Economic Commission for Europe

ECONOMIC SURVEY OF EUROPE IN 1992-1993

Prepared by the
SECRETARIAT OF THE
ECONOMIC COMMISSION FOR EUROPE
GENEVA

NEW YORK, 1993

NOTE

The designations employed and the presentation of the material in this publication do not imply the expression of any opinion whatsoever on the part of the Secretariat of the United Nations concerning the legal status of any country, territory, city or area, or of its authorities, or concerning the delimitation of its frontiers or boundaries.

UNITED NATIONS PUBLICATION
Sales No. E.93.II.E.1
ISBN 92-1-116555-5
ISSN 0070-8712

PREFACE

The present *Survey* is the forty-sixth in a series of reports prepared by the secretariat of the Economic Commission for Europe to serve the needs of the Commission and to help in reporting on world economic conditions.

The *Survey* is published on the responsibility of the secretariat, and the views expressed in it should not be attributed to the Commission or to its participating Governments.

The pre-publication text of this *Survey* was completed in March 1993 as a document for the 48th session of the Economic Commission for Europe. The final text, incorporating minor changes, will be published in mid-April 1993.

EXPLANATORY NOTES

The following symbols have been used throughout this *Survey*:

A dash (-) indicates nil or negligible;

Two dots (..) indicate not available or not pertinent;

An asterisk (*) indicates an estimate by the secretariat of the Economic Commission for Europe;

A slash (/) indicates a crop year or financial year (e.g., 1990/91);

Use of a hyphen (-) between dates representing years, for example, 1989-1991, signifies the full period involved, including the beginning and end years.

Unless the contrary is stated, the standard unit of weight used throughout is the metric ton.

The term "billion" signifies a thousand million.

References to dollars ($) are to United States dollars unless otherwise stated.

The following abbreviations have been used:

CIS	Commonwealth of Independent States
CMEA	Council for Mutual Economic Assistance
ECE	Economic Commission for Europe
EEC	European Economic Community
FAO	Food and Agriculture Organization of the United Nations
GDP	Gross domestic product
GNP	Gross national product
GSP	Gross social product
IMF	International Monetary Fund
NMP	Net material product
OECD	Organization for Economic Co-operation and Development
SDR	Special drawing rights

Eastern Europe, as employed in the text and tables of this publication, refers to the formerly centrally planned economies of Albania, Bulgaria, Czechoslovakia, Hungary, Poland and Romania, and the successor states of the Socialist Federal Republic of Yugoslavia. Czechoslovakia, which separated into the Czech and Slovak Republics at the beginning of 1993, is dealt with as a policy-making entity for 1992, but data for the two new states are shown separately where available. Among the newly-independent republics of the former *Soviet Union*, a distinction is made between the *Baltic states* (Estonia, Latvia and Lithuania) and the majority of the remaining republics which cooperate in the institutions of the Commonwealth of Independent States — the *CIS* countries.

CONTENTS

CHAPTER 1

Overview and policy issues

(i)	Introduction	1
(ii)	The western economies	2
(iii)	The transition economies	4
(iv)	The siren call of "shock therapy"	7
(v)	Popular support for the transition process	10
(vi)	Industrial policy	15
(vii)	Western support for the transition process	17

CHAPTER 2

Western Europe and North America

2.1 OUTPUT AND THE COMPONENTS OF DEMAND ... 23

2.2 INFLATION AND LABOUR MARKETS ... 39
- (i) Costs and prices ... 39
- (ii) Labour markets ... 46

2.3 TRADE AND CURRENT ACCOUNT DEVELOPMENTS ... 52

2.4 COUNTRY NOTES ... 56
- (i) The major economies ... 56
- (ii) The smaller economies ... 62

CHAPTER 3

The European transition economies

3.1 OUTPUT AND DEMAND ... 71
- (i) Overview ... 71
- (ii) Output developments ... 72
- (iii) Factors affecting output developments ... 75
- (iv) Sectoral developments ... 77
- (v) Domestic demand ... 81
- (vi) Foreign sector impact ... 85
- (vii) Short-term prospects ... 86

3.2 LABOUR MARKETS, PRICES, INCOMES ... 89
- (i) Labour markets ... 89
- (ii) Prices and inflation ... 101
- (iii) Nominal and real incomes ... 106

3.3 FOREIGN TRADE AND PAYMENTS ... 111
- (i) Foreign trade of the transition countries ... 111
- (ii) Current account, reserves and debt ... 119
- (iii) East-west trade developments ... 125
- (iv) Intra-eastern trade continues to shrink ... 135
- (v) The new foreign trade ... 138

		Page
3.4 REVIEW OF MACROECONOMIC POLICIES		142
(i)	Overview	142
(ii)	Fiscal policies: falling revenues and ballooning deficits	143
(iii)	Monetary policies: restrictive but inefficient	146
(iv)	Foreign exchange policies	152
(v)	Price and incomes policies	157
(vi)	The failure of the Russian stabilization programme in 1992	160
(vii)	The break-up of the rouble zone	169
3.5 MARKET REFORMS		173
(i)	Markets and competition	173
(ii)	Reform of the banking system	175
(iii)	Fiscal reform	181
(iv)	The dilemma of trade liberalization	184
3.6 PROGRESS IN PRIVATIZATION, 1990-1992		191
(i)	Introduction	191
(ii)	The framework of analysis	192
(iii)	Agents of privatization	193
(iv)	Establishing property rights	195
(v)	Restitution: a long detour	197
(vi)	Small privatization	199
(vii)	Fundamental changes in agriculture	202
(viii)	Commercialization of large firms	207
(ix)	The new owners	213
(x)	Privatization revenues	214
(xi)	The size of the private sector	215
(xii)	Outlook	216
3.7 COUNTRY NOTES		217
(i)	Eastern Europe: the central eastern countries	217
(ii)	Eastern Europe: the south-eastern countries	221
(iii)	The countries of the former Yugoslavia	223
(iv)	The CIS countries	227
(v)	The Baltic republics	232

CHAPTER 4

Aid and trade: western response to the transition

4.1 FINANCIAL FLOWS AND COMMITMENTS OF ASSISTANCE		237
(i)	Private capital flows	237
(ii)	Official financing	239
(iii)	Special financing	246
(iv)	Resource transfer	247
(v)	Adequacy of short-term international assistance and prospects	248
4.2 EAST-WEST COOPERATION AGREEMENTS AND MARKET ACCESS		253

CHAPTER 4

Statistical appendices

Appendix A. Western Europe and North America	262
Appendix B. Eastern Europe and countries of the former Soviet Union	275
Appendix C. International trade and payments	278

LIST OF TABLES AND CHARTS

Page

Chapter 2

Table

		Page
2.1.1	Quarterly changes in real GDP, 1991-1992	24
2.1.2	Annual changes in real GDP in western Europe and North America, 1990-1993	25
2.1.3	Major changes in nominal exchange rates against the deutsche mark, January 1992-January 1993	29
2.1.4	Changes in manufacturing production in western Europe and North America, 1990-1992	32
2.1.5	Capacity utilization rates in manufacturing industry, 1979-1992	34
2.1.6	Contribution of demand components to real GDP growth in western Europe and the United States, 1990-1992	34
2.1.7	Real private consumption expenditures in western Europe and North America, 1990-1992	35
2.1.8	Real public consumption in western Europe and North America, 1990-1992	35
2.1.9	Real gross fixed capital formation in western Europe and North America, 1990-1992	36
2.1.10	Real gross fixed capital formation, by type of asset, 1991-1992	36
2.1.11	Real housing investment in western Europe and North America, 1990-1992	37
2.1.12	Volume of trade in goods and services in western Europe and North America, 1991-1992	38
2.2.1	Unit labour costs and unit profits, 1991-1992	43
2.2.2	Contribution to the change in the GDP deflator, 1991-1992	45
2.2.3	Contribution to the change in the domestic demand deflator, 1991-1992	47
2.3.1	Changes in the volume of exports and imports, 1991-1992	52
2.3.2	Real effective exchange rates, 1985-1992	53
2.3.3	Annual changes in the terms of trade in western Europe and North America, 1990-1992	53
2.3.4	Current account balances, 1990-1992	54

Chart

		Page
2.1.1	Consumer confidence in the European Community and the United States, January 1990-February 1992	25
2.1.2	Annual growth rates of real GDP, 1980-1992	26
2.1.3	Nominal interest rates, January 1990-January 1993	27
2.1.4	Debt, interest payments and savings of private households, 1970-1992	30
2.1.5	Real housing prices, 1981-1992	31
2.1.6	Quarterly index numbers of manufacturing output, 1989-1992	33
2.1.7	Real gross fixed investment in western Europe and the United States, 1980-1992	37
2.2.1	Monthly changes in consumer prices, 1991-1992	40
2.2.2	Monthly changes in the prices of manufactured goods and services, 1991-1992	40
2.2.3	World market prices of raw materials, in US dollars and ECUs, 1991-1992	41
2.2.4	Quarterly changes in average hourly earnings in manufacturing industry, 1990-1992	42
2.2.5	Intermediate and final output prices in manufacturing industry, 1991-1992	43
2.2.6	Unit labour costs in total economy, 1982-1992	44
2.2.7	Employment growth by sectors, 1980-1992	48
2.2.8	Labour market development, 1979-1992	49
2.2.9	Quarterly standardized unemployment rates, 1990-1992	50

Chapter 3

Table

		Page
3.1.1	European transition countries: Economic activity, 1989-1993	73
3.1.2	GDP by major sectors, 1989-1992	78
3.1.3	European transition countries: Gross agricultural production, 1989-1992	80
3.1.4	European transition countries: Domestic production and absorption, 1989-1992	82
3.1.5	European transition countries: Gross fixed investment, construction, 1989-1993	83
3.1.6	European transition countries: Volume of retail sales, 1989-1992	84
3.1.7	European transition countries: Volume of foreign trade, 1989-1992	85
3.2.1	Selected transition countries: Changes in the apparent labour force	92
3.2.2	European transition countries: Unemployment, 1991-1992	93
3.2.3	European transition countries: Change in output, employment and productivity in industry	94
3.2.4	European transition countries: Registered vacancies	95
3.2.5	European transition countries: Change in employment, 1990-1992	96
3.2.6	Selected transition countries: Unemployment rates by sex	97
3.2.7	Selected transition countries: Regional dispersion of unemployment rates in September 1991 and September 1992	97
3.2.8	Selected transition countries: Change in employment by branch	98
3.2.9	Czechoslovakia and Poland: Comparison of wide and narrow measures of employment	98
3.2.10	Comparison of registration- and survey-based measures of unemployment in Hungary, 1992	100
3.2.11	Reconciliation of the two measures of unemployment in Hungary, 1992	101
3.2.12	Measures of employment change in Hungary, 1992	101
3.2.13	European transition countries: Rate of consumer price inflation, 1991-1993	102
3.2.14	European transition countries: Rate of inflation after price liberalization shocks, 1990-1992, month over preceding month	104
3.2.15	European transition countries: Nominal and real wages, 1989-1992	107
3.2.16	European transition countries: Average nominal monthly wage	108
3.2.17	Transition economies: Percentage change in real wages, 1990-1992	108
3.2.18	Transition economies: Changes in real incomes of population, 1989-1992	109

		Page
3.3.1	European transition countries: Foreign trade, by direction, 1990-1992	112
3.3.2	European transition countries: Trade balances, 1989-1992	113
3.3.3	European transition countries: Change in foreign trade values and trade balances by partner region, 1990-1992	114
3.3.4	Conflicting foreign trade data for the CIS and Russia, 1991-1992	115
3.3.5	European transition countries: Convertible currency current account balances, 1990-1993	120
3.3.6	European transition countries: Balance of payments in convertible currencies, 1990-1992	121
3.3.7	European transition countries: Foreign currency reserves and BIS deposits, 1989-1992	122
3.3.8	Eastern Europe and former Soviet Union: External debt in convertible currencies, 1990-1992	124
3.3.9	IMF quotas and shares applied to the IMF assets and liabilities of the successor states to Czecholovakia and Yugoslavia	124
3.3.10	European transition countries: Trade with the west, 1990-1992	127
3.3.11	European transition countries: Trade with the west, by major western partner	130
3.3.12	European transition countries: Trade with the west, by commodity, 1992	132
3.3.13	Eastern Europe: Imports of western semi-manufactures and engineering goods, 1989-1992	133
3.3.14	Hungary's trade with the transition economies, 1991-1992	136
3.3.15	Foreign trade of the Baltic countries, 1990-1992	141
3.4.1	Fiscal balance in transition economies, 1990-1992	144
3.4.2	Changes in money supply in transition countries in 1990-1992	146
3.4.3	Foreign currency deposits held with the banking system in transition countries	146
3.4.4	Cumulative change in real money stock and in GDP over 1990-1992	147
3.4.5	Changes in net credit emission in transition economies, 1990-1992	148
3.4.6	Monthly nominal and real interest rates in Bulgaria, Czechoslovakia and Poland, 1991-1992	149
3.4.7	Initial foreign exchange rate adjustments in transition economies	153
3.4.8	Changes in nominal and real foreign exchange rates in transition countries, 1991-1992	154
3.4.9	Changes in fuel prices in Russia, 1991-1992	157
3.4.10	Russian state budget revenues in January-September 1992	163
3.4.11	Changes in the monetary base in Russia in January-June 1992	164
3.4.12	Monetary survey of the Russian Federation, selected items, 1992	164
3.4.13	Results of convertible currency auctions in Russia in 1992	167
3.4.14	How did the new currencies fare? Exchange rates in the post-Soviet states	171
3.4.15	Monetary systems in the former Soviet republics, as of February 1993	172
3.5.1	Banking reform in transition countries	176
3.5.2	Biggest commerical banks in Russia: Assets, capital and financial ratios	177
3.5.3	"Bad loans" in the largest Polish banks, as of 30 June 1992	180
3.5.4	Selected items of budget revenues in transition countries, 1990-1992	182
3.5.5	Tax regimes in transition economies, as of 1 January 1993	183
3.6.1	The main mechanisms of privatization	192
3.6.2	The main motivations for and goals of privatization	193
3.6.3	State institutions involved in the privatization process	194
3.6.4	Restitution and compensation measures approved or planned	197
3.6.5	Restitution and compensation claims	198
3.6.6	Progress in small privatization	200
3.6.7	Sales of municipally-owned dwellings	202
3.6.8	Past and planned land ownership changes in eastern Europe	203
3.6.9	Restructuring in the former state agricultural sector	204
3.6.10	Private farms in the CIS republics	206
3.6.11	Voucher privatization programme in the transition economies	209
3.6.12	The scenario of voucher privatization in Czechoslovakia	211
3.6.13	Methods of privatization in Russia for medium-sized enterprises	212
3.6.14	Official estimates of the contribution of the private sector to GDP	215

Chart

3.2.1	Rate of inflation, 1991-1992	103
3.3.1	European transition countries: Geographical structure of exports, 1989-1992	116
3.3.2	European transition countries: Geographical structure of imports, 1989-1992	118
3.3.3	Specific western import demand facing eastern Europe and the former Soviet Union, 1990-1993	128

Chapter 4

Table

4.1.1	European transition countries: Medium- and long-term funds raised on the international financial markets, 1988-1993	238
4.1.2	European transition countries: Joint ventures and foreign direct investment, 1990-1992	239
4.1.3	Commitments of international assistance to the countries of eastern Europe and the independent states of the ex-USSR	240
4.1.4	Selected multilateral financing to the European transition countries, 1990-1992	242
4.1.5	European transition countries: IMF arrangements in force in 1992	243
4.1.6	Complementary financial assistance provided by the G-24 and EC to the European transition countries	245
4.1.7	European transition countries: Two measures of net transfer of resources and changes in reserves, 1990-1992	247
4.1.8	Total number of asylum-seekers originating from central and eastern Europe and arrived in western Europe, 1985-1992	251
4.2.1	Overview of east-west trade regimes: dates of entry into force	254
4.2.2	Improvements in access to the EC market under the EC association agreements with the former Czechoslovakia, Hungary and Poland	257

Statistical Appendices

Appendix A. Western Europe and North America

Appendix table

		Page
A.1	Gross domestic product	263
A.2	Private consumption	264
A.3	Public consumption	265
A.4	Gross domestic fixed capital formation	266
A.5	Volume of exports of goods and services	267
A.6	Volume of imports of goods and services	268
A.7	Current account balances	269
A.8	Industrial production	270
A.9	Consumer prices	271
A.10	Average hourly earnings in manufacturing	272
A.11	Total employment	273
A.12	Annual unemployment rates	274

Appendix B. Eastern Europe and the Soviet Union

B.1	Economies in transition: Selected indicators	276

Appendix C. International trade and payments

C.1	World trade: Value, by region	279
C.2	World trade: Volume change, by region	280
C.3	Western Europe and North America: Trade volume change	281
C.4	Eastern Europe and the Soviet Union: Exports by main directions	282
C.5	Eastern Europe and the Soviet Union: Imports by main directions	283
C.6	East-west trade: Value of western exports, by country of origin	284
C.7	East-west trade: Value of western imports, by country of destination	285
C.8	East-west trade: Western trade balances by western country	286
C.9	East-west trade: Western exports, imports and balances by eastern country	287
C.10	Eastern Europe and the Soviet Union: Balance of payments in convertible currencies	288
C.11	Eastern Europe and the Soviet Union: Gross debt, foreign currency reserves and net debt in convertible currencies	289

Chapter 1

OVERVIEW AND POLICY ISSUES

(i) Introduction

For the economies of Europe and North America 1992 was another disappointing year. The recovery of the United States economy from recession was very slow and, until late in the year, was also very hesitant. By the turn of the year, however, the United States recovery was beginning to appear stronger and more settled, thus providing one of the few positive developments in the ECE region. In western Europe hopes of stronger output growth proved unfounded: in the last issue of this *Survey*, which was more pessimistic than most other forecasts at the time, it was thought there might be a slight improvement from average growth of 1 per cent in 1991 to about 1.5 per cent. In the event, annual growth was unchanged and there was a sharp weakening of economic activity in the second half of the year.

In eastern Europe,[1] 1992 was another year of falling activity, the fourth in succession, the cumulative drop in output reaching just over 30 per cent. Nevertheless, there were signs that the downturn was beginning to bottom out and in a few countries there were signs of a recovery during the year. In Poland there was actually a small increase in GDP for 1992 as a whole. However, in Russia, the other CIS states and the Baltic states, the slump in output deepened through 1992. The differences among the transition economies have widened a lot: in some, substantial progress has been made in setting up the institutions of a democratic society and a market economy, and there are also hopes of an improvement in economic activity; in others, institution-building is still at a relatively early stage and subject to considerable strains from political fragility and, in some cases, from civil war and armed conflict. For many people in the eastern countries the dominant issue is not so much the "transition process" as immediate survival.

Personal insecurity, albeit at a different level, has increased sharply for many people in the western economies. At the end of 1992, 30 million people were unemployed in western Europe and North America, an increase of more than one third in two and a half years. In December, the unemployment rate averaged just under 10 per cent in western Europe compared with 8.9 per cent a year earlier. In many countries the proportion of people unemployed for over a year has risen sharply, to a third or more of total unemployment in several cases. One of the most disturbing features of the current situation is the exceptionally high rates of youth unemployment. In Spain a third of young people (under 25 years old) in the labour force are without a job and the proportion is close to 30 per cent in a number of other countries including France and Italy. The enormous economic waste and the potential for serious social problems implied by such figures is obvious. Not surprisingly unemployment began to move to the top of the economic policy agenda in 1992.

In eastern Europe unemployment is more difficult to measure than in the west, but the data available point to a rapid increase in 1992. At the end of the year some 6.5 million people were registered as unemployed, with rates ranging from 5 per cent in Czechoslovakia to 14 per cent in Poland and even more in the Federal Republic of Yugoslavia (Serbia and Montenegro). Figures for unemployment in the former Soviet Union are very unreliable and the registered figure of some 1 million is probably too low. In all the transition economies unemployment is expected to start rising rapidly as privatization gets under way and bankruptcy laws are brought into effect. Concern at the social implications of this development explains the anxiety to get social safety nets into position and the increasing willingness of western countries to support their construction with technical and financial assistance.

In other parts of the European region the "new order" has brought insecurity on a completely different scale. In the violent conflicts which have followed the break-up of the Soviet Union and Yugoslavia more than 200,000 people have been killed (of which some

---

[1] *Eastern Europe*, as employed in the text and tables of this publication, refers to the formerly centrally planned economies of Albania, Bulgaria, Czechoslovakia, Hungary, Poland and Romania, and the successor states of the Socialist Federal Republic of Yugoslavia. Czechoslovakia, which separated into the Czech and Slovak Republics at the beginning of 1993, is dealt with as a policy-making entity for 1992, but data for the two new states are shown separately where available. Among the newly-independent republics of the former *Soviet Union*, a distinction is made between the *Baltic states* (Estonia, Latvia and Lithuania) and the majority of the remaining republics which cooperate in the institutions of the Commonwealth of Independent States – the *CIS* countries.

85 per cent in Yugoslavia) and around 3 million have been displaced.[2]

The revolutions of 1989 in eastern Europe were accompanied by euphoric proclamations of a "new world order". That phrase now appears less frequently in the lexicon of international statesmen. Indeed, a west European foreign minister has recently dismissed the phrase as "Utopian folly" because it "promised more than we shall ever be able to perform". The best that can be expected is "action to avert the continuing slide into disorder".[3] A fairly large share of that disorder — actual and potential — now lies within the borders of the former communist regimes of eastern Europe and the former Soviet Union. These remarks may appear somewhat bleak but they also indicate the beginning of a more realistic appraisal of the scale of the problems in the transition economies and of what is needed to overcome them. At the end of 1989 there were excessively optimistic expectations about how long it would take and how much it would cost to create new market economies. At the same time there was a certain disengagement of the western countries from the problems of the east. The threat of nuclear war was over and the problems of transition could hopefully be left to the international financial institutions and, after a short delay, to private capital both domestic and foreign. That sanguine view is now being sharply revised, especially in the light of developments in Russia over the past year. There now seems to be a realization that the revolutions of 1989 posed a much greater challenge to the economic and foreign policies of the western countries than was previously supposed, and that is reflected in the greater urgency with which action to assist Russia is now being discussed.

(ii) The western economies

The economic situation in the western countries is now very weak and the forecasts for growth in 1993 have been lowered since last autumn. For western Europe and North America combined, the annual growth rate of real GDP appears likely to be somewhat less than 2 per cent in 1993. This outcome is slightly better than in 1992, but there is an increasing divergence between western Europe and North America.

In the United States the recovery now appears to be firmly established, even if the remarkable strengthening of activity in the second half of 1992 partly reflected special factors. Real GDP is expected to rise by some 3 per cent in 1993 (annual change).

The outlook for western Europe is instead rather sombre. Against a background of turbulent foreign exchange markets and with the German economy moving into recession, prospects for 1993 have deteriorated sharply and consumer and business confidence are now at very low levels. After growth of only 1 per cent in 1992, stagnation in west European GDP in 1993 now seems increasingly likely.

With low or zero rates of output growth there will be a further rise in the number of persons unemployed. In western Europe employment is therefore likely to continue to fall and the unemployment rate to move significantly above 10 per cent. In the United States the labour market is expected to improve in 1993, but employment growth is likely to be relatively modest.

Among the four major economies in western Europe, real GDP is likely to fall in Germany, France and Italy in 1993. Only in the United Kingdom is GDP expected to grow, albeit by a small amount, as it emerges from its longest recession in the post-war period.

Among the smaller countries the outlook is for a pervasive slow-down in economic growth, with a number of countries close to stagnation or even an absolute fall in total output. The prospects of most of these smaller economies are closely tied to the outlook for Germany.

High interest rates, resulting from the tightening of monetary policy in Germany, and the commitment of most west European countries to fixed exchange rates are major factors behind the economic downturn in western Europe. The strains of this combination increased throughout 1992 to the point where the United Kingdom and Italy decided to leave the ERM, and several other countries devalued within the system or severed unilateral exchange rate links to the ECU.

German monetary policy has been progressively eased since 15 September 1992 against a background of tensions within the ERM, but the cuts have been relatively small so that on average real interest rates are still high.

Hopes of significantly larger and more rapid cuts in German interest rates have been pinned on a successful conclusion of the so-called Solidarity Pact. The Pact, in principle, is designed to allow for the continuation of very high support payments to the east German economy. This requires the full integration of the east German *Länder* into the established revenue-sharing system between the federal government and the west German *Länder* as from 1995. At the same time the Pact is also an integral part of a broader package of measures designed to raise revenues and curb expenditures and so pave the way to a medium-term reduction in the public sector deficit.

It is also an essential part of the strategy to convince the trade unions that the severe deterioration in the east German labour market calls for wage restraint and that the insistence on a very rapid closure of the wage gap between the two halves of the country conflicts sharply not only with the objective of preserving jobs but also with the need for more stringent budgetary policies.

---

[2] Based on figures in the *New York Times*, 7 February 1993 and the weekly press release of displaced refugees in former Yugoslavia, issued by UNHCR Geneva.

[3] Speech on "The New Disorder" by Douglas Hurd, delivered at the Royal Institute for International Affairs, Chatham House, London, 27 January 1993.

However, the agreement reached in March 1993 has actually clouded the prospects for a significant reduction in the deficit. The main emphasis is on raising tax revenues *inter alia* by reintroducing the surcharge on income and corporate taxes as from 1995, while there will be only limited cuts in expenditure. The immediate impact is to increase the public sector borrowing requirement in 1993 and possibly beyond. Also, the trade unions have not yet changed their position on the question of the catch-up in eastern wages.

Thus, the Pact, as agreed, does not appear to pave the way to a substantial change in the policy mix in Germany. Given the uncertainty surrounding future budget deficits and the continuing high rate of inflation, it therefore seems likely that for the first half of 1993, at least, the Bundesbank will continue to lower official interest rates only slowly and cautiously, largely in response to the weakening of economic activity and changes in the inflation rate.

The easing of German monetary policy since September 1992 has been only hesitantly followed in other countries because of continuing speculative pressures on their exchange rates. In France there is now a sizeable margin over German short-term rates which translates into an even bigger gap in real terms, given that the inflation rate is lower in France.

In those west European countries which abandoned fixed exchange rate targets against the deutsche mark or left the ERM, monetary policy has been substantially eased since the final quarter of 1992, but the scope for further easing is now constrained because of the inflationary implications of further currency depreciation.

Although there has been some scope for monetary easing, governments in most of western Europe see hardly any room for a discretionary relaxation of fiscal policy to counter the forces of recession. Governments' financial positions generally deteriorated in 1992, mainly because of the impact of lower than expected economic growth on revenues and expenditures. Most governments will continue to allow the automatic stabilizers to work in 1993, but at the same time they are pursuing efforts to stabilize discretionary spending and reduce their borrowing relative to GDP. The pressure for fiscal restraint has been increased considerably by attempts to meet the Maastricht convergence criteria prior to European Monetary Union. None of the west European countries will be able to meet these criteria in 1993.

In the United States monetary policy was considerably eased in 1991 and 1992 and the need for further relaxation appears now to be much reduced given the strengthening of the economy in the second half of 1992.

The new administration has announced a package of measures designed to curb significantly the budget deficit in the medium term. At the same time it is also intending to provide a short-term fiscal stimulus (equivalent to about 0.5 per cent of GDP) because of concerns about low employment growth. The announced stimulus does not appear to have undermined credibility in the medium-term plan to curb the deficit; long-term interest rates have eased.

The outlook is for steady growth in the United States in 1993. Although lower interest rates have eased debt service burdens, private sector demand will continue to be restrained by balance sheet problems and by the glut in commercial real estate. The need to consolidate budget deficits will check growth in government expenditures. Given the relative strength in domestic and foreign demand, changes in the real foreign balance will act as a drag on economic activity levels.

Any forecast of central tendencies is, of course, surrounded by a margin of error. The general feature of short-term forecasts for the western economies over the past two years or so has been the presence of rather large downside risks, largely reflecting the difficulties of predicting consumer spending against a background of very high debt-income ratios and falling asset prices. Although balance sheet problems have been easing recently, they are still very important and more negative "surprises" cannot be excluded, especially in western Europe, where the deterioration in the labour market is lowering the prospects for income growth and increasing uncertainty for individuals. Another important uncertainty is how the large changes in real effective exchange rates will affect net exports and domestic activity levels in the various countries and the scope for further monetary easing in Germany in 1993.

With prospects of only low output growth and a further worrisome rise in unemployment, there would appear to be a need for closer macroeconomic policy coordination among the major industrialized economies, the general objective being to enhance the scope for output and employment growth. A step in this direction was taken by the European Community which, at the December 1992 Edinburgh summit, adopted a package of measures (the so-called "growth package") designed to stimulate investment demand. But implementation appears to be slow and the overall impact is expected to be rather modest. Moreover, closer coordination of economic policies does not currently appear to be high on the agenda of the G7.

The medium-term outlook for western Europe is also more uncertain than it appeared in 1990. One important uncertainty concerns the prospects for future monetary union. The European Council Agreement at Maastricht in December 1991 was seen as leading to the culmination of a "Europe without borders". However, the conditions laid down for economic convergence, to be met by 1997, are relatively severe and the process of meeting them is likely to lead to fairly restrictive monetary and fiscal policies. To meet the inflation requirement most countries will try to avoid further currency realignments, although this may prove difficult if persistently large government budget deficits in Germany force the Bundesbank to keep interest rates relatively high. Also, the absence of labour market indicators among the convergence criteria will not encourage initiatives by individual countries to stimulate growth. Economic growth in western Europe is therefore likely to be rather subdued for the rest of the 1990s.

This is in sharp contrast with the optimism prevailing at the end of the 1980s in anticipation of the single European market which entered into force at the beginning of 1993. The optimism for the "decade of Europe" appears to have been dampened since the Maastricht treaty was signed.

(iii) The transition economies

Output continued to fall quite substantially in most of the transition economies of eastern Europe and the former Soviet Union in 1992.[4] In eastern Europe GDP/NMP fell on average by about 10 per cent, bringing the aggregate decline since 1989 to more than 30 per cent. However, the economies of eastern Europe are becoming increasingly heterogeneous, partly reflecting the different pace and degree of success with which reforms have been introduced, and it is therefore difficult to generalize about short-run economic changes. The estimated falls in GDP last year ranged from some 27 per cent in the Federal Republic of Yugoslavia to some 4-6 per cent in Hungary, while in Poland there was actually a small increase. Data for industrial production show that a recovery began to get under way in Poland in the second quarter of 1992, while in Hungary and Czechoslovakia the decline appeared to have come to an end. However, in January 1993 industrial output began to fall again in Hungary. Nevertheless, for virtually all the east European countries, except Romania, the falls in output last year were smaller than in 1991 (although the improvement in Bulgaria was small).

In Russia, the other CIS countries and in the Baltic states the slump in production deepened considerably in 1992. In Russia NMP fell by 19 per cent, and elsewhere falls ranged from 11 per cent in Belarus, where reforms have been introduced in a limited and slow way, to more than 40 per cent in Armenia, which is at war with and subject to an economic blockade by Azerbaijan.

In the Baltic states the falls in output were even larger than in the CIS countries: 28 per cent in Estonia, 44 per cent in Latvia and 35 per cent in Lithuania. These massive cuts in production in a single year reflect the impact of the abrupt severing of trade ties with the countries of the former Soviet Union. The breakdown in a highly specialized division of labour was bound to have a larger impact on the small economies of the Baltic states. In addition, the renegotiation of trade prices, especially for Russian oil, led to a large terms of trade loss for the Baltic countries and there was also some disruption in supplies of gas and oil. Moreover, for most of the year balance of payments financing from the IMF was not available so that adjustment was particularly sharp.

In general, industry has tended to bear the brunt of the fall in output in the transition economies, although there has now been some recovery in Poland. *Agricultural output* added to the decline in 1992, falling by 9-13 per cent in most of eastern Europe and by 10 per cent or so in the CIS. The *construction* sector has continued to decline, although the first signs of recovery have appeared in Czechoslovakia. *Housing* construction appears to have declined again in most countries. Data on *services* are still limited, especially for the CIS, but their share in total output appears to be rising in eastern Europe, in some cases (Hungary, for example) quite sharply. Indeed, in eastern Europe the only sectors to show any growth at all in employment were banking, finance and business services.

Statistics on the scale of activity in the *private sector* are still very limited, but those that are available indicate a rapid growth, especially in eastern Europe and particularly in agriculture, construction, trade and other services. Their share in domestic output appears to have increased substantially in 1992, from 9 per cent in 1991 to 20 per cent in Czechoslovakia, and from 5 to 10 per cent in Bulgaria. In Hungary the share of the private sector was around 25-35 per cent of GDP last year, with 45-50 per cent in Poland, and perhaps 26 per cent in Romania. The importance of private producers has also risen in the CIS countries. The numbers, although not always comprehensive or very reliable, suggest that despite all the disruption and uncertainty in the transition economies, significant progress is being achieved in one of the key areas of the transition process.[5]

National accounts statistics showing the components of demand are still fairly limited, but there appears to have been a slight recovery or stabilization of domestic demand in Czechoslovakia and Poland, and a further slight fall in Hungary. Elsewhere the declines were much larger.

The signs of recovery in Poland and perhaps Czechoslovakia are strengthened by the decline in *fixed investment* coming to an end. Zero growth in 1992 compares with falls of 4 and 27 per cent respectively in 1991. Investment is still falling in the rest of eastern Europe although less rapidly than in 1991. In Russia and the other CIS countries the slump in fixed investment deepened last year (down by 45 per cent in Russia), which is not surprising given the economic turmoil and uncertainty arising from hyperinflation.

Data on *private consumption* are also scarce but proxy variables such as retail sales and real income per head suggest that the squeeze on consumption and material living standards has continued. However, in Czechoslovakia private consumption rose (by about 5 per cent) and also in Poland. The falls in real consumption have been particularly severe in Russia and the other CIS countries, mainly because of the squeeze in real incomes as inflation outpaced the increase in money incomes by a considerable margin. In Russia the number of people officially classified as living in poverty (i.e., below the minimum subsistence income)

---

[4] Statistics of aggregate output in the transition countries are still surrounded by a considerable degree of uncertainty. See below, section 3(ii).

[5] For a detailed discussion of privatization developments, see section 3.6. On progress in other aspects of market reform, see section 3.5.

more than doubled in the twelve months to the end of 1992, from 12 per cent to 29 per cent of the population. A similar development appears to have occurred in most of the other CIS countries. In considering the impact of the transition period on living standards and people's sense of personal security, it should also be remembered that public services — health, culture, education — have also been hard hit by tight fiscal policies, that social safety nets for the unemployed are still rudimentary in many countries, and that increasing income inequalities are raising social tensions as larger numbers of people fall below the official poverty lines. However, it should be noted that the statistical dimensions of these social consequences of the adjustment process are still very imprecise.

Unemployment began to rise sharply in eastern Europe in 1991 as industrial restructuring got under way, as bankruptcy laws were put into place, as privatization began to spread into traditional sectors of the economy, and as market forces began to influence the behaviour of the "new" entrepreneurs.

Registered unemployment continued to rise steadily in all countries of the region in 1992, with the notable exception of Czechoslovakia. In most of the east European countries rates of unemployment continued to rise sharply over the year, although more slowly than in 1991. Nearly 6.5 million people were registered as unemployed in December 1992 (compared with 5 million in December 1991), with the unemployment rate ranging between 5 per cent in Czechoslovakia and nearly 25 per cent in the Federal Republic of Yugoslavia. The largest increase was in Hungary where the rate of unemployment nearly doubled to 12.3 per cent of the labour force in December 1992. Employment in large Hungarian enterprises fell by nearly 40 per cent between 1989 and the autumn of 1992, far too large to be quickly absorbed by the private sector. In Poland, where both the rate and the level of unemployment doubled during 1991, there was some slow-down in the rate of deterioration in 1992, particularly in the second half of the year. Nevertheless, the number of unemployed exceeded 2.5 million and the unemployment rate reached nearly 14 per cent at the end of the year. In striking contrast, there was a fall in the level and the rate of unemployment in Czechoslovakia after the spring: the rate of unemployment fell to 2.6 per cent in the Czech Republic and 10.4 per cent in the Slovak Republic at the end of 1992 (down from 4.1 per cent and 11.8 per cent at the end of 1991 respectively). This substantial improvement occurred while employment in large enterprises continued to fall sharply (by some 14 per cent). Although a tightening of the eligibility rules for unemployment benefit, in January 1992, and the expansion of public employment programmes played a role, much of the fall appears to be genuine and to reflect the strong performance of the private sector, particularly in the Czech Republic.

Despite the considerable understatement of the level of unemployment in the official statistics of the former Soviet Union, unemployment was also increasing in the successor countries during 1992: in the twelve months to the end of 1992, the number of registered unemployed is estimated to have reached about 1 million. Nevertheless, unemployment was much lower than might be expected given the size of the fall in output: the unemployment rate was less than 1 per cent in December 1992 in all republics except Armenia, Estonia and Latvia, where it ranged between 3.5 and 2 per cent. In the former Soviet Union in particular and to a lesser extent in most of the east European countries, the initial response of unemployment to the decline in output was delayed because of soft budget constraints, slow privatization, and sharply falling real wages, which encouraged the hoarding of labour. This hidden unemployment is clearly reflected in large productivity losses, particularly in the industrial sector. Given the expectation of a further deterioration in output, particularly in the CIS countries, it seems likely that the rise in open unemployment will soon begin to accelerate sharply. In the absence of comprehensive social safety nets such a development would inevitably increase the dangers of social and political instability.

High rates of *inflation* are still a major problem in all of the transition economies, although its intensity varies considerably among individual countries.[6] Annual (year-on-year) inflation rates fell in a number of east European countries in 1992, and strikingly so in Czechoslovakia (from over 50 per cent in 1991 to 11-10 per cent) and Bulgaria (from 340 to 80 per cent). The rate also fell in Hungary and Poland, to 23 and 43 per cent respectively. Elsewhere inflation accelerated and was over 200 per cent in Albania, Romania and Slovenia. In the Federal Republic of Yugoslavia hyperinflation has been unconstrained since April 1992: for the year as a whole prices rose by more than 9,000 per cent and were accelerating sharply in the early months of 1993.

In Russia and the other members of the CIS both the initial impact of the January 1992 price reform and the "after shocks" were much greater than expected and proved difficult to control. There was some deceleration to mid-summer but the rate of increase then picked up again after the central bank abandoned any serious attempt to restrict the expansion of credit. In Russia and most of the CIS states prices were rising at about 25 per cent a month in the final months of 1992. In contrast to the CIS countries, the Baltic states have rapidly reduced their monthly inflation from double-digit rates at the beginning of 1992 to some 3 to 5 per cent by the end of the year, a result of effective monetary reforms and restrictive monetary policies.

The *foreign trade*[7] of the transition countries in 1992 continued on the lines of the previous two years, namely, a continuing and probably accelerating collapse in intra-eastern trade, falling trade with the western

---

[6] For a detailed discussion of inflation see section 3.2(ii).

[7] See section 3.3 for a detailed discussion of trade and payments. It must be emphasized that the data used for analysing trade developments is even more fragile than in the past.

market economies in the case of the former states of the Soviet Union, and continued significant growth in the trade of eastern Europe with the west.

In the former Soviet Republics, the disruption of inter-republican trade links, falling oil production, rapid inflation, and a monetary and payments crisis have all contributed to a large fall in trade, not only with former CMEA partners but also with the western market economies.

One of the most significant developments in trade has been the sustained growth of exports from eastern Europe to *western* markets over the last four to five years. Bulgaria, Czechoslovakia, Hungary and Poland all maintained high rates of volume growth in the first three quarters of 1992, although there was some deceleration from 1991 in Hungary and Poland. Exports of manufactures have risen faster than the growth in the total volume of total exports with consumer goods and engineering leading the way. This performance indicates that a wide range of east European goods can meet the quality requirements of western markets despite opinions to the contrary a few years ago. The attempt to offset part of the destruction of trade with other eastern countries with exports to the west also shows increasing adaptability on the part of eastern exporters and their success has provided the only source of demand growth for most of them.

Assessing the *short-run prospects* for the transition countries is extremely difficult because the problems they are grappling with are essentially structural, not cyclical. The purpose of the reforms is to transform the very institutional structures and patterns of behaviour, stability in which is necessary for building short-term forecasting models. It is therefore misleading to employ the same business-cycle vocabulary which is appropriate for the west to the situation in the transition economies.[8] As noted above, there has been an upturn of industrial production in Poland and preliminary estimates suggest a rise in GDP in 1992. Hesitant signs of an upturn in industrial output also appeared in the last quarter of 1992 in Hungary and Czechoslovakia, although, as noted already, output fell again in Hungary in early 1993. However, recovery in a transition economy must mean much more than an end to the slump in output, highly welcome though that may be. When the basic infrastructural reforms are in place a strong and sustained rate of growth, perhaps in the neighbourhood of the 7 per cent or so annual rate achieved by western Germany in the 1950s, will be required. For that demand will need to expand and there must be enough capacity to meet it without creating inflationary pressures; and that, in turn, will require a rapid growth of fixed investment both in the private and public sectors. For most of the transition economies these conditions are still far from being met: inflation rates are still being reduced, government expenditures are being curbed and the stance of macroeconomic policy in general is not yet conducive to the expansion of fixed investment and the correction of inappropriate industrial structures (see below). From this perspective, one of the most encouraging developments in Poland last year was a small rise in fixed investment in equipment which is expected to accelerate in 1993 and 1994.

However, on the basis of government forecasts, which have been repeatedly over-optimistic in the last three years, the outlook for the leading reform countries is for some recovery in 1993, albeit slow and hesitant. In Poland GDP growth of 3 per cent is currently forecast, while in Hungary the uncertainties are reflected in a range of zero to 3 per cent.

The Czech government was forecasting growth in 1993 at 1-3 per cent at the start of the year, but this is now surrounded by growing uncertainty about the size of the effects on trade between the Czech and Slovak Republics following the break-up of the Czechoslovak Federation. At first it was estimated that there might be a drop in mutual trade of some 10 per cent from last year's monthly average; this was later increased to 30 per cent. However, preliminary data for the first month in 1993 under the new regime suggest a fall of 60 per cent.[9] If this collapse in mutual trade is not contained, the Czech government's hopes of 1-3 per cent GDP growth in 1993 are likely to be disappointed while Slovak GDP, which is relatively more dependent on mutual trade than the Czech Republic, will decline much more than is currently expected.

In all the other east European economies, total output is expected to continue to fall in 1993, although, it is hoped, at a somewhat lower rate than last year.

Forecasts for the CIS countries other than Russia are not generally available, and even those for the latter may be little more than subjective opinions or hopes on the part of beleaguered officials. The stabilization of the Russian economy is still the crucial and urgent issue, and that in turn requires solutions to the institutional problems of establishing monetary control, the clarification of monetary and currency relations among the CIS members, and the creation of effective payments arrangements among them in order to arrest and reverse the disastrous decline in inter-republican trade. But progress on these matters must wait upon the resolution of even more fundamental issues of constitutional authority and popular legitimacy.

Finally, the short- and long-run outlook for the transition economies can also be significantly influenced by external developments. As already noted, some of the east European economies have benefited from rapid export growth, especially to western Europe. That in turn was facilitated by the strong import boom in Germany in 1990 and 1991, and by a lowering of import barriers in western Europe and elsewhere. Both factors now look to be less favourable in the next year or so. There is little that can be done to quickly reverse the slow-down in west European import demand, but it is possible for western policy makers to resist a reversal

---

[8] See United Nations Economic Commission for Europe, *Economic Bulletin for Europe*, Vol.43(1991), New York, 1991, pp.5-6.

[9] See section 3.3(v).

of the trade liberalization which has been achieved and to extend it as quickly as possible.

(iv) The siren call of "shock therapy"

A question which continues to figure prominently in the public discussion of the transition process is whether eastern governments should adopt a gradual approach to market reforms or go for them in one "big bang" or a dose of "shock therapy".[10] It is reckoned by many observers that the policies pursued in Poland, Czechoslovakia and Bulgaria, for example, can be categorized as "shock therapy" while those in Hungary can be described as "gradualist". In Russia the programme introduced by the Gaidar government[11] at the beginning of 1992 is widely regarded as an example of "shock therapy", while the critics in the Congress of People's Deputies are labelled as "gradualists" or "reactionaries" depending on the observer's point of view. The Markovic programme of 1989 in the former Yugoslavia is another case of "shock therapy" and the former German Democratic Republic was subjected to the most severe and comprehensive shock of all. If the metaphor of "shock therapy" is taken literally, namely, that it provides quick and lasting relief from a severe disorder, then the record is not a successful one. But, clearly, this is too simple an approach which just shows up the limitations of the metaphor.

The important distinction between the various transition economies lies not so much in whether they consciously chose "shock therapy" or "gradualism" but in the initial conditions in which they started their reforms and the particular "disorders" which had to be dealt with. By the end of 1989 the Polish economy was characterized by pervasive shortages and the imminent prospect of hyperinflation. The central plank of the government's programme in January 1990 therefore consisted of a fundamental shift in macroeconomic policy to deal with the threat of hyperinflation. To do this, swift and decisive action was necessary and both experience in other parts of the world and a large body of economic analysis would support the Polish approach as the correct one. The Balcerowicz Plan also included instant price and trade liberalization which are more controversial (see below), but it was also quite clear that the Deputy Prime Minister was not suggesting a "big bang" approach to privatization and other institutional reforms. Thus Polish "shock therapy" was successful in heading off the disorder of hyperinflation, but in curing the more fundamental, structural disorders of the Soviet-type economy progress has inevitably been more gradual. In Hungary, by contrast, there was no need for macro "shock therapy" insofar as inflation was fairly low, while the relative prominence of problems requiring structural solutions, which had been underway for many years under the previous regime, meant that a more gradualist approach was unavoidable. The Czechoslovak situation was again very different in that the new government inherited an economy with no serious macroeconomic imbalances but, unlike Hungary, had undergone little structural reform in the previous twenty years or so. The government therefore went for rapid price and trade liberalization, backed by tight controls on wage increases, but in most other respects the process of reform has been more gradual and preceded by careful preparation.[12]

The argument between "shock therapists" and "gradualists" is not in fact very helpful because the dichotomy is a false one. The ability of a country to change — and to adjust to change — depends crucially on the nature and strength of its economic, political and legal institutions, on its social cohesiveness and traditions, and on the conditions prevailing at the start of the process of reform. These factors, which obviously vary among countries, will largely determine the speed of the transition to a market economy. The pace of reform will also be greatly influenced by the ability of political leaders to establish credibility in their economic programmes and to create and sustain a popular consensus behind a positive vision of the society at which they are aiming. In seeking popular support it is also important to make an honest attempt at estimating the scale, duration and distribution of the costs of adjustment, and to prepare measures for attenuating their impact on the most vulnerable sections of the population. Since the costs of the transition will appear much more quickly than the benefits, the degree of social support, which in part will depend on the presence of safety nets, will determine the scale of the shock that can be supported without derailing the entire programme.

These considerations tend to be swept aside by the advocates of shock therapy. The discussion is also clouded by a failure to specify clearly what policies are included within the domain of the shock treatment: as noted already, there is little disagreement over the need for urgent measures to deal with the threat of hyperinflation, but the advocates of this approach also include structural measures such as trade liberalization and even privatization. There is also considerable vagueness as to when the benefits of "shock therapy" can be expected to appear. The metaphor implies a short, sharp shock with immediate effects, something not too far removed from the text-book, neoclassical abstraction of instantaneous and costless adjustment to a new equilibrium. That should not be taken too literally, but can a three-year transition recession still qualify as shock treatment? The language of shock

---

10 Perhaps it is a sign of changing expectations that the more exuberant "big bang" has been pushed aside in the last year or so by the more gloomy "shock therapy".

11 For a discussion of the Russian programme see section 3.4 below.

12 The discussion of "shock therapy" is frequently confused by the use of conventional terms in innovative ways. This is particularly the case with "convertibility" the rapid adoption of which is usually seen as an important element of a shock programme. Governments frequently imply that their currencies are convertible while retaining an array of controls on access to foreign currency. In Romania, for example, current account convertibility was introduced in November 1991 but import licensing was not abolished. "Shocks" are not always what they are claimed to be.

therapy is fuzzy[13] but it clearly promises a more rapid adjustment than is possible in practice and as a result must bear some responsibility for the current disappointment and disillusion in many of the transition countries (see below).[14]

"Shock therapists" also tend to overstate the extent to which their recommendations are based on experience and rigorous research. The argument, for example, that trade and price liberalization cannot be introduced gradually is simply not true. In western Europe wartime price and other controls were phased out in line with supply-side improvements[15] so as to avoid setting off an inflationary spiral. The western economies have also liberalized their international trade in a long, gradual process which began in 1947 and is still not complete. Unilateral trade liberalization can also be successfully introduced in a gradual manner, as Israel showed in the 1960s.[16] The important requirements of such policies is not so much that they are gradual but that they are credible: as long as everyone is confident that the government's course is feasible and will be maintained — a confidence which can be greatly strengthened by dated targets — then the signals for changing behaviour and reallocating resources will begin to operate from the moment that the policy is implemented.

Those who advocate "shock therapy" appear to believe that it is possible to bring about a radical change in attitudes, expectations, and behaviour in a very short time. It is difficult to find much evidence for this. Many economists and most businessmen would probably agree with Alfred Marshall[17] that it is continuous, small incremental changes that keep the world moving: any large shock tends to increase (Keynesian) uncertainty and, thus, to paralyse action and propagate wait-and-see attitudes. This was evident, for example, not only with the large increases in oil prices in the 1970s but also with the large *fall* in the oil price in 1986.[18] "Big bangs" may also have a large and unexpected fallout which may be very costly: the protracted recession in the Scandinavian countries, the United Kingdom and, until recently, the United States, is to a significant degree an unforeseen consequence of the rapid deregulation of financial markets in the 1980s.[19]

What does not seem to be widely appreciated is that in democratic societies "big bang" policies are the rare exceptions rather than the rule. The general tendency is to "gradualize" as many adjustment problems as possible. Western governments normally pursue macroeconomic adjustment within medium-term programmes for controlling monetary aggregates and reducing budget deficits, although even they can run into social resistance if meeting a specific target imposes too heavy costs on sections of the population (meeting the Maastricht targets for economic convergence is creating such problems in some west European countries). Gradualism at the level of microeconomic adjustment is so common that it is hardly ever remarked upon. The completion of the European Community's single market by the end of 1992 was essentially a five-year programme of micro reforms intended to round off a process of structural change which had been under way for decades — and at the beginning of 1993 a number of objectives has still not been met. The introduction of noise limits on commercial aircraft, of catalytic converters on motor cars, and the phasing out of CFCs, are just a few of the thousands of other cases where governments seek to internalize or eliminate gradually specific externalities arising from private activity: those who suffer from the nuisance obviously want it removed as quickly as possible while the producer, who must bear the cost of its elimination, will want to postpone action for as long as possible. Bargaining takes place to reconcile these two positions and the outcome is usually an agreed date for the removal of the nuisance, not so soon as to impose too large and sudden a reduction in the producer's economically viable capital stock but not so distant as to make the policy incredible.[20] The results are not always optimal and there is always a danger that well-organized producer interests will have a disproportionate influence on the outcome. Nevertheless, the principle is clear that the nature and distribution of adjustment costs is a legiti-

---

13   The lack of a theoretical foundation is evidenced by the frequent resort of "shock therapists" to a bewildering array of metaphors. Some are used to describe their own recommendations — "big bang", "shock therapy", "cold turkey", and "jump start". But most of them are polemical devices to discredit contrary arguments — "you don't pull a tooth slowly" or "amputate a leg bit by bit" and "you can't cross a chasm in two jumps", etc. The comparison of the transition process to an attempt to "turn fish soup into fish" comes from a completely different school and would appear to be a cry of despair against the second law of thermodynamics. A moment's reflection will show that none of these metaphors connect with the reality of the transition process: they are therefore unable to illuminate it.

14   A common recommendation, especially from western and "international" advisers, is that eastern governments should, for example, privatize "as quickly as possible". Such statements have little operational significance.

15   See United Nations Economic Commission for Europe, *Economic Survey of Europe in 1989-1990*, New York, 1990, pp.21-22.

16   A World Bank study of 17 countries (D. Papageorgiou, et al., *Liberalising Foreign Trade*, Blackwell, 1991, 7 volumes.) has concluded that a bold trade liberalization tends to be more sustainable than one which is staged. However, Professor Greenaway has criticized this study for making extravagant claims for the generality of its results and has pointed out that there is much greater diversity of experience in the 6 volumes of country studies than is conceded by the authors of the summary volume. See David Greenaway, "Liberalising Foreign Trade Through Rose Tinted Glasses", *The Economic Journal*, 103, (416), January 1993. See also, Dani Rodrick, "Closing the productivity gap: does trade liberalization really help?" in G.K. Helleiner (ed.), *Trade Policy, Liberalization and Development*, Oxford: Clarendon Press, 1992.

17   To emphasize the role of continuity in his analysis, Marshall chose the motto *natura non facit saltum* for the title page of his *Principles of Economics*, published in 1890.

18   See United Nations Economic Commission for Europe, *Economic Survey of Europe in 1986-1987*, New York, 1987, pp.14-18.

19   See section 2.1 below.

20   Thus, the Oil Pollution Act of August 1990 in the United States set a limit of 15 years to phase out single-hulled tankers using United States' ports. Last December the EC agreed to advance its target date for the reduction of CFC use by 75 per cent to the end of 1993 and to phase it out entirely by 1996.

mate concern of democratic politics and, as a result, adjustments will tend to be gradual.[21]

In general, open democratic government must mean, among other things, adopting a pace of policy making which gives time for genuine discussion and for the electorate's fears and interests to be listened to and taken into account. This is a basic principle, but it applies with special force in the transition countries where the issues for debate are not simply banausic economic details but fundamental questions about the structure of society and the nature of changes in the political and ethical values previously associated with collectivism. These countries are engaged not only in creating market economies but also in building democratic institutions and developing democratic practices for the resolution of conflicting interests. In theory it is possible to separate the economic and political programmes but in practice the two are inextricably intertwined. "Shock therapists" tend to ignore these questions: their approach tends, in effect if not by intention, to be autocratic and often arrogant in suggesting that "there is no alternative" to their proposals. On the contrary, there is always an alternative: that is what market economics and democratic politics are all about. But it is the job of the economic adviser to estimate the costs of the various alternatives so that responsible choices can be made by those who must either bear the costs directly or face the political consequences of underestimating them.

Somewhat curiously, "shock therapists" tend to be more worried about a reversion to the old command economy, or a diversion to some "third way" between the latter and the market, than the risk of losing popular support for nascent democratic institutions and the difficult transition to a market economy. This is surprising because the revolutions of 1989 in eastern Europe and the rejection of the coup plotters in 1991 in Russia clearly demonstrated a massive disdain for the old regimes and a strong will for democracy and the market economy. The subsequent disillusionment and weakening support for the reforms (discussed below) reflects not so much a nostalgia for the past as disappointment at the level and duration of the costs of transition and, in particular, fear of unemployment and disquiet at what appears to be a highly uncertain future. But the policy response to this situation should not be obsessed with cutting off all possible retreat to the past but on maintaining a popular consensus behind a transparent programme to establish the market economy and democratic practice.

The argument between shock therapists and gradualists diverts attention from the central issues of policy making for the transition. These are, first, to create a credible programme for the transition and second to create and maintain popular support for the duration of the programme. The two are related, of course.

Credibility has several dimensions. But the crucial ones involve the government being able to define a clear set of objectives, to demonstrate a clear understanding of what is required to reach them, and to convince economic agents and the electorate at large that the government possesses − or will soon acquire − the necessary economic instruments to implement its programmes.[22]

The drive towards the market economy, therefore, needs to be spelled out in terms of a series of intermediate objectives (institutional reforms etc.) and the interdependencies between them set out in a transparent and comprehensible manner. This is related to the issue of sequencing. The literature on sequencing highlights the fact that the probability of success in achieving a given objective may be increased by following one sequence of measures rather than another − in the worst case an incorrect sequencing would destroy all chance of success. The insight is a useful one but the theory does not provide operational formulae for policy makers in the transition economies.

In practice, "correct" sequencing will have to be a matter of pragmatic judgement, made in the light of each country's individual circumstances although there may exist a few general rules derived from experience. Thus, for example, the experience of some of the transition economies already suggests that it may be unwise to go for rapid price liberalization when the authorities do not possess effective instruments of monetary control. The literature on sequencing may also be misleading insofar as it suggests alternative, linear sequences of policy actions. The transition process, with economic and political programmes of reform intertwined, is highly complex and will frequently require simultaneous action on a number of interdependent issues for significant progress to be made. The task is therefore to identify key sets of interdependencies which in turn would help to identify those subsets or menus of reform[23] which should be given priority in the transition programme.

The need for simultaneous action on a number of fronts is of course one reason why some advisers propose a "big bang" or "shock therapy". However, needs and possibilities are not always congruent, and this advice overlooks the fact that different elements of the reform will have their own subsets of requirements and may require very different periods of time for their implementation. This implies that the need is for the simultaneous *completion* of a set of reforms rather than a simultaneous *start* on them. It is perhaps not surprising that rapid price and trade liberalization have

---

21 The economic adviser to the Indian Finance Minister has said that India's reform was going ahead "at a gradual pace" because "there was no other way of carrying out change *in a large democracy* such as India", (italics supplied). *Financial Times*, 6 November 1992.

22 Credibility is not necessarily the same as determination. Determination without a coherent programme and effective instruments is mere bluster, while the determination of a government that has worked out a coherent programme and secured the necessary instruments for its implementation is self-evident.

23 An analogy with this problem is that of interdependent or complementary investment decisions. See, Hollis B. Chenery, "The Interdependence of Investment Decisions", in Moses Abramovitz et al., *The Allocation of Economic Resources*, Stanford, 1959 and Jan Tinbergen, *Development Planning*, London, 1967, chapter 7.

figured so prominently in the reform programmes recommended and adopted in the eastern countries because these are among the few reforms which *can* be implemented quickly. However, the results do not suggest that this was an obviously desirable step to have been taken.[24] Indeed rapid price and trade liberalization may be destabilizing in a transition economy with inelastic supply responses, which are partly due to institutional rigidities, and with inadequate instruments and institutions for creating and maintaining macroeconomic stability.

Perhaps the principal reason why many observers, including the authors of this *Survey*, are sceptical about the feasibility of a rapid advance to the market economy on "shock therapy" lines, is that the major part of the transition reforms consist of institutional and structural changes. This is not just a question of passing laws and creating new legal institutions but also of developing the informal codes and standards of behaviour, including managerial and business practices, which underpin the workings of all effective institutions. The institutional capital of a country also includes the accumulation of skills and experience. Economists of the Austrian School, most prominently Ludwig von Mises and Friedrich von Hayek, are nowadays frequently quoted to explain the failures of central planning in eastern Europe and the former Soviet Union but are rarely invoked when the transition process is being discussed. Yet one of their key points is highly relevant to the attempt to construct a market economy: not only did they emphasize the primordial importance of the institutional infrastructure for the effective operation of the market but they also stressed the fact of the evolutionary development of institutions and behaviour over extended periods of time. Institutions in a legal sense can be easily replicated but the development of institutional effectiveness is a much slower process of learning and behavioural adaptation. Those who believe that the former centrally planned economies can be "shocked" into market economies would appear to have overlooked some of the most fundamental requirements for a market economy to function effectively.[25]

Many of these points congregate on the question of privatization. "Small" privatization (shops, services, small-scale enterprises, etc.) has developed rapidly and proved very popular in the countries which have implemented it. But the difficult problems arise with the privatization of the large — and by western standards, extremely large — state enterprises. In many respects "large" privatization (large state-owned enterprises) is the core of the restructuring problem and one of the most crucial stages in the transition to the market economy. But a rapid change of ownership of these enterprises will not automatically solve their problems or even assist in the creation of an efficient market economy. To sell such enterprises without restructuring them would undermine the creation of a competitive environment, although there are doubts that many of them would actually interest private buyers in their present state.[26] Questions of competitive structure, of effective corporate governance etc., all need to be settled before privatization takes place, otherwise the new private owners will probably resist any belated attempt to correct for market failure. A slower pace for "large" privatization, giving time for such questions to be settled, as well as for the improvement of managerial skills, the corporatization of enterprises, and the adoption of commercial patterns of behaviour, may thus be more effective in achieving the ultimate objective of an efficient and sustainable market economy.[27] The case for gradualism has been put very succinctly by the Prime Minister of the Czech Republic, Mr. Vaclav Klaus: asked whether he had plans to "speed up the privatization process" he replied that "privatization is a major social and multidimensional process. It cannot be speeded up. It is something which has its own dynamics and there are so many dimensions to it that I don't believe it can be accelerated."[28]

(v) **Popular support for the transition process**

The widespread and enthusiastic support for radical political and economic change that followed the revolutions of 1989 has now given way to a large measure of disillusionment and discontent as the costs of adjustment have proved to be much greater than expected.

Some of the reasons for this falling off in popular support were suggested in the previous section: a failure to construct coherent and transparent programmes for the transition and to accompany them with serious efforts to explain them to the electorates and to build a sustainable consensus in support of the reforms over the medium term. Expectations about the probable speed of reform — and of improvements in living standards — were greatly exaggerated at the end of 1989. At the same time, the stabilization policies adopted and the rapid introduction of price and trade liberalization, in the absence of adequate monetary and fiscal institutions and endemic microeconomic distortions, led to a much deeper and longer recession than any eastern political leader or policy adviser expected. Consequently, not only have expectations of improved

---

[24] See Michael Bruno, "Stabilization and Reform in Eastern Europe. A Preliminary Evaluation", IMF, *Staff Papers*, vol.39, No.4, December 1992, p.752.

[25] See United Nations Economic Commission for Europe, *Economic Survey of Europe in 1989-1990*, New York, 1990, pp.22-23.

[26] See section 3.6(viii) below.

[27] Although the situation of very large state-owned enterprises figures prominently in the discussion of privatization, it should be emphasized that the problems mentioned above — of corporate governance, the improvement of management skills, the adoption of commercial behaviour, etc. — apply to most firms of all sizes where there is — or will be — a separation of management and ownership.

[28] *Financial Times*, 11 January 1993, in a special supplement sponsored by the Czech Ministry of Foreign Affairs. In the same interview Mr. Klaus also criticized those who mixed up the short-term and long-term dimensions of issues — "short-term gains may well be long-term losses and therefore we are wary of economic myopia".

living standards been rudely disappointed, but the costs of adjustment are so much greater than expected that living standards are perceived to have fallen sharply for large sections of the population. Growing resentment at this outcome is increased by the tendency of shock therapists, as well as many political leaders, to ignore the democratic requirement to explain and to allow time not only for discussion but also for people to adjust psychologically to the new orientations of economic and social policies.[29]

In Hungary, 4 million voters turned out in March 1990 for the first free elections in 40 years, but within six months the Hungarian Democratic Forum (MDF) was losing by-elections on a 30 per cent turnout and in 1991 only 8 per cent of the electorate was bothering to participate in some by-elections. By the middle of 1992 opinion polls were recording widespread dissatisfaction with the economic (87 per cent) and political (72 per cent) situations.[30] A similar degree of discontent was apparent at the start of 1992 in Poland: 87 per cent were dissatisfied with the political situation and 80 per cent with their standard of living.[31] The Hungarians put most of the blame for their situation on their government, but rising economic and political dissatisfaction in Poland has been accompanied by falling confidence in most institutions, the police and the military being the principal exceptions.

Frustration is also considerable in Russia. In a poll of Russians in October 1992 only 5 per cent were satisfied with the economic situation, 16 per cent with their standard of living, and 10 per cent with the political situation.[32] Moreover, a large majority was dissatisfied with the political *system* and there is little enthusiasm for the main political leaders. Only the church and the military received majority approval from those polled. These results do not necessarily imply that Russians are indifferent to efforts to introduce more democratic ways of government, to extend the domain of civil society, to develop an independent and objective media network, and so on. Earlier polls indicate significant support for these objectives, although there is a large proportion of people with little understanding of the points at issue — the "don't knows". However, the grinding costs of economic disruption and the impact of the January 1992 price liberalization appear to be behind the rapidly growing majority of Russians who would prefer greater social security to more political freedom. Most Russians face the prospects for 1993 with considerable trepidation.

Popular attitudes to specifically economic reforms are complex and often contradictory. In eastern Europe a majority of the population is by and large in favour of the market economy but when questions get closer to their personal situation the responses are more guarded. A succession of polls in Hungary suggested that while a majority supports the reform process, those who support privatization *and* are willing to accept the costs (in terms of unemployment and increased inequality of incomes) are in a minority which is getting smaller.[33] So-called "small" privatization is generally popular but there is considerable opposition to the transfer to private ownership of large enterprises, large estates and other items of wealth.[34] In April 1992 a majority of Hungarians supported the idea of a market economy, but only 35 per cent wanted an untrammelled market system, while a similar proportion wanted to preserve positive elements from the former system.[35] Similarly in Poland, only a small minority hankers after the old system of central planning, narrow wage differentials etc., but a large majority in January 1992 was in favour of retaining specific features of the former regime, the most prominent being the state welfare system (92 per cent) and full employment (82 per cent).[36] There is still a lot of resistance to privatization in Poland, especially among the workers, a majority of whom would prefer continued state ownership or some form of control by employees. Plans for a free distribution of stock in 400 large Polish enterprises was abandoned in early 1992 because of political opposition.

The most positive attitudes towards the market economy in general and privatization in particular are to be found in the Czech Republic, and such attitudes appear to have strengthened over the last two years, in contrast with developments in other east European countries. In Slovakia, however, there is much greater fear about the consequences of privatization and in 1992 this unease was a source of support for those parties promising a change in economic policies and a more moderate pace of privatization than in the Czech Republic.

The situation in east Germany is, of course, a special one, not least because of the considerable transfer payments from the western part of the country which have supported disposable incomes in the east. Ac-

---

29 That people usually take some time to recognize a radical change in their environment is often seen more clearly by non-economists. It " ... is not only a state of affairs, but a process of gradual realization. First one has to get used to the idea of it. The idea then has to become part of everyday life. Then rules can change, rules of behaviour, of language, of expectations". The writer is actually referring to the approach of war in Croatia but her description probably applies to the reaction of most people to severe shocks. See Slavenka Drakulic, *Balkan Express. Fragments from the Other Side of War*, London, Hutchinson, 1993, p.18.

30 Judith Pataki, "Hungarians dissatisfied with political changes", *RFE/RL Research Report*, 6 November 1992, pp.66-70.

31 Michael Deis and Jill Chin, "Roundup: Life in Poland", *RFE/RL Research Report*, 22 May 1992, pp.62-63.

32 Mark Rhodes, "Political Attitudes in Russia", *RFE/RL Research Report*, 15 January 1993.

33 E. Hann and M. Laki, "The Hungarian Public on the Advance of Private Economy", *Acta Oeconomica*, 44 (1-2), 1992, pp.191-200.

34 Loc.cit.

35 Yudit Kiss, "Privatization in Hungary — Two years later", *Soviet Studies*, 44(6), 1992, pp.1015-1038.

36 Michael Deis and Jill Chin, loc.cit.

cording to one opinion poll only 15 per cent of east Germans considered themselves less well off in summer 1992 compared with their situation before unification and 42 per cent said they were better off. But the share of the population with a favourable opinion of the market economy dropped between the spring of 1990 and the summer of 1992 from 77 per cent to 44 per cent. The proportion of persons supporting the closure of non-viable firms has fallen and support for state intervention to restructure firms has increased. These changes in attitude appear to reflect the personal stress associated with systemic change, a perceived lack of personal security, and the sudden collapse of the former organization of private lives, which has led to considerable disorientation about the future.[37]

What conclusions can be drawn from such surveys of public opinion? The contradictions and qualifications surrounding general statements of support for the market system should not be surprising to those familiar with the western market economies: voters have frequently and notoriously voted simultaneously in favour of increased public expenditure while resisting any increase in taxes; they may also support government programmes to restructure national industries but oppose any change in their own, especially if their own jobs are at risk. The eastern electorates are just as contradictory: at one level they complain that there is too much change while at another they object that there is too little (often when they see familiar faces from the old *nomenklatura* still occupying positions of power and influence). For the individual, actual or imminent private cost will nearly always loom larger in his or her calculations than any notional social benefit. The problem of overcoming such resistance was recognized a long time ago. When Macchiavelli declared that " ... there is nothing more difficult to take in hand, more perilous to conduct, or more uncertain in its success than to take the lead in the introduction of a new order of things",[38] he identified two reasons for failure: first, the determined resistance of those benefiting from the status quo and, second, the lukewarm support from those who stood to benefit from the reform. The latter's pusillanimity was due in part to fear of the former's hostility but also to uncertainty about the outcome, "men ... never really trusting new things unless they have tested them by experience".[39]

In modern political economy the frustration of socially desirable reforms is most often explained by the potential losers, those with an interest in the *status quo*, being able either to out-vote the potential gainers, or to out-manoeuvre them through superior lobbying power.[40] However, neither of these possibilities appears to fit the situation in the transition economies. The status quo, to a large extent, has already lost the vote in favour of reform in general and, anyway, many of those who did *well* under the old system appear to be doing well under the new. So why should an increasing majority in the eastern countries be reluctant to support the specific measures of reform which, it believes, will benefit the population and the economy as a whole? A recent attempt to answer this question[41] focuses on the *ex ante* uncertainty concerning the distribution of *individual* costs and benefits: if *individuals* are uncertain as to whether *they* will benefit from a reform, there may be a bias against the change even if it is certain that the majority will benefit. Individual uncertainty can therefore distort the aggregate vote − the majority might vote against itself.[42]

Creating and maintaining support for the transition

How should governments respond to the ambiguous and weakening support for the transition, as described above? One response might be to dismiss the evidence, either as being inaccurate, a common retort of politicians when the results are not congenial, or as an inevitable short-run reaction to the initial costs of otherwise sound policies. In both cases the recommendation would be to push ahead with the chosen policy and, in the second case, patiently wait for the return of public support when the promised benefits of the policies begin to appear. The analysis of individuals' *ex ante* uncertainty, referred to above, might appear to justify such a course. However, there are a number of difficulties with this response, especially in the context of the transition countries. One is the scale and complexity of the reforms being undertaken, more or less simultaneously, which ensure that a large majority of the population will be adversely affected by the changes in the short run − and the potential resistance is therefore very large. Another is the considerable uncertainty surrounding the time required to produce the promised benefits − there is a good chance that they will arrive too late to save the government from electoral defeat. The prospect of the latter will, in turn, be an incentive to abort or weaken the reform. Finally, the data given above indicate that there is already a widespread loss of patience with governments and, presumably, less willingness to take at face value the promises and predictions of policy makers. The governments of the transition economies are therefore unlikely to succeed by simply "charging on". Instead they will have to secure a large and broad-based measure of popular support if the transition to the market econ-

---

[37] See Renate Köcher, "Die Ostdeutschen frösteln in der Freiheit", *Frankfurter Allgemeine Zeitung*, 9 September 1992.

[38] Niccolò Macchiavelli, *The Prince*, Penguin Books, 1961, chapter VI, p.51.

[39] Ibid.

[40] The second possibility may in turn derive from an asymmetry in the distribution of costs and benefits, a situation which is often used to explain the persistence of import restrictions whose removal would produce a net social benefit.

[41] Raquel Fernandez and Dani Rodrik, "Resistance to Reform: Status Quo Bias in the Presence of Individual-Specific Uncertainty", *The American Economic Review*, 81 (5), December 1991, pp.1146-1155.

[42] *Ex ante* hostility to reforms can therefore quickly change to *ex post* support once they have been "tested by experience" and seen to work. Fernandez and Rodrik, loc.cit., p.1154, point out that radical trade reforms introduced by autocratic regimes in the face of opposition or little support have subsequently proved to be popular. The examples they give are South Korea (early 1960s), Chile (1970s) and Turkey (early 1980s).

omy is to be achieved within a framework of democratic institutions.[43]

However, not the least of the problems in most of eastern Europe and the former Soviet Union is that political leaders and policy makers have invested few resources into building up a national consensus behind well articulated *programmes* of reform. As was remarked in this *Survey* three years ago,[44] the objective of creating a market economy is a general statement of principle which is vague both about the precise nature of the objective and the route to achieve it. It is therefore not very difficult for opinion polls to find considerable support "for the market economy", although, even at this level of generality, majority support appears to be lacking in Russia and other former republics of the Soviet Union.[45] The difficult task is to draw up a programme of action in which all the component parts are clearly related, in which a realistic timetable for achievement is set out, and where estimates of the inevitable costs are laid out together with policies to attenuate their impact on specific groups in society. That done, the programme must be explained and "sold" to the electorates if the required consensus for each stage of the reform is to be created. Hardly anything of this sort has been carried out systematically in the transition economies.[46]

Instead, government policies often appear to be characterized by a large measure of improvisation and by series of ad hoc measures which may or may not be coherent. Moreover, communication with electorates has been hampered by the lack of broadly-based parties in most of the transition economies and in many cases by weak news media. Parliaments are also, for the most part, composed of myriads of small parties supporting inevitably fragile coalition governments, which tend to be subject to a constant barrage of conflicting criticism from within and without the ranks of the government. It is thus perhaps not very surprising that popular support for the details of reform has been weakening. In the middle of last year over half of Hungarians felt that their government was not very close to the people and that the distance was increasing.[47] In Russia, where a number of radical changes were introduced last year, there was virtually no attempt by the Government to launch a propaganda campaign to explain and gather popular support for the Gaidar programme. The contrast with the recent efforts of the President of the United States to explain and build support for *his* economic programme could not be more stark.

It was the fashion in a few western countries in the 1980s to deride the notion of consensus since, it was claimed, the search for it was an obstacle to action by governments and rulers who knew best what to do. Such an autocratic approach may be sustained for a while if governments have large parliamentary majorities and are subject to few constitutional restraints by the legislature. But it is difficult to see how the transition process can be sustained in emerging (or re-emerging) democracies without a large and sustained measure of popular approval. A consensus has to be created and maintained to carry out the programme of reform over a fairly long period and, almost certainly, beyond the life of the government that initiates it.

The need for consensus on a coherent programme for transition is essential because, as mentioned already, many of the institutional changes which constitute the building blocks of the entire process are not details which can be left to economic technicians but matters which will influence the type of market economy and, indeed, the nature of the society that will emerge at the end of the process. Means and ends cannot be easily separated when the objective is a radical systemic change. Consensus on the institutional structure of the market economy is thus important for future societal stability because it will help to underpin the removal of ideology from day-to-day economic decision-making. But it will also help to insulate the institutions of civil society from the transitory popularity of individual governments.[48]

Another reason for eastern policy makers to pay much more attention to explaining and "selling" their programmes for economic reform in a broader institutional and social framework is the widespread anxiety and disorientation in the eastern countries that many observers report. This refers not only to economic security but also to the *ennui* associated with a collapse of settled values and patterns of behaviour, as well as the problems of "identity" that arise when states break up and ethnic tensions are revived. The scale of these problems varies, of course: they seem to be marginal in the Czech Republic but considerable in Russia and the former German Democratic Republic. Probably most periods of rapid systemic change give rise simultaneously to raised expectations and to widespread anxiety, and so the apparent contradiction between a majority being both in favour of change and of wanting to hold it back is not surprising. But that is the point where governments must provide a vision of the way ahead and a credible set of policies for moving forward. A distinguished French historian has argued that "the attitudes of individuals and groups of individuals to their own situation in society and the conduct these attitudes dictate are determined not so much by actual

---

[43] Machiavelli was not impressed by the efficacy of persuasion and thought that force was the more effective instrument for changing the order of things.

[44] United Nations Economic Commission for Europe, *Economic Survey of Europe in 1989-1990*, New York, 1990, pp.14-15.

[45] See above, and also, Commission of the European Communities, *Central and Eastern Eurobarometer*, Brussels, February 1993.

[46] There was, of course, the Balcerowicz Plan for Poland in January 1990 and the four-year Kupa programme for Hungary in February 1991, but the former was inevitably dominated by the severe stabilization problems facing Poland at the end of 1989 and the latter was largely a set of macroeconomic targets rather than an articulated programme of action. The Czechoslovak *Scenario for Economic Reform* (1990) corresponds most closely to the suggestion made here.

[47] Judith Pataki, loc.cit., p.67.

[48] This is not to say that institutional questions can be removed from politics, only that the prospects for stability are greatly strengthened if most of the people have confidence in most of their institutions most of the time.

conditions as by the image in the minds of the individuals and groups".[49] It is the task of the policy maker and, above all, of the political leader, to alter that "image" in such a way as to support the dynamics of the transition process and the building of a democratic, market-based society. Failure in this regard will not only compromise the transition but is likely to enlarge the constituency for ethnic chauvinism and other outlets for frustrated expectations and unrelieved anxieties.

The difficulties facing the governments of the transition economies are considerable and building a consensus for the transition is essentially a task for national politics. But insofar as western governments and the international financial institutions have some influence through the provision of financial and technical assistance, they can encourage the preparation of transparent programmes of structural change and, following the example of the Marshall Planners, even make them part of the conditionality of aid. Specialized technical assistance in this area can also be very useful.[50] However, one important step would be for western and international donors to insist on discussing programmes not only with governments but also with opposition parties and other social groups. In this regard, the excessive confidentiality which frequently surrounds governments' discussions with the Bretton Woods institutions[51] is counter-productive: not only does it encourage many unfounded allegations about these institutions *and* western governments, but no-one has ever demonstrated that a lack of openness improves the quality of economic policy making. Greater transparency is also desirable within the transition countries, especially on issues such as the privatization of state enterprises.

The role of safety nets

Despite the manifest economic and political failures of the former communist regimes and, in eastern Europe, their association with Soviet hegemony, it would be wrong to assume that the populations of these countries have rejected every single aspect of the former system. The opinion surveys discussed above are clear, that while rejecting central planning and all its coercive features, a significant proportion of the population in eastern Europe wishes to retain what it regards as positive elements of the previous system, notably its provision for welfare and job security.[52] This possibly reflects a more egalitarian preference than is the case among west Europeans, who in turn tend to be less tolerant of large income differences and more welfare-minded than Americans. However, it is more likely that the main reason is the anxiety referred to already and the real fear not only of losing one's job but of not being able to function adequately in the more competitive environment of the prospective market economy. It is clear from the opinion surveys that support for the reform process and for privatization is strongest among the young, the better educated, and the better off; and that fear of reform and resistance to privatization is greatest among the workers, especially in large state enterprises, the less educated and those at the bottom of the income scale. The task of policy is therefore to ensure that the energy of the first group is not frustrated by the fears of the second.

Clearly, not all of the perceived "positive elements" of the former system can be retained in the transition process: some of them are simply inconsistent with a market economy, while the scale on which welfare was previously provided is probably impossible to maintain without putting intolerable strain on general government budgets. Nevertheless, an effective safety net does have to be in place if the reform process is not to be blocked by those who think they might lose. So far this has not been a major issue, but it seems likely to become so as the major hurdle of privatizing the large state enterprises is approached.

The reorganization or reconstruction of social safety nets involves a number of detailed issues but two points can be stressed here. One is that safety nets cover a wide range of situations of which unemployment is only one — from the perspective of minimizing interference with labour market incentives, unemployment benefit should be clearly separated from other types of benefit.[53] Secondly, insofar as many welfare functions in the centrally planned economy were carried out by enterprises, especially in the states of the former Soviet Union, the reorganization of welfare systems is closely connected with the restructuring of state enterprises: before the latter are privatized the former will need to be nationalized.[54]

The proponents of "shock therapy" did acknowledge the need for adequate social security nets, but their assessment of what would be adequate was based on an overoptimistic view of the speed and likely costs of transition. The depth and length of the "adjustment recession" are so much greater than expected that the ability of governments to support adequate safety nets has been severely eroded. If the minimum level of

---

49 George Duby, quoted in Linda Colley, *Britons: Forging the Nation 1707-1837*, Yale University Press, 1992, p.43.

50 For example, the BBC has helped Radio Russia to produce a 26-part radio serial on "The Free Market Economy" and Radio Free Europe has produced a series on democracy and other subjects, sometimes in response to specific requests from eastern governments.

51 However secrecy is not just a problem with these institutions. The European Parliament complained last year about the secrecy which surrounded the Commission's negotiations on the Association Agreements with Czechoslovakia, Hungary and Poland.

52 An opinion poll carried out in April 1991 asked the question "What kind of a society should be built in the Russian Federation?". Twenty nine per cent of Russians answered "socialism", 3 per cent "capitalism" and 56 per cent "a society combining the best features of both social orders", *Izvestiya*, 24 April 1991, p.1.

53 See below, section 3.2(i).

54 In the Soviet-type economy the "factory was less a specialized institution and school of modernity than a functionally diffuse neopatriarchal provider of houses, vacations, medical attention, food, and, to some extent, social activity for its workers". Ken Jowitt, *New World Disorder — The Leninist Extinction*, Berkeley: University of California Press, 1993, pp.289-290.

provision falls below popular perceptions of what is acceptable and if international assistance is not sufficient to raise the threshold level of support, then the consensus for maintaining the transition process is likely to fall apart.

In preparing to cope with the social consequences of "large" privatization, one of the first steps for government to take is to make some estimate of the distributional impact of market reform on individual industries and enterprises. Some work has already been done which, for some countries, should help to identify the extent to which particular sectors will be "shocked" by the onset of market pressures.[55] Such estimates will not only provide a guide to the likely scale of government expenditure on unemployment benefits but, just as important, will indicate the priority areas for retraining programmes and other forms of adjustment assistance aimed at integrating workers into the market economy. The emphasis should be on augmenting human capital rather than the preservation of redundant physical capacities. As "large" privatization proceeds, this should be reflected by a falling share of enterprise subsidies in public expenditure and a rising share for active labour market policies.

Placing retraining in the context of the social safety net emphasizes the need for an integrated set of policies designed to assist the process of structural adjustment rather than just to provide a static net to catch the casualties. Coordinating safety net and active labour-market policies in coherent programmes of industrial restructuring should make it easier to attract western financial assistance to ease the pressures on the pace of government deficit reduction and would provide a sharper focus and a clearer set of priorities for foreign technical assistance. But this also raises the larger issue of industrial policy.

(vi) Industrial policy

In the 1980s there was a widespread reaction in western Europe and North America against the idea that the state had a major role to play in the running of the economy. Some of the critics were extreme, virtually claiming that state intervention could never do good and would invariably do harm. Behind the polemics was a serious argument about the feasibility of effective fiscal policy which had its roots in the macroeconomic failures of the 1970s. The conclusion of this, often bitter, debate appears to be that Keynesian intervention in the economy *can* influence the level of activity but that the constraints on such action are much greater than in the 1950s and 1960s.[56]

Parallel with the attacks on Keynesian macroeconomics was a similar assault on the efficiency of government industrial policies, whether targeted at whole sectors or individual enterprises. This attack was also a mixture of polemics and serious debate. The traditional case for state intervention rested on the belief that it was capable of correcting a range of market imperfections. The principal ones are (a) positive or negative externalities which would lead to suboptimal or excessive levels of investment in the activities which generated them; (b) the presence of economies of scale which might lead to reductions in competition and even to monopoly; and (c) imperfections in financial markets which might lead to suboptimal investment in risky or very long-term projects.

The intellectual attack on industrial policy was rooted in the theory of rent-seeking[57] and the economics of bureaucracy. State intervention will anyway tend to create distortions (by increasing taxes to pay for the intervention), but the political economy of rent-seeking emphasizes the diversion of resources into lobbying and other activities designed to influence government action in favour of one group of producers or another. Bureaucrats may be fooled or overwhelmed by the arguments and pressures of sectional interests, and they may even ally with them in order to increase their own budgets and power.

Again, the conclusion of the argument appears not to deny a role for state intervention but to underline the need to assess the costs more carefully and comprehensively and to be on guard against the distortionary effects of rent-seekers and bureaucratic empire builders. It also warns against the facile assumption that market failure can always be turned into government success. Measures to overcome such problems might include a clear institutional framework for conducting policy, an insistence on detailed and transparent programmes, and provision for effective parliamentary scrutiny and independent auditing. These, it will be recalled, are among the desirable characteristics of the programme for transition suggested as necessary in the previous sections.

The recent renewal of interest in industrial policy in western Europe and, especially, in the United States is much more focused on the externalities that might encourage economic growth rather than on the spotting of "winners" or "national champions" in specific industries or products. In particular, positive externalities are believed to exist in business investment, especially in machinery and equipment;[58] in public infrastructure

---

55 See, for example, Gordon Hughes and Paul Hare, "Industrial Policy and Restructuring Eastern Europe", Centre for Economic Policy Research, *Discussion Paper Series No.653*, London, March 1992.

56 This conclusion emerges from a large body of empirical research which showed that Ricardian equivalence was not perfect in practice. Another constraint on *national* fiscal policy arises from the greater openness and interdependence among the industrial economies. This suggests that the scope for fiscal policy will be greater if there is coordination among the major economies. See, G.D.N. Worswick, "The scope for macroeconomic policy to alleviate unemployment in western Europe", United Nations Economic Commission for Europe, *Discussion Papers*, Vol.2 (1992), No.3, New York, 1992.

57 For a collection of papers on this see J.M. Buchanan, R.D. Tollison and D. Tullock (eds.), *Towards a Theory of the Rent-Seeking Society*, Texas A and M University Press, 1980.

58 See J. Bradford De Long and Lawrence H. Summers, "Equipment Investment and Economic Growth: How Strong Is The Nexus?", *Brookings Papers on Economic Activity*, 2:1992, pp.157-211.

investment which can raise the productivity and rate of return of private investment[59] (a "crowding-in" effect); in research and development; and in education and training (or human capital investment). This analysis leads to policy recommendations which include tax relief (or subsidies) on fixed investment and saving, subsidies and increased expenditure on education and training, and tax credits for research and development. The new United States Administration's plan to promote technology and growth also includes proposals for (i) a network of manufacturing "extension services" to encourage a more rapid diffusion of new technologies and management techniques among small and medium-sized enterprises and (ii) the creation of regional networks of companies and research institutes which could cooperate in developing new products and markets. Some of these proposals are inspired by programmes which have already been implemented in western Europe (see below).

Many of these ideas are relevant to the problems of industrial structure and regeneration facing the eastern transition economies. Western experience in this domain, especially at the regional level in western Europe, could prove to be a rich source of relevant technical assistance and provide a basis for future cooperation. However, to be effective, both industrial policy and technical assistance from abroad need to be coordinated within a broad framework in which the government sets out its ideas on the broad sectoral composition of output ("high tech" versus heavy industry), desirable market structures (competition policy), and the scale of capital spending required including public infrastructure investment. Again, one returns to the need for coherent and transparent programmes as part of the effort to build efficient market economies.

The spirit of the new approach to industrial policy is caught in a comment by Nobel Laureate James Meade, whose commitment to the market economy is unquestioned: " ... the price mechanism is the worst possible form of economic system except the others ... Two hearty cheers for the Price Mechanism".[60] However, many policy advisers to the transition economies are still hostile to the whole idea of industrial policy which they see as working *against* rather than *with* market forces. To some extent this hostility reflects an extrapolation of ideas from a number of countries in the west in the 1980s when the emphasis was on deregulation and the removal of obstacles to competition. However that experience, the removal of marginal frictions in otherwise predominantly market economies, is not very relevant to the current problems facing the east. Moreover, the rejection of industrial policy as a matter of principle overlooks the fact that the governments of the transition economies are being forced into adopting industrial policies *de facto*.

A purely free market approach to restructuring is attractive in that it would avoid too much reliance on the state bureaucracy which many people in the transition economies either distrust or judge to be incompetent to carry out an effective industrial policy. However, such an approach is probably impossible to follow in that many enterprises would probably find no buyers in their present state while the alternative of a massive jump in unemployment would be politically unacceptable. Moreover, as noted already, there is a strong case for breaking up the large state-owned enterprises before privatization in order to ensure competitive market structures in the longer term.

Consequently, many eastern governments and their bureaucracies are already involved in trying to identify which enterprises — or parts of conglomerates — could survive in the market without any government help, those which could survive after a period of "care and restoration", and those which should be liquidated as soon as possible. This is proving to be a difficult task. The problems include bureaucratic rivalries, limited managerial competence in assessing the potential viability of enterprises in a market environment, the politicization of the selection process for deciding which enterprises should survive or be liquidated, shortages of good managers, and so on. There are no easy solutions, no "big bangs" on the agenda, as is shown by the experience of Poland which, since 1989, has experimented with six types of industrial policy in an attempt to overcome a variety of problems, such as those mentioned above, and to reconcile a number of conflicting problems and interests.[61] In Germany the *Treuhand* has also been forced into the active management of the reconstruction of east German industry. Fears of regional de-industrialization have led the Federal Government to support Treuhand policies to preserve a core of local industrial activities.[62]

One approach to industrial policy, which has had considerable success in western Europe, is the development of regional partnerships between enterprises, business associations, and the local or regional government. If the local government is large enough to be responsible for economic strategy and major infrastructure investment within its region, powerful mechanisms can be set up for promoting regional development and providing effective support services for individual businesses.[63] Among the key conditions for success in such regional schemes are the creation of a vision of future development behind which a consensus can be created of all the interested parties; the need

---

59 See Alicia H. Munnell, "Infrastructure Investment and Economic Growth", *Journal of Economic Perspectives*, Volume 6, No.4, Fall 1992, pp.189-198.

60 J.E. Meade, *The Intelligent Radical's Guide to Economic Policy*, London: Allen and Unwin, 1975, p.123.

61 For a most useful account of these policies and the pressures which drove the government from one to another, see Ben Slay, "Evolution of Industrial Policy in Poland since 1989", *RFE RL Research Report*, Vol.2, No.2, 8 January 1993.

62 Large-scale, regional industrial decline tends to meet considerable political and social resistance in democratic societies. On this point see Martin Bangemann, *Mut zum Dialog*, Bonn, 1992, p.20.

63 For a succinct survey of four highly successful programmes of this type (Catalonia, Lombardia, Limburg and Hamburg) see Coopers and Lybrand, *Lessons from Continental Europe: Promoting Partnership for Local Economic Development and Business Support in the UK*, Business in the Community, London, 1992.

for the involvement of all concerned in the drawing up of strategic plans; the use of formal (business) planning techniques to define strategies clearly; and the direction of effort towards a few key goals based on a realistic assessment of available resources.[64] These requirements are clearly very similar to those suggested above for the broader programmes of transition.

However, at present, most of the transition economies are grappling with a considerable array of problems which, as the Polish experience suggests, could easily end in policy grid-lock. These are precisely the circumstances where an appropriate amount of well-targeted and timely assistance from abroad could make a considerable difference to the chances of success or failure.

The above discussion suggests a number of obvious targets for foreign assistance which could directly ease the problems of industrial restructuring. However, it would be a mistake to focus only on industrial or micro-targets. In the discussion of "shock therapy" above, it was suggested that rapid price liberalization might be a very risky policy if effective fiscal and monetary institutions were not in place. In fact, fiscal and monetary instruments in most of the transition economies are still very imperfect. On the one hand, tax collection systems are being created but they are still a long way from being comparable to those in western Europe. The burden of adjusting government budget deficits therefore falls almost entirely on expenditure. On the other hand, monetary policy is not very effective in controlling private sector behaviour when the commercial banking system is still rudimentary and when high levels of interenterprise debt allow most state enterprises to escape from the reach of the monetary authorities. However, the one point in the economy where the monetary authorities can be relatively effective is in deciding whether or not to finance the government deficit. Consequently the entire burden of macro stabilization policy falls largely on government expenditure. This is obviously a very dangerous tendency because, if it is allowed to continue, the resources for expenditure on social safety nets will be diminished just at the moment when the need for them is going to increase considerably. There will also be little chance of getting a long-term programme of public infrastructure investment under way which is needed to encourage a revival of private investment and thus a restructuring of productive capacities. Instead, the further cut in demand is likely to deepen or prolong the recession, discourage private investment (both domestic and foreign) and to make it even more difficult for governments to resolve the dilemmas of industrial restructuring. Moreover, further reductions in public expenditure will become increasingly difficult in the face of rising unemployment and growing popular resistance to more cuts in social programmes.

The appropriate targets for foreign assistance should therefore include the public sector. Foreign grants and loans would not only remove a growing constraint on infrastructure investment and restructuring but would also help to dampen the inflationary consequences of deficits as well as boosting reserves of hard currency. They would also help to contain growing discontent and potential opposition to the transition process. However, for assistance to be targeted, targets must first be selected; and for assistance to be effective the targets must be coherently related in a clearly articulated programme. The continuation of aid for any reasonable time will only be likely if it is made contingent on agreed conditions and targets being met. These can best be specified in such a programme: the conditions need not be too rigid — as long as the transition process is kept on course the aid will be well spent.

(vii) **Western support for the transition process**

The basic strategy envisaged by western governments for assisting the eastern countries in their transition to decentralized market economies was for an initial burst of technical assistance, backed up by improved market access for eastern exports to western markets and by IMF short-run financing and World Bank structural adjustment loans, which, it was hoped, would open the way for private investment to take over most of the task of economic restructuring. Technical assistance was needed to help in the creation of the institutional structures of a market economy and in training people to operate it effectively. Without such assistance it was thought that the capacity of the former centrally planned economies to absorb financial aid and attract foreign private investment would be limited. There was — and remains — considerable agreement on the need for such technical assistance. However, the ECE secretariat, in one of the earliest analyses of the subject, stressed that the relative *proportions* of technical assistance and financial aid to the eastern countries should be the reverse of those in the Marshall Plan which was "long on grant aid and short on technical assistance".[65] There was no suggestion that grant aid and credits were not important and a number of priority areas for such help, such as infrastructure investments, were suggested.[66] In addition, it was stressed that assistance would be most effective if, following the Marshall Plan model, it was targeted within coherent programmes of adjustment drawn up by the recipients themselves. The need was also identified for an overall coordinating function to be entrusted to an international institution which would help to clarify the needs for assistance, monitor progress and provide a forum for the exchange of ideas and experience.[67]

---

[64] Ibid., p.19.

[65] See United Nations Economic Commission for Europe, *Economic Survey of Europe in 1989-1990*, New York, 1990, p.15.

[66] Loc.cit., p.24.

[67] See below, chapter 4 for a discussion of some of these problems and a presentation of the available data.

If an attempt is made to assess the assistance effort in the light of these suggestions, the conclusions are not very favourable.

The first point to stress is that it is virtually impossible to obtain a clear cut, statistical picture of current assistance efforts. The "large" numbers which are frequently quoted in the press are usually commitments (of various degrees of strength and often tied to purchases from specific countries), and usually refer to a cumulative total since 1989 or 1990. Concessionary loans are mixed up with bilateral and other credits at market rates of interest, and commentators may or may not include estimates of the value of debt rescheduling in the numbers they quote.[68] The OECD maintains a data bank on financial flows, including development assistance to eastern Europe, and was also asked by the 1992 Washington and Lisbon Conferences to provide an information clearing house on requests for and offers of assistance to the CIS. However, the information provided by donors is neither complete, timely nor, in some cases, arranged in accordance with standard definitions of assistance. The latest available estimates of total gross financial flows, including genuine aid (i.e., grants and concessionary finance), are for 1990.

The reasons for this state of affairs are not very clear. To some extent they may simply represent the teething problems of organizing a new monitoring capability covering a large number of donors. The number of ministries and institutions involved in assistance within individual donor countries is also larger than those normally concerned with development assistance and this too complicates and delays the reporting to international institutions. But there is also an element of competition among donors to support the most prestigious projects and those with spin-offs for their own exporters — and this may tend to slow down the provision of timely data.

But whatever the reasons may be, if the basic function of monitoring is still in such a rudimentary state, it is hardly surprising that there has been growing criticism at the lack of international coordination of assistance to the transition economies.[69]

Poor monitoring does not simply undermine effective coordination among donors but can lead to unnecessary acrimony between donors and recipients. An example of this is the controversy over how much of the G-7's $24 billion aid package for Russia, announced on 1 April 1992, was actually delivered. Estimates have ranged from $8.5 billion to more than $24 billion, while the Managing Director of the IMF has stated that the G-7 obligation was "essentially" met.[70] Part of the argument arises from the protagonists referring to different parts of the package[71] but there still seems to be large differences in the numbers quoted for bilateral credits. It is very difficult to sort all this out when the data come from a variety of sources, not all of them easily accessible or referring to the same period of time. These are not inconsequential squabbles among economists. A large discrepancy between what is claimed to have been given and what is asserted to have been received could lead to a deterioration in a country's standing in relation to its creditors and so affect the flow of future finance. Disagreement may also damage the credibility of western donors in the eyes of the eastern electorates whose continued support for the transition programmes is essential for their success.

Donor credibility is also at risk from the increasing tendency to refer to virtually all financial flows to the transition countries, from outright grants to loans at market rates of interest, as "assistance". Credibility is stretched even further when such an aggregate is favourably compared to the scale of Marshall Aid which in fact was largely composed of grants.

An accurate monitoring of the levels and composition of financial flows to the transition economies is therefore urgently needed as an essential step in improving coordination among the providers of assistance.

In the area of technical assistance there is no doubt that a considerable amount of activity is under way and that it has expanded rapidly in the last two years. The European Community runs a large programme and there are important national efforts. Obviously it is very difficult to assess the economic impact of such assistance: it cannot be easily quantified, except as lists of projects, and the effects are often diffused in ways which are also difficult to measure. Anecdotal evidence suggests that a lot of this assistance is proving to be very useful but, as the programmes have grown, they too have been subject to the criticism of poor coordination among donors and questions have been raised about their cost in relation to their presumed benefit to the recipient countries.[72]

One of the most important aspects of the coordination of assistance is matching it with the needs of the transition economies. This does not mean financing any good project that happens to be presented to the donors but of identifying the key areas and problems which are constraining the development of private sector activity and the growth of the market economy. The increasing references to "well-targeted" aid, especially in relation to the problems in Russia, suggest that donor governments are preparing to put greater em-

---

68 See below, section 4.2.

69 See the comments of the United States Treasury Secretary, Mr. Brady, at the 1992 annual meeting of the EBRD (*Financial Times*, London, 13 April 1992). Also, United States General Accounting Office, *Former Soviet Union Assistance by the United States and Other Donors*, Washington, D.C., December 1992.

70 IMF, *Survey*, Washington, D.C., 22 February 1993.

71 Mr. Camdessus said "... frankly, there has always been confusion about the composition of the package", ibid.

72 On the EC programmes of assistance see the Court of Auditors, "Annual report concerning the financial year 1991 together with the institution's replies", *Official Journal of the European Communities*, C330, vol.35, 15 December 1992. The OECD has also criticized western technical assistance to the former Soviet Union for its relative neglect of certain sectors and for an unbalanced distribution between individual republics. *Financial Times*, 14 October 1992. (This latter report does not appear to be publicly available.)

phasis on this. Nevertheless, as noted above in the context of industrial policy, an efficient selection of targets is difficult without a broader framework for the transition in each country. One way of providing such a framework is to follow the example of the Marshall Plan in insisting that the recipient countries draw up their own assessment of their needs within a 5-10 year programme.[73] (Some of the poorer transition economies may need assistance to do this in order to avoid being left out in the distribution of resources.)

As for the actual flows of financial resources to the transition countries, not a great deal can be added at the present time to the picture presented in the last edition of the *Economic Bulletin for Europe*,[74] (for the reasons noted above).

The hopes for large inflows of *private foreign capital* to the transition economies have still not materialized for most of the countries concerned. Net inflows of foreign direct investment *have* risen, from about $0.6 billion in 1990 to just under $3 billion last year.[75] This is an encouraging improvement, but the amounts involved are still very small in relation to the $28 billion or so flowing to developing countries and world-wide flows of some $200 billion.[76] Moreover, the bulk of foreign direct investment (some 90 per cent) is concentrated in Hungary and the Czech Republic. FDI also increased in 1992 in Poland, Romania and Slovenia although the amounts still remain relatively small.

Hungary and the former Czechoslovakia are also the only transition countries which were able to borrow significant amounts over the last three years on the international capital markets.

The outlook for foreign private investment in the eastern countries remains uncertain. The well-known deterrents to investment include political uncertainties, which are considerable in some countries, incomplete market infrastructures and delays in resolving basic problems such as property rights. The much delayed recovery of activity in the eastern economies together with a deepening recession in western Europe, has led to western companies reassessing their investments in the east and to an increasing tendency to postpone new investments.

Thus, most of the transition economies are dependent for outside funds on official "assistance", multilateral, and government-guaranteed bilateral credits. As was emphasized above, it is extremely difficult to make a reliable estimate of the total flow of these resources — reporting is inadequate and some institutions are hesitant about releasing data on annual disbursements as opposed to cumulative commitments.

The ECE secretariat has estimated that the aggregate flow to eastern Europe in 1991 was probably of the order of $20 billion.[77] For 1992, there appears to have been a fall in multilateral financing, from some $8.5 billion to less than $5 billion, with special financing (debt write-offs, rescheduling, deferrals and arrears) accounting for another $5 billion or so in 1992. There is great uncertainty over what happened to officially guaranteed bilateral credits, so it is difficult to say whether the total funds available to eastern Europe increased or not last year.

In addition to these funds, the countries of the former Soviet Union received some $14 billion in special financing and Russia received perhaps $12 billion in bilateral credits and a further $1 billion from the IMF.

Altogether, perhaps some $40 billion or so of financing was made available to all the transition countries last year. Again, the uncertainty surrounding these figures must be emphasized. It should also be underlined that only a small proportion of this money consists of grant aid or concessionary finance as defined by the OECD's Development Assistance Committee. Bilateral credits are debt creating and carry market rates of interest. About one half of the above total is accounted for by debt rescheduling and other forms of "special financing". This gives important short-term relief, but of course the debt and the obligation to service it remains. Also, given the importance of conditionality, it should be noted that very little of the finance made available is directly focused on and integrated with specific programmes of reform. Finally, the numbers quoted here should be seen in the perspective of the scale of *net* government transfers in Germany to the new, eastern *Länder*: $75 billion in 1991 and $96 billion in 1992.

When debt servicing and other income payments are set in the balance against capital inflows, there was actually a net outflow of resources from most of the east European countries in 1992. Foreign resources have therefore done little to support consumption and investment, which fell in most of the transition countries, while the rise in imports was made possible by increased export earnings. For Russia, estimates for the first three quarters of 1992 indicate an *inflow* of resources of some $4 billion, which reflects new bilateral credits and special finance being partially offset by large amounts of debt servicing.

An increasingly worrisome aspect of the programmes for financial assistance is their uneven distribution. The countries which have had most success in attracting private capital are also those which have had least difficulty in meeting the requirements for short-run

---

[73] This point has been frequently stressed in previous issues of this *Economic Survey of Europe* and the *Economic Bulletin for Europe*.

[74] United Nations Economic Commission for Europe, *Economic Bulletin for Europe*, vol.44(1992), New York, 1993.

[75] Newspaper reports often give much larger figures than these, but the statistics for FDI are frequently subject to the same confusions as those noted above for financial assistance. Announcement of projects and intentions to invest may refer to an aggregate to be spread over several years and, for various reasons, the actual investment may be much less than originally announced. The above figures are also on a balance of payments cash, not accrual, basis. See below, section 4.2.

[76] Data for 1991. Some four fifths of annual FDI flows are among the developed market economies.

[77] United Nations Economic Commission for Europe, *Economic Bulletin for Europe*, vol.44(1992), New York, 1993, p.8 and chapter 3.

balance of payments support from official sources. However, for other countries the problems of obtaining timely support appear to be considerable. Despite estimates that they would need some $20 billion of external support in 1992, the successor states to the Soviet Union *excluding* Russia obtained very few commitments of help and even less in actual disbursement. The Baltic states were only able to draw on their first IMF credits in late 1992; commitments by the G-24 have so far fallen below target and disbursement is scheduled only for the early months of 1993. The problems surrounding this *short-term assistance* — significant long-term development assistance is still a long way off — are complex: they include delays on the part of donors in meeting their pledges, but the countries themselves are often unable to meet the standard requirements, laid down by the international financial institutions and individual donors, for obtaining funds. Also their commitment to reforms appears to be hesitant in some cases, although this is sometimes a reflection of the enormous difficulties they face and doubts about their own ability to overcome them. Whatever the explanation in individual cases, delayed and inadequate assistance simply makes their problems worse.

Within eastern Europe the problems of Albania, Bulgaria and Romania are particularly difficult, although their specific situations vary considerably. All three are among the poorest countries in Europe and the circumstances surrounding their attempts to build market economies have been very unfavourable. They all started with severe macroeconomic imbalances and Bulgaria suffered more than most transition countries from the collapse of CMEA trade. The effects of the Gulf war were also negative.[78] They are now suffering considerable additional costs as a result of the war in the former Yugoslavia and by their compliance with UN sanctions on trade with the Federal Republic of Yugoslavia. Nevertheless, they have made progress in their reforms, although, not surprisingly, less than in the countries which started earlier. However, pledges of assistance have not only fallen short of IMF estimates of their needs but disbursements have repeatedly been less than promised and subject to considerable delay. The international community has also been dilatory in taking steps to share the burden of UN sanctions in an equitable manner.

The problems of the Balkan states in obtaining assistance are by no means entirely located on the side of the providers of funds, but many of the domestic difficulties are rooted in the conditions noted above. These countries not only need generous financial assistance but they also need greater understanding of their difficulties in meeting the standard pre-conditions for obtaining it. In considering the needs of the Balkan states, west European governments might recall the fact that the flow of emigrants from these countries has been increasing rapidly. At the end of 1992 they accounted for some 36 per cent of all asylum seekers from the transition countries in western Europe, or 62 per cent if those from the former Yugoslavia are excluded.[79] Moreover, they also accounted for an estimated $1 billion out of total west European government expenditure on asylum seekers of around $8 billion in 1992. The problem for western governments is therefore becoming one of *where* to spend the money rather than *whether*.

The long recession in the market economies has not only made western governments more cautious about increasing financial assistance to the transition countries, but has also left them more willing to respond to protectionist pressures from particular interests in their own countries. One of the most positive developments in the last few years has been the ability of several eastern countries to partly offset the collapse in CMEA trade by boosting their exports to the west. In this they were helped by the boom in western, particularly German, import demand in 1990-1991 and by the reduction of western trade barriers, prominently exemplified by the Association Agreements between the European Community and Poland, Czechoslovakia and Hungary.[80] The Agreements represent an improvement on the previous situation, but their commitment to open markets and free trade is heavily qualified by a gradual relaxation of restrictions on "sensitive" products, which happen to be those where the current comparative advantages of eastern Europe are relatively concentrated — and by the anti-dumping and safeguard clauses which are heavily biased against the interests of exporters. These escape clauses also create considerable uncertainty for foreign and domestic investors in eastern Europe and therefore undermine the incentives to action which would speed up the integration of the eastern countries into the international economy.

The fears of many observers now appear to have been confirmed by the application of the safeguard clause against steel imports from the east and the subsequent imposition of anti-dumping duties on steel tubes from Czechoslovakia, Hungary and Poland. Proposals by the EC to introduce tariff quotas on a range of products and threats of further action in other sectors point to a significant retreat from the commitment to open markets and liberal trade policies. (It should be stressed that the European Community is not alone in its qualified commitment to liberal trade policies.) This is a particularly threatening development for the eastern countries, where the few signs of a recovery in output have been largely based on export growth. Western trade restrictions not only put that recovery at risk but also threaten to undermine the reform pro-

---

[78] On the effects of the Gulf crisis of 1990 and the ensuing war, see United Nations Economic Commission for Europe, *Economic Bulletin for Europe*, Vol.42(1990), New York, 1990, pp.38-40.

[79] Based on Intergovernmental Consultations on Asylum, Refugee and Migration Policies in Europe, North America and Australia, *Asylum-seekers in Western Europe (EC and EFTA) in 1992*, Geneva, 18 December 1992, and direct communications. These estimates understate the true numbers because they exclude some 146,000 Albanian migrants in Greece and perhaps 100,000 ethnic Turks who have left Bulgaria for Turkey. Asylum-seekers in western Europe from the transition countries totalled nearly 1 million (of which 42 per cent were from the former Yugoslavia) out of a grand total of 2.6 million, see chapter 4, table 4.1.8.

[80] Agreements have also been reached with Bulgaria and Romania and they come into effect at the beginning of July and May respectively.

grammes as well. They also raise questions about the credibility of western commitments to supporting the reform process in the east, as has been frankly acknowledged by one of the EC Commissioners.[81]

Helping Russia

The considerable problems facing economic reform in Russia have been greatly exacerbated by unsettled questions of political authority and legitimacy. The constitutional crisis in the country, which has now come to a head, has thrown into stark relief the seriousness of the economic problems and the inability of the authorities to respond effectively. The spectre of economic and political instability in Russia spilling over into the countries of eastern Europe has alarmed governments in both east and west. This in turn has led to calls for emergency action on the part of the G-7 to help Russia. Certainly, speedy action in a number of areas could be helpful and a number of proposals have been made, including direct food aid and loan guarantees for the construction of housing for the Russian army. However, it is not clear that an "emergency" programme is the most appropriate response. A warning was sounded in this *Survey* a year ago that "the danger is that the growing introspection of some of the leading western countries will only be broken by the onset of a major crisis in one of the transition economies and that this could lead to hasty measures which would probably be less than optimal".[82] That danger is now imminent and is increased by disagreements among the western governments as to the scope for effective action: some argue for rapid action but others warn of the dangers of wasting resources, of giving money to people "with holes in their pockets". One of the worst outcomes would be a compromise which produced a list of short-term measures, hastily put together with a maximum of publicity, but with nothing to address the chronic underlying problems of monetary instability and structural change. What is needed therefore is a policy which can reconcile the demands for action with the sceptics' fears of wasting money. At the same time, every effort needs to be made to avoid yet another "false start" on the road of Russian reform: there have been too many since 1985 and the political costs in disillusion and apathy on the part of the Russian people are already considerable.

On the basis of the arguments set out in the sections above and in previous issues of this *Survey*,[83] what appears to be needed is a commitment to a long-term programme of reform on the scale of the post-war European Recovery Programme (ERP). On the one hand, the western countries would commit themselves to a significant and sustained level of technical and financial assistance (aid) over a period of, say, 8 to 10 years, while the Russian government would undertake to draw up a long-term programme of structural reform and accept that continued aid would be contingent on intermediate targets being met. However, at present the monitoring of economic developments in Russia (as well as in other transition economies) is seriously hampered by the lack of reliable statistics. This is one of the problems that requires emergency action.[84]

Considerable efforts would be necessary to win support from the Russian population. Recent suggestions by western economic advisers to the Russian government that western aid should be seen to be benefiting the population ties in well with the discussion above on building popular support for the reforms.

The commitment of both western governments and a Russian administration to such a programme could have the same powerful effect on confidence and expectations as the Marshall Plan produced in western Europe, and not least in western Germany, in the late 1940s. The announcement that the G-7 would be prepared to make such a commitment could have more influence on domestic developments in Russia than attempts at more direct intervention, which are likely to stand little chance of success in any case.

However, although Russia is of great strategic importance for economic and political stability in eastern Europe and the rest of the world, a major commitment to Russian reform should not "crowd out" the needs of the other transition economies. Even those regarded as being in the van of the transition process are still facing considerable problems which, left unsolved or solved without sensitivity to what is politically feasible, could lead to serious setbacks. Some countries, as noted above, are lagging behind in the transition and are virtually being overlooked by western policy makers. A much broader, regional perspective of the transition process is needed rather than the current tendency for assistance efforts to treat each eastern country in isolation. This tendency probably contributed to the serious underestimation of the effects of the sudden rupture of trade relations, first among the members of the CMEA, then among the republics of the former Soviet Union, and of the former Yugoslavia, and now, it appears, between the Czech and Slovak Republics. In the cases of the former USSR and CSFR and, in the future, with the states of the former Yugoslavia, multilateral efforts to restore trade and avoid further sudden disruptions, if necessary through the creation of payments unions, should be made with international support. A regional perspective is also

---

[81] Sir Leon Brittan, the EC Trade Commissioner, is reported as believing that the EC is reneging on its promises to eastern Europe in threatening to limit severely steel imports into the Community. *The Times*, 20 March 1993. The Industry Commissioner, Herr Martin Bangemann, however, has said, referring to the possible loss of 50,000 jobs in the steel industry, that "we cannot sustain these socially without the help of central and eastern European countries. We must be sure the EC market doesn't collapse". *Wall Street Journal*, 18 February 1993.

[82] United Nations Economic Commission for Europe, *Economic Survey of Europe in 1991-1992*, New York, 1992, p.10.

[83] United Nations Economic Commission for Europe, *Economic Survey of Europe in 1989-1990*, New York, 1990, pp.13-17, and the *Economic Bulletin for Europe*, vol.43(1991), New York, 1991, pp.7-9.

[84] On this problem see, *inter alia*, United Nations Economic Commission for Europe, *Economic Bulletin for Europe*, Vol.43(1991), New York, 1991, pp.4-5. The difficulties are also discussed at various points in the present volume – for a specific illustration, see section 3.3(i) below.

desirable for other reasons, not least in the preparation of major infrastructure projects.

An approach to the transition problems of Russia and eastern Europe, similar to the Marshall Plan or European Recovery Programme, would imply the need for a serious reappraisal of the current methods of assessing the requirements for and coordinating the delivery of assistance. It would have to include a review of the criteria on which the international financial institutions determine their assistance, and if governments think that greater flexibility in this area would damage the credibility of these institutions then alternative arrangements should be considered. In particular the problems of the coordination of assistance, which appear to be institutional, would need to be addressed jointly by the leaders of the G-7 and of the transition countries.[85]

The secretariat's earlier suggestion that a programme along the lines of the European Recovery Programme was needed to deal with the long-run problems of transition was dismissed on a variety of grounds: that the world was now very different from the 1940s, that no single coordinating authority could now oversee the programmes of a considerable number of powerful international organizations, and that the existing institutions and arrangements were more than capable of carrying out the job. But most of this is beside the point. The essential thing about the Marshall Plan was its clear and long-sighted vision of the donor's self-interest in supporting strong democracies and the reconstruction of market economies in western Europe and its unequivocal commitment to ensuring that they were established. The Marshall Plan, it should be remembered, also followed nearly two years of economic turmoil in western Europe and increasing complaints in the United States that billions of dollars were disappearing into western Europe without any tangible effects on reconstruction. There are signs that a reappraisal of western self-interest in a similar achievement in Russia and eastern Europe may now be under way. That provides some ground for optimism.

---

[85] One west European proposal, which is based on a recognition of these problems, calls for the appointment of an EC Commissioner with special responsibility for Central and Eastern Europe and the CIS and the preparation of a White Paper setting out an integrated approach to the problems of the former communist countries. The Action Committee for Europe (Chairman: Vicomte Etienne Davignon), *Proposals for Increased Assistance to Eastern Europe*, Brussels: September 1992.

Chapter 2

WESTERN EUROPE AND NORTH AMERICA

Economic conditions have deteriorated significantly in western Europe in the course of 1992 against a background of turbulence in foreign exchange markets. In contrast, in North America the recovery gained considerable momentum in the second half of the year. Inflationary pressures abated further in the western market economies, reflecting sluggish demand and reduced cost pressures. Employment fell in western Europe and the unemployment rate rose to nearly ten per cent at the end of 1992. Only small gains in employment were posted in the United States in 1992: the rate of unemployment started to fall in autumn. The short-run outlook is rather bleak for western Europe and real GDP is projected to broadly stagnate in 1993 compared with the previous year. In the United States the recovery is expected to continue although at a more moderate rate than that recorded during the second half of 1992.

2.1 OUTPUT AND THE COMPONENTS OF DEMAND

Moderate economic growth in the industrialized countries

Economic developments in the western market economies of the ECE region were marked by diverging cyclical changes in North America and western Europe in 1992. Against widespread expectations to the contrary, economic growth did not strengthen in western Europe in the course of the year; instead, economic conditions deteriorated significantly. In contrast, in North America the recovery gained momentum in the second half of the year, although the characteristic buoyancy of previous cyclical upswings was clearly lacking. For the industrialized countries as a group the strengthening of economic growth in North America, however, has been more than sufficient to offset the deceleration of economic activity in western Europe and Japan.

Hopes for a pervasive strengthening of economic growth in the industrialized countries in the course of 1992 had been temporarily nourished by the strong rebound in economic activity in the first quarter of the year (table 2.1.1). This, however, largely reflected special factors, *inter alia* the mild climate which favoured construction activity. In western Europe economic growth started to slow down again in the second quarter. Domestic demand weakened and this, in turn, affected demand for foreign goods, tendencies which, given the close trading links between west European countries, became mutually reinforcing in the second half of the year. The disappointing performance in the second quarter led to a more pessimistic re-assessment of short-term prospects. Consumer confidence, which was already fragile, started to fall markedly (chart 2.1.1) and this dampened the propensity to spend. Also business confidence deteriorated further and investment plans were curtailed.[1] The downturn continued in the second half of the year: Germany moved into recession; overall economic activity levels started to fall in France and Italy and stagnated at a low level in the United Kingdom. Economic activity also became increasingly sluggish among the smaller economies, of which some have been in recession since 1991.

Short-term economic indicators point to a severe downturn in industrial activity in the final months of the year, a decline which has continued in early 1993. Against a background of very high and rising unemployment, bleak sales prospects and persistent turbulence in the foreign exchange markets, consumer and business confidence in western Europe had fallen close to record low levels at the beginning of 1993 (chart 2.1.1).

In the United States, a recovery has been underway since the second quarter of 1991, but growth was very uneven and hesitant until the second quarter of 1992. In the third and fourth quarters of 1992, however, there was a pronounced acceleration in the pace of economic activity (table 2.1.1), reflecting strong growth in household demand and fixed investment as well as a contin-

---

[86] As reported in Commission of the European Communities, *European Economy*, Supplement B (monthly).

TABLE 2.1.1

Quarterly changes in real GDP, [a] **1991-1992**
(Percentage change over preceding period)

	1991				1992			
	QI	QII	QIII	QIV	QI	QII	QIII	QIV
France	0.1	0.7	1.0	0.2	0.8	0.2	0.5	-0.5
Germany [b]	1.7	0.8	-0.2	-0.3	2.0	-0.3	-0.5	-1.0
Italy	0.4	0.5	0.2	0.6	0.6	0.2	-0.6	..
United Kingdom	-0.5	-0.7	0.2	-0.1	-0.7	-0.2	0.1	0.2
Four countries above	0.5	0.4	0.3	-	0.8	-	-0.1	
United States	-0.8	0.4	0.3	0.1	0.7	0.4	0.8	1.2
Canada	-1.4	1.3	0.1	-	0.2	0.1	0.4	..
Japan	1.4	0.7	0.4	0.5	1.0	-0.2	-0.4	..
Seven countries	-0.1	0.5	0.3	0.2	0.8	0.1	0.3	..

Source: National statistics.

[a] At market prices; data are seasonally adjusted.
[b] West Germany.

uing favourable export performance. Consumer confidence, which has been very volatile over the last two years, rose markedly in the final months of 1992 (chart 2.1.1). Business confidence also strengthened.[87] However, there was a deterioration in consumer and business confidence in January and February 1993 against a background of adverse changes in a number of short-term economic indicators. Market sentiment, at the time of writing, is interpreting this latter development as a temporary setback rather than a reversal of the improvement in confidence.

The overall rate of economic growth also strengthened in Canada in the second half of the year, largely a reflection of stronger gains in private consumption and robust export demand from the United States.

For the industrialized countries in aggregate, real GDP rose by 1.5 per cent in 1992 compared with a moderate increase of 0.5 per cent in 1991 (table 2.1.2). This slight improvement in the average performance, however, is entirely due to stronger growth in North America, where real GDP, after falling in 1991, rose by some 1.5 per cent in 1992. In contrast, the annual rate of economic growth in western Europe was only 1 per cent in 1992, slightly less than in the previous year. But, as already noted above, this masks a sharp slowdown in the second half of the year. In Japan, where there is a pronounced economic downturn, real GDP rose by 1.3 per cent in 1992, a significant deceleration from a growth rate of 4 per cent in 1991. In fact, this was the lowest Japanese growth rate since the recession of 1974, when output actually declined. Chart 2.1.2 illustrates the varying strength of the international growth cycle between 1980 and 1992.

Although there has been a pervasive slow-down in economic growth in western Europe in 1992 there remains, nevertheless, a great diversity in the performance of individual countries (table 2.1.2). Only France, Ireland, Norway and Turkey had higher growth in 1992 than in 1991, the rate of expansion being much stronger in the three smaller economies. Finland, Sweden, and the United Kingdom experienced a second year of severe recession. For the United Kingdom this is the most severe downturn in the post-war period. Cumulative output losses of about 10 per cent in Finland in 1991-1992 were larger than those recorded in the depression of the 1930s. Switzerland has also experienced two years of falling GDP, but its recession is much more moderate when compared with the other three countries.

The diverging developments in western Europe and North America reflect to a large degree differences in the stance of monetary policy and in the timing of the business cycle. In the United States the significant easing of monetary policy in 1991 was continued in 1992 and this finally led to a more pronounced demand response in the second half of the year. In western Europe the economic downswing which started in 1989 was temporarily attenuated by the stimulus to demand stemming from German unification. This impulse petered out in the course of 1991 and was supplanted by the contractionary effects of tight monetary policy in Germany. High German interest rates have reduced the scope for an easing of monetary policy in most of western Europe, given the constraints imposed by exchange rate objectives. This has been a major factor restraining economic activity since the second half of 1991. Against a background of large public sector deficits in many countries, the scope for expansionary fiscal policy has been closely circumscribed. Also, in North America and several countries of western Europe private sector spending continues to be restrained by the need to reduce the large debt burdens accumulated during the 1980s.

Economic developments in the course of 1992 were overshadowed by the turbulence in the west European foreign exchange markets, which drove the Exchange Rate Mechanism (ERM) to the brink of collapse and led to the suspension of unilateral exchange rate pegs in Scandinavia. Realignments within the ERM and the

---

[87] As measured by the Purchasing Managers' Index.

CHART 2.1.1

Consumer confidence in the European Community and the United States, January 1990-February 1993

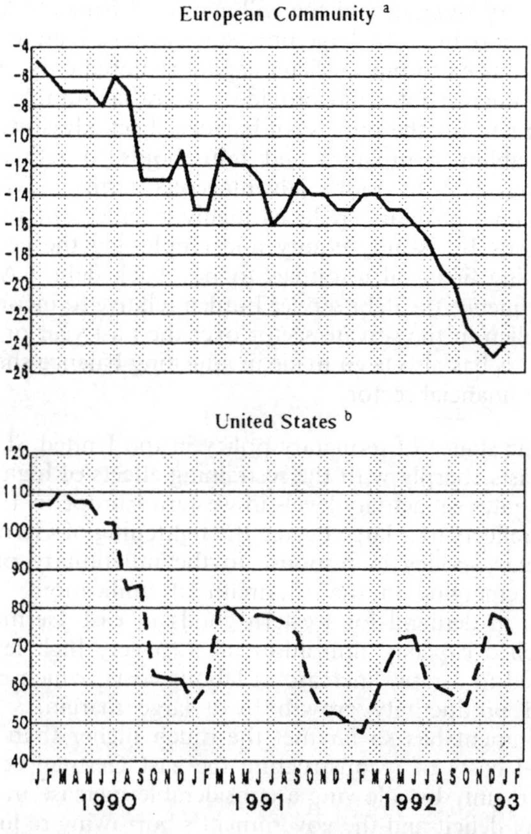

European Community [a]

United States [b]

Source: Commission of the European Communities, *European Economy*, Supplement B, monthly; United States Department of Commerce News, *Survey of Current Business* and Consumer Research Center, New York.

[a] Net balances between respondents giving positive and negative answers to several specific questions. For details see any edition of the source.
[b] Conference Board Index of consumer confidence, 1985 = 100.

TABLE 2.1.2

Annual changes in real GDP in western Europe and North America, 1990-1993
(Percentage change over previous year)

	1990	1991	1992	1993 [a]
Western Europe	2.8	1.1	1.0	-
4 major economies	2.7	1.2	1.0	-0.5
France	2.2	1.2	1.8	-0.5
Germany	5.1	3.7	1.5	-1.0
Italy	2.2	1.4	1.2	-0.5
United Kingdom	0.5	-2.2	-0.6	1.0
13 smaller economies	3.2	0.9	1.0	1.0
Austria	4.6	3.0	1.8	0.5
Belgium	3.7	2.1	1.2	0.5
Denmark	2.0	1.3	1.0	1.5
Finland	0.3	-6.4	-3.5	-
Greece	-0.2	1.8	1.2	1.0
Ireland	7.1	1.7	2.7	2.5
Netherlands	3.9	2.2	1.7	0.5
Norway	1.8	1.9	2.9	2.5
Portugal	4.4	2.0	1.9	1.0
Spain	3.6	2.4	1.2	0.5
Sweden	0.7	-1.8	-1.5	-1.5
Switzerland	2.3	-0.1	-0.6	0.5
Turkey	9.2	0.3	5.4	5.0
North America	0.7	-1.2	2.0	3.0
United States	0.8	-1.2	2.1	3.0
Canada	-0.5	-1.7	1.0	2.5
Total above	1.6	-0.3	1.6	2.0
Memorandum item:				
Japan	4.8	4.0	1.3	1.5
Total above including Japan	2.1	0.5	1.5	1.5

Source: National statistics and ECE estimates.

[a] Forecasts rounded to the nearest 0.5 percentage point.

floating of several currencies in the final months of the year led to significant changes in exchange rates and, consequently, in price competitiveness, which will influence the patterns of trade and growth of west European countries in the course of 1993.

Monetary policy

In the United States monetary policy continued to sustain the recovery with a further cut in the discount rate from 3.5 per cent to 3 per cent at the beginning of July, its lowest rate since the first half of the 1960s. This further easing of monetary policy, which had already been considerably relaxed in the course of 1991, was a response to an increasing sluggishness in the pace of economic activity in the second quarter of 1992.

The discount rate has since remained unchanged, but the federal funds rate, i.e., the rate at which commercial banks borrow excess reserves from each other, fell below 3 per cent in the autumn. Over time the generous supply of liquidity has led to a fall in short-term interest rates from more than 8 per cent in spring 1991 to some 3 per cent at the beginning of 1993 (chart 2.1.3).

The prime rate charged by US banks on short-term business loans, however, only fell from about 9 per cent to 6 per cent over the same period. These wider margins between prime rates and short-term rates reflect the attempts by banks to restore their profitability and more cautious lending policies.

The steep fall in the "headline" short-term interest rates therefore tends to exaggerate the stimulus to economic activity provided by monetary policy: real interest rates facing consumers and enterprises are still relatively high in the United States and have fallen much less than nominal short-term money-market rates. Moreover, investment decisions are in general influenced more by long-term interest rates, though these are, of course, only one among several factors which determine investment behaviour. Whereas short-term interest rates can be directly controlled by monetary policy, this is not the case for long-term rates, which also reflect factors such as expectations of inflation, output and profits.

Long-term rates have, indeed, fallen in the United States over the past two years, although the decline has been much less than that for short-term rates. This reflects to a large degree the normalization of the term structure of interest rates which tends to pivot on the

CHART 2.1.2

Annual growth rates of real GDP, 1980-1992
(Percentage)

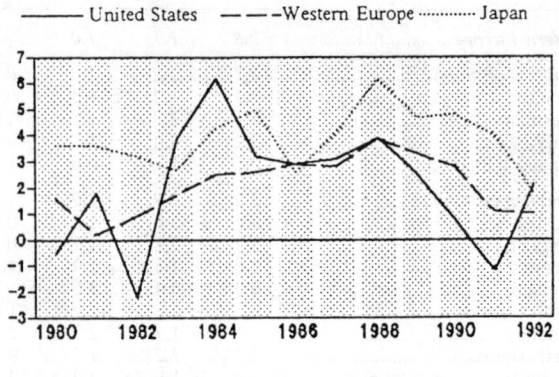

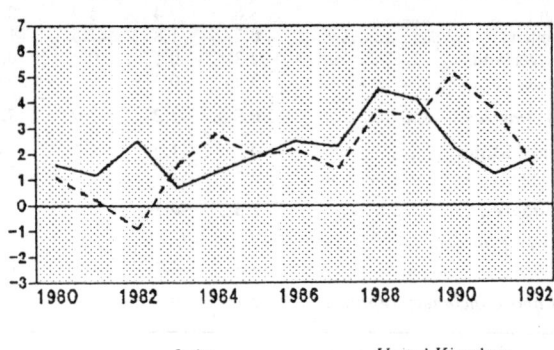

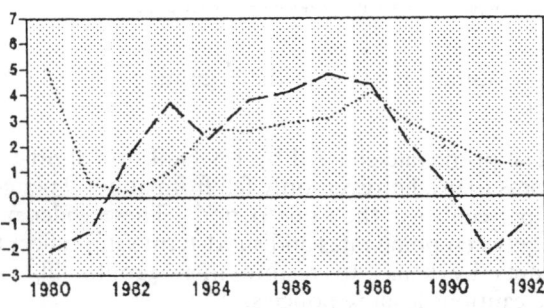

Source: National statistics.

long-term end. Economic reasoning leads to the expectation that in general short-term rates will be lower than long-term rates, but this shape of the yield-curve may be reversed if short-term rates are driven up as part of counter-inflationary policies. The fact that long-term rates have not fallen more strongly therefore partly reflects this concentration of policy changes on the short-end of the term structure. It may also reflect earlier pessimism about high budget deficits and expectations that inflation may rise once the rate of expansion becomes more vigorous. However, these fears seem to have abated since the new administration announced its budget plans and in February the long-term government bond yield fell to its lowest level since 1979.

Another, and probably more important, reason why the decline in nominal interest rates failed to generate a pronounced demand response is the high level of debt prevailing in the private sector. Demand for credit has been very sluggish and the willingness of banks to lend has been reduced by large provisions against loan losses and the need to meet stricter capital requirements. But lower interest rates did provide a more favourable environment for the process of balance sheet adjustments by enabling companies and households to refinance existing debt at more advantageous terms. Many companies have also reduced highly leveraged financial positions by issuing equity and employing the corresponding financial resources to pay back debt. As a consequence the debt service burden relative to incomes has fallen in the private sector over the past year or so. Progress has also been made in adjusting balance sheets in the financial sector.

The stance of monetary policy in the United States contrasts sharply with the restraining effects of high interest rates on activity levels in western Europe in 1992. This reflects to a large degree the tightening of German monetary policy in response to the inflationary pressures emerging in the aftermath of unification: the boom in demand for western goods in east Germany propelled capacity utilization rates to very high levels and led to a rapid increase in labour costs, which outpaced productivity growth by a large margin. The government has so far met the much higher than expected costs of reconstructing the east German economy mainly by allowing a considerable increase in the budget deficit and the government's borrowing requirement.

Hopes that inflationary pressures would abate and that the Bundesbank would have scope to lower interest rates in the second half of 1992 did not materialize. On the contrary, the Bundesbank raised the discount and Lombard rates to record levels in July 1992, only two weeks after the US discount rate had been reduced to its lowest level for more than two decades. These increasing divergences in monetary policy led to a substantial increase in the difference between short-term interest rates in the two countries from the beginning of 1991, a gap which began to narrow somewhat only in autumn 1992, when German interest rates started to fall (chart 2.1.3).

High German interest rates spilled over to other member countries of the Exchange Rate Mechanism (ERM) of the European Monetary System (EMS) because the rules of the system permit only very limited exchange rate fluctuations and because members were unwilling to consider a realignment. Given that financial markets do not expect a devaluation of the deutsche mark, German interest rates provide in general an effective floor to interest rates in the other ERM countries. Other west European countries outside the EMS which had chosen the DM or the ECU as a nominal anchor were confronted with the same problem and also tried to maintain the existing parities. But for many of these countries this was also an inappropriate response to their specific, domestic economic situations.

CHART 2.1.3

Nominal interest rates, January 1990-January 1993
(Percentages)

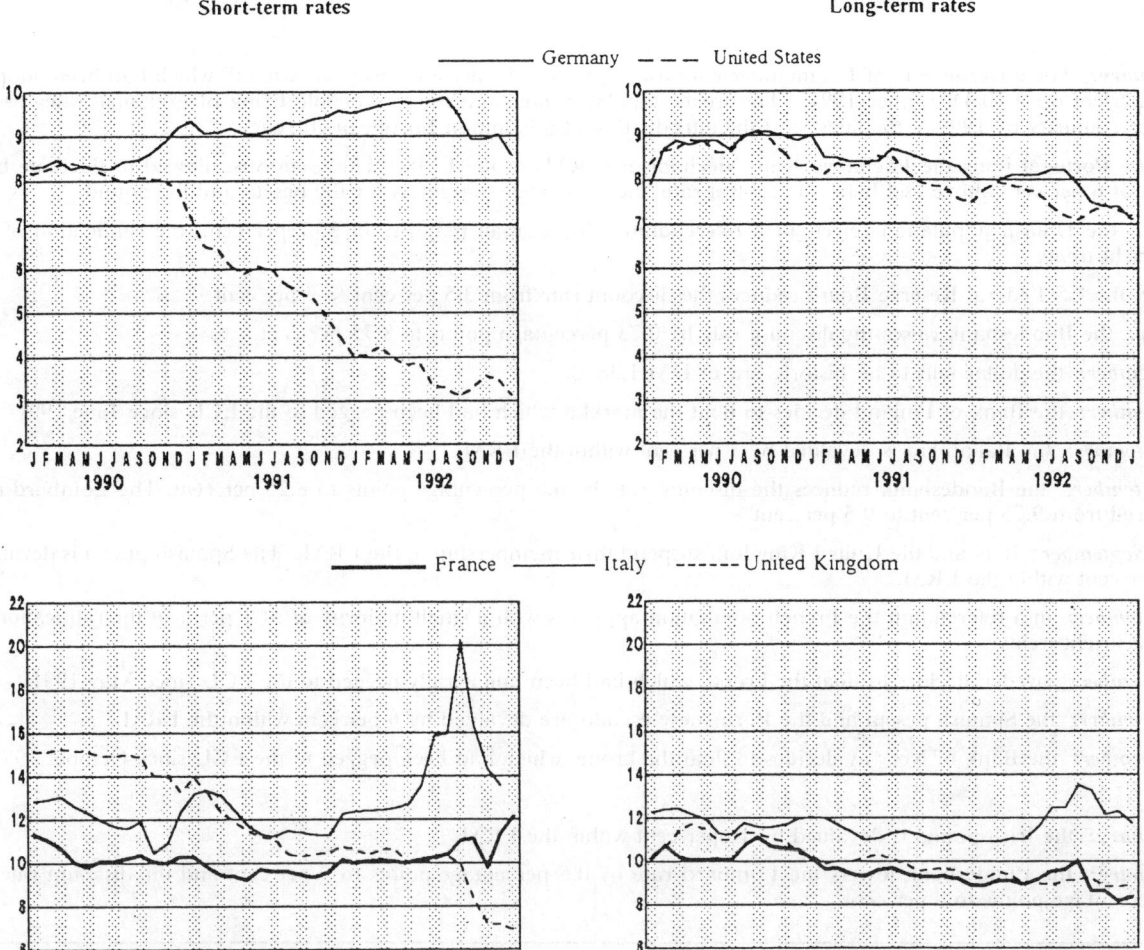

Source: OECD, *Main Economic indicators*, Paris.
Note: Monthly rates are averages of daily rates.

Financial markets became increasingly doubtful that the prevailing parities within the ERM could be sustained given the differences in cyclical positions and the differential changes in fundamental economic variables which had occurred since the last realignment in 1987. Tensions in west European foreign exchange markets started to build up gradually in the second quarter and were then accentuated, first, by the depreciation of the dollar, which started in the second quarter of 1992 and, later on, by the negative vote in the Danish referendum on the Maastricht treaty. These strains intensified considerably during the summer against a background of considerable uncertainty about the outcome of the corresponding referendum scheduled in France for 20 September. Heavy selling of some currencies led to a crisis in the EMS, which led first to the devaluation of the lira on 14 September and a small concessional cut in key lending rates in Germany, which nevertheless marked a turning point for west European interest rates. This was followed only a few days later by the devaluation of the peseta and the decision of Italy and the United Kingdom to suspend membership in the ERM and float their currencies. Further devaluations within the ERM were agreed upon for the peseta and escudo in November 1992 and the Irish pound at the end of January 1993, but overall the foreign exchange markets still appear to be highly unstable. German money market rates have been falling since September 1992 but prevailing pressures on exchange rates have not allowed all countries to lower their interest rates to the same degree. This was notably the case for the French franc. In contrast, the severing of the exchange rate constraint has allowed a sharp fall in short-term interest rates in the United Kingdom, accompanied by a large depreciation of sterling. Short-term interest rates have remained very high in Italy (chart 2.1.3). The objective of reducing tensions within the ERM was seen as a major factor behind the cuts in the German discount and Lombard rates at the beginning of February. (A chronology of the principal monetary policy events in 1992 and the first two months of 1993 is provided in box 2.1.1.)

> **BOX 2.1.1**
>
> *Principal monetary policy events from February 1992 to February 1993*
>
> *1992*
>
> *7 February:* The governments of EC member countries sign the "Treaty on European Union" which had been adopted at Maastricht on 9-10 December 1991. The treaty stipulates *inter alia* the irrevocable fixing of exchange rates of EC member countries by 1999 at the latest and the introduction of a common European currency.
>
> *6 April:* Portugal joins the Exchange Rate Mechanism (ERM) of the EMS. The escudo is allowed to fluctuate by a margin of 6 per cent above and below its central rate, the same wide margin as for the peseta and the pound.
>
> *2 June:* the Danish population rejects in a referendum, with a small majority of 50.7 per cent, the ratification of the Maastricht treaty.
>
> *2 July:* the US Federal Reserve Board reduces the discount rate from 3.5 per cent to 3 per cent.
>
> *17 July:* the Bundesbank raises the discount rate by 0.75 percentage points to 8.75 per cent.
>
> *2 September:* the dollar falls to an historic low of DM 1.3870.
>
> *8 September:* the Bank of Finland decides to float the markka which had been pegged to the ECU since June 1991.
>
> *13 September:* the Italian lira is devalued by 7 per cent within the ERM.
>
> *15 September:* the Bundesbank reduces the discount rate by 0.5 percentage points to 8.25 per cent. The Lombard rate is lowered from 9.75 per cent to 9.5 per cent.
>
> *16/17 September:* Italy and the United Kingdom suspend their membership of the ERM. The Spanish peseta is devalued by 5 per cent within the ERM.
>
> *20 September:* in a referendum the French population approves with a small majority of 51.1 per cent the ratification of the Maastricht treaty.
>
> *19 November:* Sweden decides to float the krona, which had been unilaterally pegged to the ECU since May 1991.
>
> *21 November:* the Spanish peseta and the Portuguese escudo are devalued by 6 per cent within the ERM.
>
> *10 December:* the Bank of Norway decides to float the krone, which had been pegged to the ECU since October 1990.
>
> *1993*
>
> *30 January:* the Irish pound is devalued by 10 per cent within the ERM.
>
> *4 February:* the Bundesbank lowers the Lombard rate by 0.5 percentage points to 9 per cent and the discount rate by 0.25 percentage points to 8 per cent.

Outside the ERM the crisis in the foreign exchange markets has considerably affected the Nordic countries, all of which had pegged their currencies to the ECU. Given persistent and strong speculative runs on their currencies against a background of deep recession and large financial imbalances, the Finnish authorities severed the link of their currency with the ECU in September, a move which was followed by Sweden in November, and then by Norway (see box 2.1.1).

The various realignments and changes in currency regimes which have occurred since September 1992 have changed substantially the pattern of exchange rates in western Europe. Table 2.1.3 shows the major changes in the nominal exchange rates of west European currencies against the deutsche mark between August 1992 (the month before the crisis broke) and January 1993. The depreciation of the US dollar against the deutsche mark in the second and third quarters was reversed later on in the year. The total impact of these depreciations on export performance will be influenced by the geographical orientation of trade flows in the various countries and by future changes in relative production costs. It is likely that the recent turbulence in the foreign exchange markets and the ensuing uncertainty about future exchange rates have tended to reinforce a wait-and-see attitude among economic agents.

Fiscal policy

The basic objective of fiscal policy has continued to be the consolidation of public sector finances. In fact, several countries adopted stringent measures in the course of 1992 designed to contain a further rise in their budget deficits. But the dominating feature has been to accept the working of the automatic stabilizers in an increasingly recessionary environment and the ensuing adverse impact on fiscal deficits, which in some cases have proved to be much larger than projected. Only in France, Norway and the United Kingdom was fiscal policy set to spur economic activity in 1992. But against a background of a large increase in the public sector borrowing requirement, the UK government tightened expenditure controls in the second half of the year. In western Europe pressures to curb deficits and outstanding debt relative to GDP have been considerably intensified with the adoption of the Maastricht convergence criteria.

TABLE 2.1.3

Major changes in nominal exchange rates against the deutsche mark, January 1992-January 1993
(Appreciation (-), depreciation (+) in per cent)

	January 1992-August 1992	August 1992-January 1993	January 1992-January 1993
Italy	+ 0.7	+ 21.3	+ 22.1
United Kingdom	+ 1.4	+ 13.8	+ 15.4
Finland	+ 0.9	+ 22.6	+ 23.6
Greece	+ 7.6	+ 7.8	+ 16.0
Norway	+ 0.4	+ 7.8	+ 8.2
Portugal	-0.5	+ 4.4	+ 3.8
Spain	+ 1.2	+ 10.5	+ 11.8
Sweden	+ 0.1	+ 23.3	+ 23.4
Memorandum item:			
DM per US dollar	-8.0	+ 11.4	+ 2.5

Source: United Nations Economic Comission for Europe, Division for Economic Analysis and Projections, based on OECD, *Main Economic Indicators* Paris, various issues.

Note: Comparisons are based on average monthly exchange rates in national currency units per deutsche mark.

Private sector indebtedness

An important factor behind the protracted weakness in economic activity in a number of countries has been the rapid rise in private sector debt relative to incomes, a development which occurred mainly during the second half of the 1980s. Since the deterioration of economic conditions around 1990, high debt service burdens have been restraining consumption and investment expenditures and, in some countries, this has offset the potential stimulative effects of lower interest rates. Indeed, the widespread failure of economic forecasters to anticipate such behaviour on the part of economic agents helps to explain why forecasts of growth have turned out to be too optimistic over the past year or so.

Current debt problems have been greatly influenced by the fact that the boom in real estate investment in the 1980s has turned into a serious slump. The surge in demand for real estate pertained to both private housing and commercial property, i.e., office space, commercial and industrial buildings etc. The importance of the debt problem varies considerably among the industrialized countries; among the larger countries it appears to be a problem mainly in the United States, the United Kingdom and Japan but more recently a deep crisis in the real estate sector has also been emerging in France.[88] Among the smaller west European countries the debt problem looms particularly large in Scandinavia. Switzerland has experienced a severe downturn in the real estate market since 1990 but the economic importance of this event, and in particular, the spill-over to the banking sector, is much smaller in comparison with the Scandinavian countries.

The surge in real estate investment was to a very large extent financed by loans from the banking sector. In other words, the counterpart of the acquisition of assets (private property) was a considerable increase in financial liabilities. Indeed, in the balance sheet of a household a private home is in general the most important single asset, while the associated mortgage constitutes the lion's share of total debt outstanding.

The rapid expansion of household debt relative to disposable incomes in North America, the United Kingdom and Scandinavia during the 1980s is shown in chart 2.1.4. The steepening of the debt curve occurred in general between 1983 and 1985, although in the United Kingdom there had already been a striking change in the slope of the curve in 1980. The rise in the debt ratio abated in the late 1980s and has since fallen in the Scandinavian countries;[89] in the United States it started to level off in 1991; and the indications are that a decline occurred in the United Kingdom in the course of 1992. But in Canada the indebtedness ratio continued to rise in 1992, reflecting increased mortgage financing.[90] There has been a considerable variation among countries in the level of the debt ratio, the peak values ranging from 84 per cent in Finland[91] to more than 170 per cent in Norway. Chart 2.1.4 also shows the development of debt relative to income in Germany, where there have been no signs of distress in the household sector and the real estate market. There is, indeed, a striking contrast in the level and time pattern of change. It can be seen that the German debt ratio rose steadily from about 50 per cent in the mid-1970s to a peak of about 75 per cent in the mid 1980s, a value which is much lower than in the other countries.

A general feature is for rising debt ratios to be accompanied by falling saving ratios. Debt consolidation, in turn, is reflected in a rise in saving ratios.

There are a number of factors which played a role in the rapid expansion of credit in the 1980s, although their relative importance varies from country to country. In general, the changes in the credit market can be largely attributed to the deregulation of financial markets which facilitated borrowing. Also tax systems may have favoured borrowing, for example by allowing tax deductibility on interest payments. Demographic changes also appear to have contributed to the pressures in the housing market, notably in North America. Other factors may have been inflationary expectations, for which real estate was seen as a hedge,[92] and the

---

[88] For France see: "Recession immobilière et crise bancaire", *Lettre de l'OFCE*, Paris, No.109, 27 November 1992; and for Switzerland, *Commission pour les questions conjuncturelles: Rapport du décembre 1992*, supplement to *La vie économique*, No.1, 1993.

[89] In Norway the rise in the debt ratio came to a virtual halt in 1987 on account of the adverse impact of the oil price decline on the Norwegian economy and the policy responses to cope with this shock.

[90] See Statistics Canada, *Canadian Economic Observer*, December 1992, p.13.

[91] The household debt data for Finland used in chart 2.1.4 exclude loans from subsidiaries of foreign banks and therefore understate the level of household debt. According to the Ministry of Finance the indebtedness ratio was some 84 per cent in 1989; the pattern of change, however, is not affected. Ministry of Finance, *Economic Survey 1992*, Helsinki, 1992, p.44.

[92] In Canada financial deregulation had occurred already in the 1970s and the rise in indebtedness in the second half of the 1980s is largely attributed to demo-

CHART 2.1.4
Debt, interest payments and savings of private households, 1970-1992
(As percentages of disposable income)

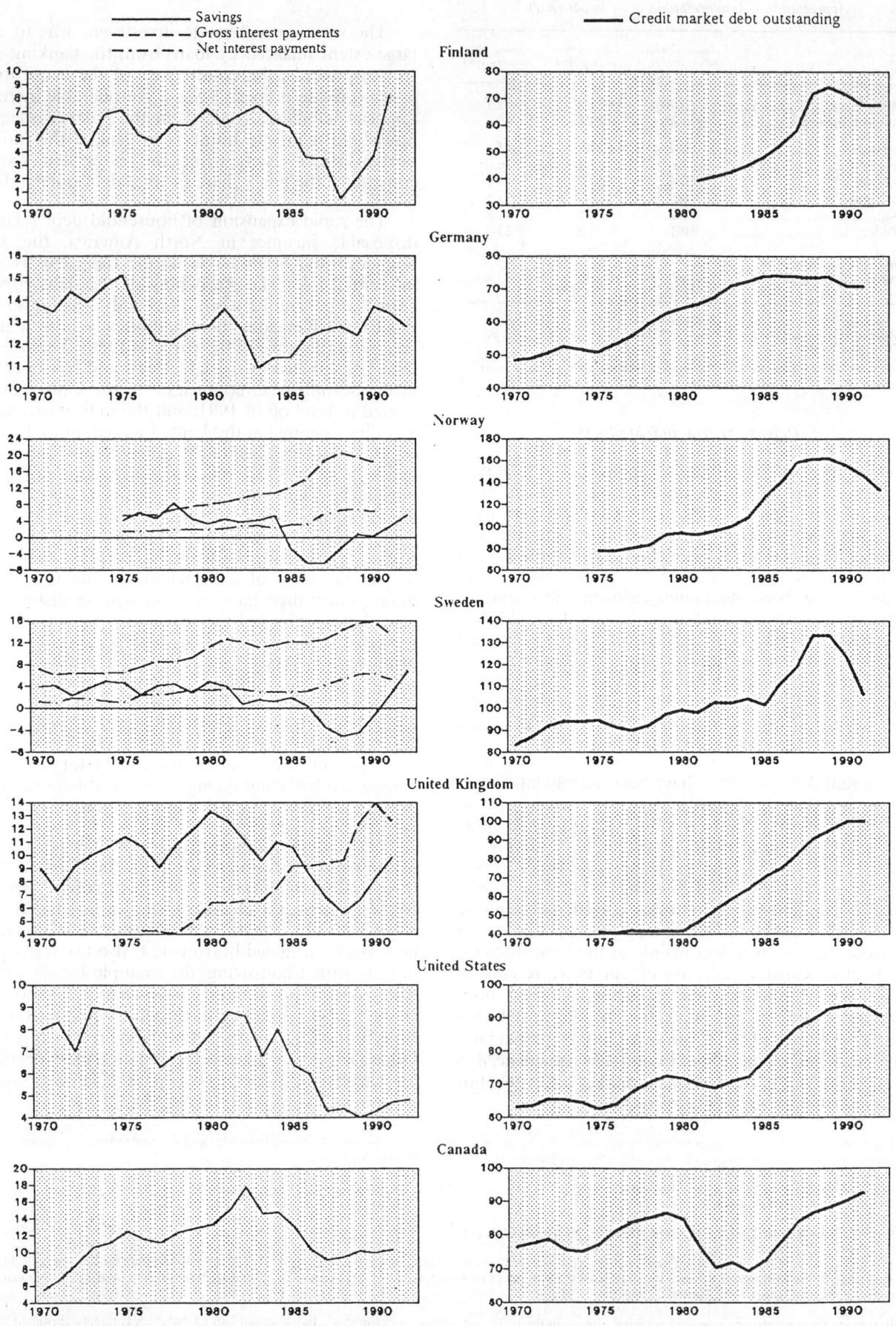

Source: UN-ECE/DEAP based on national statistics.

CHART 2.1.5

Real housing prices, 1981-1992.
(Indices, 1981 = 100)

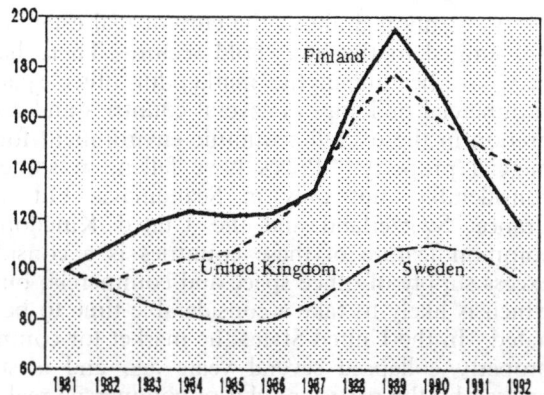
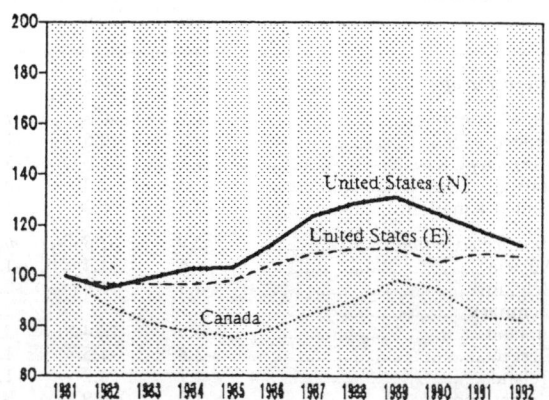

Source: UN-ECE/DEAP based on national statistics.
Notes: Data for 1992 are averages of the first three quarters for Canada, Finland, Sweden and the United States and of the first six months for the United Kingdom. Real housing prices were obtained by deflating the nominal house price indices with the consumer price index. Data pertain in general to existing houses and flats. Data for Canada cover new houses only. For the United States separate series for average prices of new housing units (N) and for existing housing (E) are shown.

sustained period of economic expansion which bolstered real incomes and consumer confidence and provided a conducive environment for borrowing. Against a background of increased competitive pressures arising from deregulation many banks were also prepared to accept a greater exposure to risk — by softening terms and standards of lending — in order to maintain or increase market shares. Many of these factors were mutually reinforcing.

The surge in the demand for housing led to significant increases in the value of property, thus stimulating speculative investment in anticipation of expected capital gains, which gave a further boost to real estate prices. Between 1986 and 1989 real housing prices rose at an annual rate of some 15 per cent in Finland and the United Kingdom. In Sweden the corresponding increase was somewhat lower (some 10 per cent) but still considerable. Increases in property values were also significant in North America (chart 2.1.5) and there have been large price increases in other countries such as France, Norway and Switzerland. The appreciation of property values was tantamount to an increase in collateral and this supported not only the growth of mortgage lending but also of consumer credit. In fact, it is more than likely that many banks provided loans on the assumption that property prices and hence the value of the collateral would continue to rise. The windfall gains in wealth from rising property prices may have been a factor in the fall in saving ratios noted above.[93]

Rising debt levels are a cause of concern because of the increased risk of default and the ensuing adverse repercussions on the real economy and the banking system if a critical default level is exceeded.[94] The very rapid debt expansion in the 1980s and the ensuing record high debt levels were in general accompanied by considerably higher income gearing, that is, the share of disposable incomes devoted to debt service payments rose markedly (chart 2.1.4). Evidently, such a situation is in general only sustainable in a favourable macroeconomic environment.

Rising interest rates, reflecting the tightening of monetary policy, and the downturn in economic activity, with its associated deterioration of income prospects, quickly revealed the financial vulnerability of many households from 1989 or so.[95] The increasing burden of debt service payments on disposable incomes, reflecting also the rising importance of variable rate loans in house financing, depressed the demand for both consumer goods and for real estate. Rising saving ratios and falling debt ratios are indicative of the balance sheet adjustments which began in the late 1980s and the priority which households attached to paying back debt rather than increasing spending when interest rates started to fall again (chart 2.1.4).

The debt adjustment process has been aggravated by occasionally steep falls in real estate prices, reflecting the normalization of prices after a period of overheating, large excess supply arising from speculative overbuilding, and forced selling on account of defaults

---

graphic developments (the maturing of the baby-boom generation) and inflationary expectations. See Marie-Claude Monplaisir, "Developments in the balance sheet of the household sector over the past two decades", *Bank of Canada Review*, July 1992, pp.3-14.

[93] See Erkki Koskela, "Household saving in Finland: Why does it fluctuate?", *Bank of Finland Bulletin*, Vol.66, No.9, September 1992, pp.7-11. Koskela observes a clear negative relationship between changes in house prices and the savings ratio in Finland over the period 1970-1990.

[94] This critical level depends on a number of variables and is impossible to determine *ex ante*. See E.P. Davis, "Rising sectoral debt income ratios: A cause for concern?", *BIS Economic Papers*, No.20, June 1987, Bank for International Settlements, Basle.

[95] The situation was somewhat different in Norway, where the credit boom had already foundered in 1986. The backdrop to this was the fall in oil prices and the ensuing policy adjustments by the government.

TABLE 2.1.4
Changes in manufacturing production in western Europe and North America, 1990-1992
(Percentage change over previous year)

	1990	1991	1992 a
Western Europe	2.2	-0.6	-1.0
4 major economies	1.9	-0.3	-1.5
France	0.2	0.5	-1.5
Germany	5.5	3.0	-2.0
Italy	-0.3	-2.3	-0.5
United Kingdom	-0.5	-5.2	-1.0
13 smaller economies	2.9	-1.3	0.5
Austria	8.6	1.4	-1.0
Belgium	4.3	-1.6	-0.5
Denmark	..	..	..
Finland	-0.5	-10.4	3.5
Greece	-2.8	-1.5	-1.0
Ireland	4.7	3.2	11.0
Netherlands	4.0	-0.9	-
Norway	-	-1.5	1.5
Portugal	5.0	-1.8	-4.0
Spain	-0.1	-1.3	-1.0
Sweden	-2.7	-8.5	-3.0
Switzerland	2.6	0.6	-
Turkey	9.5	1.9	5.5
North America	0.5	-2.5	2.0
United States	0.9	-2.3	2.0
Canada	-5.1	-6.6	-
Total above	1.4	-1.5	0.5
Memorandum item:			
Japan	4.6	2.1	-6.0
Total above, including Japan	2.0	-0.8	-1.0

Sources: National statistics; ECE secretariat estimates.

a Preliminary estimates rounded to the nearest 0.5 percentage point.

(chart 2.1 5). Real housing prices fell by 40 per cent in Finland between 1989 and 1992 and by more than 20 per cent in the United Kingdom. Property prices also fell in other countries (France, Norway,[96] Sweden, Switzerland). For a given nominal value of debt, this asset price deflation has squeezed the net worth of households and is probably behind the rise in the propensity to save (chart 2.1.4) In some cases the fall in real estate prices has gone so far as to create "negative equity", i.e., a situation where the outstanding mortgage exceeds the market value of the property. In the United Kingdom more than 1.5 million households found themselves in such a situation in the second half of 1992.[97]

The important questions, of course, are how far indebtedness will have to fall before households feel reassured that their financial position has become sustainable again and how long the adjustment process will take. However, the debt burden which households regard as "reasonable" is not known. Even average debt levels of the past may be a poor guide insofar as the adverse experience of the current crisis may have led to a "burned fingers syndrome"[98] which may entail a more cautious borrowing and spending behaviour for some time to come.

As mentioned above the current real estate crisis is not limited to the household sector. There has also been overheating in the market for office space and commercial buildings in several countries and a deterioration in the overall economic environment for commercial property. Prices have fallen, in some cases considerably, not only in Scandinavia but also in France, Switzerland and the United Kingdom. A common feature of these countries is a considerable excess supply of office space and commercial buildings which is likely to take a considerable time to be eliminated. In the United States the market for commercial property collapsed several years ago and is still depressed by the persistent glut of commercial real estate.

The slump in the real estate sector has inevitably spilled over into the balance sheets of the banking sector: there has been a considerable rise in non-performing loans and the fall in property prices has meant a reduction in the value of the collateral against which these loans were provided. The necessary process of balance sheet adjustment in the banking sector (the writing-off of losses; making provisions for non-performing loans, strengthening the capital base, etc.) has led to more cautious lending policies and a tightening of the terms and standards for granting credit. Weak balance sheets have led banks not to pass on immediately or entirely the reductions in the cost of funds resulting from the recent easing of monetary policy. In several countries banks have instead widened their lending margins.[99] The problems of the banking sector are particularly severe in the Scandinavian countries where huge losses from real estate loans have affected the major banks and required substantial measures of government support.

Industrial output and capacity utilization

Against a background of weak demand, both abroad and at home, *manufacturing output* in western Europe fell by nearly 1 per cent in 1992. In 1991 average output levels had already declined by 0.6 per cent (table 2.1.4). The downturn gathered pace in the final quarter of 1992 when output fell at an annual rate of more than 5 per cent (chart 2.1.6). Manufacturing output fell in all four major economies in 1992: the last time this happened was in 1981. Chart 2.1.6 illustrates the increasing sluggishness of manufacturing output in western Europe in the course of 1992 and the sharp downturn in Germany and some other countries in the second half of the year. With the downturn of economic activity the demand for investment goods has

---

96 Housing prices had fallen in real terms by about one third in 1992 compared with their peak in the second quarter of 1988. See, Norges Bank, *Economic Bulletin 1992*, Vol.LXIII, No.4, p.236.

97 *Financial Times*, London, 7-8 November 1992.

98 See Economic Commission for Europe, *Economic Bulletin for Europe*, Vol.44(1992), New York, 1993, p.4.

99 For the United States see, "Recent Developments Affecting the Profitability and Practices of Commercial Banks", *Federal Reserve Bulletin*, July 1992, Vol.78, No.7, pp.459-483.

CHART 2.1.6

Quarterly index numbers of manufacturing output, 1989-1992
(1985 = 100)

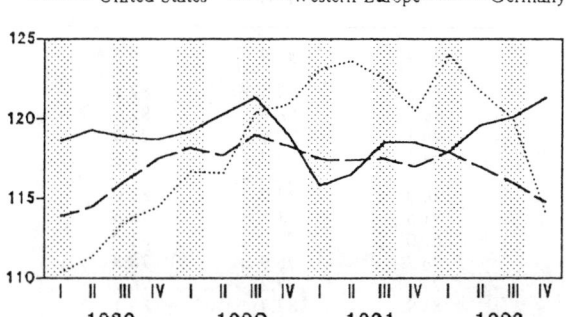

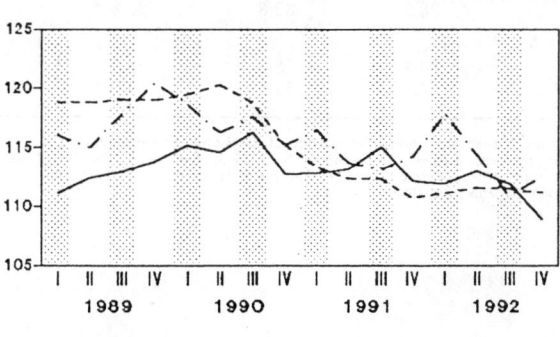

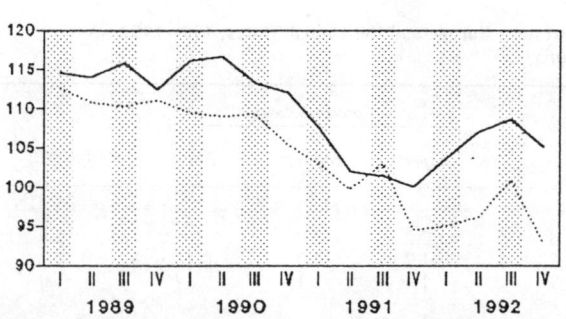

Source: UN-ECE DEAP based on national statistics.

weakened considerably. Among the smaller economies the buoyancy of the Irish manufacturing sector stands out — a reflection of strong export growth by foreign subsidiaries in "high-tech" industries. The strong rebound in manufacturing output in Finland is also based on export growth and reflects the strong gains in price competitiveness resulting from the sizeable devaluation of the currency in November 1991 and from wage restraint. Branches mainly oriented towards the domestic market saw their production levels falling. There were significant declines in total manufacturing output in Portugal and Sweden in 1992.

The pervasive sluggishness of manufacturing output has persisted over the last few years in western Europe (chart 2.1.6). In about half of the west European countries average output levels in 1992 were lower than in 1989. Among the four larger economies, the only exception is Germany. Among the smaller economies, manufacturing output in Finland, Spain, Sweden, Portugal was below the levels of 1989, while in Norway output more or less stagnated between 1989 and 1992.

Capacity utilization rates in manufacturing declined in the course of 1992. In a number of countries they were heading towards the low levels reached during the recession of 1981/82 (table 2.1.5). The depth of the Finnish recession is reflected in the low level of capacity utilization, which in the second half of 1992 was significantly lower than in the recession of 1981-1982. The 1992 data for most west European countries refer to autumn 1992 but, given the deterioration in economic conditions since then, capacity utilization rates will have fallen further, although the data for most countries are not yet available. In Germany, for example, utilization rates had already fallen to some 80.5 per cent in December compared with 83 per cent in October.

In the United States manufacturing output rose by 2 per cent in 1992, although this was not enough to offset the decline of 2.3 per cent in 1991. Utilization rates had fallen in the first quarter of 1992 and the moderate upturn in output growth led only to a slight reduction in spare capacity in the course of 1992. Compared with the previous peak (table 2.1.5) there appears to be ample spare capacity, which should help to contain inflationary pressures in the short term. In Canada the stagnation of output in 1992 as a whole masks a steady strengthening in activity since the second quarter of 1992; however, total output remains relatively depressed given the steep fall in 1991.

The components of demand

In western Europe the relative importance of the major components of demand did not change significantly in 1992 as compared to 1991. Table 2.1.6 shows their contribution to the overall growth in real GDP. The strong deceleration in economic growth in 1991 was due to the faltering of private consumption and the fall in investment, which was only partly offset by rising net exports. Growth in domestic demand remained weak in 1992 and was broadly unchanged from the previous year. However, there was actually an absolute fall in total domestic demand in several of the smaller countries.[100] Consumption was even less of a support to overall output growth in 1992 than in 1991 and investment remained depressed. Changes in the real foreign balance, in the aggregate, have continued to support overall economic activity, although its contribution to growth was smaller than in 1991.

Among the four major economies total domestic demand growth slowed down sharply in 1992, especially in Germany and Italy. In France growth in domestic demand remained subdued and broadly unchanged from 1991. Domestic demand stagnated at a low level in the United Kingdom, following a sharp fall in 1991. Only in the United Kingdom were changes

---

[100] This was the case for Denmark, Finland, Ireland, Sweden and Switzerland, where there had already been an absolute decline in total domestic demand in 1991.

TABLE 2.1.5
Capacity utilization rates in manufacturing industry, 1979-1992
(Per cent)

	Previous peak 1979-1980 a	Previous low 1982-1983 b	Quarterly rates c		
			1990	1991	1992
Western Europe	..	..	*85.4*	*82.0*	*80.2 **
France	85.3	81.1	86.5	83.5	82.1
Germany	86.0	75.3	89.5	86.6	82.9
Italy	77.3	69.1	79.1	77.1	75.3
United Kingdom	87.6	73.0	84.0	77.7	78.5
Austria d	86.0	79.0	86.0	83.0	82.0
Belgium	78.8	74.6	81.5	78.5	76.4
Finland e	90.1	86.5	87.0	79.0	81.0
Greece	..	..	76.7	76.9	78.8
Ireland	68.1	56.8	77.4	77.2	76.7
Netherlands	83.0	75.8	85.9	83.7	83.1
Portugal	..	..	80.6	79.0	76.9
Spain	..	..	80.2	77.7	76.4
Sweden	83.7 f	79.7	87.4	82.3	81.0 *
Switzerland	87.0	80.1	88.5	82.8	80.8 *
North America					
United States	87.3	70.0 g	80.8	78.2	78.2
Canada	87.5 *	65.2 g	74.2	72.6	74.0 *

Sources: Belgium, France, Germany, Greece, Ireland, Italy, the Netherlands, Portugal, Spain, United Kingdom: Commission of the European Communities, *European Economy*, supplement B, November 1992; table 1 (data refer to January, April, July and October); Sweden: *Allman Manadsstatistik*, No.2, 1993. Canada and the United States: OECD, *Main Economic Indicators*, Paris; Austria: Oesterreichisches Institut für Wirtschaftsforschung, *Monatsberichte No.1*, 1993, and previous issues; Switzerland: Bundesamt für Statistik, *La vie économique*, No.2, 1993, and previous issues; Finland: *Bank of Finland Monthly Bulletin No.1*, 1993.

a Previous peak. Quarterly high; for the United States monthly high.
b Previous low. Quarterly low; for the United States monthly low.
c Quarterly rates are seasonally adjusted (Sweden: unadjusted).
d Date refer to November of each year.
e Only semi-annual data are available.
f No data are available for 1979.
g Break in series.

TABLE 2.1.6
Contribution of demand components to real GDP growth in western Europe and the United States, 1990-1992
(Percentage points)

	Domestic demand					Foreign balance			
	Private consumption	Public consumption	Fixed investment	Stocks	Total	Exports	Imports	Net	GDP
Western Europe									
1990	1.9	0.5	0.6	-0.2	2.8	2.2	-2.2	-	2.9
1991	1.0	0.3	-0.2	-0.4	0.7	1.7	-1.3	0.4	1.1
1992	0.8	0.3	-0.2	-0.2	0.7	1.4	-1.1	0.3	1.0
United States									
1990	0.8	0.5	-0.4	-0.5	0.4	0.8	-0.3	0.5	0.8
1991	-0.4	0.2	-1.3	-0.3	-1.8	0.6	-	0.6	-1.2
1992	1.5	-0.1	0.8	0.3	2.5	0.7	-1.1	-0.4	2.1

Source: ECE secretariat calculations based on national and international sources.

Note: Differences between totals and sum of components are due to rounding.

in real net exports a drag on total domestic activity levels.

In the United States the fall in real GDP in 1991 was largely accounted for by the faltering of private consumption and investment. The resulting fall in domestic demand was partly offset by favourable changes in real net exports (table 2.1.6). The increase in real GDP in 1992 reflects in the main the strengthening of domestic demand. The growth contribution of exports rose slightly, but this was more than offset by the strong growth in imports: in contrast to 1991, changes in net exports subtracted from output growth.

The rate of growth of *real private consumption* remained rather subdued in western Europe in 1992. For the year as a whole there was an increase of about 1 per cent, compared with an increase of 1.5 per cent in 1991 (table 2.1.7).

The major factors behind the sluggishness of household expenditures have been low income growth, on account of broadly stagnating or falling employment levels, and higher savings ratios. Against a background of high and rising unemployment, consumer confidence was depressed and in many countries households decided to save a larger part of their disposable incomes than in 1991. High real interest rates made consumer credit very costly and this has dampened demand for

TABLE 2.1.7

Real private consumption expenditures in western Europe and North America, 1990-1992
(Percentage change over previous year)

	1990	1991	1992
Western Europe	**3.1**	**1.6**	**1.3**
4 major economies	*3.0*	*1.5*	*1.0*
France	2.9	1.5	1.9
Germany	5.4	3.6	0.9
Italy	2.8	2.8	1.7
United Kingdom	0.7	-2.0	-0.3
13 smaller economies	*3.1*	*1.9*	*1.9*
Austria	3.8	2.4	2.5
Belgium	2.6	2.7	2.2
Denmark	0.6	1.2	1.5
Finland	0.2	-3.7	-5.3
Greece	2.0	1.2	1.3
Ireland	1.1	0.5	3.5
Netherlands	4.1	3.3	1.7
Norway	2.9	-0.3	1.5
Portugal	5.3	4.9	4.0
Spain	3.7	3.0	2.6
Sweden	-0.3	1.0	-1.5
Switzerland	1.5	1.5	-0.3
Turkey	10.4	2.1	10.6
North America	**1.1**	**-0.7**	**2.1**
United States	1.2	-0.6	2.2
Canada	0.9	-1.7	0.5
Total above	**1.9**	**0.2**	**1.8**

Source: National statistics.

TABLE 2.1.8

Real public consumption in western Europe and North America, 1990-1992
(Percentage change over previous year)

	1990	1991	1992
Western Europe	**2.5**	**1.9**	**1.6**
4 major economies	*2.2*	*1.9*	*1.5*
France	1.8	2.9	2.0
Germany	2.4	0.5	2.5
Italy	1.3	1.7	0.7
United Kingdom	3.2	2.7	0.5
13 smaller economies	*3.0*	*1.8*	*1.6*
Austria	1.2	2.6	1.5
Belgium	1.0	1.1	-0.1
Denmark	-0.4	-0.2	0.7
Finland	4.4	2.3	-0.1
Greece	0.6	-0.7	0.7
Ireland	3.5	1.7	1.7
Netherlands	2.1	1.6	0.7
Norway	2.0	2.3	2.7
Portugal	1.5	3.0	3.1
Spain	4.2	4.4	3.7
Sweden	2.9	0.2	0.9
Switzerland	4.7	2.8	2.5
Turkey	17.0	0.9	1.9
North America	**2.8**	**1.3**	**-0.1**
United States	2.8	1.2	-0.3
Canada	2.9	1.9	1.6
Total above	**2.7**	**1.6**	**0.5**

Source: National statistics.

consumer durables. Fragile financial positions have also been a tight constraint on the expenditures of many households who need to reduce their financial liabilities. In Finland, Norway, Sweden and the United Kingdom this has entailed a sharp rise in the saving ratio over the last few years, which has been an important factor behind falling or sluggish consumption expenditures (see above).

Among the various countries there was a conspicuous slow-down in the growth of consumer demand in Germany and Italy in 1992. Household expenditures declined further — albeit by a very small amount — in the United Kingdom, reflecting a further significant rise in the savings ratio. Consumer demand was still relatively robust in a few countries in 1992, notably in Austria, Ireland, Spain and Portugal. And household demand was remarkably buoyant in Turkey. In contrast, there was a further significant fall in real consumption in Finland against a background of soaring unemployment and austerity policy, which led to a decline in real incomes. There was also a decline in real private consumption in Sweden, but this was entirely due to a large rise in the saving ratio.

In the United States, the consumption expenditures of private households picked up in 1992, following a decline in the preceding year (table 2.1.7). Falling inflation was a major factor behind the growth in real disposable incomes. The average household saving ratio remained broadly stable between 1991 and 1992 but there was a marked fall in the second half of 1992, which supported higher spending levels. Household spending rose notably for consumer durables, which was partly related to the strong rise in investment in single-family housing which led to a concomitant increase in the demand for furniture and household equipment. Demand for consumer credit remained subdued and total instalment credit outstanding at the end of 1992 was lower than one year earlier. Against the background of a still high debt-service burden, the current savings ratio still appears to be rather low, particularly when seen in a somewhat longer perspective (chart 2.1.4). It is therefore unlikely that a fall in the savings ratio will support consumer demand in 1993, which instead will have to rely on higher income and employment growth.

A bleak situation in the labour market has been depressing personal income growth in Canada and this, together with high indebtedness, has restrained consumers' expenditure, which grew only moderately in 1992.

The ongoing consolidation of public sector budgets is mirrored in a further slow-down in the growth of *public consumption*, which in western Europe rose by only 1.5 per cent in 1992 (table 2.1.8). Restraints on the purchase of goods and tighter limits on public sector employment levels have been among the factors behind this modest growth. Overall this demand component contributed 0.3 percentage points to real GDP growth in 1992, slightly less than in 1991. In the United

TABLE 2.1.9

Real gross fixed capital formation in western Europe and North America, 1990-1992
(Percentage change over previous year)

	1990	1991	1992
Western Europe	*3.0*	*-0.8*	*-1.1*
4 major economies	3.3	-0.2	-0.3
France	2.9	-1.2	-1.9
Germany	8.7	6.5	1.7
Italy	3.3	0.9	-0.1
United Kingdom	-3.1	-10.1	-2.0
13 smaller economies	2.2	-1.9	-2.6
Austria	5.8	4.9	2.3
Belgium	8.3	0.3	1.4
Denmark	-0.9	-4.2	-10.0
Finland	-4.9	-19.8	-14.9
Greece	4.8	-2.0	0.6
Ireland	9.5	-5.5	0.5
Netherlands	3.6	0.1	0.9
Norway	-26.6	1.0	5.6
Portugal	5.9	2.6	1.8
Spain	6.9	1.6	-2.7
Sweden	-0.5	-8.3	-8.7
Switzerland	2.6	-2.5	-6.7
Turkey	14.0	-0.4	1.3
North America	*-2.9*	*-8.0*	*5.0*
United States	-2.8	-8.5	5.6
Canada	-3.9	-3.7	0.7
Total above	*-0.2*	*-4.5*	*1.9*

Source: National statistics.

TABLE 2.1.10

Real gross fixed capital formation, by type of asset, 1991-1992
(Percentage change over previous year)

	Machinery and equipment		Construction	
	1991	1992	1991	1992
Western Europe	*-1.1*	*-2.1*	*-0.3*	*0.5*
4 major economies	-0.2	-1.8	-	2.2
France	-2.5	-4.3	0.6	2.3
Germany	9.1	-2.1	4.1	5.5
Italy	0.7	-0.5	1.2	0.3
United Kingdom	-11.9	0.6	-8.4	-1.4
13 smaller economies	-3.4	-2.8	-0.7	-2.8
Austria	3.8	-1.0	5.8	5.0
Belgium	-1.9	0.7	2.0	3.8
Denmark	3.1	-17.9	-10.9	-2.5
Finland	-26.3	-9.0	-15.4	-12.9
Greece	3.3	4.5	0.8	-3.0
Ireland	-6.0	-1.5	-5.0	1.0
Netherlands	1.8	-0.8	-2.2	-
Norway	-4.7	17.3	5.0	-2.5
Portugal	1.0	4.8	4.5	2.5
Spain	-2.5	0.2	4.2	-3.4
Sweden	-11.9	-13.0	-4.8	-5.1
Switzerland	-1.2	-5.4	-3.1	-7.7
Turkey	..	..	..	..
North America	*-4.0*	*7.2*	*-11.6*	*3.3*
United States	-4.6	7.3	-12.3	4.0
Canada	1.1	6.4	-6.6	-1.4
Total above	*-2.7*	*2.7*	*-6.9*	*1.9*

Source: National and international statistics.

States real government purchases[101] fell by 0.3 per cent in 1992, a reflection of cuts in defence expenditure, which more than offset higher spending on other items.

There was a further weakening of *fixed capital formation* in western Europe in 1992. Total expenditure fell in real terms by about 1 per cent compared with the previous year, when there was a decline of 0.7 per cent (table 2.1.9). Real expenditures on machinery and equipment fell by 2.2 per cent, while construction investment grew only marginally in 1992 (table 2.1.10). There has been a marked slump in fixed investment over the last two years in a number of countries, reflecting the combined impact of cyclical factors and structural imbalances. The investment downturn has been particularly steep in Finland where aggregate fixed investment in 1992 was some 10 per cent lower than in 1980. The large cuts in investment over the past three years has effectively led to a reduction in the productive capacities of industry.

High interest rates, increasing margins of spare capacity and a deteriorating outlook for sales and profits led many firms to curb investment plans in the course of last year. Against such a background the dominating motive of business investment has tended to shift from increasing capacity to rationalization and modernization. In general, there was no significant support for total investment from the public sector where the prevailing concerns were focused on large fiscal deficits.

The current weakness of expenditure on machinery and equipment may also reflect to some degree a natural pause, when seen against the sustained period of growth in the second half of the 1980s (chart 2.1.7) Investment in machinery and equipment rose on average by nearly 40 per cent in western Europe between 1985 and 1990, a sign of the efforts undertaken in the business sector to improve competitiveness in anticipation of the completion of the internal market of the EC by the end of 1992.

Sluggish construction investment reflects only partly the reduced importance of increasing capacity during a cyclical downswing. Rather, in many countries, construction has been depressed for some time because of the severe imbalances which have emerged after a period of speculative building and overheating in both the markets for housing and commercial buildings. This applies particularly to France, the United Kingdom, the Scandinavian countries and Switzerland (see above). The slump in housing in several west European countries is shown in table 2.1.11.

In the United States fixed investment began to recover in 1992 after two consecutive years of decline (table 2.1.9). Industrial investment was stimulated by a more favourable outlook for profits. A large increase in purchases of EDP equipment was the major factor behind the upturn of investment in producers' durables (table 2.1.10). Construction investment also rose after several years of decline (chart 2.1.7), although this im-

---

[101] Note that in the United States this demand component also includes public investment expenditure.

CHART 2.1.7

Real gross fixed investment in western Europe and the United States, 1980-1992
(Indices, 1980 = 100)

——— Total — — — Machinery and equipment ········· Construction

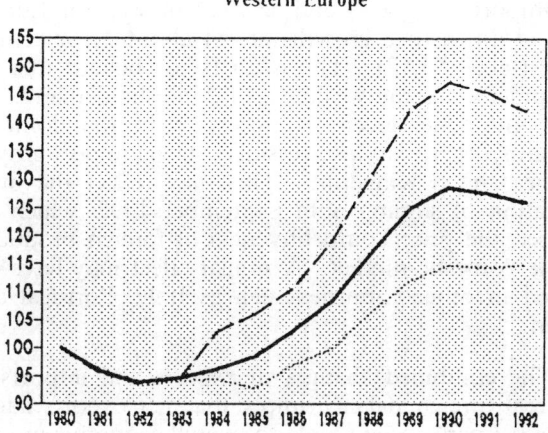

Source: UN-ECE/DEAP based on national and international statistics.
Note: Data for the United States cover private sector investment only.

provement hides a further fall in expenditure on industrial structures, a reflection of a persistent surplus of office space and other commercial buildings. The major factor behind the growth in construction investment was strong demand for single-family houses which, stimulated by low mortgage rates, increased by more than 20 per cent in 1992 compared with 1991. Total residential investment, however, rose less strongly because of the continuing imbalance in the market for multi-family housing (table 2.1.11).

Against a background of weak domestic demand, *real exports of goods and services* were the mainstay of overall economic activity in western Europe in 1992. For the year as a whole exports increased by about 4 per cent, which compares with less than 1 per cent growth in real total domestic demand (table 2.1.12). The figures for the year as a whole, however, hide an increasing sluggishness of foreign demand in the second half.

Foreign demand was dampened by the erosion of price competitiveness due to the large appreciation of west European currencies *vis-à-vis* the US dollar in the second and third quarters of 1992. Exports of many countries were restrained by the faltering of domestic demand in Germany. West German imports (excluding intra-German trade) rose by only 1.5 per cent in 1992 following very strong growth (annual increases of more than 9 per cent) in the two preceding years.

The impact of the various exchange rate changes within and outside the ERM in the final quarter of 1992 probably had little impact on export performance in 1992, but the effects are likely show up in the course of 1993 provided the gains in price competitiveness from devaluation are not offset by domestic inflation.

TABLE 2.1.11

Real housing investment in western Europe and North America, 1990-1992
(Percentage change over previous year)

	1990	1991	1992
Western Europe	0.5	-2.8	..
4 major economies	0.8	-1.6	..
France	-1.5	-1.2	1.5 *
Germany	7.8	4.2	6.7
Italy	2.7	1.8	..
United Kingdom	-11.1	-20.2	-0.5
9 smaller economies	-2.8	-10.5	-4.1
Austria	..	..	..
Belgium	10.8	-3.9	7.7
Denmark	-11.7	-11.5	-4.0
Finland	-6.7	-22.3	-14.5
Ireland	..	..	..
Netherlands	-3.5	-7.6	-0.5
Norway	-20.7	1.0	-18.4
Sweden	3.0	2.2	-6.0
Switzerland	..	..	..
North America	-9.1	-12.5	12.6
United States	-9.1	-12.6	13.4
Canada	-9.1	-11.8	5.3
Total above	-5.7	-8.9	..

Source: National statistics.

Among the four larger economies the favourable export performance of France is noteworthy since it reflects large improvements in the competitiveness of French industry. West German exports to the rest of the world (i.e., excluding transactions with east Germany which are included in the west German export figures) were very sluggish and rose by only 0.5 per cent in 1992. Given the strong specialization of German

TABLE 2.1.12

Volume of trade in goods and services in western Europe and North America, 1991-1992

(Percentage change over previous year)

	Exports		Imports	
	1991	1992	1991	1992
Western Europe	4.9	3.9	3.7	3.0
4 major economies	5.5	3.8	4.3	3.3
France	3.6	4.7	2.9	1.4
Germany	12.8	3.4	11.7	2.8
Italy	-0.8	4.5	2.9	3.0
United Kingdom	0.2	3.2	-3.1	6.5
13 smaller economies	4.2	4.1	2.7	2.5
Austria	8.2	4.0	8.9	4.1
Belgium	3.1	1.1	2.7	1.7
Denmark	7.9	3.7	4.9	0.2
Finland	-6.7	9.1	-12.1	0.4
Greece	16.4	6.4	13.2	3.8
Ireland	5.7	9.7	1.7	4.5
Netherlands	4.6	3.5	3.7	1.5
Norway	6.3	6.1	1.3	1.0
Portugal	1.9	3.5	6.9	5.6
Spain	6.6	5.8	8.9	6.7
Sweden	-2.2	1.2	-5.9	0.5
Switzerland	-0.7	4.1	-1.7	-3.2
Turkey	11.4	8.7	-2.2	9.2
North America	4.7	6.2	0.4	8.5
United States	5.8	6.0	-0.1	9.5
Canada	0.5	7.0	2.2	4.3
Total above	4.9	4.7	2.5	4.8

Source: National statistics.

industry in machinery and equipment, the downturn of the international investment cycle has hit German exports relatively more than those of other countries. The below average growth of the United Kingdom's exports in 1991 and 1992 partly reflects the stronger orientation to markets in North America in comparison with other west European countries.

Although the international environment has been weak, several smaller countries were able to achieve relatively strong export growth in 1992. This was particularly so for Finland, Greece, Ireland, Norway, Spain, Switzerland and Turkey (table 2.1.12). The precise reasons for this favourable performance are not always easy to disentangle. In the case of Finland the increase in exports reflected sizeable gains in price competitiveness which arose from the devaluation in late 1991, while Ireland benefited from the strong demand for technology-intensive products produced by foreign-owned firms. Norwegian exports were mainly supported by strong demand for oil and gas and Spanish exports appear to have been favourably influenced by the considerable inflow of foreign direct investment since it joined the EC. Finally, Swiss exporters appear to have increased their share of foreign markets by accepting lower profit margins. This has also been a factor in the relatively robust export growth of the Netherlands in the last few years.

Imports of goods and services in western Europe slowed down to a growth rate of about 3 per cent in 1992 (table 2.1.12). The main factor behind this development was the weakness of domestic demand. At the same time, given the recession in the domestic economy, there was a surprisingly strong import growth in the United Kingdom. This may reflect the overvaluation of sterling until September 1992, which increased the price competitiveness of foreign producers in the domestic market. A similar effect may have occurred in other countries such as Italy, Portugal and Spain.

The combined changes of exports and imports in the real foreign balance imply that net exports added some 0.3 percentage points to overall economic growth in western Europe in 1992, slightly less than in 1991. The growth contribution originating in the real foreign account varies considerably among individual countries both with regard to sign and size. Among the four major economies, changes in the real foreign balance were a drag on output growth only in the United Kingdom. Among the smaller countries this was true for Belgium, Portugal, Spain and Turkey.

In the United States real exports of goods and services rose by 6 per cent in 1992, broadly the same as in the previous year. This robust growth reflects partly the beneficial effects of the depreciation of the dollar in the course of 1992 which supported sales in western Europe, and partly favourable demand developments in important markets, such as Latin America and Canada. The recovery of domestic demand led to a considerable surge in the volume of imports, which rose by 9.5 per cent in 1992 following stagnation the previous year. Consequently the deficit in the real foreign balance increased to the extent that the change in net exports subtracted nearly half a percentage point from overall economic growth of 2.1 per cent. In 1991 the change in real net exports had added 0.6 percentage points to total output growth.

The recovery of US domestic demand was a major factor behind the large increase in Canadian exports in 1992 (table 2.1.12). The positive impact on overall domestic activity (a contribution to GDP of about 2 percentage points) was largely offset by the acceleration in import growth.

2.2 INFLATION AND LABOUR MARKETS

(i) Costs and prices

Inflationary pressures in the western countries of the ECE region abated further during 1992, reflecting the combined effect of the continued weakness of economic activity, tight monetary conditions, wage moderation and, to a lesser extent, an improvement in the terms of trade. Measured by the *consumer price index*, the upturn in the rate of inflation which emerged in the late 1980s was reversed in 1991 and by the end of 1992 rates had fallen in some countries to levels not seen since the 1960s. The annual rate of inflation in 18 western economies of the ECE region[102] fell from 5.3 per cent in 1990 to 4.5 per cent in 1991 and 3.3 per cent in 1992 (Appendix table A.9). Thanks to a sharper fall in the high inflation countries, the convergence towards lower inflation rates in the region continued during 1992. The rate of inflation accelerated only in Germany and Austria, particularly during the first half of the year, due to reunification induced demand pressure in the former and continued strong economic activity in the latter. Disinflation was most significant in the United Kingdom, Finland, Sweden and Switzerland, where economic activity shrank further in 1992. Despite relatively stronger output growth there was also a marked disinflation in North America, a reflection of a significant moderation in the rate of increase in unit labour costs. In the fourth quarter of 1992 year-on-year rates of inflation were at their lowest since the first quarter of 1987 in North America (2.9 per cent) and since the second quarter of 1988 in the 16 west European countries taken together (3.5 per cent).

"Underlying" inflation rates,[103] which refer to consumer prices excluding seasonal and volatile food and energy prices, on the other hand, have remained rather sticky at around 4 per cent for western Europe as a whole (chart 2.2.1). Among the west European countries the "underlying" rate fell in the second half of the year in France, the United Kingdom, Switzerland and probably, given the considerable weakening in the growth of their unit labour costs and/or profits, in Finland, Sweden and Italy. In North America and particularly in Canada, the down-trend in "underlying" rates, which started in the last quarter of 1990, gathered momentum mainly as a result of a significant moderation in unit labour costs. However, despite its faster rate of deceleration, the "underlying" rate of inflation was also still higher (3.6 per cent) than the rise in the total index in North America (2.9 per cent) because of the dampening effect of falling food and energy prices on the latter.

Within the west European consumer price index, changes in the *prices of manufactured goods* accelerated during the first half of 1992 but by mid-year had resumed the down-trend which had started in late 1991 (chart 2.2.2).[104] In the last quarter the annual rate of increase was 1.5 per cent reflecting to a large extent the unusually heavy degree of price discounting in the "sales" in many countries. In contrast the rate of change in service prices stabilized at around 5 per cent throughout most of the year. Thus, the increase in the prices of manufactured goods fell from 2.7 per cent in 1991 to 2.2 per cent in 1992 while service prices accelerated from 4.2 per cent to 4.8 per cent (averages for six European economies). This difference in the rates and direction of price changes reflects the much larger impact of falling raw material prices on the production costs of manufactures, as well as the fact that most services are more labour intensive and much less subject to international competition than are producers of manufactured goods. The rate of change in the prices of manufactured goods accelerated also in the United States, in fact much more strongly than in western Europe: from 0.8 per cent in December 1991 to 3.2 per cent in July 1992. This partly reflects higher energy prices which, in western Europe, were partially offset by the weaker dollar. As in western Europe, US manufactured goods prices resumed their down-trend from mid-year and for the year as a whole, the increase was 2.4 per cent, down from 3.3 per cent in 1991. While both the direction and rates of changes in manufacturing prices in western Europe and the United States were similar during 1992, service price inflation in the United States, in contrast to western Europe, decelerated and, for the year as a whole, increased by just under 4 per cent (down from 5.1 per cent in 1991). This reflects a much faster rate of moderation in unit labour costs in the United States, which in turn was due to large productivity gains in the service sectors.

World commodity prices

Significantly weakened world demand together with large increases in production, mainly due to producing countries' need for foreign exchange but also to their response to repeatedly over-optimistic forecasts of growth in the industrial countries, created huge excess supplies particularly for commodities sensitive to cyclical conditions. *Non-energy commodity prices*, after

---

[102] Excluding Turkey where the annual rate of inflation averaged 70 per cent in 1992 despite some deceleration during the second half of the year.

[103] Weighted average, excluding Italy, Finland, Sweden and four south European countries.

[104] Weighted average of France, Germany, Belgium, Denmark, Norway and Switzerland.

CHART 2.2.1

Monthly changes in consumer prices, 1991-1992
(Percentage change over corresponding month of the preceding year)

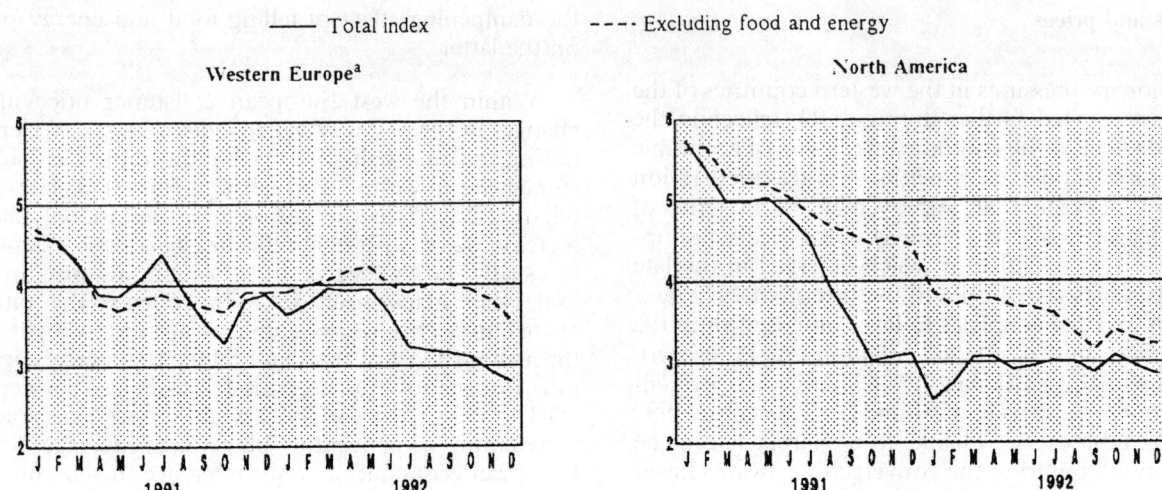

Source: National sources.

[a] Weighted average of 10 countries, excluding Italy, Finland Sweden and four South European countries.

CHART 2.2.2

Monthly changes in the prices of manufactured goods and services[a], 1991-1992
(Percentage change over corresponding month of the preceding year)

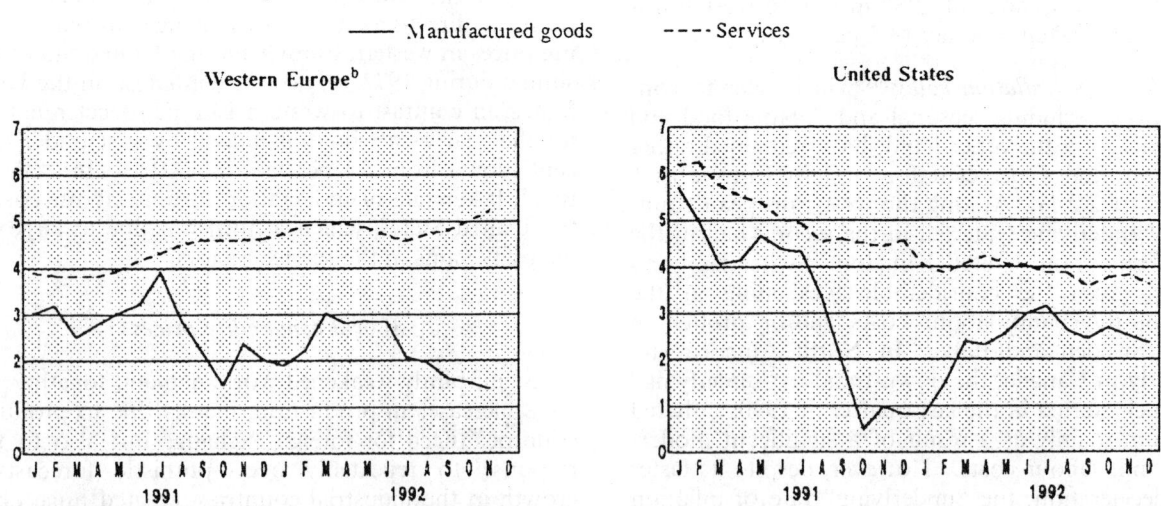

Source: National sources.

[a] Components of the consumer price index.
[b] Weighted average of France, Germany, Belgium, Denmark, Norway and Switzerland.

CHART 2.2.3

World market prices of raw materials, in US dollars and ECUs, 1991-1992
(December 1990 = 100, semi-logarithmic scale)

——— US dollars   ---- ECUs

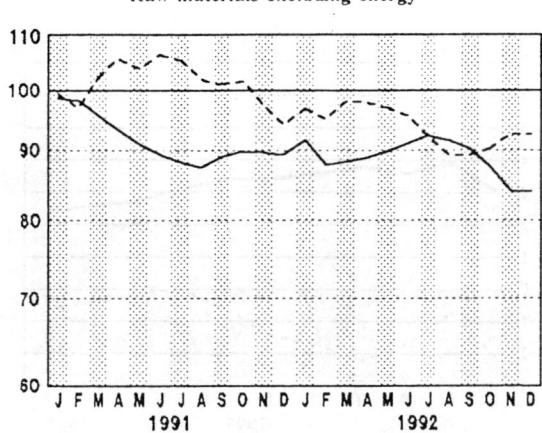

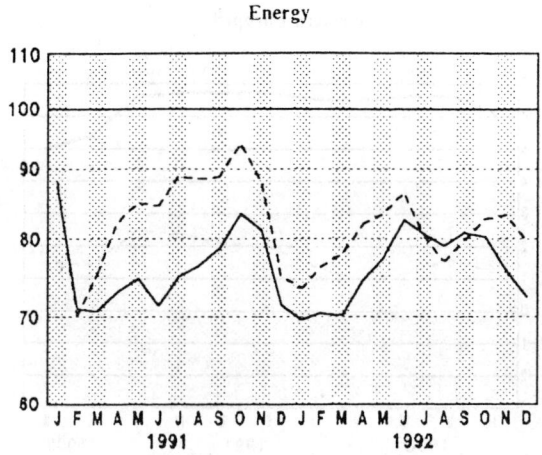

Source: The dollar index is published in Hamburg Institute for Economic Research (HWWA), *Intereconomics*, Hamburg (bi-monthly). The conversion to ECUs was made by the ECE secretariat on the basis of the US dollar-ECU exchange rate published in IMF, *International Financial Statistics*, Washington, D.C., (monthly).

reaching a cyclical peak in September 1990 fell continuously for nearly a year (chart 2.2.3). During the last quarter of 1991 and the first half of 1992 non-energy prices stabilized somewhat but only to resume their down-trend during the late summer. In the closing months of 1992, non-energy prices had fallen to levels not seen since mid-1987, a reflection of the further weakening of industrial activity in most of the importing countries and, probably to some extent of the significant rebound of the dollar. The HWWA non-energy world commodity index[105] in December 1992 was 6.2 per cent lower than a year earlier in dollar terms and 1.7 per cent lower in ECUs. However, given the strong depreciation of the dollar during the second and third quarters, non-energy prices for the year as a whole fell by 2.7 per cent in dollar terms and 7.2 per cent in ECUs (1992 compared with 1991). Industrial raw materials, especially basic metals and minerals, suffered most, partly due to massive exports from Russia. Thanks to record harvests in grains, particularly in the United States, and the collapse of international commodity agreements (such as those for cocoa, coffee and sugar), food prices also fell considerably during the second half of 1992.

Oil prices started the year at less than $17 a barrel for Brent crude. During the spring, in anticipation of a strong recovery in demand, prices reached $20 a barrel in May. Until late October prices fluctuated around that level, which was not far from OPEC's target price ($21). However, with the absence of recovery in industrial demand, unseasonably warm weather, and continued over-production by some members of the OPEC cartel, prices fell sharply during the closing months of 1992 and were less than $17 a barrel by mid-January 1993. Despite large fluctuations during the year, *world energy prices*[106] for 1992 as a whole were virtually the same as in 1991 in terms of dollars while in ECUs they fell by 4.7 per cent.

Average hourly earnings, input and output prices in manufacturing industry

Reflecting significantly weaker labour markets and the maintenance of tight monetary policies in most countries, the *rate of increase in average hourly earnings* decelerated considerably in each quarter of 1992 both in western Europe and North America (chart 2.2.4). In western Europe average earnings increased by 5 per cent in the third quarter of 1992 compared with 6.7 per cent during the same quarter of 1991. In North America the rate fell from 3.4 to 2.4 per cent during the same period. In western Europe the rate of growth in average hourly earnings also fell below 2.5 per cent in Finland and Norway in the third quarter of 1992. In real terms (deflated by the change in consumer prices), average hourly earnings fell in Finland, Italy (during the second half of 1992) and the United States (for the sixth consecutive year albeit at much lower rates than in the previous two years). On the other hand, in most west European countries, and despite a relatively strong moderation in nominal earnings growth, real earnings increased significantly (notably in Germany, the United Kingdom, Austria, Belgium, Ireland, Sweden and

---

[105] Produced by the Institute for Economic Research, Hamburg, this index weights world market prices (in dollars) by the relevant commodity shares in total imports of the western industrialized countries in 1974-1976.

[106] The weight of oil in the total energy component of the HWWA index is 91.3 per cent.

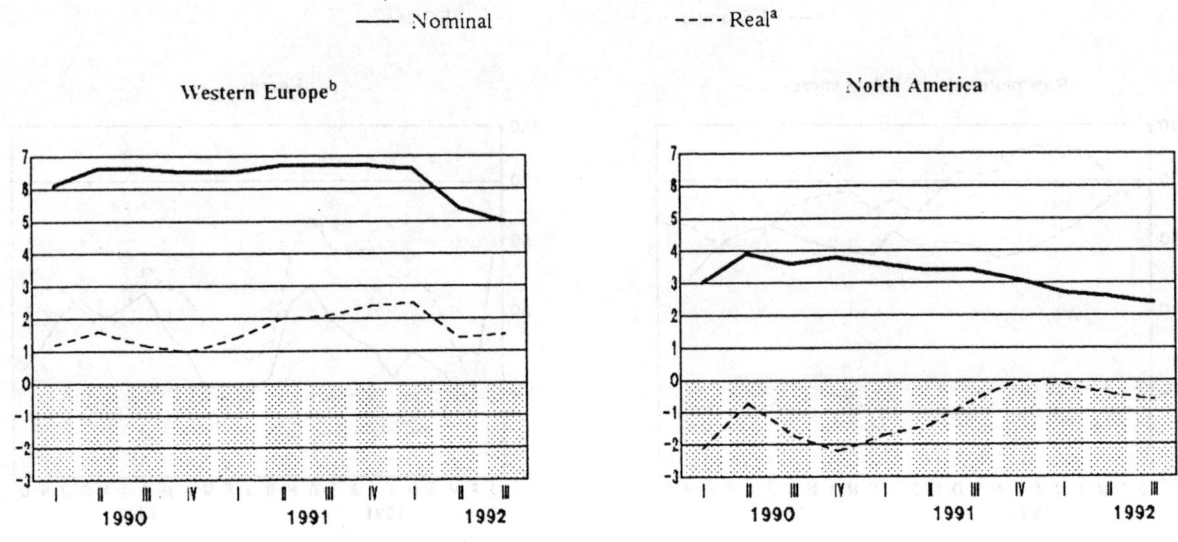

CHART 2.2.4

Quarterly changes in average hourly earnings in manufacturing industry, 1990-1992
(Percentage change over corresponding quarter of the preceding year)

Source: OECD, *Main Economic Indicators*, January, 1993 Paris; national statistics.
[a] Deflated by consumer price indices.
[b] Weighted average of 13 countries excluding the four South European countries.

Switzerland). Thus, in western Europe as a whole, the increase in real average hourly earnings was just below 2 per cent during the first three quarters of 1992, as much as during the same period of 1991. However, given the large falls in employment and the subsequently large productivity gains in most countries, manufacturing industries' unit labour costs moderated also in all these countries with the exception of Germany, where earnings growth outpaced weaker productivity gains during 1992.

Manufacturing industry's *raw material and intermediate goods prices* continued to fall in 1992: by 0.9 per cent in western Europe and 0.5 per cent in North America (chart 2.2.5). The larger fall in western Europe was mainly due to the strong depreciation of the dollar during the second and third quarters *vis-à-vis* most of the west European currencies. (In the United States there was a small increase in input prices during those two quarters.) In western Europe manufacturing industries' input prices increased throughout the year only in Denmark, Finland and to a much lesser extent in Ireland. However, input prices increased strongly also in the United Kingdom and Italy in the fourth quarter due to the large depreciation of the pound and the lira following their withdrawal from the ERM in September.

The growth of manufacturing industry's *output prices*, which moderated significantly during 1991 in western Europe and particularly in North America (where they actually fell during the last quarter), accelerated slightly during the first half of 1992. However, reflecting the significant weakening of consumer demand and the strong deceleration in labour and material costs of production, the rate of change in output prices resumed their down-trend in western Europe while they more or less stabilized at around 1.5 per cent in the United States. The change in output prices accelerated slightly in December and early 1993 in the United States mainly due to the strengthening of demand. In western Europe, despite the pressure of the dollar's appreciation on the input prices, the growth in output prices remained stable at just above 1 per cent, a reflection of a significant moderation in unit labour costs in general and, in some countries, reduced profit margins. Thus, for the year as a whole output prices increased by only 1.7 per cent in western Europe and 1.2 per cent in the United States compared with 2.5 per cent and 2.2 per cent respectively in 1991. In Canada the output prices increased by less than half a percentage point in 1992 after falling some 6 per cent in 1991. However, during the second half of the year there was a noticeable acceleration in Canadian output prices, in contrast to most west European countries and the United States.

Even though total manufacturing profits shrank in real terms in 1992 and the number of business closures and bankruptcies rose sharply because of the significant weakening of demand (particularly for durables and investment goods), profit margins probably increased throughout most of the year in the majority of the western industrialized countries, thanks to the strong deceleration in unit labour costs and the fall in material input prices.

Labour and non-labour unit costs in total economy

Reflecting the increased slack in the labour markets inflationary pressures arising from *wage and non-wage labour costs* continued to moderate in 1992. Labour

CHART 2.2.5

Intermediate and final output prices in manufacturing industry,[a] 1991-1992
(Percentage change over corresponding month of the preceding year)

——— Intermediate product prices   - - - - Final output prices

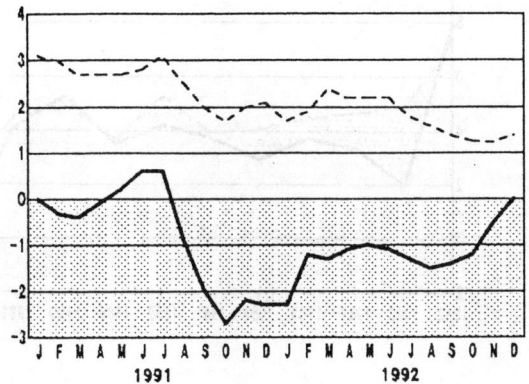

Western Europe[b]

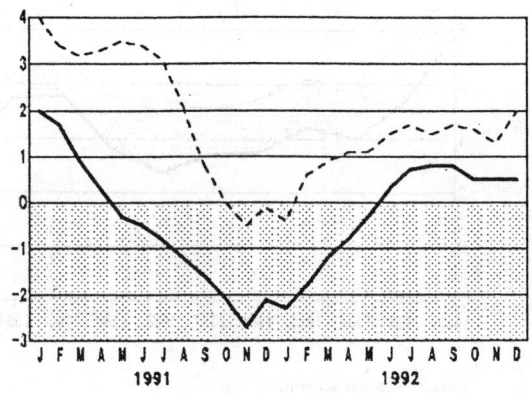

United States

Source: National sources.
[a] For definition of indices, see United Nations Economic Commission for Europe, *Economic Survey of Europe in 1982*, New York, 1983, pp 32-33.
[b] Weighted average of 12 countries, excluding Italy and four South European countries.

costs per employee in western Europe increased on average by 5.3 per cent in 1992 compared with 6.4 per cent in 1991 and 7.0 per cent in 1990 (chart 2.2.6). Despite this general down-trend of the last two years, in some countries labour costs per employee were still increasing well above the annual domestic inflation rate in 1992: most notably in Sweden (7.8 per cent), the United Kingdom (7.0 per cent) and, to a lesser extent, in Belgium and Ireland (both 5.7 per cent). Nevertheless, the rate of change accelerated in 1992 only in Ireland and was unchanged only in France (4.6 per cent) and Germany (5.2 per cent). In contrast, there was a marked deceleration in the rate in Finland, Norway and, albeit at higher rates, in Italy and Switzerland. In 1992, not only did wage and non-wage labour costs continue to moderate but, unlike 1991, they were also accompanied by large gains in *labour productivity*, a result of the more rapid decline in employment in most west European countries during 1992. West European labour productivity increased on average by 1.5 per cent in 1992 compared with just under 1 per cent in 1991. The largest gains were in France (2.4 per cent), Ireland, Finland, Norway and Sweden (between 3-3.5 per cent). Productivity growth slowed down only in Germany, Austria and Denmark: in the former two there was still a relatively high growth of employment, particularly during the first half of the year, and a sharp slow-down in the output growth during the second.

Consequently the average rate of change in west European *unit labour costs* slowed down considerably in 1992 (3.7 per cent compared with 5.5 per cent during the previous two years, table 2.2.1). They accelerated only in Germany and only in a few countries did the rate of increase remain above 4.5 per cent (the United Kingdom, Austria, Germany and Sweden). In Finland,

TABLE 2.2.1

Unit labour costs and unit profits, 1991-1992
(Annual percentage change)

	Unit labour costs		Unit profits [a]	
	1991	1992	1991	1992
France..........................	3.8	2.2	2.9	4.7
Germany......................	4.1	4.6	1.9	3.2
Italy.............................	7.6	4.1	6.2	6.5
United Kingdom [b]	8.5	5.4	1.7	4.8
Austria.........................	5.3	5.1	1.2	2.7
Belgium........................	4.0	3.9	1.4	1.0
Denmark	1.5	1.4	4.6	2.8
Finland.........................	7.8	-3.0	-4.9	9.8
Ireland.........................	3.4	2.7	3.3	3.0
Netherlands.................	3.5	2.4	2.3	3.4
Norway [b]	2.0	0.1	1.8	-2.9
Sweden........................	8.3	5.3	6.9	-9.8
Switzerland..................	7.6	3.5	4.4	-1.9
Western Europe [c]	**5.5**	**3.7**	**2.9**	**3.5**
United States...............	4.2	0.8	2.6	5.7
Canada [b]...................	4.7	1.7	-3.2	-1.9

Source: National accounts.
[a] Operating surplus including capital consumption per unit of real gross value added.
[b] Based on data for three quarters.
[c] Weighted average of 13 countries.

unit labour costs, which had soared by nearly 8.5 per cent during the previous two years, declined by 3 per cent in 1992: output fell by 3.5 per cent but there was a nearly 7 per cent decline in employment and negligible growth in wages. In Norway unit labour costs virtually stabilized at 1991 levels (due mainly to faster output growth), whereas in France, Italy, Sweden and Switzerland the rate was about half of that in 1991. *Real*

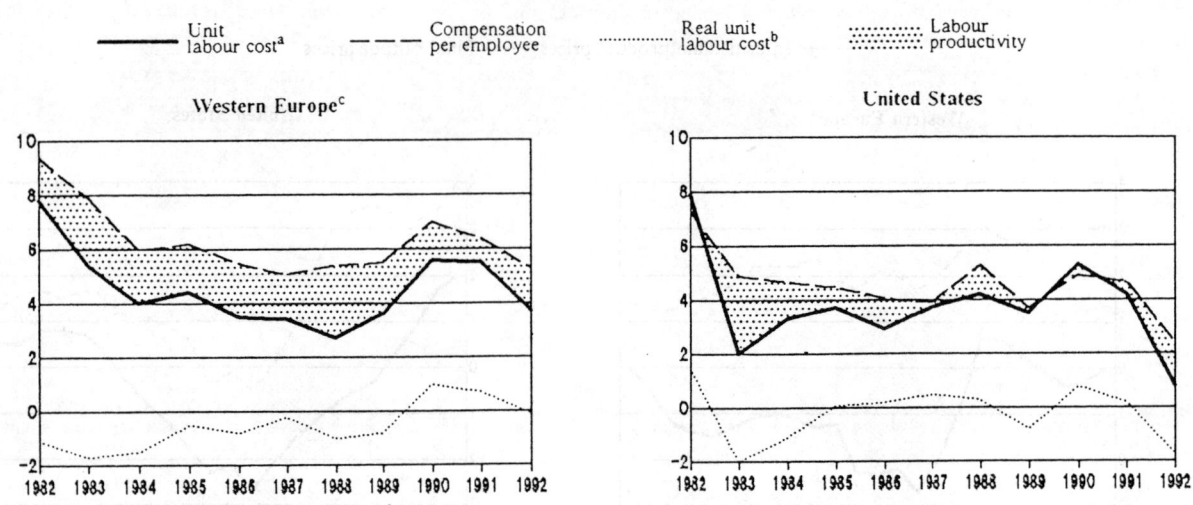

CHART 2.2.6

Unit labour costs in total economy, 1982-1992
(Annual percentage changes)

Source: National accounts.
a Compensation of employees per unit of real gross value added.
b Deflated by implicit price deflator for GDP.
c Weighted average of 13 West European countries excluding four South European countries.

unit labour costs[107] were stable for western Europe as a whole and increased only in Austria, Belgium, Switzerland, Sweden and Norway. In Norway this was due to a decline in the GDP deflator as there was virtually no change in nominal unit labour costs. Real unit labour costs fell by 4 per cent in Finland and by about 1 per cent in France, Italy and Ireland.

For western Europe as a whole *unit profits* (the share of gross operating surplus in value added) increased by 3.5 per cent in 1992 compared with 2.9 per cent in 1991. However, while unit labour costs moderated everywhere except in Germany, the direction and size of change in unit profits in 1992 varied considerably between countries. They grew strongly and at much higher rates than in the previous two years in France, the United Kingdom, Italy, Finland, Germany, Austria and the Netherlands, but they fell in Norway, Sweden, and, to a much lesser extent, in Switzerland.

In North America unit labour costs weakened much more than in western Europe. They increased only by 1.7 per cent in Canada and 0.8 per cent in the United States, thanks to a sharp acceleration in productivity (mainly due to falling employment in Canada and output recovery in the United States) and a further moderation in wage and non-wage labour costs (3.5 and 2.4 per cent respectively). Real unit labour costs increased by just under 1 per cent in Canada but fell by nearly 2 per cent in the United States. On the other hand, unit profits in 1992 shrank further in Canada and rose sharply in the United States (5.7 per cent, the largest increase in the last eight years).

The sources of inflation

In 1992, the rate of change in the *GDP deflator*, the broadest measure of inflation, which reflects domestic cost pressures, decelerated both in western Europe and North America (table 2.2.2). In western Europe, the average rate of increase fell to 3.8 per cent in 1992, after accelerating for the previous four years and reaching 4.8 per cent in 1991. In North America the disinflation, which had already started in 1991, was much stronger: down to 2.5 per cent in 1992 compared with 3.9 per cent in 1991 and 4.3 per cent in 1990. There was some acceleration in the GDP deflator in Germany, Austria, Belgium and Ireland, but a strong disinflation in the Nordic countries, Switzerland and Canada. In the majority of countries this favourable price performance was mainly due to a significant moderation in the growth of unit labour costs which compensated for the larger increase in unit profits (notably the case in France, Italy, the United Kingdom, Finland, the Netherlands and the United States). In Norway, Sweden, Switzerland and Canada, not only did unit labour cost growth weaken significantly but unit profits also fell, thereby pulling down the GDP deflator. Nevertheless, for western Europe as a whole, the rise in unit profits contributed more than one third of the total increase in the GDP deflator in 1992 compared with less than a quarter in 1991. In the United States more than two thirds of the increase in the GDP deflator was due to the large increase in unit profits. On the other hand, the impact on prices of net indirect taxes, which had increased considerably in 1991 in some countries (namely Germany, Italy, the United Kingdom, Sweden and Canada), moderated in 1992, and their contribution

---

[107] Nominal unit labour costs deflated by the GDP deflator.

TABLE 2.2.2

Contribution to the change in the GDP deflator, 1991-1992
(Percentages)

	Change in the GDP deflator [a]	of which due to:			Unit profits [c]	Unit indirect taxes net of subsidies
		Unit labour costs				
		Total	Compensation per employee [b]	Labour productivity		
France						
1991	3.0	2.0	2.4	-0.4	1.0	-
1992	3.0	1.1	2.4	-1.3	1.6	0.3
Germany						
1991	4.2	2.2	2.8	-0.6	0.7	1.3
1992	4.5	2.5	2.9	-0.4	1.1	1.0
Italy						
1991	7.3	3.4	3.7	-0.3	2.9	1.0
1992	5.4	1.8	2.3	-0.4	3.0	0.6
United Kingdom [d]						
1991	6.8	4.8	5.5	-0.6	0.5	1.5
1992	5.2	3.1	4.0	-0.9	1.4	0.7
Austria						
1991	3.4	2.8	3.3	-0.5	0.4	0.2
1992	4.6	2.7	2.8	-0.1	0.9	0.9
Belgium						
1991	2.5	2.1	3.0	-0.9	0.5	-0.1
1992	3.0	2.0	3.0	-0.9	0.4	0.5
Denmark						
1991	2.4	0.8	2.0	-1.2	1.5	0.2
1992	1.8	0.8	1.6	-0.9	0.9	0.1
Finland						
1991	2.3	4.3	3.6	0.6	-1.6	-0.3
1992	1.0	-1.7	0.3	-2.0	3.0	-0.3
Ireland						
1991	3.2	1.7	2.7	-0.9	1.3	0.2
1992	3.6	1.3	2.9	-1.5	1.2	1.1
Netherlands						
1991	3.0	1.8	2.3	-0.5	0.9	0.3
1992	2.7	1.3	2.0	-0.7	1.3	0.1
Norway [d]						
1991	1.9	1.0	2.6	-1.5	0.7	0.1
1992	-1.0	0.1	1.8	-1.7	-1.1	0.1
Sweden						
1991	8.2	5.1	4.9	0.2	1.8	1.3
1992	0.4	3.3	4.8	-1.4	-2.5	-0.4
Switzerland						
1991	6.1	4.6	4.6	-	1.5	-0.1
1992	1.8	2.1	3.1	-0.9	-0.6	0.3
Western Europe [e]						
1991	4.8	2.9	3.4	-0.5	1.1	0.8
1992	3.8	2.0	2.8	-0.8	1.3	0.5
Canada [d]						
1991	2.7	2.6	2.7	-0.1	-1.1	1.2
1992	0.9	1.0	2.0	-1.0	-0.6	0.5
United States						
1991	4.0	2.5	2.8	-0.3	0.9	0.6
1992	2.6	0.5	1.4	-0.9	1.8	0.3

Source: National accounts. Small discrepancies are due to rounding.

a GDP at market prices.
b Wage and non-wage labour costs per person employed.
c Includes capital consumption.
d Based on data for three quarters.
e Weighted average of 13 countries.

to the increase in the GDP deflator was about one percentage point only in Germany, Austria and Ireland. Thus, domestic cost pressures weakened in most of the developed market economies of the ECE region during 1992 mainly because of increased productivity accompanied by weaker labour costs. With few exceptions, unit profits put more pressure on prices than in 1991 and in some countries (France, Italy, Finland and the United States) they contributed significantly more than unit labour costs to the overall increase despite the much lower share of profits in value added compared with the share of labour.

The change in the *domestic demand deflator*, which takes into account the terms of trade effect on the overall rate of domestic inflation, decelerated during 1992 both in western Europe and North America (table 2.2.3). Domestic inflation accelerated only in Austria, the Netherlands and, to a much lesser extent, in France.

In most west European countries and the United States there was a small but favourable terms of trade effect on domestic prices in 1992. The change in import prices put upward pressure on the domestic price level only in Finland, Switzerland and, to a much lesser extent, in Austria and Canada. In Finland all of the increase in the domestic demand deflator was due to the large devaluation of the markka.[108] Falling effective exchange rates were also the main reason for the unfavourable terms of trade effect in Canada and partly in Switzerland.

The effect of changes in suppliers' prices[109] on domestic inflation rates was favourable for Canada and the United States and for a few European countries (Italy, the United Kingdom, Finland, and Sweden). Given the important effect of the varying composition of imports both in terms of goods and of country of origin on the change in suppliers' prices, it is difficult to generalize about the reasons for the substantial disparity in supplier price pressures on individual countries. Significantly weaker food prices were probably an important factor for the United Kingdom, Finland and Sweden, where food has a large share in total imports whereas in Italy, the United States and Canada there appear to have been price reductions by exporters designed to maintain market shares in these large economies, a common practice during periods of weak world trade accompanied by turmoil on the foreign exchange markets.

Notwithstanding the favourable terms of trade effect, the main factor behind domestic price disinflation in the majority of western countries during 1992 was the significantly smaller contributions of domestic cost pressures.

(ii) Labour markets

Reflecting sluggish output growth or recession in most of the western economies of the ECE region during the last two and a half years, *employment*[110] stagnated in 1992, with employment in western Europe falling by 0.4 per cent and increasing by 0.5 per cent in the United States (Appendix table A.11). These rates of change compare with zero growth, and a decline of 1.6 per cent, respectively, during 1991. Given the usual lagged response of employment to changes in output, the labour markets of individual countries in 1991 generally reflected the sharp differences in output growth in 1990. To a large extent, this desynchronization of national growth cycles ceased in early 1991 in western Europe. Consequently, employment in 1992 fell or stagnated almost everywhere in western Europe except in Germany, Italy, Austria, the Netherlands and Turkey where employment increased albeit at much slower rates than in 1991 (in which year there was a significant acceleration in output growth). In 1992, 700,000 jobs were lost in western Europe as a whole. However, excluding the countries where employment rose, the total job losses were nearly 1.5 million during 1992, almost half being in the United Kingdom. The largest relative declines in employment were in Finland (7 per cent), Sweden (4 per cent), the United Kingdom (2.5 per cent) and Switzerland (2 per cent), where economic activity has deteriorated significantly during the last two years.

In the United States, where the recession officially ended in the second quarter of 1991, there was a jobless recovery during 1992: a 2.1 per cent increase in the total economy output created only a 0.5 per cent rise in employment. However, employment started to rise significantly in early 1993: in February, non-farm payroll employment rose by 365,000, the biggest monthly increase since 1988.

Labour force survey data on *employment by sectors* (chart 2.2.7) show that, unlike previous downturns, not only industry but also many branches of services (especially private and non-financial services) have suffered retrenchment during the current recession in the United Kingdom, Finland, Sweden, Switzerland, the United States and Canada. Except in the United States, total service employment has fallen for several quarters in these countries. Nevertheless, industry (especially durable consumer goods, investment goods and construction) has continued to lose a far larger proportion of jobs than the service sector in all countries, notwithstanding the fact that total industrial employment has fallen less than in the previous downturn in the United States, Canada, France, Italy, Belgium and Switzerland. This recent convergence of employment changes in services and industry can be explained to a large extent by the persistently high real interest rates in western Europe during the recent slow-down. Dur-

---

[108] The markka was devalued by 12.3 per cent in November 1991. Due to the further deterioration of the economic situation during 1992, it came under renewed speculative pressure, and in September 1992 it was allowed to float. During the 18 months to the beginning of 1993, the markka depreciated by more than 25 per cent against the ECU.

[109] The prices of imports in terms of the national currency of the country of origin.

[110] Employment in this section, unless otherwise stated, is defined in national accounts terms.

TABLE 2.2.3
Contribution to the change in the domestic demand deflator, 1991-1992
(Percentages)

c	Change in domestic demand deflator	Changes in GDP deflator excluding exports [a]	Import prices Total	Import prices Exchange rates [b]	Export prices of suppliers
France					
1991	2.9	3.0	-0.1	0.5	0.5
1992	3.1	3.1	-	-0.7	0.7
Germany					
1991	4.2	3.6	0.6	0.3	0.3
1992	4.1	4.5	-0.4	-0.9	0.5
Italy					
1991	6.5	6.5	-	0.3	-0.4
1992	5.2	9.1	0.1	0.6	-0.5
United Kingdom [d]					
1991	6.1	6.6	-0.5	-0.1	-0.4
1992	4.5	4.8	-0.3	0.9	-1.2
Austria					
1991	3.4	3.3	0.4	0.2	0.2
1992	4.1	3.8	0.3	-0.9	1.1
Belgium					
1991	2.8	2.2	0.6	0.1	0.6
1992	2.8	2.9	-0.1	-1.3	1.3
Denmark					
1991	2.9	2.5	0.4	0.5	-
1992	2.3	2.8	-0.5	-0.8	0.3
Finland					
1991	3.3	3.1	0.2	1.0	-0.7
1992	1.5	-0.2	1.7	3.3	-1.6
Ireland					
1991	5.9	5.1	0.8	0.8	-0.1
1992	4.3	4.5	-0.2	-2.0	1.8
Netherlands					
1991	3.3	3.3	-	0.4	-0.4
1992	3.6	4.3	-0.7	-1.2	0.5
Norway [d]					
1991	3.9	3.5	0.4	0.8	-0.4
1992	3.4	3.7	-0.2	-0.5	0.3
Sweden					
1991	8.3	8.2	0.1	0.1	-
1992	-	0.8	-0.7	-0.4	-0.4
Switzerland					
1991	5.7	5.5	0.2	0.6	-0.4
1992	2.8	1.9	0.9	0.6	0.4
Western Europe [e]					
1991	4.6	4.5	0.1	0.3	-0.2
1992	3.7	3.9	-0.2	-0.2	0.1
Canada [d]					
1991	2.9	3.4	-0.5	-0.4	-
1992	1.0	0.6	0.4	1.6	-1.3
United States					
1991	3.8	3.9	-0.1	0.1	-0.2
1992	2.4	2.6	-0.2	0.3	-0.5

Source: National accounts and IMF, *International Financial Statistics*, Washington, D.C. Small discrepancies are due to rounding.

a Calculated as the residual of the change in the domestic demand deflator minus the contribution of the change in import prices.
b Based on nominal effective exchange rates.
c These are the prices of imports in terms of the national currency of the country of origin.
d Based on data for three quarters.
e Weighted average of 13 countries.

CHART 2.2.7

Employment growth by sectors, 1980-1992
(Annual percentage changes)

——— Total   — — — Industry   ······· Services

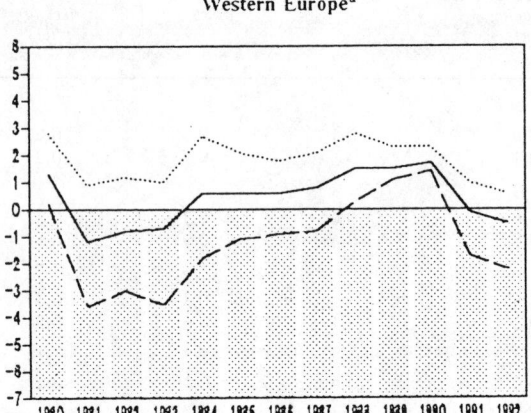

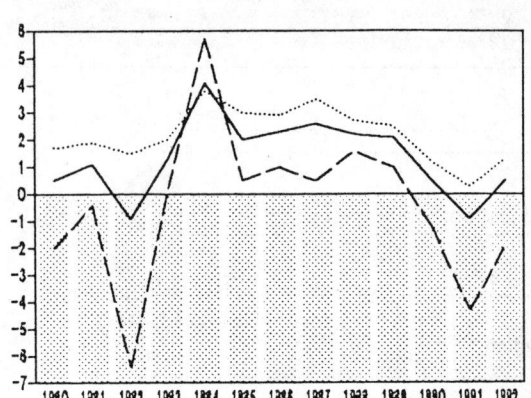

Source: OECD, *Quarterly Labour Force Statistics*, Paris, various issues.
a Weighted average of France, Germany, Italy and the United Kingdom.

ing the recession of the early 1980s, the squeeze was to a large extent limited to the tradeable goods producing sectors of the economy, that is, industry and only those services closely linked with the tradeable goods sectors (particularly wholesale trade and transport). In the current recession, not only have most west European currencies been overvalued due to ERM membership or formal links to it, but real interest rates have also been at record high levels. This sharp increase in the cost of borrowing, in addition to tighter constraints on credit, hit also those enterprises which mainly serve the domestic market. A large proportion of these enterprises are usually small in size and in the past have been a stabilizing factor during downturns and an important source of new jobs during the recovery and subsequent expansion.[111] During the current downturn they have also shed labour. Furthermore, in response to increased exposure to foreign competition through deregulation and sharply increased foreign direct investment in many service branches, labour productivity in most services has risen significantly during the last few years thanks to new cost-cutting office technologies. These technologies have enabled the service sector to restructure like manufacturing did during the 1980s, and its employment elasticity has consequently increased. Normally, during an economic recovery, productivity increases sharply in industry through higher utilization of existing capacity. That is, manufacturing industry, in particular, does not immediately create new jobs within the sector during the early phases of recovery but generates higher disposable incomes for those in work which, in turn, increases demand, *inter alia*, for services and hence new service jobs.[112] At present, this demand for services, at least partially, is being met by increasing productivity and not immediately by employing new staff, as in the past. This may explain to a large extent the slower than usual recovery in the US labour markets during 1992, where investment in labour-saving capital equipment has increased sharply during the last decade or so.[113]

In the US economy, employment and output growth during the last decade have been much more closely synchronized than in western Europe (chart 2.2.8). This may be explained to some extent by the behaviour of labour costs. That is, in the US, earnings seem to adjust more quickly to market clearing levels. During the 1979-1992 period, labour cost per person employed in the US increased persistently at lower rates than in the four large economies of western Europe. Labour cost growth slowed down steadily in the United States, even during periods of accelerating growth in output and employment (1983, 1984 and 1987). In western Europe, however, the down-trend of the first

---

111 In the United States, small businesses were responsible for almost all the 20 million new jobs created during the 1980s. According to the US Small Business Administration, companies with fewer than 20 employees created 4 million jobs between 1988-1990. Companies employing over 100 lost 1.2 million jobs. *Wall Street Journal*, 4 February 1993. In the United Kingdom, between 1980 and 1992, the service sector, which includes most small businesses, was the exclusive generator of new employment in the economy, producing more than 1.5 million net new jobs. During that period small business increased its share in total employment from 59.6 per cent to 71.6 per cent, while manufacturing declined from 30.3 per cent to 21.2 per cent. According to official estimates, small businesses created around 1 million jobs in the second half of the 1980s. *Financial Times*, 6-7 March 1993.

112 For example in January 1993, the average manufacturing work week in the United States rose to 41.4 hours (the highest level since 1966), which increased overtime payment significantly.

113 During 1992, productivity in the US non-farm business sector rose 2.7 per cent, the biggest increase for 20 years. "In the fourth quarter, overall productivity actually grew slightly faster than manufacturing productivity, suggesting that service companies were improving their efficiency even faster than manufacturers." *Financial Times*, 8 February 1993.

CHART 2.2.8

Labour market development, 1979-1992

(Percentages)

Source: National accounts; OECD, *Quarterly Labour Force Statistics*, various issues, Paris.
a Annual percentage growth rates.
b Rate of unemployment.
c Weighted average of France, Germany, Italy and the United Kingdom.

half of the 1980s was reversed in 1988 when employment started to recover relatively strongly. This may help to explain the increasing long-term trend of unemployment rates in western Europe. After seven years of uninterrupted economic expansion in the four largest European economies, the unemployment rate in 1990 (7.6 per cent) was much higher than its previous trough in 1979 (5.3 per cent), whereas in the United States the rate of unemployment in 1989 was below its level in 1979 (5.2 compared with 5.8 per cent).

In December 1992, the average *rate of unemployment* in western Europe was 9.8 per cent compared with 8.9 per cent in December 1991 and its previous low of 8 per cent in the fourth quarter of 1990. For the year as a whole, the rate of unemployment in western Europe in 1992 averaged 9.4 per cent (Appendix table A.12). In the United States the rate reached its peak in June (7.6 per cent) and then, reflecting the stronger recovery in output growth, it fell gradually to 7 per cent in February 1993, which was still much higher than its previous low point of 5.2 per cent in the second quarter of 1990. In Canada the unemployment rate was higher than in most west European countries (11.4 per cent in December 1992). Among the western ECE countries it was only in the Netherlands that the average rate of unemployment fell in 1992 compared with 1991. However, the unemployment rate in the Netherlands increased significantly during the second half of the year, from 6.1 per cent in July to 7.3 per cent in December, whereas in both Portugal and Italy the rate remained rather stable, at around 4 per cent and 10 per cent respectively during 1991-1992. During the last two and a half years the most rapid increases in unemployment rates were in Finland, Sweden, Switzerland and to a much lesser extent in Canada and the United Kingdom, albeit at very different levels (see chart 2.2.9). Unemployment at the end of 1992 was highest in Spain (19.5 per cent), Ireland (16.6 per cent) and Finland (15.1 per cent), while the lowest rates were in Austria, Portugal and Switzerland (about 4 per cent).

The total number of people registered as without work in the western countries of the ECE region reached 30 million persons in December 1992 compared with 27.7 million in December 1991 and 22.4 million in June 1990 (its lowest level during the present cycle); in other words, the numbers unemployed have increased by some 7.5 million or by more than one third during the last two and a half years.[114] Not only have the numbers and the rates of unemployment reached very high levels, but also the average *duration of unemployment* and, hence, the share of those who have been unemployed more than one year (long-term unemployment) has increased significantly. The share of long-term unemployed in 1992 was one third or more of total unemployment in France, Italy, the United Kingdom, Sweden, Denmark and probably in Turkey. The share was more a than half in 1992 in Ireland, Spain and the Netherlands, despite the fall in the average unemployment rate in the latter.

Unlike previous downturns, *male unemployment rates* have increased faster than *female rates*, albeit at lower levels: in the European Community, the seasonally adjusted male unemployment rate had increased to 9.9 per cent in December 1992 compared with 6.5 per cent in June 1990 (its lowest rate in the current cy-

---

114 These estimates exclude eastern Germany where the numbers unemployed reached 1.1 million at the end of 1992 compared with 142,100 in June 1990.

CHART 2.2.9

Quarterly standardized unemployment rates, 1990-1992
(Per cent of labour force)

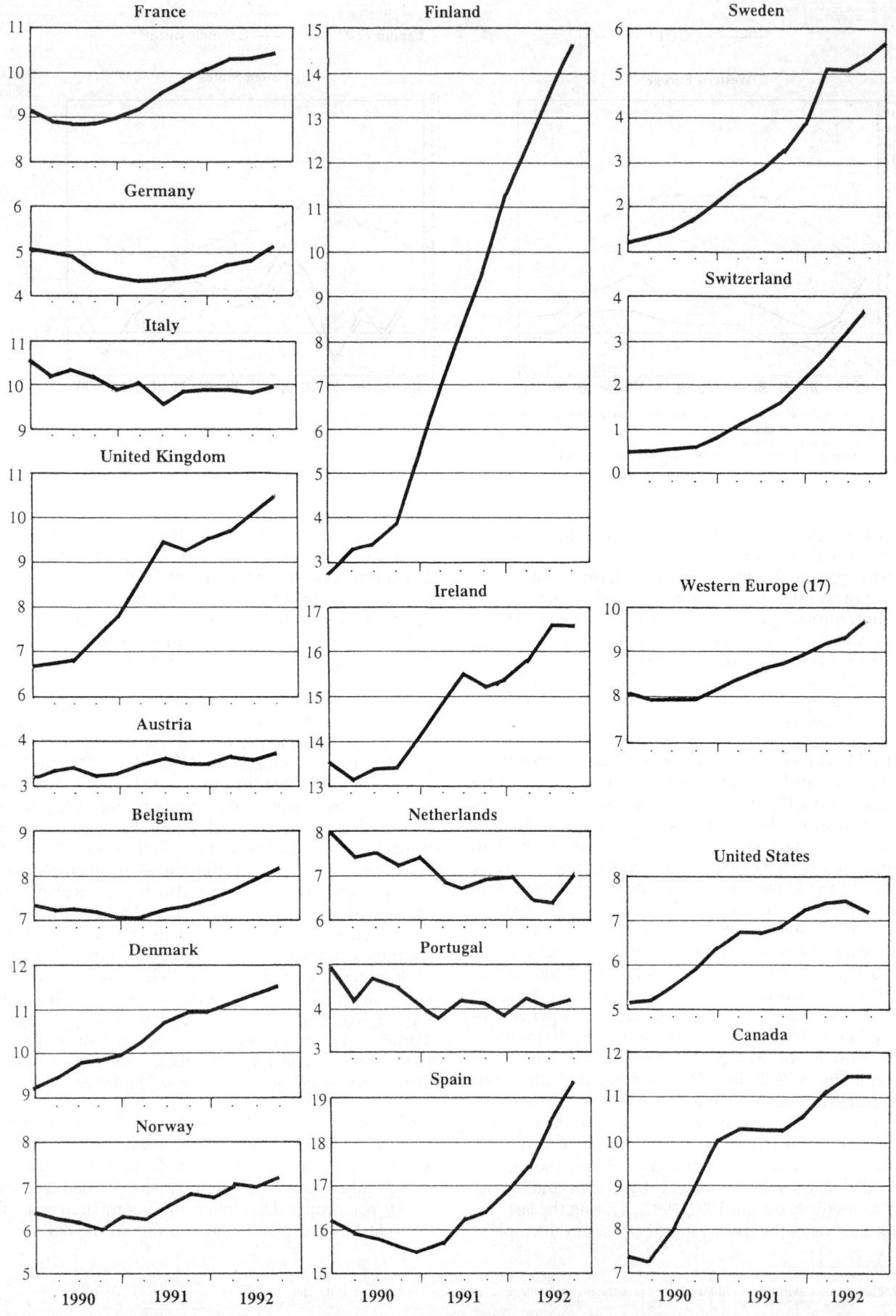

Source: OECD, *Press release: Standardized unemployment rates*, Paris, various issues; national sources.

cle).[115] During the same period the female rate has risen from 11 per cent to 11.8 per cent. The number of persons unemployed in western Europe increased by about 10 per cent in 1992, the increase for males (13 per cent) being more than twice that for females (6 per cent). Nevertheless male rates of unemployment were lower than female rates in all countries except in the United Kingdom and the United States.[116]

Part- and short-time employment, which were already rising during the 1980s, mainly in the service sectors, have risen significantly during the last two and half years and have also spread to some branches of manufacturing as a form of temporary lay-offs with pay. For example in west Germany where the recession has deepened significantly and the unemployment rate has soared during the early months of 1993, the number of workers on short shifts was 1 million in February 1993, up from 650,000 at the end of 1992. In the United States, the recovery of employment in February was largely due to part-time service jobs.

One of the most disturbing features of the current labour market situation is that the unemployment rate for persons under 25 years of age has risen considerably during the last two and a half years. The *youth unemployment rate* in western Europe reached nearly 20 per cent at the end of 1992, up from 15.5 per cent in mid-1990. The highest youth unemployment rates were in Spain (33 per cent), France, Italy, Ireland (about 28 per cent in all three) and, probably, Turkey. The share of young persons in total unemployment has thus reached disturbing levels in most countries. In Italy half of all the unemployed (that is more than one million persons) were younger than 25 years. In the United Kingdom, Ireland, Spain, Sweden and Canada the share of youths in total unemployment was about one third or more. This huge army of unemployed young persons is not only an enormous economic waste but, unlike the elderly unemployed in general, is also a major factor behind the serious aggravation of a range of social problems.

Given the present record low levels of consumer confidence and the relatively weak economic fundamentals in general, particularly in the large west European economies, the expected upturn is likely to be much more gradual than previous recoveries. Cyclical rates of unemployment can therefore be expected to increase further and put additional pressure on budget deficits, which have already soared in the last two years due to higher interest rates, foregone tax income and increased unemployment-related expenditures.

---

115 Eurostat, *Unemployment*, various issues, Luxembourg.

116 In these two countries, this peculiarity can be explained to some extent by the fact that a large proportion of the females in these economies are employed in cyclically less sensitive service sectors such as health and education. Furthermore, part-time, low-paid, low-skill jobs, which are unattractive to unemployed males receiving benefits, are taken by female entrants, particularly in the United Kingdom where benefits are based on flat rates.

2.3 TRADE AND CURRENT ACCOUNT DEVELOPMENTS

After slowing down to a growth rate of about 3 per cent in 1991 the increase in the volume of world trade picked up in 1992 to about 4.5-5 per cent. This improvement was largely due to the strengthening of economic activity in North America which led to a considerable boost to import demand and more than offset the sluggish growth of imports into western Europe and Japan. For the western industrialized countries combined, import volumes rose by about 4.5 per cent in 1992 compared with about 3 per cent the previous year. The volume of exports is currently estimated to have risen by about 3.5 per cent in 1992 (table 2.3.1).

As in 1991, world trade was also supported by buoyant demand for foreign goods originating in developing countries, viz. South-East Asia, China, Latin America and the Middle East. The continuing adjustment crisis in the transition economies of eastern Europe and the former Soviet Union was reflected in a further sharp fall in their aggregate imports in 1992.

Although changes in international trade and payments in 1992 were largely the result of differential movements in demand in the various countries, diverging developments in relative price competitiveness over the past year or so also played a role.

Exchange rate changes

The US dollar appreciated in the first quarter of 1992 against the ERM currencies reflecting market sentiment that the monetary authorities would not lower interest rates any further; but given the weakening of economic growth in the second quarter and the widening interest differentials, notably against assets in deutsche mark, the dollar depreciated strongly in the second and third quarters of 1992 against a background of mounting tensions in the ERM. This fall in the dollar was fully reversed in the final quarter when it appreciated against a background of more favourable economic growth in the United States and falling short-term interest rates in western Europe: the dollar's average exchange rate against the deutsche mark in December 1992 was the same as in January 1992. The dollar continued to appreciate against the deutsche mark in January and February of 1993. Thus, the deterioration in international competitiveness of many west European countries on account of the depreciation of the dollar in the second and third quarters of 1992 has to be set against the beneficial effects of dollar appreciation in the first and from the final quarter of 1992.

In marked contrast the realignments in the ERM and especially the switch to a floating exchange rate regime in a number of west European countries (see section 2.1 above) altered significantly the pattern of nominal and real effective exchange rates in the final quarter of 1992.

Real effective exchange rates in the various west European countries had diverged, on occasion considerably, since 1987 (the date of the last realignment in the ERM) and the summer of 1992, largely because of differential price and cost movements (table 2.3.2). Thus a sizeable real appreciation of the exchange rate eroded competitiveness in a number of countries (notably Italy, Portugal, Spain and the United Kingdom) and this was increasingly reflected in the level and rate of change in industrial activity and in the foreign balances. Among the four major west European econo-

TABLE 2.3.1

Changes in the volume of exports and imports, 1991-1992
(Percentage change over previous year)

	Exports 1991	Exports 1992	Imports 1991	Imports 1992
Western Europe	2.4	2.9	3.7	2.9
4 major economies	1.8	2.8	4.7	3.7
France [a]	3.1	4.8	2.1	2.1
Germany	1.5	2.2	13.1	2.5
Italy	0.7	1.9	3.0	4.6
United Kingdom	1.7	2.8	-2.8	6.0
13 smaller economies	3.4	2.9	2.2	1.8
Austria [b]	5.7	6.1	1.3	3.0
Belgium [b]	3.8	-0.3	3.9	0.9
Denmark	6.4	3.6	4.5	2.8
Finland	-8.8	8.6	-16.7	-1.8
Greece [a]	14.5	6.2	12.7	3.3
Ireland	5.5	9.8	0.7	4.5
Netherlands [a]	5.4	1.7	3.1	-
Norway	-0.8	6.1	1.9	3.3
Portugal [b]	1.1	8.9	7.1	13.4
Spain	12.0	0.3	11.8	7.1
Sweden	-2.0	1.0	-6.0	-
Switzerland	-1.4	4.4	-1.4	-4.0
Turkey [c]	6.7	3.3	-3.9	1.6
North America	5.2	6.2	1.0	9.8
United States [d]	6.5	5.8	0.6	10.8
Canada [a]	1.0	7.2	2.4	5.5
Total above	3.1	3.7	2.9	5.0
Japan	3.0	0.8	3.0	-0.9
Total above including Japan	3.1	3.4	2.9	4.4

Source: National statistics.

Note: Data for Germany cover only west German foreign trade.

[a] 1992: January-November compared with same period of preceding year.
[b] 1992: January-September compared with same period of preceding year.
[c] 1992: January-October compared with same period of preceding year.
[d] Merchandise trade according to the national accounts.

TABLE 2.3.2

Real effective exchange rates
(Index 1987 = 100)

	1985	1987	1990	1991 December	1992 June	1992 December
France	98.1	100	94.9	91.5	91.4	95.2
Germany	84.0	100	103.4	105.5	106.3	113.6
Italy	98.5	100	109.8	115.1	117.4	99.1
United Kingdom	107.5	100	110.9	117.1	121.0	106.2
Austria	96.6	100	97.6	94.4	91.8	95.9
Belgium	95.8	100	99.2	98.7	94.8	100.7
Denmark	85.7	100	100.2	96.5	96.5	101.4
Finland	106.4	100	105.3	87.2	84.5	74.5
Greece	104.4	100	110.0	113.5	110.9	118.2 [a]
Netherlands	91.6	100	95.8	93.9	93.4	96.5
Norway	97.6	100	98.2	96.7	95.4	95.6
Portugal	102.4	100	112.1	121.5	133.9	133.5 [a]
Spain	98.7	100	118.7	121.0	123.9	114.3
Sweden	97.1	100	111.5	104.7	100.2	87.1
United States	146.2	100	91.4	82.0	84.1	87.9
Canada	104.7	100	116.8	118.0	113.1	108.4
Memorandum item:						
Japan	77.3	100	88.9	99.3	102.4	108.0

Source: IMF, *International Financial Statistics*, Washington, D.C. (monthly).

Note: A fall in the index is equivalent to an improvement in price competitiveness and vice versa.

[a] November 1992.

mies only France improved its international competitiveness. A similar favourable development occurred in a number of smaller economies. In Finland the devaluation of the markka against the ECU by 12 per cent led to a considerable improvement in competitiveness as from the end of 1991.

The potential impact of these various exchange rate changes on the pattern of international competitiveness can be gauged by comparing real effective exchange rates in December 1992 with those in June 1992 and earlier (table 2.3.2). Finland, Italy, Spain, and the United Kingdom have experienced sharp gains in international competitiveness. In contrast, Germany and those countries which maintained their parities against the deutsche mark suffered a deterioration in their international competitiveness. Thus, in France the competitive gains achieved since 1990 were completely wiped out in the final quarter of 1992.

The real effective exchange rate of the US dollar also edged upward in 1992, but overall the competitive position of the US economy still appears to be quite strong, especially if compared with 1985, when the dollar was regarded as considerably overvalued.

These changes in real effective rates will have a potentially important influence on the relative trade performance of west European countries in 1993. But a depreciation may provide only a temporary gain in competitiveness, which in the long run can be offset by changes in relative costs and prices. The immediate effect of a depreciation is a deterioration in the terms-of-trade and this will lead in the short run to an adverse change in the trade account. The gains in competitiveness will tend to support nominal net exports only in the medium term (the J-curve effect). These effects

TABLE 2.3.3

Annual changes in the terms of trade [a] in western Europe and North America, 1990-1992
(Percentage change over previous year)

	1990	1991	1992
Western Europe	0.6	-0.3	1.0
4 major economies	1.0	-0.4	1.8
France	-0.3	0.6	1.1
Germany	1.4	-2.3	3.0
Italy	3.2	2.2	1.8
United Kingdom	1.6	-0.2	1.5
13 smaller economies	-	-0.4	-0.1
Austria	2.3	-3.5	-1.6
Belgium	-0.3	-1.8	1.6
Denmark	1.3	1.0	1.6
Finland	-2.8	-2.3	-3.4
Greece	-0.2	-0.5	-
Ireland	-4.7	-3.0	1.0
Netherlands	1.2	-	-1.2
Norway	-5.6	-0.1	-4.7
Portugal	0.1	0.8	3.5
Spain	0.9	1.5	3.7
Sweden	-	1.0	1.0
Switzerland	1.8	2.7	-1.4
Turkey	4.7	-5.4	3.3
North America	-2.4	1.1	0.7
United States [b]	-2.4	2.0	0.7
Canada	-2.1	-1.5	0.1
Total above	-0.3	0.1	0.9
Japan	-6.0	8.5	8.8
Total above including Japan	-0.8	0.9	1.7

Sources: National statistics.

Note: Data for Germany cover only west German foreign trade.

[a] Ratio of unit value index of exports to net of imports.
[b] Based on implicit deflators for merchandise trade from the national accounts.

TABLE 2.3.4
Current account balances, 1990-1992
(Billion US dollars and per cent)

	1990	1991	1992	Per cent of GDP 1991	Per cent of GDP 1992
Western Europe	-11.6	-52.3	-64.1	-0.7	-0.8
4 major economies	-7.6	-58.1	-67.2	-1.2	-1.2
France	-9.7	-5.9	2.8	-0.5	0.2
Germany [a]	47.1	-19.8	-25.1	-1.3	-1.4
Italy	-14.7	-21.1	-23.9	-1.8	-1.9
United Kingdom	-30.2	-11.3	-20.9	-1.1	-2.0
13 smaller economies	-4.1	5.8	4.0	0.3	0.1
Austria	1.2	-0.2	-0.3	-0.1	-0.2
Belgium	4.6	4.6	4.2	2.3	1.9
Denmark	1.3	2.2	4.5	1.7	3.2
Finland	-6.9	-6.6	-5.1	-5.3	-4.6
Greece	-3.5	-1.5	-2.0	-2.2	-2.6
Ireland	1.4	2.6	3.4	6.1	7.1
Netherlands	10.7	9.1	8.4	3.2	2.6
Norway	3.9	5.0	3.6	4.7	3.2
Portugal	-0.2	-0.7	0.1	-1.0	0.1
Spain	-15.6	-16.0	-23.4	-3.0	-4.1
Sweden	-6.8	-3.3	-3.8	-1.4	-1.6
Switzerland	8.6	10.2	14.7	4.4	6.1
Turkey	-2.6	0.3	-1.3	0.2	-1.1
North America	-112.4	-29.2	-86.2	-0.5	-1.3
United States	-90.4	-3.7	-62.5	-0.1	-1.1
Canada	-22.0	-25.5	-23.7	-4.3	-4.2
Total above	-124.0	-81.5	-149.3	-0.6	-1.0
Memorandum item:					
Japan	35.8	72.9	117.6	2.2	3.2
Total above including Japan	-88.2	-8.6	-31.7	-0.1	-0.2

Sources: National statistics; IMF, *International Financial Statistics*, February 1993, Washington, D.C.; OECD, *Main Economic Indicators*, February 1993, Paris.

[a] Data as from the second half of 1990 include the five new Länder.

work in the opposite direction in countries with appreciating currencies (such as France and Germany).

Trade volumes

Against this general background the annual rate of growth in the volume of west European *exports* was about 3 per cent in 1992 (table 2.3.1). There was, nevertheless, a much more favourable export performance in a number of countries — a reflection, in part, of improved competitiveness (France, Finland) or a favourable product mix, notably in Ireland (high-technology goods) and Norway (oil and gas).

Hardly any country, however, escaped from the pervasive weakening of export growth in the second half of the year, due to the strong cyclical downturn in western Europe. The importance of intra-regional trade means that the weakening in domestic demand and exports has been mutually reinforcing. The strong demand impulse originating from German unification, which had given a considerable boost to the exports of a number of west European countries in 1991, has petered out. The marked weakening in the growth of German import demand appears to have been an important factor in the deteriorating export performance of Belgium, Denmark and the Netherlands in 1992. Austria and France appear to have continued to benefit from the high level of German import demand until the first half of 1992.

Exports from the United States grew quite strongly in 1992, largely a reflection of the earlier strong real depreciation between 1985 and 1987, which continued at a more moderate rate until about 1990. This considerable improvement in competitive position was broadly preserved in 1992. The strengthening of economic activity in the United States spilled over into stronger demand for Canadian goods and this, together with the real depreciation of the Canadian dollar, was the main factor behind the strong rise in Canadian exports in 1992.

The growth in the volume of west European *imports* in 1992 was 2.9 per cent, the same as for exports (table 2.3.1). The considerable variance among the various countries reflects in the main the relative strengths of domestic demand. The major exception is the United Kingdom where imports rose strongly although the economy remained in deep recession. This was partly the result of the sizeable real appreciation of sterling (see table 2.3.2) which favoured the substitution of domestic goods by foreign goods. Import demand was also buoyant in Portugal and Spain in 1992 as a whole, although there was a marked slow-down in the second half of the year.

There was a considerable rebound in the demand for foreign goods in the United States as economic activity gathered pace in the second half of the year: real merchandise imports rose by about 11 per cent in 1992 compared with 1991, when they virtually stagnated.

Current account balances

Current account developments in the industrialized countries in 1992 mainly reflected the strength of domestic demand relative to foreign demand. Favourable changes in the terms of trade helped to improve the current account position in most countries (table 2.3.3) although these were partly offset by demand movements and other factors.

The combined current account deficit of the industrialized countries, which had been influenced in 1991 by one-time transfers from countries in the Middle East to defray the costs of the Gulf conflict, rose to 0.2 per cent of GDP in 1992 (table 2.3.4). This reflects larger deficits in western Europe and North America which, however, were largely offset by the rise in the Japanese current account surplus to a record level.

In Germany the current account had moved from surplus into deficit in 1991, largely because of the demand effects of unification, which compressed the trade surplus, and special transfers to the United States and the Soviet Union. The trade surplus started to rise again in 1992, reflecting in the main the sluggishness of imports, favourable changes in the terms of trade and a smaller net outflow of unilateral transfers. But these developments were more than offset by a large deficit in services (which had been in surplus in 1991). The current account deficit consequently rose to 1.4 per cent of GDP in 1992.

A favourable competitive position in foreign markets and listless domestic demand were the main factors behind the swing into surplus of the trade balance in France in 1992. In consequence, the current account, which had been in deficit since 1978, also recorded a small surplus.

In the United Kingdom the current account deteriorated strongly in 1992, the deficit rising to 2 per cent of GDP. The unusual strength of import demand during the recession and a weak export performance led to a considerable increase in the merchandise trade deficit which was aggravated by a smaller surplus in the invisibles account. The latter had been influenced in 1991 by financial contributions of countries in the Middle East towards UK expenditures related to the Gulf conflict.

There was also a further rise in the Italian current account deficit, to nearly 2 per cent of GDP in 1992. This was largely due to the deterioration in international competitiveness which followed the real appreciation of the lira until summer 1992.

In the United States depressed domestic demand, robust export growth and improved terms of trade led to a considerable reduction in the merchandise trade deficit in 1991. Financial contributions from other countries to cover the costs of the operations in the Persian Gulf led to a net inflow of unilateral transfers, the first in the post-war period. As a result, the current account deficit nearly disappeared in 1991. However, in 1992 the merchandise trade deficit rose again, mainly the effect of the surge in import demand, and unilateral transfers moved back into sizeable deficit. These were the major factors behind the renewed deterioration in the current account in 1992, the deficit increasing to 1.1 per cent of GDP.

In Canada the favourable development in trade with the United States provided the largest contribution to the rise in the merchandise trade surplus in 1992, but this was largely offset by larger deficits in the services and investment accounts. As a result the current account deficit has remained at a very high level, equivalent to more than 4 per cent of GDP.

In Japan the current account surplus reached a new record of about $117.5 billion or 3.2 per cent of GDP in 1992. This mainly reflected a sharp rise in the merchandise trade surplus: the weakness in domestic demand relative to foreign demand was amplified by a considerable improvement in the terms of trade by nearly 9 per cent in 1992. There had been a similarly large gain in the terms of trade in 1991. In addition the deficit in non-merchandise transactions in 1992 was much lower than in 1991 (table 2.3.4). Although the current account surplus was at a record level in 1992, this was not the case when it is expressed as a ratio of GDP, which peaked at 4.3 per cent in 1987. The lower ratio in 1992 reflects the very rapid growth in output during recent years.

Among the smaller west European economies with a current account surplus in 1991, only Denmark, Ireland and Switzerland had further improvements in 1992, largely because of favourable developments in their merchandise trade. In Ireland the surplus in merchandise trade rose to about 12 per cent of GDP in 1992. Adverse changes in the trade account played an important role in the widening of current account deficits in Austria, Greece, Spain and Sweden in 1992. In Finland the large current account deficit was somewhat reduced in 1992: there was an increase in the surplus in merchandise trade but this was substantially offset by rising net interest payments on the foreign debt, which, in turn, were partly a reflection of the depreciation of the markka. Finally, there was a swing into a small current account surplus in Portugal in 1992, while strong import demand led to a sizeable swing into deficit in Turkey.

Short-term prospects

The prospects for international trade in 1993 are difficult to assess at the time of writing. The more solid upswing in North America should continue to stimulate world trade, although the direct impact on western Europe is likely to be rather small. Against a background of weak domestic demand, intra-regional trade flows are likely to remain subdued in western Europe in 1993. Considerable improvements in international competitiveness should help to sustain export growth and higher activity levels in Finland, Italy, Sweden and the United Kingdom. Import demand is expected to remain depressed in the eastern transition economies especially because of tight current account constraints. The entering into force of the Single Market at the beginning of 1993, and of the European Economic Space later on, should in principle provide a strong stimulus to international trade, but for most of 1993 this is likely to be offset by a further weakening of the economic growth cycle in western Europe. A major uncertainty is the future of the Uruguay Round and the continuing risks of an outbreak of tit-for-tat trade conflicts which could seriously hamper the growth of world trade.

2.4 COUNTRY NOTES

(i) The major economies

Germany

In western Germany the economic situation started to deteriorate after the first quarter of 1992, and in the second half of the year the economy moved into recession, real GDP falling by 0.5 per cent between the second and third quarters and by another 1 per cent in the final quarter. The downturn has been particularly severe in manufacturing industry, where output declined by some 5 per cent between the third and final quarters of 1992 against a background of sharp declines in new orders from the domestic economy and abroad.

Total employment fell by 0.3 per cent in the fourth quarter compared with the same period in 1991, the last year-on-year decline in total employment having been in 1984. The number of persons unemployed rose to 2.3 million at the end of January 1993, an increase of some 20 per cent over the same month in 1992.

The contraction of economic activity reflects the significant weakening of both domestic and foreign demand. Private consumption was restrained by a significant slow-down in the growth of disposable incomes, largely a reflection of the increasing sluggishness of employment. The abolition of the solidarity tax in the second half of the year had no significant impact on consumer spending, which may be explained by the decline in consumer confidence. Given the increase in VAT rates at the beginning of 1993, many households brought forward purchases and this led to a temporary strengthening of household demand in the final quarter. Business investment in machinery and equipment fell strongly in an environment characterized by large increases in spare capacity and shrinking profits. In contrast, construction investment held up much better largely on account of the strong demand for housing, which has been stimulated by interest rate subsidies.

Exports provided little support to economic activity in 1992. The east German import boom has petered out[117] and demand from abroad was affected by the significant weakening of economic activity in western Europe in the second half of the year and by the deterioration in German price competitiveness on account of the real effective appreciation of the deutsche mark. There was a sharp fall in export volumes in the final quarter of 1992.

Consumer price inflation remained high in 1992 mainly reflecting the strong growth in prices for services and the upward pressure on rents on account of the housing shortage, which has been accentuated by a large increase in the number of immigrants in recent years. The rise in the rate of VAT by one percentage point from the beginning of 1993 was the main factor which pushed the inflation rate above 4 per cent in January; inflationary pressures appear to have abated somewhat in February.

Monetary policy in 1992 was aimed at keeping money supply growth, as measured by M3, within a range of 3-5 per cent between the final quarters of 1991 and 1992. However, the actual increase was 9.4 per cent. There are various reasons for this overshooting. One important factor appears to have been the inverted yield curve, which stimulated a shift in portfolios towards term deposits which are part of M3. Another major factor was the strong demand for credit, partly reflecting subsidy schemes to stimulate construction in east and west Germany. Speculative capital inflows and intervention to support weak ERM currencies also added to the strong growth of M3 in the final quarter of 1992.

Concerns about monetary expansion and inflationary pressures led to a further tightening in monetary policy in July 1992. The subsequent reduction of the discount and Lombard rates in connection with the devaluation of the Italian lira on 15 September marks the turning point for German monetary policy in 1992. Official interest rates were lowered again in February 1993, but in the meantime the Bundesbank had eased progressively the terms for its securities repurchase agreements (the instrument of open market policy). As a result, short-term interest rates fell from a peak of 9.9 per cent in August 1992 to 8.4 per cent in February 1993. Long-term rates, which had already started to decline, fell from 7.7 per cent to 6.9 per cent over the same period. This policy was continued in early March with a further cut in the tender rate for securities repurchase agreements from 8.5 per cent to 8.25 per cent against a backdrop of deepening recession.

The financial position of the public sector continues to be strained by support payments for the reconstruction of the east German economy. These payments[118] corresponded to some 5.5 per cent of west German GNP in 1992 and are unlikely to decline significantly in the medium term. The total net borrowing requirement of general government rose from DM 89.1 billion in 1991 to DM 93.1 billion in 1992, although the ratio of net borrowing to GDP fell slightly from 3.2 per cent to 3.1 per cent. But these figures exclude the off-budget

---

117 It should be recalled that economic transactions with east Germany are recorded as foreign transactions in the west German national accounts.

118 These payments are net of federal tax revenues and social security contributions collected in east Germany.

borrowing of the Treuhandanstalt, the railways and the postal services, which in total accounted for another 2 per cent of GDP in 1992. A large increase in the general government budget deficit is projected for 1993, largely reflecting the impact of the recession. The financial position of the government sector will be strained increasingly from 1995 when the debt accumulated by the Treuhand will be taken over and the moratorium on the debt of the east German housing stock will expire. Additional financial burdens will arise from the reform of the railway system in east and west Germany and from the need to fully integrate the east German regional authorities into the constitutional revenue-sharing schemes (*Finanzausgleich*) between the federal government and the west German regional authorities. This function has been fulfilled so far by the German Unity Fund which will be phased out at the end of 1994. All these changes will require further fiscal tightening in west Germany.

Against this backdrop the government has been seeking to forge a "solidarity pact" with west German regional authorities, employers and trade unions. The basic idea is to develop a coherent and credible strategy to combine the need for very large transfer payments to east Germany with the requirement of significantly reducing the budget deficit in the medium term.

The federal government has adopted such a medium-term plan — "federal consolidation programme" — which envisages reducing the general government deficit to DM 100 billion through a combination of expenditure restraints and tax increases by 1995. An important step forward was made in mid-March when the federal government and the regional authorities agreed on the modalities of further financial support for the east German economy: *inter alia* these involve, as from the beginning of 1995, new rules for sharing tax revenues, the re-introduction of a 7.5 per cent surcharge on income and corporate taxes, and higher taxes on private wealth. But a number of important details remain to be settled.

France

Real GDP in France rose by 1.8 per cent in 1992 but growth was very sluggish in the second and third quarters and real GDP fell by 0.5 per cent between the third and final quarters of 1992. At the time of writing the economy appears to be on the brink of recession.

Domestic demand was very weak in 1992, although private consumption was relatively robust, reflecting higher demand for consumer durables, notably motor cars. The volume of retail sales declined for the second consecutive year.

Against a background of increasing margins of spare capacity and high real interest rates, there was a further fall in real fixed capital formation in 1992. Investment fell in all major sectors, but there was a striking fall of more than 10 per cent in industry, where there had already been a cut of 8.5 per cent in 1991. Business confidence deteriorated in the second half of the year as earlier expectations of a strengthening of export growth failed to materialize. Construction investment was depressed, reflecting high interest rates and the collapse of the real estate boom, which has left behind a glut of office space and commercial buildings in several regions. Exports of goods and services rose strongly in 1992 as a whole but exports actually fell between the first and second halves of the year.

Against a background of weak domestic demand there was a further slow-down in the rate of inflation. There has been a significant deterioration in the labour market in the past two years, which was only partly cushioned by special labour market measures introduced by the government. Total employment fell in 1992, considerable job losses in industry being only partly offset by increased demand for labour in other sectors. This continued a development which was already under way in 1991.

The rise in unemployment since the beginning of 1991 continued in 1992. Nearly 3 million persons were seeking jobs at the end of December 1992, corresponding to an unemployment rate of 10.5 per cent.

Developments in the external sector of the economy were more favourable. Improved industrial competitiveness was a major factor behind a robust export growth for manufactured goods which, combined with gains in the terms of trade, led to a swing in the trade balance for manufactures from the traditional deficit to a surplus in 1992. Largely as a consequence of this swing, the current account also moved into a small surplus of 0.2 per cent of GDP. But export growth was increasingly affected in the final quarter by the slow-down of demand in major foreign markets and by the strong appreciation of the franc against the lira, peseta, sterling and other west European currencies since September 1992. The appreciation of the currency could constitute, as for Germany, a major constraint on export growth in 1993.

The stance of monetary policy was very tight in 1992 and high real interest rates had an increasingly dampening effect on domestic activity. The official discount rate has remained unchanged at 9.5 per cent since 1989. The determination of the authorities to maintain the parity against the deutsche mark allowed for only a temporary fall in short-term interest rates in the wake of the September crisis in the ERM. Short-term interest rates fell to 9.8 per cent in November, a decline of more than 1 percentage point compared with October. This led to a corresponding narrowing of the interest rate differential with Germany, but as a result of renewed speculative selling of the franc in December the decline in short-term interest rates was more than reversed. A further tightening of monetary policy drove short-term rates above 12 per cent in January 1993, while German short-term rates fell below 9 per cent. Monetary policy was slightly eased in February, but short-term rates have remained very high. To underpin their full commitment to a monetary policy oriented towards price stability, the government announced at the beginning of 1993 that the Banque de France would be granted independence in the near future.

The financial position of the public sector weakened considerably in 1992, largely because tax revenues were much lower than projected on the basis of earlier optimistic assumptions about economic growth. The state budget deficit for 1992 is currently estimated at some FF 230 billion or 3.2 per cent of GDP; when the budget for 1993 was adopted last December, the deficit for 1992 was still being estimated at about FF 188 billion. The general government deficit, which includes the financial balances of the social security system and of the regional authorities, rose to 3.7 per cent of GDP in 1992 compared with 2.2 per cent in 1991. This was much worse than expected and the deterioration in the overall economic environment will require a review of the 1993 budget adopted in December 1992. At that time, the government's objective was to reduce the deficit to FF 165.5 billion, some 2.2 per cent of GDP.

Prospects for economic growth in 1993 have deteriorated significantly since the final quarter of last year. According to a business survey by the Banque de France, industrial output fell in all the major industrial sectors in January 1993, a reflection of weak demand at home and abroad. Business and consumer confidence is very low. Given the severe deterioration in the financial position of the public sector, fiscal policy is likely to be tightened. The outlook will therefore depend very much on the scope for an easing of monetary policy in 1993.

Italy

After a strong rebound in the first quarter of 1992 the rate of economic expansion slackened considerably in Italy in the second half of the year. The backdrop to this development was the introduction of tough fiscal policies and the turbulence in financial markets. Business and consumer confidence deteriorated strongly after the summer. For the year as a whole economic growth slowed down to 1.2 per cent, mainly supported by relatively robust private consumption and rather favourable export growth. But these yearly averages mask the faltering of domestic and foreign demand in the second half of the year.

Industrial output fell in 1992 for the third consecutive year, the prolonged weakness partly reflecting the loss of price competitiveness due to the real effective appreciation of the lira prior to the crisis in the ERM in September 1992. Falling capacity utilization rates together with high real interest rates and deteriorating sales prospects led to a pronounced fall in industrial investment. For the economy as a whole there was a decline in expenditure on machinery and equipment and a slight gain in construction investment.

Private consumption turned increasingly sluggish in the final months of 1992 as fiscal austerity and the deterioration in the labour market led to downward revisions of expected income.

The relatively favourable export performance reflects in the main a strong rise in exports to markets outside the OECD area. The drag of imports on domestic activity levels abated in the second half of the year; in the event, changes in the real foreign balance made a small positive contribution (0.2 percentage points) to GDP growth in 1992, which compares with a negative contribution of 1 percentage point in 1991.

The weakening in economic activity led to a deterioration in the labour market in the second half of 1992. Employment in industry fell strongly as large companies accelerated labour shedding, while in the service sector employment stopped rising. The rate of unemployment rose above 11 per cent in the second half of 1992.

Nominal wage growth slowed down considerably in 1992 as a result of wage restraint in both the private and public sectors. The inflation-indexing of wages *(scala mobile)* was suspended in December 1991 and formally ended in an agreement which the government, trade unions and employers signed on 31 July 1992. At the same time the government promised to freeze public sector tariffs. The abolition of wage indexing should help to contain inflationary pressures and to improve the price competitiveness of industry.

Consumer price inflation decelerated in 1992, mainly reflecting reduced labour cost growth and the overall slack in the economy. But the strong depreciation of the lira in the final quarter of 1992 has led to higher import prices and the pass-through effect is likely to drive up the inflation rate in 1993. The impact of higher import prices was mitigated in the final quarter of 1992 by the overall sluggishness of demand: the annual inflation rate fell to 4.7 per cent in December 1992 compared with 4.9 per cent in November. The current account deficit deteriorated further to some 2 per cent of GDP in 1992, reflecting in the main a larger trade deficit and rising net interest payments on the growing foreign debt.

Downward pressure on the exchange rate led to a substantial tightening of monetary policy between June and September, the discount rate being raised progressively from 12 per cent to 15 per cent. The rate on interbank sight deposits rose temporarily above 20 per cent in September. The suspension of membership in the ERM on 17 September 1992 and the subsequent flotation of the lira led to a sizeable depreciation against the deutsche mark and the dollar. The freeing of the lira from the narrow exchange rate constraint has since allowed a gradual easing in the stance of monetary policy. Against a background of recessionary tendencies and favourable inflationary trends, the discount rate was reduced in several steps to 11.5 per cent in February 1993, the same level as in November 1991. Short-term interest in the interbank market have fallen from some 20 per cent in September to less than 14 per cent in January. Against a backdrop of increasing political uncertainties, downward pressure on the lira re-emerged in February. Exchange rate considerations will limit the scope for further interest rate cuts given the need to contain the inflationary pressures which will emerge from the already large depreciation. The short-term interest rate differential with Germany fell from 7.4 percentage points to some 4.5 percentage points between October 1992 and January 1993, whereas the inflation differential was only 1 percentage point in December 1992. This suggests that there is still a considerable risk premium on lira-denominated assets, a risk which is also reflected in very high long-term interest rates.

In the second half of 1992 fiscal policy embarked upon a major programme to curb the huge budget deficit and reduce public debt in the medium term. The backdrop to this move was the growing crisis of confidence in the international financial markets triggered by a large budget overrun in the first half of the year and lingering doubts about the possibility of the country being able to meet the Maastricht convergence criteria.

The new government adopted a package of measures in July 1992 designed to reduce expenditures and enhance revenues. The net impact of these measures, which include privatization revenues from selling shares in state companies, was intended to reduce the net borrowing requirement to 10.2 per cent of GDP for 1992, as compared with 10.7 per cent in 1991. In the event net borrowing corresponded again to about 10.7 per cent of GDP largely because of delays in the privatisation programme.

The budget for 1993, passed by parliament in late December 1992, aims at budget savings corresponding to some 6 per cent of GDP, but the contractionary impulse is largely offset by the rise in interest payments on existing debt. The government's total borrowing requirement is projected to decline to some 9.5 per cent of GDP in 1993. The budget projections, however, were based on assumptions about economic growth in 1993 which now look rather optimistic. It is therefore likely, that the target for borrowing will be missed unless further corrective measures are taken. The European Community has recently approved a medium-term ECU 8 billion balance-of-payments loan for Italy. The allocation of the various tranches, however, is contingent upon the fulfilment of the stringent fiscal targets.

At the time of writing the short-term outlook for the Italian economy appears to be deteriorating against a backdrop of mounting political uncertainty. There has been a widespread freezing of public sector construction projects against a background of corruption charges. Forecasts for economic growth in 1993 have been reduced sharply pointing to a small decline of GDP or at best broad stagnation.

United Kingdom

In the United Kingdom the hoped-for recovery did not materialize in 1992. The fall in real GDP appears to have bottomed out after the second quarter but the recovery has remained elusive. Real GDP increased only slightly in the third and fourth quarters. Manufacturing production remained listless through 1992 and, for the year as a whole, output fell for the third consecutive year. A moderate upturn in activity in the first two quarters was almost completely reversed in the second half of 1992. Manufacturing output in the final quarter of 1992 was 7.5 per cent below its peak in the first quarter of 1991. Against this backdrop the number of firms working at below capacity has increased further. Construction activity continued to decline against a background of large excess supply in the market for commercial real estate and falling demand for private housing. Prices of houses and commercial property continued to fall.

Business and consumer confidence strengthened markedly in the second quarter but fell back to very low levels in the second half, when it became clear that earlier signs of an upturn were but a false dawn.

Domestic demand remained sluggish and the small increase which occurred was fully met by imports. In fact, real imports of goods and services rose by 5 per cent in 1992 which is unusual for an economy in recession.

Consumers remained very cautious in their spending behaviour, a reflection of the strong rise in unemployment and the need to pay back high levels of outstanding debt. Lower interest rates have reduced the debt service burden but many households are in a debt trap, in that falling house prices have led to a sharp reduction in their net wealth. About 1.5 million households in the second half of 1992 were faced with a situation where the value of their mortgage was higher than the current market value of their house (i.e., the so-called "negative equity" situation).

Balance sheet restructuring in the industrial and commercial sector led to falling investment and labour shedding. Access to external financing is made more difficult because the banks, whose own balance sheets have weakened, have tightened the terms of lending.

The labour market steadily deteriorated in the course of 1992. Falling employment was the major factor behind the rise in unemployment, which was close to 3 million at the end of 1992, corresponding to an unemployment rate of 10.5 per cent. Inflationary pressures abated in the course of 1992, largely a reflection of depressed demand and moderate growth in unit labour costs. The year-on-year inflation rate fell below 2 per cent in January 1993. But import prices started to rise more strongly in the final quarter of 1992 on account of the sharp depreciation of sterling: this will put upward pressure on domestic prices in 1993 unless offset by gains in productivity or a reduction in profit margins.

The withdrawal of sterling from the ERM in September 1992 has allowed a considerable easing in monetary policy. Base lending rates were cut in four steps, from 10 per cent in September to 6 per cent at the end of January 1993, against a background of continuing weakness in output. But the scope for a further relaxation of monetary policy is now highly constrained because lower interest rates would risk a further depreciation of sterling and additional inflationary pressures to those already emerging from higher import prices.

The government's target is to keep year-on-year inflation, measured by the retail price index excluding mortgage interest payments, within the range of 1 to 4 per cent. At the beginning of 1993 inflation was close to the upper end of this range, although the "headline" rate of inflation was in the lower half. Cost pressures arising from the depreciation of sterling have started to appear. There was a large rise in import prices in the final quarter of 1992 which led to a marked acceleration in the rate increase of manufacturing input prices. In the short run, this rise may be offset to some extent by improved productivity growth and reduced wage cost pressures. But the rate at which higher import prices will be passed through to domestic output prices in 1993 will also depend on the credibility of the anti-inflationary stance of monetary policy. Even if inflation is rising above the target range, however, there is likely to be some hesitation in raising interest rates before a recovery is well established.

Sterling has depreciated by some 15 per cent since September and this considerable gain in price competitiveness should not only allow a boost to exports but also enable domestic producers to capture a larger share of the domestic market. However, export prospects have been weakened by the overall slow-down in the world economy. Manufacturing goods exports strengthened in the final quarter but it remains to be seen whether this favourable trend will continue in 1993. The extent to which the gains in price competitiveness can be preserved will also depend on future changes in domestic costs relative to those in other countries.

A striking feature of 1992 was the increase in the current account deficit to some 2 per cent of GDP, largely because of the exceptional strength of import demand during a period of a recession. Forecasts point to a further increase in the deficit to some 3 per cent of GDP in 1993. There are concerns that a strong recovery would lead to a further strengthening of import demand with a consequential rise in the current account deficit which would constrain higher output growth.

Various fiscal measures were adopted in the spring and autumn to support economic activity, but their potential stimulus was rather limited. A striking development was the severe deterioration in the financial position of the public sector in 1992: the public sector borrowing requirement rose to some 6.5 per cent of GDP in 1992 compared with 2.8 per cent in 1991. A further deterioration is projected for 1993.

It is generally acknowledged that the rise in the government budget deficit largely reflects the impact of cyclical factors on government outlays and revenues. Nevertheless, although it cannot be measured accurately, the indications are that the increase in the structural deficit is far from negligible. The large budget deficit has exhausted any scope for expansionary fiscal policy. In his March budget statement the Chancellor announced a small increase in taxation for the coming year but with larger increases for subsequent years. In doing so he appears to have more or less followed the advice of the panel of independent economic advisers who recommended that tax increases, while necessary to reduce public borrowing over the medium term, should not be implemented until the recovery is well established.

The prospect is for only modest output growth in 1993. Given the depressing effects of the debt overhang on spending propensities in the private sector, any recovery will have to be largely export-led. The business climate seems to have improved somewhat since late 1992 but overall sentiment is still rather gloomy. Lower interest will help the restructuring of balance sheets but growth in domestic demand will continue to be sluggish. Exports will benefit from the depreciation, but the overall rate of expansion will be dampened by weakening activity levels in foreign markets. The sharp deterioration in the terms of trade will be a major factor behind a further increase in the current account deficit, reflecting the typical J-curve effect.

United States

In the United States the rate of economic growth strengthened considerably in the second half of 1992 and the recovery appears now to be well established, although the risk of setbacks can never be excluded. Real GDP rose at an annual rate of 4.8 per cent in the final quarter of 1992 but short-term economic indicators for the first two months of 1993 suggest that this strong rate of expansion has not been sustained.

A major factor behind the more favourable performance in the second half of 1992 was rising consumer confidence and the consequent strengthening of personal consumption. Business investment in machinery and equipment was very buoyant after the first quarter, reflecting the improved outlook for profits. Exports, which had faltered in the second quarter, also rose vigorously. Low mortgage rates stimulated investment in single-family housing although the two-digit growth rates should be seen against the large declines in preceding years. Output growth remained subdued in industry and capacity utilization rates showed little increase in the course of the year. Inflationary pressures have continued to subside partly because of low growth in unit labour costs.

The unemployment rate peaked at 7.7 per cent in June 1992 and has since declined steadily to 7 per cent in February 1993, its lowest level since November 1991. The response of employment to strengthening output growth has been rather weak in 1992, a reflection of a significant acceleration in productivity growth. The cautious hiring policy may have also been influenced by the prevailing uncertainty about future sales prospects and the deepening recession in western Europe and Japan. Employment rose in manufacturing industry between November 1992 and January 1993 but the preferred response of companies to higher demand appears to have been to increase the number of average weekly hours worked, which rose to its highest level for more than 20 years in January. Preliminary estimates point to a surprisingly large rise in non-agricultural payroll jobs by 365,000 in February compared with the preceding month; but these estimates tend to be subject to large revisions. An earlier estimate of an increase in payroll jobs by more than 106,000 in January was later revised to an increase of only 44,000.

Lower interest rates have helped the restructuring of balance sheets in the private sector. Household debt levels relative to income have fallen, but they are still quite high and restraining expenditures. The decline in commercial and industrial loans outstanding bottomed out in August but net borrowing has remained sluggish. Though lower interest rates have provided a more conducive environment for borrowing, the non-price terms of lending in the banking sector have remained rather tight, a reflection of cautious lending policy and attempts by banks to restore their profitability. In fact, there was a considerable rise in bank profits in 1992.

Monetary policy had already been relaxed considerably in the course of 1991 and was eased further in July 1992, when the discount rate was cut to 3 per cent, its lowest level for nearly thirty years. The discount rate has since remained unchanged, but the stimulus provided by lower interest rates may have eventually started to show up more clearly in the second half of 1992.

The stimulus provided by monetary policy to economic activity is not reflected in the Federal Reserve's target variable: in the fourth quarter of 1992 M2 was only 1.9 per cent higher than in the same period of 1991, well below the target range of 2.5 per cent to 6.5 per cent. An important factor behind the sluggish growth in M2 has been the shift in private sector portfolios towards financial assets which are not included in M2. The reason for this was the steepening of the yield curve, which made these assets more attractive than those which are included in M2. The substantial easing of monetary policy is clearly discernible in the strong growth of M1 which rose by 14 per cent between the fourth quarters of 1991 and 1992; this was the largest increase since 1986. The scope for a further easing in monetary policy now appears to be rather limited, although the Federal Reserve can be expected to remain cautious about any tightening before a sustainable recovery is clearly confirmed in the economic statistics. This cautious attitude is also reflected in the slight reduction in the target range of money supply growth by half a percentage point, to 2 to 6 per cent between the final quarters of 1992 and 1993.

The major concern of fiscal policy remains the need to curb the large budget deficit. This amounted to $290.2 billion or 4.9 per cent of GDP for fiscal year 1991/92 and a further rise to more than 5 per cent of GDP is projected for fiscal year 1992/93. A major factor in the recent rise of the deficit has been the impact of the recession on revenues and transfer payments and the strong growth in mandatory payments for the government's health care programmes, which more than offset the spending restraint imposed by the budget agreement of November 1990 and the associated Budget Enforcement Act.

Continuing rapid growth in spending on health care programmes has been a major factor behind the further deterioration in the budget deficit, which under unchanged policies has been projected by the Congressional Budget Office to rise to nearly 7 per cent of GDP in 2003.[119] The new administration has announced plans to curb the budget deficit from some $300 billion in the current year to $206 billion or 2.5 per cent of projected GDP in 1997. The measures, which have still to be passed by Congress, aim at budget savings of nearly $500 billion over the period 1994-1997, which in broadly equal parts, should result from expenditure cuts and additional revenue from higher income and corporate tax rates and a new energy tax. Some offset to the expenditure cuts is provided by additional funds of $160 billion made available for public infrastructure projects and investment tax incentives over the target period. At the time of writing, the net impact of this mixture of contractionary and expansionary measures on activity levels is not clear, but the overall effect appears likely to be contractionary. Against a background of sluggish employment growth, the new administration has also proposed a short-term fiscal stimulus of $30 billion designed to create some half a million jobs within a year. The need for this stimulus, however, is now controversial, given the solid growth performance in the second half of the year: there are fears that it could trigger an increase in long-term interest because of a more pessimistic assessment in financial markets of prospects for budget deficit reduction and inflation. However, long-term bond yields fell in February to their lowest levels since 1979.

Canada

Economic growth picked up in Canada in the third quarter of 1992 and strengthened further in the final quarter. These changes reflected in the main a recovery in private consumption, business fixed investment, and strong export growth. Labour market data appear to confirm the recovery in the final quarter. But compared to previous cycles, the recovery has been very slow so far. Real GDP is estimated to have increased by only about 1 per cent in 1992, insufficient to offset the output losses in 1990 and 1991.

Private consumption has remained subdued against a background of low growth in real disposable incomes and the need to pay back the large debts accumulated in the second half of the 1980s. Total fixed investment edged up in 1992 following two years of decline. There was a strong rise in expenditure on machinery and equipment, reflecting higher profits and improved export prospects. Private housing started to recover stimulated by low mortgage rates. But non-residential investment continued to decline reflecting the overbuilding in commercial real estate.

Exports rose strongly and were the major driving force of economic growth in 1992. Their buoyancy was largely due to more solid growth in the United States which led to higher imports from Canada.[120] Exports were also supported by gains in price competitiveness on account of low wage growth and the sizeable depreciation of the Canadian dollar against the US dollar in 1992. The upturn in domestic demand led to a strengthening of import growth, which was, however,

---

119 Congress of the United States, Congressional Budget Office: *Reducing the Deficit: Spending and Revenue Options*, February 1993, Washington, D.C.

120 Some three quarters of all Canadian merchandise exports are shipped to the United States.

insufficient to offset the stimulus to domestic activity from higher exports.

The surge in exports to the United States was the major factor behind the considerable rise in the trade surplus in 1992. However, this was more than offset by larger deficits in services and investment income. As a result the current account deficit remained very high, equivalent to more than 4 per cent of GDP, in 1992.

The labour market deteriorated in 1992. For the year as a whole employment fell compared with 1991, reflecting lethargic output growth and efforts to improve productivity. Employment increased slightly after the first quarter but for most of the year this was insufficient to prevent a further rise in unemployment. The unemployment rate was 11.5 per cent in December 1992, this compares with 10.3 per cent twelve months earlier.

Inflationary pressures weakened considerably in 1992: the annual rise in the consumer price index fell to 1.5 per cent from 5.6 per cent in 1991. This improvement reflects the combined impact of favourable changes in unit labour costs and sluggish domestic demand, which more than offset the rise in import prices resulting from depreciation.

Monetary policy was eased further in 1992 to sustain the fragile recovery. Short-term interest rates fell under 6 per cent between June and September 1992, their lowest level in about 20 years. Against a background of turbulence in the foreign exchange markets, concerns about the outcome of the constitutional referendum in October, and a further deterioration in the financial position of the public sector, there was considerable downward pressure on the exchange rate between September and November and this led the central bank to raise short-term rates. Monetary policy has been progressively eased again since December 1992, partly in response to strong capital inflows. At the beginning of March the commercial banks' prime rate fell to its lowest level since 1973. Nominal long-term interest rates also fell in 1992, but given the sharp deceleration in inflation, real interest rates have actually increased and are currently at a very high level. This may help to explain the muted response of investment to the improving economic environment.

The considerable achievements in budget consolidation in the second half of the 1980s have been completely reversed since 1990, largely because of the recession and too optimistic forecasts of revenues. The general government net borrowing requirement rose to Can$ 41.1 billion equivalent to 6.1 per cent of GDP in 1991. The target for reducing the deficit in 1992 was not met and instead there was a further rise to Can$ 43.9 billion or 6.4 per cent of GDP. Successive deficits have led to a considerable increase in public debt, which currently stands at 96 per cent of GDP. Debt service payments have narrowed the room for manoeuvre of fiscal policy: interest payments on the public debt accounted for nearly 20 per cent of current government expenditures in the third quarter of 1992.

Against a background of concern about the severe deterioration in the financial position of the public sector, the government adopted a mini-budget at the beginning of December 1992 designed to curb expenditures in the rest of the fiscal year 1992/93 (which ends at the end of March) and for 1994 and 1995. At the same time the government introduced measures which aim *inter alia* at stimulating small business investment and employment growth. At the time of writing, the budget for fiscal year 1993/94 has not yet been presented to Parliament but the authorities have already indicated that they will introduce new measures to curb the deficit.

The recovery is projected to continue in 1993, but the rate of expansion will remain moderate. The government expects an increase in real GDP of 2.5 per cent. Tighter fiscal policies, high debt levels in the private sector and small employment gains will restrain the growth in domestic demand. A sustained recovery will therefore largely depend on a continuation of favourable export growth.

(ii) The smaller economies

Austria

Private consumption and construction investment were the mainstays of economic growth in 1992. Export growth slowed down considerably in the second half of the year, a reflection of the real effective appreciation of the schilling and, especially, the cyclical downturn in the west German market, which accounts for some 40 per cent of Austrian exports. The export slow-down was an important factor behind the decline in industrial output in the second half of the year. Revenues from tourism stagnated in real terms, reflecting the weakening of income growth abroad and the appreciation of the schilling. The boom in commercial and industrial real estate petered out in the second half of the year. The labour market deteriorated in 1992 with industrial employment continuing to fall, although for the year as whole this fall was still offset by increased demand for labour in other sectors. The unemployment rate rose to 8.4 per cent in January 1993 compared with 7.8 per cent a year earlier. Inflationary pressures intensified in 1992 as a result of large increases in wage costs and higher indirect taxes. As in 1991, there was a small deficit in the current account in 1992.

Fiscal policy remained restrictive in 1992. The basic objective of monetary policy continues to be the maintenance of a stable exchange rate of the schilling against the deutsche mark, a policy which requires official interest rates to remain broadly in line with German rates.

Economic growth is expected to slow down further in 1993, to the border of stagnation reflecting to a large degree the deterioration in the external environment. A large decline is projected for industrial investment while private consumption and construction should continue to support economic growth. Cyclical factors will entail a larger budget deficit than was projected in the 1993 budget plan adopted in 1992. In January 1993 the government adopted a package of measures designed to stimulate investment and exports, but this is not expected to have a significant impact.

Belgium

The annual growth rate of real GDP slowed down to 1.2 per cent in 1992 following an increase by 2.1 per cent in 1991. Private consumption grew little in spite of increases in real disposable incomes: consumer confidence deteriorated and there was a further rise in the savings ratio. Public consumption fell, though only slightly, a reflection of continuing efforts to consolidate public finances. Business fixed investment declined for the second consecutive year. There was a marked rebound in residential construction, which may have been partly a result of lower mortgage rates.

The weaker growth in consumption and investment was offset by changes in stocks and the contribution of total domestic demand to overall economic growth was therefore unchanged from 1991. The marked slow-down in economic growth in 1992 was therefore entirely due to adverse changes in the real foreign balance.

The labour market has been deteriorating since the second half of 1991 but the rise in unemployment accelerated in the final quarter of 1992. The unemployment rate rose to 8.6 per cent in December 1992 compared with 7.8 per cent twelve months earlier. The annual inflation rate fell to 2.4 per cent in 1992, from 3.2 per cent in 1991. There was a further rise in the current account surplus, partly because of an improvement in the terms of trade.

The stance of monetary policy continues to be determined by the exchange rate target. As was the case for the Netherlands guilder, the Belgian franc was also left unscathed by the turbulence in the foreign exchange markets. Monetary policy has been progressively eased since autumn 1992, sometimes ahead of the relevant decisions of the Bundesbank. But, against a background of heightened concerns in the financial markets about regional political tensions, the Belgian franc weakened within the ERM at the beginning February and this forced the central bank to raise short-term interest rates.

The attempt to reduce the government budget deficit and to curb the very high level of public debt has been intensified in order to meet the Maastricht convergence criteria. But the target to reduce the general government budget deficit in 1992 was missed, because of the slow-down in economic growth and higher interest rates. The budget deficit rose from 6.5 per cent of GNP in 1991 to 7.0 per cent of GNP in 1992. The original target, fixed in August 1992, was to reduce the deficit to 5.2 per cent of GNP in 1993, but the sharp deterioration in the economic environment will make this target very difficult to achieve. In October 1992 the government introduced further measures to curb expenditures and raise revenues, including the privatization of state assets, but these are insufficient to prevent a further large overshooting of the budget target. More retrenchment measures are likely to be introduced in spring 1993.

Netherlands

Economic growth in the Netherlands slowed further to 1.7 per cent in 1992. Economic activity became increasingly sluggish after the first quarter of 1992 and industrial output fell in the second half of the year. The slackening of economic activity reflects a weakening of both domestic demand and real net exports. Growth in private consumption was tempered by marginal gains in real disposable incomes, while stringent budgetary policies repressed public consumption.

Fixed capital formation provided only little support to overall economic activity. Against a background of falling rates of capacity utilization and deteriorating sales and profit prospects, private non-residential investment rose by only 1 per cent in 1992. Residential investment, which had fallen sharply in 1991, picked up slightly in spite of high real interest rates. This can be partly explained by favourable weather and increased access to subsidies for owner-occupied housing. Export growth lost considerable momentum, a reflection of the slow-down in demand from western Europe, which was accentuated by the changes in exchange rates in the final quarter of 1992. Import growth also slackened in the wake of subdued domestic demand. The contribution of changes in the real foreign balance to growth weakened in 1992, while the current account surplus fell slightly, to 3 per cent of GDP.

Inflationary pressures have been relatively high since 1991, reflecting strong wage growth and increases in excise taxes. The rise in the consumer price index slowed down in the final quarter of 1992, partly because of the reduction in VAT by 1 percentage point at the beginning of October.

The labour market deteriorated significantly in the second half of 1992 and the unemployment rate rose from 3.8 per cent in July to 4.6 per cent in December. Total employment fell slightly in 1992 compared with 1991.

The Netherlands guilder was left unscathed by the turbulence in the foreign exchange markets and monetary policy has been eased since September 1992 broadly in line with changes in German monetary policy.

The major preoccupation of fiscal policy for several years has been to consolidate the public sector finances: the progress achieved has more or less met the government's targets. The tight policy stance was maintained in 1992 but the weakening of economic growth and the working of the automatic stabilizers were partly responsible for a rise in the general government net borrowing requirement from 2.5 per cent of GDP in 1991 to 3.2 per cent.

Real GDP is forecast to rise by only 0.5 per cent in 1993, reflecting a further slackening of domestic demand and export growth. The sharp deterioration in the economic environment has already been straining public sector finances and the government has introduced measures to ensure that the budget deficit target for 1993 will be broadly met. Given the deterioration in the labour market the government has also been seeking wage restraint in the private sector. Trade unions and employers agreed in November 1992 to postpone new wage negotiations until February 1993.

Switzerland

Contrary to earlier expectations the Swiss economy failed to emerge from recession in the second half of last year: instead, the cyclical downturn deepened with real GDP falling at an annual rate of more than 1 per cent in the final two quarters of 1992. This deterioration reflected for the most part a marked decline in exports, which had been a major support to domestic activity in the first half of 1992.

Domestic demand continued to fall in the course of 1992, although the rate of decline diminished considerably in the final quarter. For the year as whole real GDP fell by 0.6 per cent. Private consumption remained depressed given the virtual stagnation of real disposable incomes. Against a background of high interest rates, large margins of spare capacity and a slump in the real estate market, there was a fall in real fixed investment by nearly 7 per cent. Of the major components of domestic demand only public consumption provided any support to domestic activity levels.

The demand for labour fell significantly in the course of 1992; for the year as a whole total employment fell by at least 2 per cent and the average number of persons unemployed was more than double the number in 1991. The seasonally adjusted unemployment rate rose from only 2 per cent in January 1992 to 4.2 per cent in January 1993.

Inflationary pressures have subsided against a background of weak demand and weak import prices.

The major concern of monetary policy was the reduction of inflation which was still close to 5 per cent in the first quarter of 1992 (over the same period in 1991). Against this backdrop, monetary policy was tightened in the first half of the year to counter the upward pressure on prices originating in the depreciation of the Swiss franc against the deutsche mark. The ensuing rise in short-term interest rates, however, was more than reversed in the second half of the year, when lower inflation and increased concern about the deteriorating economic situation led to a progressive easing of monetary policy. Short-term interest in January 1993 were some 3 percentage points lower than in Germany, compared with a difference of only 0.5 percentage points in June 1992. There has also been a significant fall in long-term interest rates. The budget deficit of the confederation in 1992 was double the planned figure, the net borrowing requirement corresponding to some 2.7 per cent of GDP compared with only 0.8 per cent in 1991.

The short-term outlook for the economy is rather bleak and only marginal growth, if any, is expected for 1993. A major uncertainty remains the short- and medium-term impact on business fixed investment of the negative vote in the referendum on the European Economic Space in December 1992.

Ireland

Economic growth strengthened in Ireland in 1992. For the year as a whole real GDP rose by 2.8 per cent compared with 1.7 per cent in 1991. This improvement was due to strong growth in private consumption and in exports of agricultural products and "high-tech" goods. Fixed investment virtually stagnated, but within the total there was robust growth in housing and public infrastructure investment. Strong export growth combined with relatively weak import growth resulted in a further increase in the already large surplus in foreign trade, which amounted to some 12 per cent of GDP in 1992. The current account surplus rose to some 8.5 per cent of GDP in 1992.

The labour market continued to deteriorate in 1992. Total employment remained broadly unchanged compared with 1991 (a decline of 0.1 per cent) but the growth in the labour force led to a rise in the unemployment rate from 14.9 per cent in 1991 to 16.1 per cent in 1992. The labour force growth reflects to a large degree an increase in the participation rate of females and the deterioration of the labour market in the United Kingdom, which led to a return of emigrants.

Inflationary pressures remained relatively low in 1992, largely because of the weakness of import prices.

The Irish pound came under strong speculative selling pressure in the wake of the flotation of the pound sterling in September 1992. The ensuing strong appreciation of the Irish pound led to a sizeable deterioration in the price competitiveness of Irish industry, for which the UK market is a very important outlet. Against a background of high unemployment, the central bank was forced to raise short-term interest rates to very high levels to defend the existing parities. However, the speculative pressures persisted and the authorities agreed to a devaluation of the pound by 10 per cent within the ERM at the end of January.

The general government budget deficit rose to 2.5 per cent of GDP in 1992, up from 1.9 per cent in 1991. This deterioration was partly due to cyclical factors. There was a further fall in the ratio of debt to GDP in 1992, but total government debt is still considerably higher than permitted by the convergence requirements of the Maastricht treaty. The devaluation of the pound has led to a large rise in debt service payments on the foreign currency denominated part of government debt and this will forestall a decline in the ratio of debt to GDP in 1993.

Economic growth is expected to slow down in 1993, principally because of a deterioration in export prospects. Small gains in employment will be insufficient to prevent a further rise in unemployment. Inflation is expected to edge upwards, partly reflecting higher indirect taxes. The government budget deficit is set to rise above 3 per cent of GDP in 1993.

Denmark

Economic growth remained subdued in 1992, mainly because of the severe drought, which depressed agricultural output. Real net exports remained the main support to domestic demand given that aggregate demand continued to fall. Private consumption increased, reflecting a rise in real disposable incomes, but there was a large fall in business investment. The decline in residential investment, which started in 1987, continued, the cumulative fall between 1986 and 1992 amounting to about 40 per cent.

Consumer price inflation slowed down further to one of the lowest rates in western Europe in 1992. But a major concern remains the high and rising level of unemployment. The government introduced in mid-1992 a package of labour market measures but the unemployment rate continued to rise to nearly 12 per cent in December 1992. Favourable export growth and an improvement in the terms of trade led to a further rise in the merchandise trade surplus and this was fully reflected in a larger current account surplus, which rose to about 3 per cent of GDP in 1992.

There was a small rise in the general budget deficit in 1992, to 2.4 per cent of GDP compared with 2.2 per cent in 1991. The higher deficit is largely due to the working of the automatic stabilizers. Denmark, together with Ireland, is currently the only EC member country which fulfils the Maastricht convergence criterion of a budget deficit of less the 3 per cent of GDP. The stance of fiscal policy will be slightly eased in 1993 as a result of measures to stimulate employment growth. The budget deficit is projected to rise by some 0.5 per cent of GDP in 1993.

The Danish krone remained relatively unscathed by the turbulence in the ERM last autumn. But since the flotation of the Swedish and Norwegian currencies in late 1992 and the ensuing loss in the price competitiveness of Danish exports, there has been increased speculative downward pressure on the currency and this has led to a progressive tightening of monetary policy. The foreign exchange markets calmed down in March 1993 and this allowed the central bank to reduce short-term interest rates again.

Forecasts prepared at the beginning of 1993 point to a slight acceleration in the rate of economic growth to somewhat less than 2 per cent in 1993. This improvement reflects to a large degree the strengthening of domestic demand following three consecutive years of decline.

Finland

The recession in Finland has led to a cumulative decline in real GDP of nearly 10 per cent in 1991 and 1992; this is more than during the depression of the 1930s when output fell by 6.5 per cent between 1929 and 1931. Total domestic demand fell by about 15 per cent during the last two years.

Exports strengthened considerably in 1992 as a result of the improved price competitiveness of Finnish industry: this was due to wage restraint and the devaluation of the markka by 12 per cent in November 1991 which was amplified by its flotation in September 1992.

The recovery in exports was the major cause of the growth in manufacturing output by nearly 4 per cent in 1992 (compared with 1991). Private consumption fell by about 5 per cent, the result of a combination of falling real incomes and a rising savings ratio, the latter reflecting the need to lower high debt levels. Fixed investment remained depressed, a reflection of large margins of spare capacity and considerable excess supply in the markets for housing and commercial real estate.

The situation in the labour market is very gloomy. The unemployment rate rose to 15.6 per cent in December 1992, and has increased rapidly from less than 5 per cent in the final quarter of 1990. The indications are that the labour market has continued to worsen in the early months of 1993.

Inflationary pressures have abated considerably. Depressed domestic demand has led to intensified competition and labour cost pressures have been low given the agreements on wage restraint.

Favourable export growth and falling imports led to a considerable rise in the trade surplus and this was the major factor behind a fall in the current account deficit, which remained, nevertheless, very high.

The deep recession and government support for the banking sector have led to a serious deterioration in public sector finances. The net borrowing requirement of general government was more than 8 per cent of GDP in 1992. In October 1992 the government adopted a package of measures designed to significantly reduce the structural part of the budget deficit by a combination of expenditure cuts and additional revenues over the period 1993-1995. A supplementary budget was passed at the beginning of February 1993 to ensure the financing of bank support in 1993. At the end of February the government proposed additional expenditure cuts (involving wage cuts in the public sector) to avoid a larger than planned deficit in 1993.

The stance of monetary policy was eased progressively after the floating of the markka in September 1992. Short-term interest rates fell between October 1992 and January 1993 from 13.4 per cent to 10.6 per cent. The positive margin over German short-term rates fell from 4.5 percentage points to 2 percentage points over the same period. But the scope for still lower interest rates is constrained by concerns about the inflationary pressures which a further depreciation of the currency would be likely to generate. These pressures are already considerable given the very large depreciation of the markka which against the deutsche mark, was more than 30 per cent between October 1991 and January 1993. Also, given that a large part of the business sector's debt is denominated in foreign currency, the devaluation has already led to a considerable increase in debt-service payments and this has been a factor in depressing domestic demand.

The government expects a further small fall in real GDP in 1993 compared with 1992. This outcome is the result of a forecast rise in real net exports which is nearly sufficient to offset the continuing decline in domestic demand.

Norway

Economic activity picked up in mainland Norway in 1992 but the favourable overall rate of economic growth was largely due to a vigorous expansion of oil exports.

Private consumption picked up as disposable incomes were bolstered by higher public sector transfer payments and lower taxes. Households continued to consolidate their net asset position: there was a further rise in the savings ratio and private housing investment continued to fall. There was also a large fall in manufacturing investment. The overall weakness of mainland fixed investment was, however, more than offset by vigorous capital formation in the oil sector.

Total employment started to edge up after the first quarter of 1992 but the increase was too small to absorb the rise in the labour force. Consequently, unemployment continued to rise.

The increase in consumer prices continued to slow down, a reflection of moderate wage growth and smaller increases in import prices.

The current account surplus was much reduced in 1992 compared with 1991. This reflects the combined effect of a deterioration in the terms of trade (oil prices fell compared with 1991) and, partly because of special factors, much larger outflows of investment income from the oil sector. Several years of current account surpluses have considerably reduced Norway's net foreign debt. Indeed, on current projections there will be a swing to a positive net asset position in 1994.

The Norwegian krone was not left unscathed by the turbulence in the foreign exchange markets. The heavy speculative pressures on the Finnish markka and the Swedish krona in September spilled over to Norway and forced the central bank to raise overnight lending rates by large amounts and to intervene in the foreign exchange markets to defend the krone-ECU peg. With the flotation of the Swedish krona in November 1992 these pressures on the exchange rate intensified considerably and the government decided to float the Norwegian krone on 10 December 1992.

Fiscal policy has been very expansionary since 1988 and the traditional budget surplus swung into a deficit corresponding to 0.4 per cent of GDP in 1991. The deficit rose considerably, to some 3.4 per cent of GDP in 1992 reflecting the combined impact of cyclical factors, lower net revenues from oil activity and discretionary measures to stimulate economic growth and employment. The crisis in the banking sector, caused by the collapse in the real estate market, continued to linger on and the government was forced to take further measures to strengthen the balance sheets of the commercial banks.

The outlook for 1993 is for a continuation of relatively robust growth with the oil sector remaining the major driving force.

Sweden

The recession deepened in Sweden in 1992. Private consumption declined in the wake of a significant slow-down in the growth of real disposable incomes and a large rise in the savings ratio. Cautious consumer spending was partly motivated by the desire to reduce the high debt levels incurred during the 1980s. Fixed capital formation fell for the third consecutive year, with a particularly sharp downturn for machinery and equipment against a background of large excess capacity and high real interest rates. Exports picked up slightly while imports remained depressed; the resulting change in real net exports added some 0.3 percentage points to overall economic growth but this was largely insufficient to offset the 1.5 per cent fall in domestic demand.

The steep fall in manufacturing output has not yet come to an end: output edged upwards during the first three quarters of 1992 but this was more than offset by a renewed decline in the final quarter. Output in manufacturing industry fell by nearly 15 per cent between 1989 and 1992.

The labour market continued to deteriorate. There were large cuts in employment in the business sector as enterprises sought to achieve gains in productivity. The unemployment rate rose to 5.5 per cent in December 1992 compared with 3.5 per cent twelve months earlier. Inflationary pressures continued to ease in 1992 as a result of wage restraint and sluggish domestic demand. The current account deficit in 1992 was 1.6 per cent of GDP, slightly more than in 1991.

Heavy speculative pressure on the Swedish krona, against the background of a severe deterioration in the financial position of the public sector, was a major factor in the significant tightening of fiscal policy after the summer. In September the government adopted two emergency fiscal packages, one designed to limit the much faster than projected rise in the budget deficit and the other to improve cost competitiveness in the business sector. On current projections the budget deficit for fiscal year 1992-1993 is estimated at about 14 per cent of GDP. This striking deterioration reflects not only the impact of the recession on revenues and outlays but also government support for the commercial banks, which are faced with heavy loan losses. The recently proposed budget for fiscal year 1993/94 aims to reduce the budget deficit to some 11 per cent of GDP. The government intends to eliminate the structural deficit, which currently corresponds to some 4.5 per cent of GDP, by 1998 but this will depend on a favourable economic environment.

The framework for monetary policy has changed since November 1992, when the link between the Swedish krona and the ECU was severed. Monetary policy has since been progressively eased, leading to a decline in short-term interest rates by 2.5 percentage points between November 1992 and January 1993. Long-term interest have not fallen, however, reflecting concerns about the large budget deficit and inflationary expectations. The target of monetary policy is to keep inflation within a range of 1-3 per cent as from 1995.

The recession is expected to deepen further and real GDP is forecast to fall by about 1.5 per cent in 1993. But the outlook is very uncertain: further action to curb the budget deficit may be required and the likely impact of the large depreciation of the krone on export growth and domestic inflation is difficult to gauge.

Greece

The rate of economic growth slowed down to 1.2 per cent in 1992. Manufacturing output fell for the third consecutive year. Domestic demand was very weak, a result of tight economic policies. Private consumption was restrained by the dampening effect of falling employment, higher taxes on real disposable incomes, and by falling real wages in the public sector, a reflection of the government's austerity policies. Accordingly, public consumption also remained subdued. Total fixed investment rose marginally, a robust growth of expenditure on machinery and equipment being largely offset by declining construction investment.

Exports were the main support for domestic activity, although price competitiveness has deteriorated as a result of the devaluations of the lira, the peseta and the escudo in the final quarter of 1992. There was only relatively moderate import growth in 1992, reflecting the overall sluggishness of domestic demand. The deficit in the current account rose to 2.5 per cent of GDP in 1992.

Total employment continued to fall and the rate of unemployment rose to about 10 per cent at the end of 1992.

Tight policies and weak import prices contributed to the easing of inflationary pressures, the average annual inflation rate falling to 15.9 per cent in 1992 compared with 19.6 per cent in 1991. The underlying rate of inflation slowed down much more, since significant increases in indirect taxes, designed to contain a sizeable budget overrun, were introduced in August 1992 and these were reflected in the consumer price index.

Monetary policy continued to be restrictive in order to reduce inflation and strengthen confidence in the drachma. Total credit to the private sector fell in real terms in 1991 and only increased slightly in 1992. Against a background of turbulence in the ERM, the drachma (which is not a member of the ERM) also came under pressure. Capital controls were introduced to fend off speculation and will only be lifted in June 1994.

Fiscal austerity, designed to curb the high budget deficit, is gradually bearing fruit, although earlier ambitious targets have not been met. The government is adamant that it will maintain the austerity policy which is targeted on reducing the public debt from 116 per cent of GDP to 100 per cent in 1996. The central government borrowing requirement is to be reduced from 9.6 per cent of GDP to 1.9 per cent over the same period.

Tight policies and weaker prospects for agricultural output are likely to lead to a further slow-down in economic growth in 1993.

Portugal

Economic growth slowed down markedly in the course of 1992, real GDP rising by less than 2 per cent. Manufacturing output declined for the second consecutive year despite relatively strong growth in domestic demand, which can be explained by a mismatch between the structure of domestic output and demand.

There was nevertheless a weakening of domestic demand which was partly offset by favourable changes in the real foreign balance. Private consumption continued to grow strongly, a reflection of large increases in real earnings. Fixed investment weakened against a background of high real interest rates and reduced profit margins. Real exports of goods and services picked up slightly, while import demand continued to grow strongly.

There was a small current account surplus in 1992, following a sizeable deficit in 1991. The trade deficit remained quite high, but this was largely offset by transfer payments from the EC.

The labour market was still close to full employment in 1992, but in manufacturing industry employment fell in the wake of a sharp drop in output. The unemployment rate edged up to 4.5 per cent in the final quarter of 1992.

High inflation, fuelled by strong growth in labour costs, remains a major concern of economic policy. Inflationary pressures abated slightly in 1992, a result of tight monetary policy and favourable changes import prices. In contrast, higher VAT rates added to upward pressure on prices. The annual increase in the consumer price index fell to 8.9 per cent, compared with 11.4 per cent in 1991.

Monetary policy has been kept tight to curb inflation. High real interest rates have led to a marked slow-down in the growth of bank credit to the private sector. The escudo joined the ERM in April 1992 with a fluctuation margin of 6 per cent. The turbulence in the foreign exchange markets led to a devaluation of the escudo by 6 per cent within the ERM in November 1992, which was partly the consequence of an earlier erosion of price competitiveness. The realignment occurred together with the second devaluation of the peseta, which is not surprising, given that a considerable share of Portuguese exports are shipped to Spain. The monetary authorities lifted all remaining capital controls in the course of 1992.

Following the general decline in interest rates in the international financial markets, monetary policy was eased slightly in January 1993 but the overall scope for relaxation is constrained by high inflation and the exchange rate commitment within the ERM.

Fiscal policy was highly expansionary in 1991 and the general government budget deficit rose to 6.2 per cent of GDP. The Maastricht treaty has increased the urgency of consolidating public finances and, consequently, fiscal policy was tightened in 1992. The target was to reduce the budget deficit to 5.2 per cent of GDP in 1992, but the outcome was slightly worse (5.7 per cent of GDP), partly because of the deterioration in the labour market.

The budget for 1993, adopted in September 1992, signalled a further significant tightening of fiscal policy: reductions in current government outlays, including cuts in real wages for civil servants, are the main instruments to reduce the budget deficit to 3.9 per cent of GDP in 1993. The budget assumed that real GDP would rise by 3 per cent in 1993 but growth forecasts have since been considerably reduced. Given the working of the automatic stabilizers, the budget target is unlikely to be reached unless a further tightening of policy is introduced.

Economic growth will slow down further in 1993, as a result of tight policies and the deterioration in the external economic environment. The slackening of demand should help to reduce inflationary pressures but the government has also been urging wage restraint in the private sector in order to bring inflation down to within the range of 5-7 per cent in 1993. The rise in unemployment in the second half of 1992 can be expected to continue in 1993.

Spain

There was a marked slow-down in economic growth in Spain in 1992. For the year as whole real GDP rose by about 1 per cent, far below the rate of 3 per cent expected by the government in the spring of 1992. At the end of 1992 the economy was on the verge of recession, real GDP falling by 0.2 per cent between the third and fourth quarters. Private consumption was still buoyant during the first half of 1992 supported *inter alia* by generous transfer payments, lower income taxes and a significant fall in the savings ratio. But the rapid deterioration in the economic environment, as well as higher taxes, depressed consumer demand in the second half of the year.

Public consumption continued to grow strongly in 1992 as a whole, but fixed investment fell as the boom associated with the Olympics and the World Expo came to an end. Exports of goods and services grew quite strongly, the slow-down in foreign demand for goods being largely offset by favourable growth in services. Import demand weakened significantly in the course of the year in tandem with slackening domestic demand. Changes in the real foreign balance, as in previous years, pulled down the rate of economic growth in 1992 but, given the increasingly depressed import demand, the negative impact was significantly smaller than in 1991. The current account deficit rose in 1992, largely reflecting the worsening trade deficit.

The slow-down in economic growth was accompanied by a significant deterioration in the labour market. There was a large fall in employment, which partly reflected attempts by firms to rationalize production in the face of strong labour cost pressures. The already very high unemployment rate, a major economic problem, rose to 18.3 per cent in the autumn of 1992. The annual inflation rate was 5.9 per cent in 1992, the same as in 1991, but this outcome was influenced by the increase in indirect taxes in January and August of 1992.

The peseta came under strong speculative pressures after the summer against the background of a gradual erosion of price competitiveness in previous years and adverse changes in the current account and in the budget deficit in the first half of 1992. Monetary policy was tightened to defend the exchange rate but in the event the peseta was devalued within the ERM, first by 5 per cent in September and then by another 6 per cent in November. To stem speculation against the peseta capital controls, which had been abolished earlier in the year, were partly reintroduced. Some cautious easing of monetary policy has occurred in the first two months of 1993, but the overall stance of monetary policy remains tight given the lingering tensions within the ERM and the need to reduce a high rate of inflation.

Fiscal policy was relaxed in the first half of 1992 reflecting overly optimistic expectations of economic growth. Subsequently, the budget deficit rose far above the target fixed in the government's convergence programme and fiscal policy was consequently tightened significantly in July 1992. An emergency fiscal package introduced higher rates of income tax and VAT, as well as more stringent controls on public expenditure. The budget for 1993, adopted in late 1992, introduced further restrictive measures. The total net borrowing requirement of the general government fell from 4.9 per cent in 1991 to 4.4 per cent in 1992, and the target is to reduce it to 3.6 per cent of GDP in 1993.

The outlook is for only very weak growth in 1993. Domestic demand is likely to be subdued given the continued deterioration in the labour market and the maintenance of tight policies. Given depressed demand for imports and the potential benefits of devaluation for exports, changes in the real foreign balance could provide some support to increased activity in 1993. Inflationary pressures are set to abate, given sluggish domestic demand and a slow-down in wage growth on account of the rise in unemployment. The government adopted various measures in February 1993 designed to stimulate investment and employment but these will be accommodated within the existing target for the budget deficit.

Turkey

In Turkey, GNP increased by 5.4 per cent in 1992 compared with 0.3 per cent in 1991 when output was affected by the Gulf conflict. The most dynamic component of demand was private consumption (an increase by 10.6 per cent) thanks to significant increases in wages, higher agricultural support prices and some employment gains. Both public consumption and gross fixed capital formation grew by less than 2 per cent and the foreign balance pulled down the growth in GDP by 2.2 percentage points. Industry, which accounts for about 30 per cent of GDP, was the leading sector with a growth rate of about 7 per cent between January and September. Foreign trade remained buoyant in 1992 with exports (f.o.b.) increasing by 11 per cent (mainly due to manufactures) and imports (c.i.f.) by 9.5 per cent. The trade deficit was $8 billion in 1992 compared with $7.3 billion in 1991. Official grants to compensate for losses originating from the Gulf Crisis declined by $1.3 billion in 1992 as scheduled. Thus, in spite of a strong recovery in revenues from tourism, the current account deficit in 1992 is estimated at $1.3 billion compared with a surplus of $0.3 in 1991. On the capital account, direct investments increased significantly (from $800 million in 1991 to $1.1 billion in 1992) and net portfolio investments reached some $2 billion. Thus, despite a large debt repayment, official reserves increased by $1.3 billion compared with a fall of $1 billion in 1991.

Inflation, measured by the consumer price index, remained for the fifth consecutive year at around 70 per cent due to a somewhat lax monetary policy and a particularly expansionary fiscal policy. Chronic inflationary expectations worsened and increased the pressures on the financial markets. According to preliminary estimates, the public sector borrowing requirement fell from 14.4 per cent of GNP in 1991 (its highest level in the last three decades) to 12.6 per cent in 1992, which was still well above the target of 8.8 per cent. In the 1993 budget, the public sector borrowing requirement is targeted to decline to 9 per cent of GNP. To achieve this the budget foresees a reduction in investment by state economic enterprises, a firm limit on company tax deductions and measures to improve revenue collection. Privatization, which made some progress in 1992, is expected to accelerate and to raise LT 25,000 billion, more than five times the sum realized in 1991. GNP growth in 1993 is expected to be between 4 and 5 per cent, led by exports and private investment. Import growth is expected to decelerate (9.5 per cent) and the annual rate of inflation to fall to about 50 per cent.

Cyprus

After a mild recession in Cyprus during the first half of 1991, due to the economic consequences of the Gulf war, the economy picked up strongly in 1992. GDP growth is estimated at 7 per cent, compared with 1 per cent in 1991. This acceleration reflected a significant increase in domestic demand and particularly in private consumption. There was also a sharp rise in foreign demand for services reflecting the strong upturn in tourism which accounts for about 20 per cent of GDP. Imports, driven by the upturn in activity, rose by almost 20 per cent in 1992 with the current account deficit widening to £C 200 million (about 4 per cent of GDP). The upturn in economic activity resulted in a substantial decline in unemployment. At the end of June the unemployment rate was 1.8 per cent compared with 2.4 per cent a year previously. During the second half of 1992, labour shortages had grown so acute that strict immigration rules were relaxed. The work force increased by about 5 per cent in 1992, with the arrival of about 12,000 east European and Asian workers on short-term contracts at wages well below the Greek-Cypriot average. None the less, labour costs increased significantly due to index-linked wage rises, and with the introduction of VAT in 1992, annual inflation rose to 6.5 per cent, its highest rate for nearly a decade. The budget deficit narrowed in 1992 (from 4.4 per cent of GDP in 1991 to 3.6 per cent) thanks to higher tax receipts reflecting the rise in activity. The 1993 budget proposal calls for the continuation of cutbacks in the number of public sector jobs through a freeze on vacancies. No new taxes are proposed, although the VAT rate might be increased. On the other hand, total government expenditure is budgeted at £C 1.1 billion (the first time it has exceeded £C 1 billion), due to a steep rise in infrastructure spending and social benefits. Nevertheless, inflation is forecast to fall to 4.5 per cent in 1993 as a result of a tighter credit squeeze. Demand is forecast to weaken, after a rush to buy cars and other household durables before VAT was imposed. During 1993, given the recession in Europe and the strong Cyprus pound (pegged to the ECU since mid 1992), income growth from tourism and textiles and clothing exports, the main foreign exchange earners, will fall and so the Cypriot economy is not expected to be as dynamic as in 1992.

Malta

According to the forecast based on data for the first three quarters, GDP growth in Malta in 1992 slowed down to 5 per cent compared with 5.6 per cent in 1991. Public consumption was the most dynamic component of GDP (9.5 per cent) followed by private consumption (3.1 per cent). The foreign balance also added strongly to growth in 1992. However, gross fixed capital formation was weak (0.6 per cent) and only marginally higher than in 1991. Retail price inflation was 2.5 per cent during the first nine months of 1992, the same as in 1991. Even though employment increased by 1 per cent, the unemployment rate increased to 3.9 per cent, compared with 3.6 per cent in 1991, due to faster growth in the labour force. The merchandise trade deficit narrowed by LM 24 million (11 per cent) during the first nine months of 1992, exports rising more strongly (29 per cent) than imports (11.5 per cent). According to the Central Bank, the current account deficit narrowed to LM 7.8 million during the first quarter of 1992 compared with LM 14.2 million during the same period in 1991. This improvement was due to the strong growth in exports and net income from tourism. In 1993, GDP growth may slow down further as a result of sluggish demand in Malta's main export markets for goods and tourism, namely Italy and the United Kingdom. The expected introduction of VAT and its effect on wage demands may contribute significantly to inflationary pressures in 1993. The government budget deficit has been growing since the early 1980s with increasingly buoyant expenditures (due to increasing debt servicing in addition to growing public expenditure) and decelerating growth in revenues. The pace of fiscal consolidation may slow down as a result of the Lisbon decision in late June 1992 not to include Malta in the list of countries to be considered for full membership of the EC. The budget deficit in the second quarter of 1992 was LM 23.2 million compared with a small surplus of LM 0.3 million in the same period of 1991. Although the government financed part of the deficit through bond issues, central bank financing was important in meeting the borrowing requirement.

Israel

The rapid pace of economic expansion in the last two years in Israel slowed down towards the end of the third quarter of 1992. Nevertheless, for the year as a whole, GDP increased by just below 6 per cent, as in 1991. Record growth in the tourism sector, the continued boom in construction, and mainly export-led manufacturing growth, have been the principal factors behind the strong performance of the economy during the last two years. Imports in 1992 rose by some 7.5 per cent, but thanks to a surge in exports (11.5 per cent) and the continued large flow of transfers from abroad, the current account deficit narrowed to $0.3 billion (from $0.8 billion in 1991). With a significant moderation in wages and relatively strong gains in productivity (2 per cent), unit labour costs ceased to put upward pressure on prices during 1992. In fact inflation slowed down markedly, particularly during the third quarter: consumer price inflation fell from 17.3 per cent in March to 8 per cent in September. Despite some seasonal acceleration during November and December, the average annual rate of inflation was 12 per cent in 1992 compared with 19 per cent in 1991. Despite strong growth in employment (4 per cent), the unemployment rate remained high in 1992 (11 per cent) as the labour force continued to expand albeit at much lower rates than in the previous two years due to the slow-down in immigration (75,000 in 1992 compared with 200,000 in 1991).

The 1993 budget, approved by the Knesset in November, reflects a re-ordering of priorities by the recently elected Labour government. It includes increased spending on infrastructure and education, tax cuts and a 50 per cent reduction in spending on public housing projects in the occupied territories. The reduction in the 1993 budget deficit mainly reflects reduced expenditure due to an expected fall of some 40 per cent in the rate of immigration. The deficit is forecast at 3.2 per cent of GDP compared with 6.2 per cent of GDP in 1992. GDP growth in 1993 is expected to slow down to some 4.5 to 5 per cent, largely because of a 35 per cent fall in residential building investment. Both public and private consumption is forecast to grow by 3 per cent and total investment to fall by 5 per cent. Thus, the stimulus from domestic demand will weaken in 1993: the main driving force behind growth is assumed to be a continued boom in exports and tourism. However, the deepening recession in Europe may make the above targets difficult to achieve.

Chapter 3

THE EUROPEAN TRANSITION ECONOMIES

This chapter provides an account of the principal macroeconomic developments in the transition economies in 1992 and discusses the prospects for 1993. The first sections, 3.1 to 3.3, survey recent changes in output and demand, employment, prices and incomes, and foreign trade and payments. Sections 3.4 to 3.6 review macroeconomic policies and the transformation process. Section 3.7, in a series of notes, focuses on policy issues in a country framework.

3.1 OUTPUT AND DEMAND

(i) Overview

The transformation process in the European transition countries continued under difficult conditions in 1992. Inherited distorted production structures and underdeveloped infrastructures, together with policies designed to transform formerly state-run systems into market-based economies, brought a deep recession, persistent inflation and growing unemployment. The prolonged economic downturn and high costs of the transition process have eroded the popular enthusiasm and optimism which prevailed immediately after the revolutions of 1989-1990 in eastern Europe and the break-up in 1991 of the Soviet Union.[1] Growing social, political and ethnic tensions have added to an already difficult economic situation in many of these countries.

Despite a generally gloomy economic picture, all the transition countries have gone through important political and economic transformations. The foundations for pluralistic democracies and decentralized market economies have been established. The first free elections in the post-war period have been held in most of these countries and important steps were taken to liberalize prices and foreign trade, to start the privatization process and to set up new legal and institutional structures.

Most countries of eastern Europe and the successor states of the Soviet Union experienced a third year of falling production and living standards. However, the outcomes of transformation policies differ greatly, linked with differing initial conditions, the inheritance of the past, the sequencing and speed of the reform process, the policies actually adopted and progress made in building up the infrastructure necessary for a functioning market economy. Thus, in 1992 there was a further differentiation between countries; their economic situation became much more diversified and the region is now heterogeneous.

Czechoslovakia, *Hungary* and *Poland* are fairly advanced with institutional and structural reform. In all three countries, the output contraction may be bottoming out and an upturn in production is anticipated for 1993 in the Czech Republic, Hungary and Poland. Czechoslovakia has brought inflation to almost single-digit level, maintained balanced budgets and a stable exchange rate and reduced unemployment. Output fell some 7-8 per cent in 1992, but the fall slowed throughout the year and a slight upturn was expected for 1993. However, the relatively favourable macroeconomic outlook of that economy is now clouded by the consequences of the country's division into two sovereign states, especially by the impact on trade between them.[2] In Hungary, the output fall was some 5

---

[1] *Eastern Europe* here refers to the formerly centrally planned economies of Albania, Bulgaria, Czechoslovakia, Hungary, Poland and Romania, and the successor states of the Socialist Federal Republic of Yugoslavia. Czechoslovakia, which separated into the Czech and Slovak Republics at the beginning of 1993, is dealt with as a policy-making entity for 1992, but data for the two new states are shown separately where available. Among the newly-independent republics of the former *Soviet Union*, a distinction is made between the *Baltic states* (Estonia, Latvia and Lithuania) and the majority of the remaining republics which cooperate in the institutions of the Commonwealth of Independent States – the *CIS* countries.

[2] In order to maintain trade between the two republics, a customs union was arranged, as well as a much more circumscribed intergovernmental agreement on the temporary maintenance of a common currency which broke down within just over a month. In February 1993, the two countries introduced their own national currencies. A settlement mechanism was set up which provides for the clearing of balances in convertible currencies. The negative impact of the

per cent, but the rate slowed throughout the year and a modest upturn is expected for 1993. As in Czechoslovakia, developments in external markets — especially in Germany — may determine whether this can materialize. Many company failures followed the adoption of a bankruptcy law in September 1991. The rate of inflation decelerated, but remains near 20 per cent annually, in part reflecting a large fiscal deficit. Poland is the only country where output has shown an upturn in 1992. Year-on-year inflation rates decelerated significantly, though they are much higher than in Czechoslovakia and Hungary, and the budget deficit is causing concern. The role of the private sector in Polish and Hungarian output and employment is now significant. In the external sector, all three countries may have succeeded in raising their exports to western markets substantially and also in improving their current account balances.[123]

By contrast, in *Bulgaria* and *Romania* output continued to fall in 1992 at about the same pace as in 1991 — by 22 and 15 per cent, respectively. Inflation has been running at high rates, and both countries have serious balance of payments difficulties. The basic steps of economic liberalization have been taken, but institution-building and structural change are lagging behind those in the first group of countries and the political situation is less stable. *Albania*, the poorest country in Europe, is in a critical situation. Its production and distribution systems have virtually collapsed, the transition process is at an early stage and, given large internal and external imbalances, it is unlikely that macroeconomic stabilization and economic recovery can progress without large-scale external assistance.

The violent disintegration of the former *Socialist Federal Republic of Yugoslavia* in a civil war has destroyed the bulk of the economic links within the former federation, which imposed substantial costs on the successor republics forced to look for new markets. While the economic situation of *Slovenia* is relatively better, the ex-Yugoslav republics are generally in deep recession. Unemployment is high and in some of them much of the capital stock has been damaged or destroyed by the war. In *Bosnia-Herzegovina*, the war is imposing immense human and material losses which at the moment cannot be quantified. The economy of the new *Federal Republic of Yugoslavia* (Serbia and Montenegro) is experiencing a particularly steep fall in output (some 30 per cent), largely in consequence of the trade embargo imposed by the United Nations in May 1992; it is also the only European country with hyperinflation. The consequences of the war and of the collapse of trade between the former Yugoslav republics have resulted in a sharp decline in output also in *Croatia*, which in 1992 fell some 15 per cent, to half its 1989 level.

Even though the *former Soviet Union* initiated some market reforms in 1990 and 1991, it was only after the failed August 1991 coup that these gained momentum and a more radical orientation in some successor states associated in the *Commonwealth of Independent States (CIS)*. The main impetus came from the policies of the Russian Federation, which was the first to introduce a series of economic liberalization and stabilization measures after November 1991. However, inconsistencies in the programme and in its implementation,[124] the unresolved fiscal and monetary problems of the rouble zone and the collapse of inter-republican trade led to a large output fall in 1992. Inflation continued at much higher rates than expected after the first price liberalization surge and accelerated markedly since September 1992;[125] uncontrolled increases in the money supply contributed to the fall in the value of the rouble internally and against foreign currencies. Growing internal imbalances were accompanied by declining levels of external trade affecting both the new foreign trade between the former Soviet republics and trade with other countries. Russia was unique in that it had not only a growing trade surplus with the CIS as well as a positive, though contracting, balance with the outer world. But, as in the CIS overall, falling imports disrupted the supply of production inputs. Falling living standards and the mounting difficulties of transition have taken their toll on parliamentary and popular support for radical economic change and the original reform programme was abandoned. However, the new government led by Prime Minister V. Chernomyrdin announced in January 1993 an anti-crisis plan to fight inflation, stabilize the currency and restore confidence in the economy.

An especially difficult situation has arisen in the *Baltic states*. The ambitious transformation programmes adopted by Estonia, Latvia and Lithuania after independence in autumn 1991 were constrained by the collapse of trade with the former Soviet republics and by large adverse terms-of-trade shifts resulting from sharp increases at the beginning of 1992 in the price of oil and natural gas imports from Russia. As a result, the fall in national product and trade accelerated during the year. However, in all three countries the high inflation experienced in the first months of the year appears to have abated subsequently after national currencies were introduced.

(ii) Output developments

For most east European transition countries, 1992 was again a year of substantial output contraction.

---

national currencies. A settlement mechanism was set up which provides for the clearing of balances in convertible currencies. The negative impact of the separation on trade between the Czech and Slovak Republics, and hence on output and employment, is expected to be rather important, particularly for Slovakia which is the more dependent of the two on mutual trade.

123 Conflicting trade and current-account data reported by Polish authorities make the judgement on developments in that country rather uncertain at the time of writing. See section 3.3 below.

124 See section 3.4 below.

125 See section 3.2 below.

TABLE 3.1.1

European transition countries: Economic activity, 1989-1993
(Percentage change over same period of preceding year)

	NMP or GDP [a]					Gross industrial output			1992				
	1989	1990	1991	1992	Forecast 1993	1989	1990	1991	Jan.-March	Jan.-June	Jan.-Sept.	Jan.-Dec.	Forecast 1993
Albania	11.7	-13.1	-30	-(8-10)*	0-3	5.0	-7.6	-42.5	..	..	..	..	..
Bulgaria	-0.3	-17.5	-25.7	-22.0 [b]	-4	2.2	-12.6	-23.3	-18.0	-21.2	-20	-22	..
Czechoslovakia [c]	1.4	-1.4	-15.9	-(7-8)*	..	0.8	-3.5	-24.7	-25.9	-20.2	-16.0	-11.0*	..
Czech Republic	2.4	0.8	-19.0	-7.1 [c]	1-3	1.7	-3.3	-24.4	-27.4 [d]	-21.6 [d]	-17.5 [d]	-10.6	..
Slovak Republic	1.1	-3.8	-19.3	-6.0 [c]	-3	-1.3	-4.0	-25.4	-26.4 [d]	-21.1 [d]	-17.0 [d]	-12.5 [d]	..
Hungary [c]	0.4	-3.3	-11.9	-(4-6)	0-3	-2.5	-4.5	-19.1	-19.5	-16.6	-13.1	-9.8	0-3
Poland [c]	0.2	-11.6	-7.6	0.5-2	2.0	-0.5	-24.2	-11.9	-8.6	-3.0	0.6	4.2	..
Romania [c]	-5.8	-7.4	-13.7	-15.4	-5	-2.1	-17.8	-19.6	-13.0	-20.1	-23.5	-22.1	..
Yugoslavia (SFR) [e]	0.6	-8.5	-15	..	..	0.9	-10.3	-19	..	..	..	..	..
of this:													
Croatia [e]	-1.5	-8.5	..	-24.3	15	-1.0	-11	-28.5	-28.0	-28.1	-22.4	-14.6	..
Slovenia [c]	-0.5	-3.4	-9.3	-6.5	-1	1.1	-10.5	-12.4	-13.6	-16.0	-15.3	-13.2	-4
Yugoslavia (FR) [e]	-1.9	-8.4	-11.1	-27.0	-15	1.0	-11.7	-17.4	-8.0	-10.9	-18.4	-22.9	-(15-30)
Eastern Europe	-1.4	-9.9	-14.4	-10.0	..	-0.5	-15.2	-19.6	-15.1	-14.7	-13.8	-11.8	..
Soviet Union	2.5	-4.0	..	..	..	1.7	-1.2	..	..	..	..	..	..
CIS	2.3	-3.4	-10.1	-18.5	..	1.9	-1.1	-7.8	-13.5*	-13.7	-17.4	-18.2	..
Armenia	8.4	-8.2	-11.4	-42.6	..	-8.3	-7.5	-7.7	-52.9	-51.3	-50.8	-52.5	..
Azerbaijan	-8.9	-11.3	-0.4	-28.2	..	0.7	-6.3	4.8	-20.7	-21.8	..	-24	..
Belarus	8.2	-2.4	-3.0	-11.0	..	4.6	2.1	-1.5	-10.2	-11.7	-13.8	-9.6	..
Kazakhstan	-0.1	-0.9	-10.3	-14.2	..	2.5	-0.8	-0.9	-13.5	-12.1	-14.8	-14.8	..
Kyrgyzstan	4.5	4.8	-5.2	-26.0	..	5.2	-0.6	-0.3	-11.5	-18.1	-23.2	-26.8	..
Moldova	8.8	-1.5	-18.0	-21.3	..	5.7	3.2	-11.1	-25.6	-26.5	-27.5	-21.7	..
Russia	1.6	-4.0	-11.0	-20	-(5-8)	1.4	-0.1	-8.0	-13.0	-13.5	-17.6	-18.8	-(7-10)
Tajikistan	-6.5	0.2	-8.4	-31	..	1.8	1.2	-3.6	-13.2	-16.1	-19.3	-24.3	..
Turkmenistan	-7.0	1.8	-0.6	..	..	3.3	3.2	4.8	-11.1	-16.5	-23.9	-16.7	..
Ukraine	5.0	-3.6	-11.2	-15	..	2.8	-0.1	-4.8	-15.3	-12.3	-9.8	-9.0	..
Uzbekistan	3.1	4.3	-0.9	-12.9	..	3.6	1.8	1.5	-8.4	-8.7	-9.1	-6.2	..
Georgia	-3.4	-4.3	..	..	..	0.7	-5.7	..	..	..	..	..	..
Estonia [c]	6.6	-3.6	-12.6	-28	6	0.7	-5.6	-9.0	-34	-38.6	-36	-38.9	..
Latvia	7.4	-1.4	-7.9 [c]	-43.9 [c]	..	3.1	-0.2	-0.1	-25.4	-30.9	-32	-35.1	..
Lithuania	1.6	-6.0	-6.7	-35 [f]	..	4.2	-2.8	-4.9	..	-42.7	-47	-51.2	..
Ex-GDR *Länder* [c]	2.1	-14.7	-33.9	6	..	2.3	-28.1*	-28.1 [g]	-12.1	-9.1	..	-6	..

Sources: National statistical publications and statistical office communications to ECE. "Forecasts" are a mixture of enunciated governmental assumptions entering into policy determination and independent forecasts of specialized institutes.

a Net material product (produced) unless otherwise noted.
b Sales of the material sphere excluding agriculture and the private sector.
c Gross domestic product.
d Enterprises with 25 or more employees.
e Gross material product (value added of the material sphere including depreciation).
f January-September 1992.
g Second half of 1991 compared with the same period of 1990.

Declines of 10 per cent in 1990 and 14 per cent in 1991 in real GDP/NMP were followed by a further 10 per cent drop in 1992 (table 3.1.1). However, in many east European countries the rate of contraction decelerated, and in some of them signs have appeared that the output fall may have reached its bottom. In the successor states of the Soviet Union, in contrast, the rate of contraction steepened in 1992, rising from 4 per cent in 1990 and 10 per cent in 1991 to about 20 per cent in 1992 in the CIS republics and to much sharper falls in the Baltic states.

The cumulative output contraction since 1989 in the transition economies jointly now amounts to about 30 per cent. There were substantial differences between countries, however, with the cumulative fall ranging from less than 20 per cent of GDP in Hungary and Poland to almost half in Albania and Croatia. In the CIS countries it amounted to approximately one third, and in the Baltic states to about one half. Despite serious measurement problems and doubts whether the official statistics reflect the true extent of the output decline, the depth of the recession is undeniable.[126]

---

[126] Aggregate output and other economic development indicators in all the transition countries are subject to a considerable degree of uncertainty. The traditional system of collecting statistical information from all basic economic units has ceased to operate and has been only partially replaced by new sampling methods of data collection. The disintegration of old administrative structures and the creation of thousands of new organizational units has greatly complicated the collection of reliable and comparable statistical information. The number of new firms is growing rapidly, the field of firms' activities may change and the quality of their accounting systems can differ widely. Coverage of the new and, in some countries, fast-growing private sector particularly is incomplete. Moreover, the shift to the conceptually and methodologically different framework of the SNA system of national accounts has interrupted previous historical

All transition countries except Poland registered an overall output fall in 1992, ranging from some 5-7 per cent in Czechoslovakia, Hungary and Slovenia, about 15 per cent in Bulgaria and Romania, to around 20 per cent in Russia. Particularly sharp declines of output occurred in the Federal Republic of Yugoslavia, in the Baltic states and in some of the Asian CIS countries.

A substantial output contraction was inevitable as market-oriented reforms eliminated inefficient and uncompetitive production lines. But the actual fall surpassed expectations. The reasons for this are examined in more detail below, but it can be noted at this stage that the historically unprecedented rapid change in political and economic systems, coupled with the abrupt dismantling of the CMEA and intra-CIS trade networks, together with the lack of preparedness of economic policy-makers and economic agents for sudden changes on this scale, were the basic factors behind the larger-than-expected output losses. Intercountry variations resulted from different combinations of these factors. At the risk of some oversimplification, the following groups of countries can be distinguished:

– *Czechoslovakia, Hungary* and *Poland* are the countries which started earliest and have progressed most rapidly with the transformation process. They are also the first ones to show signs of a slight upturn in output or a substantial deceleration in its rate of decline. In Poland, where the cumulative fall in GDP was 18 per cent between 1989 and 1992, output began to grow compared with year-earlier levels after the second quarter of 1992; full-year GDP is estimated to have increased by 0.5-2 per cent over 1991.[127] In Czechoslovakia and Hungary, total output continued to fall in 1992, but at substantially lower rates as the year proceeded; preliminary estimates indicate GDP falls of 7 and 5 per cent, respectively. In 1992, the improvement was stimulated *inter alia* by rising exports in Hungary and Poland and by a revival of domestic demand in Czechoslovakia. In Poland, which had experienced its largest output decline in 1990 as the first transition country to implement a comprehensive stabilization and liberalization policy package ("shock therapy"), the fall of output slowed in 1991 and seems to have bottomed out in 1992. The early start of the transformation effort, and the already large and growing share of the private sector in total output, are probably key factors in this development. In Hungary, the cumulative GDP fall amounted to 19 per cent between 1989 and 1992. The largest fall in Hungarian output had occurred in 1991 (12 per cent), the year the CMEA market collapsed. A slower pace of adjustment was grafted in Hungary onto a system which had long since been allowed to depart a considerable distance from the central planning model – particularly in the decade which preceded the collapse of communism. The recession was thus less severe than in most other east European countries, even though stabilization and restructuring were pursued simultaneously. Czechoslovakia experienced its largest output fall (16 per cent) in 1991, in the wake of the introduction of the first far-ranging market-oriented reforms. Domestic imbalances in Czechoslovakia had been less acute than in the other east European countries at the outset of transition, and external indebtedness was low. Strong macroeconomic stabilization policies accompanied the reforms. The ensuing output contraction was none the less the largest among the three countries (some 23 per cent in 1990-1992).

It seems that the better results of this group of countries were caused either by the long adjustment process which had preceded the final dismantling of central planning (Hungary and Poland) or by the better initial conditions for macroeconomic stabilization (Czechoslovakia). It is also notable that, in spite of the many differences between them, they inherited a less centralized and more diversified economic system than those elsewhere in the region. It is thus likely that the demolition of central planning caused fewer and less severe disruptions in production and supply than was the case in Albania, Bulgaria, Romania and the former Soviet Union. Moreover, these three countries were able to increase their exports to western markets substantially and rapidly, aided by close geographical and cultural affinities with western Europe and, more concretely, by the association agreements with the European Community. The collapse of traditional markets and the sharp compression of domestic demand none the less gave rise to larger-than-expected output falls.

– In *Bulgaria* and *Romania*, preliminary estimates indicate a 15 per cent fall in GDP in 1992 in addition to the sharp contractions already experienced in 1990 and 1991 by both countries. The cumulative contraction since the start of the recession was, respectively, almost 50 and 35 per cent, i.e., around twice as big as in the first group of countries. The consequences of the collapse of trade with the former CMEA countries have been particularly severe for Bulgaria, due to its high trade exposure in those markets and especially in the former Soviet Union. Moreover, since Bulgaria was unable to service its large external debt, the inflow of external finance and hence imports of material inputs and energy were limited. In contrast, Romania was less dependent on eastern markets, had a very low initial foreign debt and has oil and natural gas resources which can meet a large part of its energy needs. Despite this, the output fall, which started in 1988, continued in 1992 for a fifth consecutive year. The sudden collapse of a highly centralized command and control system in these countries, coupled with large structural imbal-

---

series. It has also led to numerous related problems – such as the need to overhaul accounting procedures, to restructure the whole statistical collection system, to take account of changes in enterprise coverage, etc. Distortions are also appearing due to the large shifts in relative prices – an inevitable part of the transition process but one which cannot always be fully captured in the statistics. The relatively large size of the "shadow economy" in the European transition countries, which is by definition unrecorded, often leaves a worse impression of economic developments than is actually the case. The incidence of such shortcomings varies between countries and may reduce the value of intercountry comparisons.

[127] It should be noted, however, that in NMP terms, Polish output had contracted by some 24 per cent between 1978 and 1982 and that resumed growth in the 1980s had not made up for this loss. Hence in 1992 the level of output was some 19 per cent smaller than in 1978. In Czechoslovakia, NMP in 1988 was 24 per cent above the 1978 level; the contraction in 1990-1992 reduced it to 10 per cent below the 1978 level. The same comparison for Hungary shows GDP 17 per cent above and 5 per cent below 1978 in 1988 and 1992, respectively.

ances, had resulted in severe disruptions of domestic supply and trade patterns. Their damaging effects on output were aggravated by weak export performance and severely limited access to external finance. Moreover, various external shocks (including the trade embargoes on Iraq and the disintegration of and civil war in the former Yugoslavia) reinforced these adverse developments. However, as macroeconomic policies were tightened, the contraction of domestic demand has played an increasing role in the output decline in both of them.

– *Albania* is a special case and the economic crisis there clearly deteriorated further in 1992. No meaningful 1992 statistics are available for Albania, but the evidence suggests that the GDP contraction (10 per cent in 1990 and 30 per cent in 1991) continued in 1992, albeit at a slower pace.[128] The collapse of domestic output has been so deep that no recovery is likely without large-scale external assistance. For this reason the G-24 countries have agreed to provide emergency aid to Albania to cover food needs until the 1993 harvest and to make available inputs for selected enterprises.

– Developments in the dissolved federation of *Yugoslavia* indicate a growing diversity between individual republics. While in 1991 the largest drop in output occurred in *Croatia*, due to the civil war, the 1992 industrial decline in this republic decelerated strongly – particularly in the last quarter of the year. The output contraction in *Slovenia* is the smallest of all republics of former Yugoslavia, and the GDP decline decelerated from 9 per cent in 1991 to 6½ per cent in 1992. On the other hand, a substantially accelerated output decline can be observed in the new *Federal Republic of Yugoslavia*, where the GDP fall in 1992 is estimated at some 27 per cent.

– In 1992, the GDP decline in the *Commonwealth of Independent States* as a whole accelerated to around 20 per cent (18½ per cent for NMP), with a slightly larger contraction in Russia. This was almost twice as high as in 1991, and the cumulative decline leaves total output some one third below its 1989 level – a bigger output loss than in the first group of east European countries but smaller than that in the south eastern countries of eastern Europe. In Belarus, Kazakhstan and Ukraine, the 1992 NMP and industrial output declines were significantly smaller than the CIS average. In Moldova, and to an even greater extent in Armenia, Tajikistan and Azerbaijan, armed civil or inter-republican conflict was associated with much bigger than average output losses. The output fall was also generally higher in the Asian countries other than Kazakhstan.

In Russia, the output decline in 1992 accelerated during the course of the year. This occurred despite the expansionary monetary and fiscal policies in the second half of the year which breached the policy stance that was initially maintained after the January 1992 price liberalization. In Belarus, Moldova and the Ukraine, in contrast, the decline showed some deceleration in the second half of the year. But this appears to have been partly due to seasonal factors and the evidence of the final months is not encouraging.

– In the *Baltic states* the GDP decline, estimated at some 40 per cent in 1992 for the three countries jointly, was much steeper than in most CIS countries. There was a marked acceleration in the drop during the course of the year, reflecting the fact that these three countries had to pay much higher prices for Russian fuel and other goods from the beginning of 1992 as well as suffering supply interruptions at various points during the year.

(iii) Factors affecting output developments

Numerous factors contributed to the large falls in output in the European transition economies. The leading role among them was played by the distorted structures inherited from the past, the implementation of radical economic reforms and the collapse of intra-CMEA trade in 1991. Weak supply side responses to the new economic environment and changed demand also played a role. In the former Yugoslavia and Soviet Union, the collapse of inter-republican trade was an important additional factor.

Short-term output changes in countries for which quarterly or monthly data are available indicate that the steepest falls in industrial output followed the introduction of radical economic reforms, but in many cases this also coincided with the adverse effects of external shocks. Poland experienced a very sharp drop in industrial output in 1990 under the impact of "shock therapy" stabilization policies. The substantially smaller decline of 1991 was influenced by the collapse of the CMEA market. In Czechoslovakia, the deep industrial slump in 1991 was caused mainly by radical economic reforms coinciding with the loss of traditional CMEA markets. In Bulgaria and Romania, the industrial slump was deeper and appears to be lasting longer; in both countries, supply constraints resulting from reduced capacity to import contributed significantly to the fall. Bulgaria, because of its high dependence on that market, was particularly affected by the demise of the CMEA. In the successor states of the former Soviet Union, the reform process did not start until 1992; its contractionary effect was exacerbated by the breaking of traditional inter-republican and interenterprise supply linkages. This was reflected in the acceleration of the industrial decline during 1992.

The simultaneous impact of several adverse developments caused a deeper recession than might have been expected from transition alone. Country-specific features also played an important role. Judging from the fact that the largest losses of output coincided with the introduction of radical economic reforms (the year

---

[128] The data on economic development in Albania are highly spotty. In 1991, gross industrial output is estimated to have declined by about 40 per cent. An unweighted average of physical output data for 34 industrial commodities indicate a decline of more than 60 per cent in the first half of 1992 compared with the same period of 1991. Preliminary indications suggest that output performance improved somewhat in the second half of 1992.

1990 in Poland, 1991 in other east European countries, and 1992 in the CIS countries and the Baltic states), it seems likely that tight stabilization policies which strongly compressed domestic demand were responsible for a large part of the output contraction.[129]

The rapid collapse of CMEA trade in 1991 exacerbated the impact of radical economic reforms both on the demand side (loss of traditional export markets) and the supply side (constraints arising from reduced imports of fuels and raw materials). Some estimates suggest that about half of the 1991 output fall was due to the collapse of the CMEA.[130]

The economic downturn in the transition economies was, however, not just due to stabilization policies and the dismantling of the CMEA. Radical transformation and a rapidly changing economic environment made a part of the capital stock obsolete. General uncertainty, falling investment, slow progress in privatization, frequent changes in legislation, lack of the institutions necessary for a market system to function, the continuance of plan-oriented management and the monopolistic position of many enterprises weakened the stimulus for adaptive supply-side responses of the enterprise sector to the new environment. In 1992, a severe drought reduced harvests and agricultural production, which also contributed to the overall output decline.

Among the factors which determined the size of output losses in the CIS area was probably the differing degree of commitment to reform in the various republics.[131] The continuation of direct controls (state orders, subsidies, price controls) in certain of them, notably in Belarus and Ukraine, appears to have succeeded in maintaining output for a time, but these palliatives are reaching their limits as the resulting budgetary deficits rise to unsustainable levels.

Supply side factors exerted strong downward pressure on output in 1991 in the CIS countries and to these were added very strong constraints on demand through largely uncompensated price rises and falling government expenditures in 1992. Output has been hampered above all by the breakdown of the traditional centrally administered supply system. Most of output overall, and particularly of producer goods, was centrally distributed in the former Soviet Union. In contrast with Hungary, where the central supply organization had been abolished after 1968, Poland, where it disappeared in the early 1980s, and Czechoslovakia, where the supply system was also gradually decentralized over the last decade, the Soviet State Procurement Committee *(Gossnab)* preserved its functions until the break-up of the Soviet state. Only a few large firms − notably in the automobile industry and some defence and space establishments − had the right to conclude contracts directly with suppliers. As in Albania, Bulgaria and Romania, therefore, the vast majority of former Soviet firms thus had neither sales nor buying departments nor any of the legal and financial infrastructure necessary to ensure the purchase of materials or the disposal of products. The problems of establishing nationwide links with other enterprises, and lack of knowledge of the specification and cost of possibly more competitive goods which might be available from alternative suppliers, have been impossible to resolve in the relatively short time since the Soviet *Gossnab* was abolished.[132]

These already severe problems were aggravated by various trade restrictions imposed in 1992 by the new independent states. The break-up of the Soviet banking system resulted in delays in settling accounts between enterprises in different CIS countries (see section 3.2.4 below). The potential impact of these problems is indicated by the fact that in 1990 the ratio of interrepublican shipments to NMP for the Soviet Union as a whole was 26 per cent.[133] Since that year, trade between the Soviet successor states has fallen substantially. Following an estimated decline of more than 25-35 per cent in 1991,[134] the fall in real terms between 1991 and 1992 is estimated at around a further 25 per cent.[135] Delivery shortfalls in 1992 were apparently particularly big with regard to fuel, other raw materials and industrial inputs. Deliveries of coal, crude petroleum, petrol and fuel oil fell 20-50 per cent below 1991 levels. Deliveries of rolled metal, timber, cement, tractors, cars and lorries were cut by 15-60 per cent. In some cases this freed supplies for export to non-CIS countries; this seems to have been the factor which allowed Russia to raise its export volumes of crude oil and, though only marginally, of natural gas to non-CIS customers.

In addition to the effects of falling intra-CIS trade, the effects of lower imports from countries outside the former Soviet Union were also substantial. In 1990, the imports of the former Soviet Union amounted to nearly 16 per cent of NMP. By 1992, the Soviet successor

---

[129] One study indicates that in Czechoslovak industry the main factor in the decline was the rapid contraction of domestic demand. See R. Vintrova, "Macroeconomic Analysis of Transformation in the CSFR", *Forschungsberichte*, Vienna Institute for Comparative Economic Studies, No.188, January 1993.

[130] For example, see estimates for Czechoslovakia presented by E. Sujan in *Ekonom*, No.4, 1992; see also the interview with V. Dlouhy, *Ekonom*, No.8, 1992.

[131] One problem in cross-country comparisons of output performance may be that the pace of reform of statistical reporting systems, and in particular of price indices for calculating volume changes, has also been uneven.

[132] State procurement organizations have not been fully reconstituted in any of the Soviet successor states. But even in Russia, where decentralization has advanced the farthest, public bodies handling sales for particular industries and some state procurement agencies (for the purchase of goods for state requirements and of agricultural products) still exist. Indeed, *Gossnab SSSR* was partially replaced in Russia by the Ministry for Trade and Material Resources, designed to facilitate the supply of inputs for the production of the state's own requirement of goods. This ministry was finally abolished on 1 October 1992. See *Ekonomika i zhizn'*, No.40, October 1992, p.16.

[133] This ratio was much lower for Russia (15 per cent of NMP), but twice that high in Ukraine, Belarus and Moldova and over half of NMP in most other republics.

[134] United Nations Economic Commission for Europe, *Economic Bulletin for Europe*, Vol.44(1992), New York, 1993, p.85.

[135] Direct communication to the ECE secretariat by Russian Federation Goskomstat. See also section 3.1(vi) below.

states' imports from outside the former Soviet Union were one half of their 1990 US dollar value; of these, imports from eastern Europe, which had accounted for nearly one third of the above total in 1990, had fallen by over two thirds since that year (table 3.3.1 below). The attempts of former CMEA countries in eastern Europe to switch their exports away from the Soviet successor states to hard-currency markets could not be compensated by a switch of the Soviet successor states' to imports of production inputs from the industrialized market economies. Reduced purchases of goods for further processing and engineering goods in particular appear to have been mainly responsible for the overall fall in CIS imports from the west in 1992.

Most of the successor states — Armenia, Belarus, Moldova, Georgia, the Baltic republics and the economically backward states of Central Asia, where imports from abroad and from other former Soviet republics combined were equivalent to between 63-77 per cent of NMP in 1990 — were particularly vulnerable to disruptions in trade flows. This factor should have affected economic growth less in Azerbaijan, Kazakhstan and Ukraine where import dependency was smaller in 1990 (46-55 per cent of NMP) and notably in Russia where it was lowest of all (32 per cent). One study estimated that a 10 per cent fall only in intra-CIS trade would reduce NMP by between 3-4 per cent in Russia and Ukraine and 9-10 per cent in Belarus, Tajikistan, Turkmenistan, Uzbekistan and the Baltic states.[136] The actual decline, as noted above, was considerably bigger than this.

Ranking output performance (NMP change) for the 13 former Soviet republics for which full-year 1992 data are available against overall import dependency ratios (i.e., trade with other former Soviet republics and trade with countries outside the former Soviet Union) in 1990 indicates that performance in 1992 was much better than might have been expected in Belarus and Uzbekistan. It was considerably worse in the countries engaged in armed conflict (Armenia, Azerbaijan and Tajikistan), and also in Russia. Although differences in the reliability of output data, as well as other elements such as the appropriateness of economic policies and the extent of armed conflict, may have reduced the value of such comparisons, the rapid re-establishment of intra-CIS trade links would appear to be a priority task in removing supply-side constraints on output in most if not all the Soviet successor states. This task was in fact heavily stressed at a meeting of the Russian cabinet in February 1993.[137]

On the *demand* side, inflation and investor uncertainties affected a wide range of domestic demand categories in the CIS countries. Falling real incomes constrained the level of private consumer expenditures. Changes in price relatives following price liberalization and new criteria of choice offered by marketization have all disrupted long-standing expenditure patterns of private consumers. In 1992, falling real incomes limited demand for all but basic consumer necessities. The fall appears to have been less a result of deliberate policy than a by-product of the pricing behaviour of the large number of monopolistic producers and the lack of a hard budget constraint in the former Soviet Union. Changes in public consumption patterns have also contributed to declining output; the decline of defence-related government expenditures has weakened the demand for heavy industrial production and this was reinforced by very big, and again largely unanticipated, declines in the level of investment. In addition, the continuing slump in the demand of former CMEA member countries of eastern Europe for CIS goods other than fuel was not offset by any rise in CIS trade with the market economies.

(iv) Sectoral developments

Industrial production has borne the brunt of the recession in eastern Europe, with contractions of gross output substantially larger than those of GDP/NMP since 1989 (see tables 3.1.1 and 3.1.2). Over the period 1989-1992 the fall of industrial output in eastern Europe as a whole was over 40 per cent. In several countries industrial output levels in 1992 are below those of 1975. There has been considerable intercountry variation in industrial performance. In the fast-reforming states, the first signs of stabilization or of an upturn are now appearing. In Poland, industrial production began to grow in the second quarter of 1992 and the full-year outcome was a rise of some 4 per cent. In Czechoslovakia and Hungary, the industrial decline decelerated and signs of stabilization in the level of production have appeared. However, for the whole 1992 the fall was around 10 per cent in both countries. None the less, it appears that the industrial slump may be bottoming out in those two countries also and a weak recovery could come about in 1993. In Bulgaria and Romania, industrial output continued to fall in 1992 at approximately the same rate as in 1991 (over 20 per cent) and there was no sign that the decline had begun to flatten out. In 1993 the contraction of industrial output will probably continue, although at a decelerating rate. The successor republics of former Yugoslavia are a special case, since in most of them industrial performance was strongly influenced by the consequences of civil war, the disintegration of the country and the consequent loss of traditional markets. Industrial output had contracted strongly in 1991 (by 19 per cent for the entire area of the old Yugoslav federation, with falls ranging from 12 per cent in Slovenia to 28 per cent in Croatia). In 1992, the sharpest fall probably occurred in Bosnia-Herzegovina, where the economic system collapsed in consequence of the war but for which no statistics are available, and in the new Federal Republic of Yugoslavia where a 23 per cent industrial contraction was registered — partly due to the effect of the trade sanctions imposed by the United Nations in May 1992. A further substantial fall is likely in 1993. In Croatia and Slovenia, the contraction de-

---

[136] See United Nations Economic Commission for Europe, *Economic Bulletin for Europe*, Vol.44(1992), New York, 1993, p.85.

[137] See *Izvestiya*, 19 February 1993.

TABLE 3.1.2
GDP by major sectors, 1989-1992
(Percentage shares)

	Agriculture	Industry	Construction	Services	Other and residual	Total
Albania						
1989	32.3	44.8	..	22.9	..	100.0
1990	35.9	41.8	..	22.3	..	100.0
1991	39.6	38.2	..	22.2	..	100.0
Bulgaria						
1989	10.9	59.4	..	29.7	..	100.0
1990	17.7	51.3	..	31.0	..	100.0
1991	12.9	50.0	..	37.1	..	100.0
Czechoslovakia						
1989	9.9	43.2	8.9	37.7	0.3	100.0
1990	8.4	41.9	9.5	39.8	0.4	100.0
1991	..	..				..
Hungary						
1989	13.7	31.1	7.4	34.4	13.3	100.0
1990	12.5	26.7	6.0	38.5	16.2	100.0
1991	10.0	26.8	5.3	43.8	14.1	100.0
Poland						
1989	12.9	44.1	8.2	33.7	1.1	100.0
1990	8.3	46.1	8.7	35.6	1.3	100.0
1991	6.9	40.2	10.2	40.7	2.0	100.0
Romania						
1989	13.9	52.9	6.3	26.9	..	100.0
1990	18.0	48.3	5.7	28.1	..	100.0
1991	18.5	43.6	5.0	32.9	..	100.0
Slovenia						
1989	4.4	39.8	4.6	46.0	5.3	100.0
1990	4.7	33.4	4.3	57.3	0.3	100.0
1991	5.5	38.5	3.7	56.1	-3.8	100.0
Armenia						
1989	14.4	50.2	21.2	14.2	..	100.0
1990	17.3	45.4	25.4	11.9	..	100.0
1991	25.7	48.3	14.6	11.4	..	100.0
Russia						
1989	..	..	..	..	..	..
1990	..	..	..	..	..	..
1992	10.9	37.5	11.0	16.5	24.1	100.0
Ex-GDR Länder						
1989	..	..	..	..	..	..
1990	..	..	..	..	..	..
1991	1.3	37.5	..	42.7	18.4	100.0

Sources: Annual statistical yearbooks and direct communications from governments. World Bank, *Trends in Developing Economies 1992*, Washington, D.C., 1992, p.462.

Notes: Sectoral shares were calculated on the basis of current prices. Services include producer services (trade, transport and communications) and consumer services. However, there all some differences in coverage of sectors between countries. Thus, construction is included in industry in Bulgaria and in the ex-GDR.

celerated throughout the year, but output still fell by 13-15 per cent in 1992; the fall is not expected to bottom out until the second half of 1993.

No systematic analysis of *structural change* in the transition countries' industrial sector is yet possible. Few comparable data on output in 1992 by individual industrial branch for eastern Europe, and none at all for the states of the former Soviet Union, are available. Moreover, not only is the time scale since the introduction of market mechanisms still too short, but the changes so far have taken place under conditions of sharply falling output and various shocks of both internal and external origin.

Production changes in the east European *energy* sector were generally smaller than those for industry as a whole; paradoxically, this was due mainly to lack of progress in reducing the energy intensity of output; the energy intensity of NMP in fact increased in most east European countries. Production was also stimulated by the substitution of domestic energy sources for more expensive and often unavailable imports — particularly from the former Soviet Union. *Metallurgy* was particularly affected by depressed domestic demand (especially the sharp falls in investment and the consequent contraction in engineering production) and lost CMEA and CIS export markets; on average, the contraction was greater than that of total industrial output, steel output having virtually halved between 1989 and 1992. The fall was most severe in Romania (a 60 per cent drop). Czechoslovakia registered a smaller contraction than the other east European countries due to increased exports to western markets which partially mitigated other contractionary factors. *Engineering,* which had been the most dynamic industrial branch in the past, suffered most in the majority of European transition countries, again reflecting the collapse of the former CMEA markets (most of the east European countries were exporters of engineering products, particularly to the former Soviet Union) and a more than proportional fall in investment demand. Difficulties in the conversion of the defence industry added to a critical situation. However, on the positive side, production of private motor cars recovered in Czechoslovakia and Poland. The *light and food-processing* industries were depressed strongly by the severe compression of domestic demand in the initial stage of the reform process, though in some east European countries output in these branches has begun to recover partially as the squeeze eased. This was the case in Poland, mainly in 1991, and in Czechoslovakia in 1992. Generally good performance was recorded in the wood, paper and printing industries.

In contrast with eastern Europe, the gross industrial output decline of more than 18 per cent in the CIS countries in 1992 was smaller than the fall in NMP (table 3.1.1). In Belarus and Ukraine, the industrial output decline was less than half that recorded in Russia; much steeper falls were registered in some of the Central Asian CIS republics, and in Armenia over half of output was lost. Apart from Armenia, the Baltic states suffered the sharpest industrial output declines, ranging from 35 per cent in Latvia to 51 per cent in Lithuania.

Drops in Russian production of oil of 14 per cent and in coal and electricity of 5 per cent were not large in relation to the overall production decline, and natural gas output was maintained close to year-earlier levels. This suggests that in Russia at least, as in eastern Europe, the energy intensity of the economy has not yet begun to fall. Moreover, energy shortfalls were a prime constraint on output from the supply side in other CIS countries, which were affected by falling output in Russia and most other CIS fuel producers. However, Turkmenistan actually increased its oil and gas output (by 16 and 2 per cent). Overall performance data are unavailable for the *engineering* branch, but in Russia the investment slump in the region has contributed to falls of 20-50 per cent on half of engineering product lines. Output falls in the production of tractors and other farm machinery, machine tools, forging, heavy metal processing equipment and, in Russia, large diameter pipes and, especially, electricity transmission gear were well above the average. In contrast, the decline in passenger vehicle output was rather small. Russian data also indicate that declining investment and construction activity was accompanied by big falls in the output of construction machinery and building materials; bricks, cement and roofing materials fell by 8-10 per cent and other materials by as much as 20-40 per cent, the production of this industry as a whole having contracted to the level of the early 1970s. *Ferrous metals* also recorded higher than average rates of fall. In Russia, these branches were badly affected by the unavailability of raw materials and other inputs from Ukraine. In the *chemicals* branch, output losses appear to have been below the all-industry average in the case of resins, fibres and some basic chemicals but were well above it for fertilisers. With regard to *consumer goods,* the output decline in the CIS as a whole, Moldova, Belarus, Russia, Turkmenistan and Uzbekistan, was less than for total industry; in Ukraine, the fall was in line with the overall industrial contraction. Output of industrial consumer goods other than food generally appear to have fallen by somewhat less than production overall — Armenia is the only exception. Even so, in the CIS as a whole and in Russia, production of textiles, clothing and footwear declined by up to one third. Food industry output fell by less than industrial production in Russia, Moldova and Armenia only (though its absolute decline was still very large in the latter).

Construction activity has continued to decline in most east European countries, although it has generally fallen less than industrial output. The first signs of recovery in this sector have appeared in Czechoslovakia, where construction output rose substantially in 1992. In Hungary, construction stagnated, and in Poland it declined by 3 per cent in that year. However, the situation in housing construction remains critical and most countries reported further declines in the number of dwellings constructed in 1992, after a sharp decline in housing construction in 1991. It should be noted that private enterprises have already gained a strong position in the construction sector in most east European countries. Few data are available on developments in the CIS countries, but the investment slump has severely curtailed construction activity. Housing construction was 20-50 per cent lower than in 1991 in Belarus, Moldova and Russia, but it rose in Ukraine.

In *agriculture*, developments in eastern Europe have been adverse in 1992 (table 3.1.3). Sharp shifts in the domestic terms of trade (slowly growing prices for farm products and rapidly growing prices of farm inputs), disruption caused by privatization (involving the break-up of cooperatives and often disputes about the ownership of land), as well as the depressed state of domestic demand have all affected output. A severe drought in 1992 reduced harvests; grain production declined by about one third in eastern Europe as a whole — in Czechoslovakia, Hungary, Poland and Romania by 14, 37, 28 and 36 per cent respectively. Gross agricultural output appears to have declined by 9-13 per cent in most east European countries. In Hungary and the Federal Republic of Yugoslavia the fall amounted to more than 20 per cent. In Albania, where output dropped by 24 per cent in 1991, food shortages have become widespread and a further fall is likely to have occurred in 1992 in the wake of spontaneous land distribution, acute shortages of inputs and the breakdown of the distribution, marketing and transport systems.

Most east European countries increased their imports of agricultural products and foodstuffs in 1992. The poor harvest further aggravated the unfavourable financial situation of many farmers and increased pressures on governments to provide direct support to agriculture. The Hungarian and Polish governments have already introduced a system of preferential credits, price supports and increased protection from imports for domestic agriculture.

Agricultural output in the CIS area declined by 10 per cent in 1992 — in Russia by somewhat less but by

TABLE 3.1.3

European transition countries: Gross agricultural production, 1989-1992
(Annual percentage change)

	1989	1990	1991	1992
Albania	10.7	-6.9	-24	..
Bulgaria	1.2	-6.0	-6.4	-12.9
Czechoslovakia	1.7	-3.9	-8.4	-11.8
Czech Republic	2.3	-2.3	-8.9	-11.8
Slovak Republic	0.6	-7.2	-7.4	-11.9
Hungary	-1.3	-3.8	-5	-22.7
Poland	1.5	-2.2	-2.0	-11.9
Romania	-5.0	-2.9	1.2	-9.2
Yugoslavia (SFR)	5.0	-4.8	7.6	..
of this:				
Croatia	4.5	-3.2	..	..
Slovenia	0.3	1.7	-3.3	-13.0
Yugoslavia (FR)	5.2	-7.0	9.7	-22.0
Eastern Europe	0.8	-3.7	-0.9	-13
Soviet Union	1.3	-2.9	..	..
CIS	1.5	-2.6	-6.9	-10.0
Armenia	-18.7	-11.4	11.0	..
Azerbaijan	-8.9	-0.1	-	..
Belarus	8.9	-8.7	-4.9	-16.0
Kazakhstan	-7.3	6.8	-8.0	..
Kyrgyzstan	2.5	1.3	-8.0	..
Moldova	5.2	-12.8	-11.0	-18.0
Russia	1.7	-3.6	-4.5	-8.0
Tajikistan	-10.8	2.8	-10.0	..
Turkmenistan	0.3	7.0	-2.0	..
Ukraine	5.1	-3.7	-13.2	-11.0
Uzbekistan	-4.3	6.3	-5.0	..
Georgia	-5.5	6.9	..	..
Estonia	7.6	-13.1	-20.7	-21.4
Latvia	3.9	-10.2	-3.5	-12.0
Lithuania	1.8	-9.0	-4.6	-29.5

Sources: ECE secretariat Common Data Base, based on published national statistics and communication to the ECE secretariat; USSR Goskomstat, *Soyuznye respubliki. Osnovnye ekonomicheskie i sotsialnye pokazateli*, Moscow, 1991, p.150; CIS Statistical Committee, *Ekonomika Sodruzhestva nezavisimykh gosudarstv v 1992 godu*, Moscow, 29 February 1993, p.30.

as much as 16 per cent in Belarus[138] and 11 per cent in Ukraine. Grain output increased sharply in Russia and, surprisingly, in Belarus, but fell in Ukraine and was over a third down in Moldova due to a severe drought. Other crops were generally smaller than in 1991, reflecting shortages of machinery and especially fuel, which occurred partly, as in eastern Europe, because of an unfavourable shift in the agricultural terms of trade with industrial suppliers. Inadequate supplies of fodder and other inputs gave rise to relatively larger declines in livestock output and difficulties in wintering livestock under cover due to heating fuel shortages was a factor in the substantial declines in herd sizes. It should be recalled that, in the past, agricultural production under the inherently inefficient and low-productivity conditions of central planning was expanded with the assistance of huge fiscal transfers from other sectors to levels which could not possibly be profitably sustained under market pricing. Falling real incomes have reduced demand for food to the minimum required to satisfy basic nutritional needs and it is unlikely that production will begin to recover until a new balance is established between production costs and real demand.

Information on developments in the *service sector* is limited but it is likely that services in eastern Europe have proved more resilient to the adverse economic climate than have the goods-producing sectors.[139] As a result the share of the service sector in eastern European GDP has increased in most countries for which information is available (table 3.1.2). However, there are large differences between countries. In Albania the share of services is still low. Substantial increases in the share of services in GDP were recorded for Bulgaria and Hungary. In Slovenia, their share in 1990 already exceeded 50 per cent; a comparison of the GDP decline in 1992 (6.5 per cent) and that of goods-producing sectors (more than 10 per cent) suggests that the share of services increased again in that year. It should be remembered, however, that the share of services in GDP calculated on the basis of data at current prices is strongly influenced by the different price movements of services relative to those of goods. Generally, prices of service activities have been rising more slowly than those of goods. Moreover, due to the incomplete statistical recording of private activities, the share of services is likely to be underestimated. For 1992, data on employment changes indicate that in most of eastern Europe the share of services increased. Indeed, the only sectors to show any growth in employment were banking, finance and business services.

Data on the services sector in the CIS countries are very scanty. A comparison of NMP produced and GDP for Russia indicates that non-material services (i.e., excluding goods transport, communications and trade) provided some 22 per cent of GDP at current prices in 1991, and only 13 per cent in 1992. This probably reflects the fact that the services sector has found it more difficult to increase prices than producers of goods. Even so, for the CIS countries taken together, the *volume* of NMP produced in 1992 is also reported to have fallen less than GDP, which should indicate a more rapid contraction in the service sphere. However, in view of the uncertainties about the accuracy of national accounts estimates in these countries no strong conclusions can be drawn from this observation.

The limited information available on production changes in the *private sector* indicates a dynamic performance in eastern Europe — particularly in agriculture, construction, trade and other services. As a result, the share of private activities in domestic production increased substantially in 1992.[140] In Bulgaria and

---

[138] CIS data. A report in the Belarus press gives a decline of 10 per cent. See *Sovetskaya Belorussia*, 26 January 1992.

[139] Preliminary estimates of basic macroeconomic variables by the Czechoslovak Federal Statistical Office for the first three quarters of 1992 support this: NMP (excluding non-material services) was 14.7 per cent lower than in the same period of 1991, whereas the GDP fall over the same period was only 10.1 per cent. In Romania, GDP in 1992 fell by 15 per cent while the volume of services rendered to the population was only 2.3 per cent less than in 1991.

[140] For a detailed discussion of progress in privatization, see section 3.6 below.

Czechoslovakia, where the private sector had been negligible in 1989, its share of GDP amounted to about 5 and 8 per cent, respectively, in 1991. In 1992, it is reported to have risen to 10 per cent in Bulgaria and 20 per cent in Czechoslovakia. In Hungary, the corresponding share was about 18-27 per cent in 1991, and it is likely to be around 25-35 per cent in 1992. In Poland, the private sector generated 45-50 per cent of GDP in 1992, and a 26 per cent share has been estimated for Romania in the same year (see table 3.6.14).

The share of private producers in total output also rose in the CIS countries in 1992, though it remains far below those in much of eastern Europe. In Russia, it may have been approaching 10 per cent of *industrial* output by the end of the year,[141] though it was considerably bigger in agriculture (35 per cent) even in 1991.[142] The bulk of investment in the CIS countries in 1992 (85-90 per cent) was still made by state-owned enterprises (including public utilities) and other organizations (including collective farms), most of which was financed from their own resources and less than a quarter from state funds.[143] The increasing pace of privatization in 1993 should considerably raise the still generally small shares of the private sector in domestic economic activity.

(v) Domestic demand

Developments in *domestic absorption* in 1992 in the transition countries can be assessed only tentatively since national accounts statistics for most of them are unavailable. While in 1991 restrictive domestic stabilization policies combined with external adjustment contributed to a sharp fall in domestic demand in all *east European countries* — and in Albania, Bulgaria, Czechoslovakia and Romania, the restrictions on domestic demand were particularly strong — developments in 1992 differed significantly by country (table 3.1.4). In Czechoslovakia and Poland, preliminary estimates point to a slight recovery or stabilization of domestic demand, while in Hungary there was probably another fall. Steeper falls can be surmised for Bulgaria and Romania.

No systematic information is available on changes in aggregate domestic demand in 1992 for the republics of the *former Soviet Union*. Apart from late publication, analysis is hampered by the continuing lack of adequate national accounts data and apparent inconsistencies between the partial indicators which have been published. In Russia, for example, the 19 per cent GDP decline was reported to be accompanied by 40 and 45 per cent falls in real incomes and gross fixed investment respectively. If these are taken as approximations for private consumption and net investment, respectively, this would suggest a rise in the joint share of government consumption, net exports and stockbuilding in GDP. Given the 62 per cent fall in defence procurement[144] and other government expenditure cuts and the deterioration in the balance of trade with non-CIS countries as published by Russian Federation Goskomstat in 1992, only stockbuilding could account for the difference. In 1991, inventory growth had absorbed 10 per cent of Russian GDP; this share would appear to have risen substantially in 1992. The big question mark here, however, is the true size of the Russian surplus on trade with other CIS countries which may, according to some reports, have taken up a considerable amount of the otherwise unexplained difference between output and apparent demand changes (see section 3.1(vi) below). Even so, one of the effects of the slow adjustment of monopolistic state enterprises faced with only soft budget constraints to falling demand and changes in its structure could be the piling up of large quantities of unsaleable goods.[145] This is another example of the "inertia" on the part of producing enterprises in the face of rapidly-changing conditions which has been noted recently by Russian researchers.[146] This could also explain the relatively smaller impact of the transition recession on gross industrial output in the CIS countries by comparison with experience in most east European countries; in the latter, as noted above, the decline in gross industrial output has been smaller than that of total value added (GDP).

Very sharp falls of *investment* occurred in eastern Europe in 1990-1991 when investment demand suffered from economic austerity policies and economic uncertainty (table 3.1.5). The decline did not begin until 1991 in the former Soviet Union but was substantial in that year. Much statistical information for 1992 is still lacking, but investment probably fell in most east European economies. It may have stabilized in Czechoslovakia and Poland, which would be another indication of an impending turnround. In Hungary, investment declined by more than GDP (by about 8 per cent) in 1992. Further large contractions appear to have occurred in Bulgaria and Romania. The steepest falls in 1992 occurred in the republics of the former Soviet Union, in part as a result of growing uncertainties, the unstable economic environment and the constraints imposed on entrepreneurs by still incomplete institutional reform. In Russia, the fall in gross fixed

---

[141] Based on a figure of 6.7 per cent for the beginning of October 1992. See Russian Federation Goskomstat, *O razvitii ekonomicheskikh reform v Rossiiskoi Federatsii v yanvare-oktyabre 1992 goda*, Moscow, 1992, p.138.

[142] Private subsidiary agriculture only (i.e., excluding newly-created independent private farms). See *Narodnoe Khozyaistvo Rossiiskoi Federatsii 1992*, Moscow, 1992, p.401.

[143] CIS Statistical Committee, *Ekonomika Sodruzhestva nezavisimykh gosudarstv v 1992 godu*, Moscow, 29 January 1993, p.49.

[144] Speech by Ye. Gaidar in London, *The Independent on Sunday*, 7 February 1992.

[145] The survey of business prospects conducted by the Russian Institute of Economic Policy in January 1993, based on a sample of 167 industrial enterprises, reported that 40 per cent of them were holding abnormally high stocks of finished goods for sale, compared with 32 per cent with normal and only 18 per cent with abnormally low stocks. 10 per cent of firms did not respond. *Ekonomika i zhizn'*, No.6, February 1993.

[146] Centre for Conjunctural Economics and Forecasting, *Rossiya - 1992*, Moscow, November 1992, p.3.

TABLE 3.1.4

European transition countries: Domestic production and absorption, [a] **1989-1992**

(Annual percentage change)

Country and period	NMP produced (1)	NMP used [a] (2)	Consumption			Accumulation			Memorandum items		
			Total (3)	Personal [b] (4)	Social [c] (5)	Total (6)	Net fixed capital formation (7)	Changes in stocks (8)	Retail trade turnover (9)	Real income per capita (10)	Gross investment (11)
Bulgaria											
1989	-0.3	2.0	2.9	2.8	3.4	-0.6	-11.2	6.2	0.8	-4.8	-10.1 [d]
1990	-11.5	-9.2	-4.9	-8.6	14.1	-22.6	-38.8	-14.4	-8.7	..	-18.5
1991	-25.7	..	-33.5	..	..	-45.9	..	..	-50.4	-45*	-48.6
1992	..	..	..	..	..	..	..	..	-32.2	..	..
Czechoslovakia											
1989	0.7	3.2	3.6	1.8	7.3	0.9	-22.7	79.7	2.3	2.0	1.6
1990	-1.5	3.4	2.1	3.0	0.3	10.8	-17.9	52.0	1.3	-2.8	6.1
1991	-19.2	-31.4	-21.9	-31.9	-1.6	-83.0	-99.1	-70.4	-39.2	-26.9	-27.2
1992 [e]	-7*	..	..	5.4	..	..	..	..	9.0*	5.0	-
Hungary [e]											
1989	0.2	0.4	0.7	1.9	-5.8	1.2	7.0	-25.5	-0.2	3.4	7.0
1990	-4.3	-5.3	-3.9	-4.5	-0.6	-4.2	-8.1	-14.3	-7.6	-1.8	-8.1
1991	-11.9	-9.1	-5.3	-5.8	-2.7	-22.3	-11.6	..	-9.9	-1.7	-11.6
1992	-(4-6)	-(7-8)	..	-(2-3)	-(3-4)	..	-8	..	-5.6	-1*	-8
Poland											
1989	-0.2	0.1	-1.7	0.8	-4.2	7.1	-7.4	36.7	-2.7	6.2	-2.4
1990	-14.9	-19.8	-13.0	-15.8	3.4	-45.0	-27.8	..	-17.4	-14.9	-10.1
1991 [e]	-7.6	-1.4	3.3	7.4	-6.5	-14.2	-4.5	..	3.7	..	-4.1
1992 [e]	0.5-2	..	..	..	..	..	..	..	..	..	-
Romania											
1989	-8.0	-5.0*	0.8	..	..	-23.8	-5.8	..	-1.3	2.4	-1.6
1990	-9.9	11.1*	12.0	..	..	7.2	-65.9	..	7.3	17.9	-38.3
1991 [e]	-13.7	..	..	-20.9	2.3	-0.3	..	..	-27.7	..	-28.8
1992 [e]	-15	..	..	..	..	..	..	..	-17.5	..	-18.9
Yugoslavia (SFR)											
1989	0.6	2.1	..	1.0	-1.8	..	0.5	7.2	..	25.3	-0.5
1990	-7.6	-3.6	..	2.5	1.1	..	-7.0	-15.3	..	-21.9	-7.0
1991	-15*	..	..	..	..	..	..	..	..	-7*	..
Croatia											
1989	-1.5	..	..	..	..	..	..	..	-5.6	..	-8.7
1990	..	..	..	..	..	..	..	..	-7.4	..	..
1991	..	..	..	..	..	..	..	..	-26.5	..	..
1992	..	..	..	..	..	..	..	..	-38.9	..	..
Slovenia											
1989	-0.5	..	..	..	..	..	..	..	-11.9	..	-10.1
1990	-8.3	..	..	..	..	..	..	..	-17.6	..	-8.7
1991	-12.7	..	..	..	..	..	..	..	-1.2	..	-18.0
1992 [e]	-6.5	..	..	..	..	..	..	..	-11.8	..	..
Yugoslavia (FR)											
1989	1.9	..	..	..	..	..	..	..	-14	-30.8	8.6
1990	-8.4	..	..	..	..	..	..	..	-	-21.7	-20.1
1991	-11.1	..	..	..	..	..	..	..	-2	-5.2	-12.8
1992	-27.0	..	..	..	..	..	..	..	-52	-50.6	..
Belarus											
1989	7.9	..	..	..	..	..	..	..	9.0	8.9	10.3
1990	-3.2	..	..	..	..	..	..	..	14.7	12.0	9.0
1991	-3.0	..	..	..	..	..	..	..	-0.5	-2.2	-8.0
1992	-11.0	..	..	..	..	..	..	..	-26.1	-38.3 [f]	-15.0
Russia											
1989	1.9	2.3	..	..	..	..	..	..	8.4	6.8	4.1
1990	-4.0	-4.2	..	..	..	..	..	..	10.0	9.0	0.1
1991	-11.0	-10.2	..	..	..	..	..	..	-7.2	-10.5	-15.5
1992	-20.0	..	..	..	..	..	..	..	-39.1	-46.5 [f]	-45.0
Ukraine											
1989	5.0	6.1	..	..	..	..	..	..	7.4	8.5	3.7
1990	-3.6	-2.7	..	..	..	..	..	..	11.5	11.0	1.9
1991	-11.2	-10.8	..	..	..	..	..	..	-9.7	-3.7	-5.3
1992	-15.0	..	..	..	..	..	..	..	-22.8	-30.0 [f]	-40.0

Sources: ECE secretariat Common Data Base, based on national statistical publications or direct communications to the ECE secretariat.

[a] "Net material product produced" and "net material product used for consumption and accumulation" (material balance system), unless otherwise specified.
[b] Volume of consumer goods and material services supplied to the population.
[c] Consumption of material goods in institutions providing amenities and social welfare services.
[d] At current prices.
[e] Gross domestic product. Components of final demand correspond to GDP definition and their coverage thus differs from the components of NMP used.
[f] January-November.

TABLE 3.1.5

European transition countries: Gross fixed investment, construction 1989-1993
(Percentage change over same period of preceding year)

	Gross fixed investment [a]					Construction gross output							
									1992				
	1989	1990	1991	1992	Forecast 1993	1989	1990	1991	Jan.-March	Jan.-June	Jan.-Sept.	Jan.-Dec.	Forecast 1993
Albania	10.9	-14.8	..	..	..	7.1	-13.9	..	..	..	..	..	..
Bulgaria [b]	..	-18.5	-48.6	..	..	1.5	-19.1	-68.6	..	-11.8 [c]	-13.1 [c]	-18.4 [c]	..
Czechoslovakia	1.6	6.1	-27.2	..	..	1.6	-5.3	-34.7 [d]	-1.6	4.3	7.9	..	..
Czech Republic	1.8	6.5	-26.8	..	..	2.4	-2.6	-35.6 [d]	-10.9 [e]	-8.1 [e]	-0.9 [e]	4.8 [e]	..
Slovak Republic	1.3	5.3	-28.1	7.4	..	0.5	-10.2	-33.1 [d]	-4.2 [e]	8.3 [e]	14.2 [e]	7.1 [e]	..
Hungary	7.0	-8.1	-11.6	-8	2-6	-1.1	-15.6	-9.4	-0.8	-3.1	-9.0	-0.2	..
Poland	-2.4	-10.1	-4.1	-	..	-3.7	-10.7	-0.3	..	..	..	-3.0	..
Romania [b]	-1.6	-38.3	-28.8	-18.9	..	-4.3	-36.6	-33.0	..	..	..	..	..
Yugoslavia (SFR)	-0.5	-7.0	..	..	..	..	..	..	..	..	..	..	..
Croatia	-8.7	..	..	..	..	..	..	..	..	..	..	..	..
Slovenia	-10.1	-8.7	-18.0	..	-15.0	-7.0	-11.3	-16.6	-6.4	-10.2	-10.6	-10.7	..
Yugoslavia (FR)	8.6	-20.1	-12.8	..	..	15	-18	-4	..	..	..	-14	..
Eastern Europe	-1.5	-13.7	..	..	..	..	..	..	..	..	..	..	..
Soviet Union	4.7	0.6	-12	..	..	..	..	..	..	..	..	..	..
CIS	4.8	1.0	-11.7	-45	..	..	..	..	..	..	..	..	..
Armenia	149.1	-4.6	-35.2	..	..	..	..	..	..	..	..	..	..
Azerbaijan	-14.4	-3.6	-14.3	..	..	..	..	..	..	..	..	..	..
Belarus	10.3	9	-8	-15	..	..	..	..	..	..	..	..	..
Kazakhstan	3.5	-2.9	4.8	-(40-45)*	..	..	..	..	..	..	..	..	..
Kyrgyzstan	1.7	11.3	-12.8	..	..	..	..	..	..	..	..	..	..
Moldova	5.9	-0.5	-17.9	..	..	..	..	..	..	..	..	..	..
Russia	4.1	0.1	-15.5	-45	-50 *	..	..	..	..	..	..	..	..
Tajikistan	6.8	0.7	-14.6	..	..	..	..	..	..	..	..	..	..
Turkmenistan	2.3	7.5	11.2	..	..	..	..	..	..	..	..	..	..
Ukraine	3.7	1.9	-5.3	-40	..	..	..	..	..	..	..	..	..
Uzbekistan	-0.5	13.0	4.6	..	..	..	..	..	..	..	..	..	..
Georgia	-1.2	-14.4	..	..	..	..	..	..	..	..	..	..	..
Estonia	6.2	2.7	-12.0	..	..	..	..	..	..	..	..	..	..
Latvia	4.2	-8.2	-36.3	-53	..	..	..	..	..	..	..	-57	..
Lithuania	-1.8	-10.3	-46.0	-34.3	..	..	-15.0	-20.7	..	..	..	-28.1	..
Ex-GDR *Länder* [b]	..	6.2	17.9	..	..	..	..	..	..	..	..	..	..

Sources: National statistical yearbooks and current reporting.

a Investment in state enterprises and organizations.
b Gross fixed investment (GDP concept).
c Sales of construction enterprises.
d Enterprises with 100 and more employees.
e Enterprises with 25 and more employees.

investment volume was 45 per cent; similar falls were registered in Ukraine and Kazakhstan, but in Belarus it was only about 15 per cent. The share of gross fixed investment in Russian GDP fell by about one third in 1992, to 14 per cent. The efficiency of the investment process also suffered, as indicated by rising gestation periods. Thus, the value of unfinished investment projects in Russia at the end of 1992 was almost half as large again as total investment expenditures during the year, compared with over 90 per cent in the last years of the Soviet Union.

The share of investment in GDP has in fact fallen substantially in recent years in all the transition economies. This is a worrying development since they will all need large amounts of investment to modernize and adapt the capital stock to the new economic situation and to provide the technological impetus for economic revival. These issues, which are indeed the central challenge of transition, have not yet been properly addressed. While domestic savings and resources will have to play the main role in financing investment, foreign direct investment is an essential component of structural reform. However, except for Hungary and Czechoslovakia, it has not yet played an important role in accelerating structural change.

Real private consumption declined in most of the transition countries in 1992, but precise quantifications are generally scarce. Rough indications of the direction and relative magnitude of change can be obtained from estimates of changes in real per capita incomes and the volume of retail sales (see tables 3.1.4 and 3.1.6). However, it should be kept in mind that the data on retail sales volume have started to deviate in some countries significantly from changes in real private consumption – probably due, as noted earlier, to the fact that the growing role of the private sector and the shadow economy is not yet fully captured in official statistics.

TABLE 3.1.6

European transition countries: Volume of retail sales, 1989-1992
(Percentage change over same period of preceding year)

Country or group	1989	1990	1991	1992 Jan.-March	1992 Jan.-June	1992 Jan.-Sept.	1992 Jan.-Dec.
Bulgaria [a]	0.8	-8.7	-50.4	-30.3	-33.6	-28.3	-32.2
Czechoslovakia	2.3	1.3	-39.2	2.2	13.0	9.6	..
Czech Republic	2.1	1.9	-39.4	6.9	20.5	9.9	..
Slovak Republic	2.5	0.6	-39.2	-7.4	-2.4	9.0	15.0
Hungary	-0.2	-7.6	-9.9	-16.2	-12.0	-5.8	-5.6
Poland	-2.7	-17.4	3.7	..	..	..	..
Romania	-1.3	7.3	-27.7	-12.9	-10.5	-9.3	-17.5
Croatia	-5.6	-7.4	-26.5	-48.0	-47.2	-43.2	-38.9
Slovenia	-11.9	-17.6	-1.2	-20.4*	-19.9*	-15.6*	-11.8
Yugoslavia (FR)	-14	-	-2	-25*	-33*	-43*	-52
CIS [a]	8.4	10.5	-7.1	-53	-31	-37	-36.7
Armenia	17.4	6.3	-25.8	-77	-74	-74	-72.7
Azerbaijan	8.9	7.3	-11.8	-72	-67	-61	-63.0
Belarus	9.0	14.7	-0.5	-44	-32	-27	-26.1
Kazakhstan	8.3	8.5	-12.0	-52	-43	-41	-38.5
Kyrgyzstan	8.9	9.0	-15.8	-63	-63	-65	-67.0
Moldova	10.4	14.2	-15.0	-50	-60	-54	-49.4
Russia	8.4	10.0	-7.2	-51	-42	-39	-39.1
Tajikistan	7.0	9.8	-21.5	-66	-55	-70	-72.1
Turkmenistan	7.6	8.7	-11.6	-60	-34	-38	-38.0
Ukraine	7.4	11.5	-9.7	-52	-30	-23	-22.8
Uzbekistan	9.0	8.3	-8.0	-62	-46	-37	-31.3
Georgia	7.0	11.5	..	..	..	..	..
Estonia [a]	6.3	6.8	..	..	..	..	..
Latvia [a]	8.1	8.3	-35	..	..	..	-51.0
Lithuania [a]	10.2	3.8	-37.3	..	..	..	-52.3

Source: ECE secretariat Common Data Base, based on published national statistics and communication to the ECE secretariat.

[a] State and cooperative sectors only.

Developments in 1992 show large differences in consumption growth between the east European countries. In Bulgaria, where *total* consumption had fallen by one third in real terms in 1991, a further but much smaller fall appears to have occurred in 1992, as suggested by a 9 per cent decline in the volume of retail sales including the new private sector.[147] In Czechoslovakia, private consumption is estimated to have risen by about 5 per cent after a sharp fall in 1991. Retail sales recovered strongly from the second quarter of the year. In Hungary, falls in real private consumption of about 5 per cent annually in 1990 and 1991 appear to have been followed by a further but smaller contraction in 1992. Poland had experienced its biggest drop in private consumption in 1990 and there was already some recovery in 1991; for 1992, the Polish statistical office noted an increase in private and a fall in public consumption without giving precise figures.[148] After a 21 per cent decline in 1991, Romanian real private consumption contracted further in 1992, to judge from the 18 per cent fall in the volume of retail sales (which accelerated especially in the fourth quarter). In Croatia, the full-year contraction must also have been substantial, with a 40 per cent fall in the volume of retail sales. An even more drastic deterioration in the domestic market situation can be observed in the new Federal Republic of Yugoslavia, where the retail sales decline accelerated during the course of 1992 and resulted in a full-year fall of 52 per cent. In contrast, Slovenia reported a 12 per cent contraction in retail sales volume for the year as a whole, reflecting a decelerating fall as the year progressed. In the countries which experienced a sharp fall in real incomes and consumption, the pattern of consumption has shifted to basic necessities and the share of foodstuffs in households' expenditures increased.

A very big contraction in private consumption occurred in the CIS countries in 1992, on the scale of — and possibly larger than — that recorded in the east European countries in the first transition year (1990 or 1991). Belarus, Russia and Moldova all reported steep falls in real incomes, and the same probably holds also for most other CIS countries. These falls largely resulted from the impact of inflation, which by far outpaced the rise of nominal incomes; unemployment rose only insignificantly (but the income lag may in part reflect wage earnings not actually paid because of enterprise illiquidity). In Russia, total money incomes increased 7.5 times and the consumer price index 12 times in 1992. The resulting real income decline of nearly 40 per cent was considerably bigger than in

---

[147] A much steeper fall was registered in state and cooperative enterprises (table 3.1.6).

[148] *Rzeczpospolita*, 6 February 1993.

TABLE 3.1.7

European transition countries: Volume of foreign trade, 1989-1992
(Annual percentage change)

	Exports				Imports			
	1989	1990	1991	1992 a	1989	1990	1991	1992 a
Albania	..	..	..	..	..	..	..	..
Bulgaria	-2.3	-23.3	-30.1	1.6 +	-4.6	-23.3	-17.0	27.5 +
Czechoslovakia	-2.0	-4.2	-4.9	4.7	2.7	9.7	-38.6	15.7
Hungary	0.3	-4.3	-4.9	1-2	1.1	-3.4	5.5	-8
Poland	0.2	13.7	-2.4	9.7 +	1.5	-17.9	37.8	6.1 +
Romania	-10.8*	-46.0*	-7.0*	0.9 +	3.7*	4.0*	-17.0*	-1.4 +
Yugoslavia (SFR)	4.8	2.2	-	..	13.1	21.9	-	..
Croatia	..	..	..	-5.0 +	..	..	..	-10.4 +
Slovenia	..	..	..	8.0 +	..	..	..	0.1 +
Yugoslavia (FR)	..	..	..	-46.0 +	..	..	..	-30.4 +
Eastern Europe b	-2.8	-10.8	-8.8	..	0.7	-5.6	-12.0	..
Soviet Union CIS c	-	-13.1	1.0*	..	9.3	-1.4	-7.9*	..
Russia d	..	..	..	-26	..	..	..	-22
Russia e	..	..	..	-	..	..	..	-6 +

Source: ECE secretariat Common Data Base.

a Change in dollar *value* where *volume* information is not available (indicated by " + "). The Polish figures for 1992 refer to payments flows in convertible currency trade.
b Excluding Yugoslavia (SFR).
c Former Soviet Union for 1989-1991; CIS states for 1992.
d Russian Federation Goskomstat data.
e Russian Federation Ministry of Foreign Economic Relations data.

Belarus, where the fall was about a quarter; eleven month data for Ukraine suggest a rather smaller decline still. Quarterly data suggest that, after the very large first-quarter effects of the January 1992 price liberalization on real incomes, the depressing impact of falling real incomes on private consumption diminished only slightly during the year. The decline in the volume of retail sales, which averaged over one third for the CIS as a whole in 1992 compared with 1991, was broadly in line with real income movements. The slump in retail sales was somewhat smaller in Belarus and Ukraine, and much larger in Moldova, Armenia and some Central Asian CIS countries where retail sales fell by one half or more (table 3.1.6). Quarterly retail trade movements again show a pattern of only slightly decelerating decline in the course of the year. The fall in retail sales volume was considerably bigger than in even the most recession-affected countries of eastern Europe and suggests a sizable drop in real private consumption.

As in eastern Europe, the declines in private consumption indicated above may have been partly offset by a rise in activity of new and unrecorded private sector activities. Even so, the official report on Russia in 1992 notes that by the end of the year, 29 per cent of the Russian population was living below the minimum subsistence income, as compared with 12 per cent at the end of 1991. In other CIS countries the pattern seems to have been similar.

Altogether, the social costs of transition in terms of declining real incomes and consumption have been high and have reduced the support for radical reforms in an important part of the population. In most east European transition countries the cumulative loss of private consumption since 1989 was about 30 per cent, and considerably more than that in the CIS countries. Moreover, public services — health, culture and education — were also hit hard by restrictive fiscal policies. Accelerated income differentiation and the growing portion of population living below the poverty line has led to mounting social tensions in many countries of the region.

(vi) Foreign sector impact

Assessment of the foreign sector impact on economic developments in eastern Europe is hindered because information on volume changes in trade as well as on domestic absorption is incomplete (see table 3.1.7). Foreign trade statistics for 1992 in *value* terms indicate that most of the countries concerned were net importers; trade thus helped to maintain domestic absorption. However, the picture differs by country. In Bulgaria and Czechoslovakia imports grew much faster than exports. Preliminary estimates for Czechoslovakia indicate that domestic absorption (NMP used) actually increased in 1992, while in Bulgaria absorption must have contracted less than output. In 1991, policy in both countries had stressed external adjustment, which had resulted in substantially sharper falls in domestic demand than in GDP in that year and squeezed imports. Imports increased substantially (in value) also in Poland, though less than exports,[149] but stagnated or contracted in all other east European countries, with a particularly sharp fall in FR Yugoslavia. Only Poland and Slovenia registered significant overall export growth in 1992.

---

[149] Polish foreign trade information was particularly uncertain at the time of writing, however, and import growth may in fact have been far stronger than shown in table 3.1.7. See discussion in section 3.3(i) and footnote (199) below.

For the CIS countries as a whole, trade turnover with non-CIS countries clearly fell sharply from its 1991 level in dollar terms.[150] However, there is considerable uncertainty about the magnitude of the fall (see section 3.3(i) below). Attempts of economic agents to evade restrictions on both exports and imports, as well as the initial shortcomings of new customs-based recording systems, have probably resulted in some degree of under-reporting.[151] The CIS countries' external trade turnover (exports plus imports) is reported to have fallen by about a quarter in 1992, with imports falling faster than exports. The net surplus of over $1 billion was said to include surpluses of $1 and $3 billion by Kazakhstan and Russia. Russia and Belarus have independently reported trade surpluses of just over $3 and $0.3 billion, respectively, reflecting a sharp narrowing (from $6.4 billion) in the case of Russia and a turn-round from a deficit position in 1991 in Belarus. Most other CIS countries thus must have registered deficits in nominal terms. This may indicate that for the CIS countries other than Belarus, Kazakhstan and Russia, domestic absorption was supported by a net inflow of resources from abroad.[152] In the case of Russia, the 1992 trade surplus can be estimated to have amounted to some 4 per cent of nominal GDP.[153]

Few aggregate data are available for *intra-CIS trade*.[154] Russian exports are reported to have fallen by over and imports by under one third in volume in 1992.[155] However, published data on developments in intra-CIS trade available at the time of writing appear very uncertain and unreliable. Thus, reports on the size of the Russian trade surplus with other CIS countries vary from R58 billion (about 1 per cent of GDP in 1992)[156] to as much as R3,000 billion (about 20 per cent of GDP).[157]

(vii) **Short-term prospects**

An economic recovery may now be in sight for 1993 in some *east European countries*, but it is likely to be sluggish and uneven. Uncertainties over the course of privatization may hamper investment. Continued weakness in the world economy could reduce export growth, which is otherwise expected to be the main motor of recovery. Inflation is likely to continue on its downward course provided fiscal discipline can be maintained in the face of increasing pressure on government spending, particularly in response to the higher unemployment which is expected once large-scale privatization gets under way.

Within this general picture, individual countries' prospects differ widely. In Albania and Bulgaria, further declines in national product are likely, though the rate of decline should slacken in 1993. In the former Czechoslovakia, favourable economic conditions for the resumption of growth are now at risk under the impact of the split of the country into two independent republics. The Czech government none the less at the turn of the year still predicted GDP growth of 1-3 per cent in 1993; in Slovakia, the government anticipated a decline of 3 per cent. These forecasts, however, reflected the assumption of a relatively mild downturn in mutual trade after the separation. In the early months of 1993, this downturn was much more severe than expected.[158] In Hungary, the government expects a GDP performance of 0-3 per cent growth in 1993 and some growth is also expected in Poland. In Romania, the government anticipates a further 5 per cent drop in GDP, based on a forecast 10-14 per cent fall in industrial output and a 12 per cent recovery in agricultural production.

However, it should be kept in mind that actual developments have often been worse than the governments have expected and a pessimistic scenario in which the recession is prolonged into 1993 is quite possible. Even if there is a return to overall growth in 1993, the depth of the recession has been such that it will take many years for pre-1989 levels of output to be regained. Macroeconomic policies will have to remain tight and further progress in structural reform is required if the recovery is to be sustainable and inflation is to remain under control. Given the large external debt of most east European countries and their growing import requirements once recovery gains momentum, development strategies will have to be based on export-led growth. In this respect improved access to western markets will be important. The prospects for economic growth in the east European countries thus depend not only on the way these countries pursue their transformation efforts, but also on external factors and western support for their development. Economic

---

150 CIS Statistical Committee, *Ekonomika Sodruzhestva nezavisimykh gosudarstv v 1992 godu*, Moscow, 29 January 1993, pp.28-30.

151 United Nations Economic Commission for Europe, *Economic Bulletin for Europe*, Vol.44 (1992), New York, 1993, pp.54-55, and section 3.3(i) below.

152 Whether this was in fact the case depends, of course, on whether *real* external balances carried the same sign as the *nominal* balances reviewed here.

153 The Russian Federation Goskomstat reports a surplus in non-CIS trade of $3.1 billion or R598 billion at current prices for 1992. The latter figure constitutes 4 per cent of the nominal GDP estimate for 1992 of R15,000 billion reported in the same publication. See Russian Federation Goskomstat, op.cit., pp.7 and 39. The rouble trade values appear to have been converted at R193 per dollar, a rate which coincides with the average of the weekly market rates at the Moscow Inter-Currency Exchange auctions in 1992.

154 See discussion in section 3.3 below and notably table 3.3.4.

155 Direct communication to the UN ECE secretariat from Russian Federation Goskomstat. These volume estimates are based on measurements in physical units of a sample of commodities representing roughly 60 per cent of total flows by value. Earlier estimates published by the Russian Goskomstat showed much smaller falls (7 per cent for Russian export volume and 12 per cent for Russian imports; see Russian Federation Goskomstat, op.cit., p.39), probably reflecting errors in deflation procedures. See also the discussion by A. Granberg in *Rossiiskie vesti*, 19 February 1993, p.3.

156 Russian Federation Goskomstat, loc.cit.

157 *Izvestiya*, 19 February 1993.

158 See section 3.3(v) below.

growth and restructuring will also need to be supported by a recovery in investment. The foreign debt position of most east European countries will limit large-scale foreign borrowing and inflows of foreign direct investment are unlikely to be very large in relation to investment needs. Hence consumption growth will have to be restrained in order to provide resources for investment. Progress with privatization to give a more stable framework for investment, and reform of the financial sector to increase the efficiency of the allocation of scarce savings, will also be required for sustained investment growth.

Most of these considerations apply also to the *CIS countries*. But it should also be borne in mind that although the post-1989 fall in production in the former Soviet republics so far is not dissimilar to the declines in eastern Europe over the same period, the average 10 per cent fall in eastern Europe of 1990 was not matched until a year later in the former Soviet Union. The output declines in its successor states have thus generally been more concentrated in time. But even the best east European precedents (Hungary, Czechoslovakia and Poland where the output declines could have bottomed out or even been reversed in 1992) suggest that the output fall in the CIS countries, the Baltic states and Georgia could continue beyond the two years or so already experienced. Prospects for the Baltic economies may be more promising than for other former Soviet republics, but only because they seem to have better chances of receiving external financial assistance, both from the IMF and from individual western governments.

Quantified forecasts for the CIS economies are scarce. For Russia, the new Chernomyrdin government's economic policy programme of January 1993 aims to achieve a deceleration in the production decline and a stabilization in output during 1993. It is also intended to limit inflation to 5 per cent monthly and to keep the budget deficit within 5 per cent of GNP.[159] Another estimate foresees a GNP decline of 5-7 per cent and 12-15 per cent for industrial production.[160] The Russian Ministry of the Economy expects a similar GDP decline, a slightly smaller 7-10 per cent fall in industrial output, unchanged agricultural output and, provided that monetary and fiscal stabilization is achieved, a halt or even some reversal in the investment decline.[161] No improvement in private consumption levels is expected in these forecasts. No quantified forecasts are available for the other CIS economies.

Three broad sets of inter-related factors will determine the timing of any reversal in the precipitous economic decline in the CIS economies. The first set concerns the timing and implementation of a consistent set of stabilization objectives and further development of the new mechanisms for implementing them. The second relates to the rapid removal of often policy-determined obstacles on the production and movement of goods and services. The third concerns the speed at which economic agents at all levels can be induced to respond to the new environment so created.

In the first case, it is imperative that CIS governments succeed in creating a short- to medium-term macroeconomic policy framework within which businesses can operate with confidence. Disagreements between governments, parliaments and powerful economic interests in 1992 resulted in inconsistent policy expedients which proved a recipe for disaster. Public discontent, which has so far played a negligible role in influencing policy, could now become disruptive if production and real incomes decline much further. Coherent economic stabilization programmes must not only be announced and pursued, but a major effort needs to be made to explain to the public the reasoning behind and the results expected of such programmes.

The second set of factors concerns obstacles to business activity which can be relatively easily removed. The most important are the "beggar-thy-neighbour" restrictions on exports to "conserve national resources" introduced by some countries during 1992 and which were one reason for the continuing decline in intra-CIS trade.[162] Similarly, there seem to be few valid arguments for the maintenance of restrictions on imports from other CIS countries, especially given the remaining effects of preferences under central planning for inter-firm and inter-regional complementarities rather than competition. While structural balance of payments problems due to the Soviet successor states' high level of dependence on Russian fuel and raw materials are likely to persist, reliance on tariffs and other restraints on trade will not yield optimal solutions to such problems and can only worsen existing supply constraints. Domestically, a range of new legislation which appears to echo the distaste for private entrepreneurial activity of the *ancien régime* needs to be reviewed with a view to removing such anti-market regulations as those limiting "excess" profits and to abolishing discriminatory tax regimes imposed on private business activity.

Governments also have an active role to play with respect to the third group of factors. The foundations of institutional change have been laid. If stabilization is achieved, there will be opportunities to harness the market in the reform process. Rudimentary commodity exchanges to replace central distribution and supply have shown the capacity for rapid development and the same is true of banking, insurance and financial services. Procedures were also put in place in 1992 for an accelerated rate of privatization. But industrial policy formulation has so far been very limited. A central element in it should be to ensure, first, that new insti-

---

159 "On Financial-Economic Policy in Russia in 1993", *Izvestiya*, 26 January 1993.

160 Russian Central Bank. See *Radio Free Europe Radio Liberty Research Report*, Vol.1, No.48, 4 December 1992.

161 *The Economic and Commercial Bulletin of the Ecotass News Agency*, Pergamon Press, No.4, 18 January 1993.

162 Such restrictions should of course be distinguished from those imposed by a CIS country of origin to prevent the resale abroad at world market prices of commodities purchased at controlled prices by a CIS customer.

tutions are in fact giving appropriate signals to economic agents to stimulate dynamic output, profit and investment performance and, second, that advice and assistance are available to those firms which are prepared to respond. Nowhere is this more urgent than in the sphere of competition policy. So far, the business community has in fact already responded in many respects rationally to the new business environment: monopolistic producers have reacted to rising costs, falling real demand and the absence of a hard budget constraint by reducing output and raising prices. It now remains to be seen whether privatization will raise competition to the point where the new firms will, instead, feel impelled to lower costs and margins to increase sales. It is vital for governments to ensure that private monopolies do not replace public ones. The limited antimonopoly rules do not so far attempt to apply market share or other criteria similar to those used to assess market domination in western antitrust legislation. It will thus be important to follow up privatization with an ongoing review of enterprise behaviour and to strengthen competition rules where this appears necessary.

Beyond the above and other factors relevant for further design and implementation of the institutional and structural change and economic policies, both short- and medium-term prospects of Russia and its transformation into an open market economy depend heavily on the resolution of the political struggle between the two centres of political power, that of President Yeltsin and his team on the one hand, and that of the Congress of People's Deputies on the other, especially as behind that struggle there appear to be two quite different economic and political strategies.

3.2 LABOUR MARKETS, PRICES, INCOMES

(i) Labour markets

General labour market developments

Assessing labour market developments in the transition economies is made particularly difficult by the limited (and often varying) coverage of available statistics. In order to give some idea of the gaps left by official sources and also of the differences in unemployment rates which can be implied by the various measures of unemployment, the sources of labour statistics are discussed in box 3.2.1 and the first results of the quarterly Hungarian Labour Survey in a note at the end of this section.

As results from household labour force surveys are not yet available and, in addition, comprehensive employment statistics for 1991 are available only for Czechoslovakia, Poland and the CIS countries, it is possible to estimate the recent growth of the *labour force* only in these countries (table 3.2.1). A crude estimate of the total labour force, obtained by adding together estimates of total employment and unemployment, suggests that both the labour force and *activity rates* decreased slightly in 1990 but then increased significantly in Poland and the Slovak Republic (but not in the Czech Republic) in 1991. Such an increase in participation at a time of deep recession and sharp falls in employment is unusual, but it may merely be a statistical artefact arising from the introduction of (initially) relatively generous systems of unemployment benefits. In such circumstances, especially when labour offices do not have sufficient resources to investigate claims in detail, there will be an incentive for the economically inactive to register as unemployed. In fact, in Poland and the Slovak Republic unemployment rates increased significantly in 1991. If generous unemployment benefits are in fact the explanation for the apparent rises in participation in 1991, one would expect to see participation fall back again in 1992 following the tightening of the benefit regimes in Poland and Czechoslovakia. By contrast, in the four CIS member countries activity rates fell both in 1990 and 1991. This different behaviour of the activity rates in the CIS countries compared with Poland and Czechoslovakia is to a large extent due to the fact that in the countries of the former Soviet Union collection of unemployment statistics was introduced only in the second half of 1991. Hence until recently the level of employment was approximately equivalent to the level of the labour force. Thus, falling employment, even if it was much slower than the state of economic activity would imply, was accompanied by falling labour force and activity rates. However, recently introduced benefits systems, though not as generous or broad in scope as in eastern Europe, together with the expected improvement in the network of labour offices may increase the number of registered unemployed. Unless labour hoarding diminishes significantly and the discouraged worker effect strengthens sharply, as usually happens during recession periods in the western economies, the labour force may be expected to stop falling or even to increase in the near future.

Available *unemployment statistics,* which in all countries of the region are based on the number of registered unemployed, show that unemployment rose steadily in all countries of the region (with the notable exception of Czechoslovakia) in 1992 (table 3.2.2). In most of the countries of eastern Europe, unemployment continued to rise quickly, though at a slower pace than in 1991. In the Baltic republics and the countries of the former Soviet Union, open unemployment emerged for the first time at the end of 1991 and rose steadily throughout 1992. However, if official statistics are to be considered comprehensive, unemployment in the countries of the former Soviet Union is still at very low levels, both relative to those in market economies and to those currently experienced by the countries of eastern Europe.

The forces driving the rise in *measured unemployment* vary substantially between countries, though two generalizations may be made. Firstly, the decline in measured employment has been lower than and has generally followed the decline in measured output with a considerably longer lag than usually experienced in the west. This holds throughout eastern Europe and is also reported to be taking place in Russia.[163] This result should not be affected by the general deficiencies of labour statistics discussed in box 3.2.1, as output figures are generally compiled on a broadly similar basis to the employment figures, though it may not apply to the unobserved sectors of the economy. Therefore, there seems to be a general tendency for enterprises in eastern Europe and the former Soviet Union to "hoard" labour in the face of falling output which is clearly reflected by sharply falling productivity (table 3.2.3).

The reasons for such a tendency are not clear but the sharp contrast with the huge recent rises in productivity in the *Länder* of the former GDR suggest several possible explanations: the absence of a

---

[163] Productivity in Russian industry is estimated to have fallen by 24 per cent in 1992, according to the Macroeconlink research centre. See *Financial Times*, 6 January 1993.

BOX 3.2.1

The methods of collecting statistics on unemployment, vacancies and employment

(a) Unemployment

Unemployment statistics are usually collected in one of three different ways, the "claimant" method, the "registration" method or from household surveys. The internationally accepted definition of unemployment – the ILO definition – counts as unemployed an individual who is without work, is actively searching for work and who is available to start work. The measure of unemployment obtained from household labour force surveys is the only one which corresponds to the ILO definition and is also the only one which can be reliably used to compare the levels of unemployment between countries. The other two methods, while not comparable between countries, provide rapid information on the development of unemployment over time within individual countries but even these trends should be interpreted cautiously at times of rapid structural change.

In the claimant method (as used in the UK for example), unemployment is measured by counting the number of recipients of unemployment benefits (UB). Unemployment figures measured this way are obviously strongly affected by changes in the rules of the benefit system or by the introduction of other benefits (for example, the introduction of a sickness or invalidity benefit which may be more generous than unemployment benefit would be expected to reduce the number of claimants for UB and hence to reduce unemployment measured in this way).

However, if there are conditions that claimants must be available for work and searching for work in order to receive benefit and these conditions are enforced (which is not always the case, especially when unemployment is rising rapidly) then it is likely that the vast majority of claimant unemployed will in fact be unemployed according to the ILO definition (of course, the converse will not be true, i.e., not all the ILO-unemployed will be claimants). Therefore, in circumstances where the structure of unemployment is not changing rapidly and benefit rules are enforced, the claimant method will yield unemployment figures which will consistently underestimate the level of ILO unemployment but may mirror its progress over time fairly well. As the statistics are collected automatically in the course of processing of claims for benefit, unemployment statistics based on counting claimants are usually relatively cheap to obtain and analyse and above all are available quickly.

In the registration method (as used in most western European countries and in all of the transition countries studied here), unemployment is measured by counting the number of people who have registered with state labour offices as looking for work and who do not have work. While there is now no direct link with the benefit system, and the registration method includes those unemployed who are not eligible for unemployment benefit, most systems of UB require all claimants to register with their local labour office as a condition for obtaining benefit. Therefore there will be a substantial overlap with the claimant method and measures which change the eligibility for benefits may also affect the numbers registering with the labour offices. In addition, there are no checks that those registering as unemployed are in fact actively searching for work or are available to start work, so some of those counted as unemployed by the registration method may not in fact be unemployed according to the ILO definition. On the other hand, labour offices may not be regarded by the unemployed as being efficient in obtaining work for them and so some of the unemployed may choose not to register with the labour office.

The relationship between registered unemployment and ILO unemployment therefore varies substantially from country to country and registration-based figures should not therefore be used to compare the level of unemployment between countries. Moreover, as recent results from the Hungarian Labour Force Survey show (see the Note on this subject), changes over time in the number of registered unemployed may not reflect changes in the underlying level of ILO unemployment. As with the claimant method, the data required for compiling registration-based statistics are collected automatically by the labour offices and the statistics are therefore available quickly and relatively cheaply.

Household labour force surveys (such as those carried out in all countries of the EC every spring), where a controlled sample of individuals are asked to fill in a questionnaire on the nature of their employment status and the methods they are using (if any) to find work, provide direct measures of unemployment according to the ILO definition. Such surveys also provide detailed information on the structure of unemployment which may not be available from other sources. However, household surveys are expensive and, as it takes time to process their results, they are usually published with a considerable delay. Moreover, great care has to be taken in the design of the questionnaire and the drawing up of the sample to ensure that the results obtained are in fact representative of the whole population and not just of that part of the population which is easily contactable.

(b) Vacancies

Statistics on unfilled vacancies are provided by labour offices and measure the number of unfilled vacancies reported to labour offices on a given date. The reliability of these figures obviously depends on the extent to which labour offices are used by enterprises to fill their vacancies. Even in countries (such as Poland) where there is a legal obligation for enterprises to notify labour offices of unfilled vacancies, it is likely that only a fraction of vacancies will in fact be reported and also that this fraction will vary considerably between countries. Official vacancy statistics will therefore always underestimate the number of unfilled jobs in the economy, though they may give a good indication of changes over time when the structure of new jobs is not changing too rapidly.

However, in current circumstances, where the state sector, which used to have close links with the labour offices, is shedding jobs and the private sector, which used to have to recruit informally rather than through the official labour offices, is gaining jobs, the relationship between registered vacancies and the actual number of vacancies will still be evolving. Until the private sector has grown to such a size that it finds it difficult to fill its vacancies through informal channels, the official vacancy figures may give a misleading idea even of trends over time as they will not reflect much of the growth of the private sector.

[continued ...

comprehensive social safety net, so that most social welfare provisions are still provided by the employing enterprise and loss of employment means that access to many welfare facilities is also lost; the lack of hard budget constraints on firms so that they are not forced to get rid of excess labour by immediate financial considerations; slow progress with privatization so that firms are less concerned with increasing profits or introducing efficient production than with the welfare of workers and, finally, the sharp decreases in real wages seen in most of eastern Europe mean that the financial incentive on firms to lay off excess labour is reduced. All these factors differ between the *Länder* of the former GDR and the other transition economies, where in the former there is a comprehensive safety net, privatization is proceeding rapidly, access to credit is controlled by a

> **BOX 3.2.1** *(concluded)*
>
> **The methods of collecting statistics on unemployment, vacancies and employment**
>
> *(c) Employment*
>
> Statistics on employment are collected in two main ways: statistics from enterprises, which count jobs rather than people, and those obtained from household surveys, which count the number of people employed rather than the number of jobs. All the transition countries base their employment statistics on figures from enterprises (see below the discussion of changes in employment as revealed in the Hungarian Labour Force Surveys). West European countries tend to use a combination of both sources, with the household surveys being used to provide information on sectors of the economy not adequately covered by enterprise-based figures (e.g., the self-employed or smaller enterprises in the service sector).
>
> In the transition countries, short-term employment statistics (those produced on a monthly or quarterly basis) generally measure the number of employees (they thus usually exclude the self-employed) and are based on reports from all enterprises satisfying certain characteristics: for example, enterprises above a certain threshold number of employees (5 in Poland, 50 in Hungary and, in Czechoslovakia, 100 in 1990 and 1991 and 25 in 1992), enterprises in the state sector alone (Bulgaria, Romania) or enterprises in certain sectors of the economy (excluding private agriculture in Poland, excluding enterprises in the non-material sector in Hungary). The excluded sectors, enterprises and the self-employed are generally included in the more comprehensive annual figures but these figures are generally available only with a long delay.
>
> Moreover all these figures are generally based on figures provided from a complete list of enterprises satisfying the criteria rather than a sample of them. West European employment statistics tend to be based more on samples of enterprises but with much less restrictions on the characteristics of the enterprises sampled.
>
> In the current circumstances of the transition economies, this combination of complete enumeration of employment in a restricted set of enterprises (at least for the short-term statistics) is unfortunate. The enterprises included in the regular statistical reports tend to be the large, state-owned firms who are expected to shed considerable amounts of labour during the transition. The firms which are expected to grow and to provide the main source of new jobs are mainly small, in the private sector and are concentrated in the service sector of the economy. In addition, large enterprises are more likely to be broken up into a number of smaller firms, some of which may then be below the threshold for regular statistical reporting. The regular employment statistics are therefore likely to overstate the fall in employment significantly and may exaggerate the fall in employment even in a given set of work places, if the break-up of enterprises is taking place on a large scale. For some quantitative evidence illustrating these points, see table 3.2.9. As the table shows, the differences between the quarterly and annual figures for employment in Poland and Czechoslovakia are large and also increased rapidly between the end of 1989 and the end of 1991, probably reflecting the growth of small enterprises in the private sector.

functioning banking system and real wages have increased dramatically.

As a result of this increase in *hidden unemployment*, employment in loss-making and soon to be privatized firms is expected to fall further even if their output recovers. The first signs of this can be seen in Poland where employment in industry has continued to fall, despite an increase in industrial output in 1992. Hence, open unemployment is expected to increase further in 1993 as privatization programmes proceed and as provisions for pushing firms into bankruptcy become more widespread.

The second, more tentative, generalization on the unemployment situation in transition economies, concerns the *causes of the rise in unemployment*. In several countries of eastern Europe the continued rise in unemployment in 1992 was accompanied by stabilization or even rises in the number of *vacancies* (table 3.2.4). In contrast, in the countries of the former Soviet Union in 1992 and in eastern Europe in 1990 and 1991, unemployment rose at the same time as stocks of vacancies were falling steadily.

This suggests that the rise in unemployment in eastern Europe in 1992 was driven rather more by changes in the structure of employment and rather less by a general fall in output and the demand for labour than was the case in 1990 and 1991.[164] While this interpretation leans heavily on official vacancy statistics, which are of untested reliability, it is supported by changes in the regional structure of unemployment (see below) and by reports that the privatization process is beginning to make progress in several countries.

Country observations

In *Czechoslovakia*, unemployment actually fell substantially in 1992, with the largest part of the fall taking place in the Czech Republic. This fall in unemployment occurred at the same time as quarterly employment statistics were showing sharp falls in employment in large enterprises. Although some of the fall in unemployment was due to a tightening of the rules for obtaining unemployment benefit in January 1992 and a further part was due to the expansion of public employment programmes, much of the drop appears to be genuine and seems to reflect the strong performance of the private sector, particularly in the Czech Republic. In contrast, the regular employment figures (table 3.2.5), which exclude small enterprises and thus most of the emerging private sector, seem to greatly exaggerate the actual fall in employment. Additional evidence for this view is provided by the rise in registered vacancies in 1992, suggesting that at least some sectors of the economy were increasing their demand for labour, and the increase in the number of registered private entrepreneurs to an estimated 1.7 million (relative to an

---

[164] In terms of the Beveridge (or U-V) curve, which plots unemployment (U) against vacancies (V), a fall in overall labour demand leads to a fall in vacancies but to an increase in unemployment and thus to a movement along the curve. In contrast, a fall in the efficiency with which the unemployed are matched to the available vacancies, caused by, for example, mismatch between the skills or location of the unemployed and those of available jobs, will cause a rise in unemployment for a given stock of vacancies and thus a shift outwards of the whole U-V curve.

TABLE 3.2.1

Selected transition countries: Changes in the apparent labour force
(Levels in thousands; rates and changes in per cent)

	Total employment		Unemployment		Labour force		Activity
	Level	Change	Level	Rate	Level	Change	rate
Czechoslovakia							
1989	7 995	..	-	-	7 995	..	88.6
1990	7 686	-3.9	77	1.0	7 763	-3.0	85.1
1991	7 292	-5.1	524	6.7	7 815	0.7	85.3
Czech Republic							
1989	5 428	..	-	-	5 428	..	90.5
1990	5 214	-3.9	39	0.7	5 253	-3.3	86.7
1991	5 045	-3.3	222	4.2	5 266	0.2	86.5
Slovak Republic							
1989	2 567	..	-	-	2 567	..	84.8
1990	2 472	-3.7	37	1.5	2 509	-2.3	82.0
1991	2 247	-9.1	302	1.9	2 549	1.6	83.1
Poland							
1989	17 558	..	56	0.3	17 614	..	80.5
1990	16 474	-6.2	1 126	6.4	17 600	-0.1	80.1
1991	15 861	-3.7	2 156	12.0	18 017	2.3	81.7
Belarus							
1989	5 198	..	-	-	5 198	..	91.0
1990	5 149	-1.0	-	-	5 149	-1.0	90.2
1991	5 020	-2.5	2	-	5 022	-2.5	88.4
Moldova							
1989	2 091	..	-	-	2 091	..	87.1
1990	2 071	-1.0	-	-	2 071	-1.0	86.3
1991	2 070	-0.1	-	-	2 071	-0.1	86.2
Russia							
1989	75 168	..	-	-	75 168	..	89.8
1990	74 383	-1.0	-	-	74 383	-0.6	88.6
1991	73 809	-0.8	62	0.1	73 871	-0.7	87.8
Ukraine							
1989	25 420	..	-	-	25 420	..	88.2
1990	25 277	-0.6	-	-	25 277	-0.6	87.8
1991	24 756 *	-2.1	7	-	24 763	-2.0	85.9

Source: National statistics.

Notes: All figures refer to the end of the year. The employment measure used here is comprehensive (it includes the self-employed and employees in all sizes of enterprises), but as it is only published annually and with a long delay, figures for 1991 are available only for Czechoslovakia, Poland and the CIS countries (only four CIS countries which are ECE members are presented). Estimates of the labour force are obtained by adding the figures for registered unemployment and total employment. Activity rates are derived using the estimates for the labour force and data for the population of working age (in the territory of former Czechoslovakia 15-59 for men and 15-54 for women; in Poland 18-64 for men, 18-59 for women).

estimated total labour force of about 7.7 million) by the end of 1992. Notwithstanding, open unemployment is expected to increase again in 1993 as privatization spreads and the delayed bankruptcy law is brought into operation.

In *Hungary*, the picture is more confused. Registered vacancies are now increasing and unemployment, at least as measured by the new quarterly labour force survey, appears to be rising only slowly. However, the monthly unemployment figures are still rising quickly and the regular enterprise-based employment figures show very large falls (a drop of nearly 40 per cent between 1989 and autumn 1992) in employment in large enterprises in the material sector of the economy. Moreover, much of the rise in official vacancy figures appears to be due to the greater efficiency of the labour offices in obtaining details of unfilled posts. So far, the measures of employment based on the labour force survey show only that overall employment is falling more slowly than the narrow measure suggests and it therefore seems that there has as yet been no significant increase in the demand for labour in the small enterprise sector.[165]

Little sign of any improvement in the labour market situation is yet visible in *Poland*. Even though the registered vacancies no longer seem to be falling and in fact have even started to increase during the second half of 1992, the regular employment statistics continue to indicate that employment is falling steadily and unemployment continues to rise. Although the employment figures are still likely to exaggerate the fall in

---

[165] The "narrow measure" usually covers employees in large enterprises, but the threshold size varies between countries. See table 3.2.9 for its definition in Czechoslovakia and Poland.

TABLE 3.2.2

European transition countries: Unemployment, 1991-1992

(Thousands and per cent of labour force, end-of-month)

	Unemployment (thousands)						Unemployment rate (per cent)					
	Dec. 1991	Mar. 1992	Jun. 1992	Sept. 1992	Dec. 1992	Jan. 1993	Dec. 1991	Mar. 1992	Jun. 1992	Sept. 1992	Dec. 1992	Jan. 1993
Albania	67.0	..	..	..	..	..	..	..	..	..	..	..
Bulgaria	419.0	452.6	475.8	538.7	577	..	11.5	12.4	13.1	14.8	15.9	..
Czechoslovakia	523.7	502.5	424.0	403.1	395.1	..	6.6	6.5	5.5	5.2	5.1	..
Czech Republic	221.7	195.2	141.7	137.0	134.8	158.1	4.1	3.7	2.7	2.6	2.6	3.0
Slovak Republic	302.0	307.4	282.3	266.1	260.3	286.0	11.8	12.3	11.3	10.6	10.4	11.2
Hungary	406.1	478.0	546.7	616.8	663	700	7.4	8.9	10.1	11.4	12.3	13.3
Poland	2 155.6	2 216.4	2 296.7	2 498.5	2 509.3	2 584.0	11.8	12.1	12.6	13.6	13.6	14.0
Romania	337.4	488.2	652.2	841.7	929.0	949.8	3.0	4.4	5.9	7.6	8.4	8.5
Croatia	283.0	275	264	262.1	261.0	262.5	18.9	18.8	18.5	18.5	18.5	18.5
Slovenia	91.2	95.5	97.2	107.8	118.2	..	10.1	10.7	11.0	12.2	13.3	..
Yugoslavia (FR)	707.1	747	741	760	749	..	21.0	23.0	23.4	24.1	24.8	..
CIS	77.0	230.0	300.0	525.0	800.0	..	..	..	..	..	..	..
Armenia	..	..	14.5	35.6	56.3	..	..	..	0.9	2.2	3.5	..
Azerbaijan	3.8	7.0	7.2	6.2	6.4	..	0.1	0.3	0.3	0.2	0.2	..
Belarus	2.0	4.2	7.1	14.9	24.0	..	-	0.1	0.2	0.3	0.5	..
Kazakhstan	..	9.2	15.8	25.0	33.7	..	..	0.1	0.2	0.3	0.4	..
Kyrgyzstan	..	0.5	0.8	1.4	1.8	..	..	-	-	0.1	0.1	..
Moldova	0.1	0.3	0.6	1.6	15.0	..	-	-	-	0.1	0.7	..
Russia	60.0	118.4	202.9	367.5	577.1	628	0.1	0.2	0.3	0.5	0.8	0.8
Tajikistan	..	-	3.5	7.2	..	..	..	..	0.2	0.4	..	..
Turkmenistan	..	..	..	..	..	..	..	..	..	..	..	..
Ukraine	6.8	..	35.6	60.7	70.5	..	-	..	0.1	0.2	0.3	..
Uzbekistan	..	..	..	3.6	8.8	..	..	..	..	-	0.1	..
Georgia	..	..	..	..	..	..	..	..	..	..	..	..
Estonia	0.9	2.4	5.7	8.7	15.0	..	0.1	0.3	0.7	1.1	1.9	..
Latvia	..	2.8	8.6	17.8	31.3	36.7	..	0.2	0.6	1.2	2.1	2.5
Lithuania	4.7	5.2	7.7	11.2	20.7	..	0.3	0.3	0.4	..	..	..
Ex-GDR Länder	1 037.7	1 220.1	1 123.2	1 110.8	1 100.8	1 194.4	12.7	15.0	13.8	13.6	13.5	14.7 [a]

Sources: National statistics and direct communications to ECE.

[a] Per cent of *civil* labour force.

employment,[166] they are more comprehensive than those in Hungary or Czechoslovakia[167] and so the bias due to unobserved growth of the private sector is smaller. The greater comprehensiveness of the Polish statistics may partly explain why the fall in measured employment in Poland is lower than in Czechoslovakia or Hungary but it may also be that the restructuring of production and employment is proceeding more slowly in Poland.

In *Bulgaria*, the rise in unemployment in 1992 was considerably lower than the massive jump seen in 1991 but sharp falls in employment in the state sector continued in the first six months of 1992. In *Romania*, where unemployment was relatively low at the end of 1991, joblessness grew very rapidly in the course of 1992 and employment in the state sector fell further in the first half of 1992. Unemployment surged particularly rapidly in the last quarter of the year, coinciding with the first privatizations of large state firms. No concrete information on the recent development of employment in the private sector is available for either country.

Information is also limited for the successor republics to the *former Yugoslavia*. In all the republics, the disruptions to normal economic life caused by war, the breaking of established trade links and the displacement of populations make interpretations of even those statistics which are available difficult. In the three republics for which regular statistics are available, Croatia, Slovenia and the Federal Republic of Yugoslavia (Serbia and Montenegro), measured unemployment has risen only slowly or has even fallen despite large falls in production. In the Federal Republic of Yugoslavia, the falls in employment in the state sector have been limited by government prohibitions on making workers redundant while economic sanctions imposed by the UN are in place so that large number of workers are being paid for doing little or no work, while in Croatia and Slovenia the slow progress in privatization and the lack of an enforceable bankruptcy code have prevented a larger number of job losses.

In the *Baltic republics* and the successor states to the *former Soviet Union*, statistics on the extent of unemployment only began to be collected in the second half of 1991 or even later. Measured unemployment in

---

[166] See, for example, table 3.2.9, which compares narrow and broad measures of employment.

[167] They cover employees in all enterprises with five or more employees and exclude private agriculture.

TABLE 3.2.3

European transition countries: Change in output, employment and productivity in industry
(Annual percentage change)

	Industrial output		Industrial employment		Industrial productivity	
	1991	1992	1991	1992	1991	1992
Albania	..	..	..	..	..	..
Bulgaria [a]	-27.8	-23.1	-18.8	-19.2	-11.0	-4.8
Czechoslovakia [b,c]	-24.7	-17.3	-12.0	-13.1	-14.4	-4.9
Czech Republic [b,c]	-24.4	-17.5	-11.6	-12.9	-14.4	-5.3
Slovak Republic [b,c]	-25.4	-17.0	-12.9	-13.7	-14.4	-3.8
Hungary [b]	-18.6	-14.2	-13.0	-16.5	-6.6	2.9
Poland [b]	-15.2	0.2	-8.7	-8.1	-7.4	9.3
Romania [d,e]	-18.4	-20.1	-9.9	-12.0	-9.4	-9.3
Croatia [b]	-28.5	-22.3	-17.5	-16.2	-13.4	-7.1
Slovenia [d]	-12.4	-16.1	-10.5	-10.6	-2.2	-6.1
Yugoslavia (FR) [a]	-18.0	-11.0	-6.5	-5.9	-12.3	-5.4
CIS states:						
Belarus	-1.5	-9.6	-0.3	..	-1.2	..
Moldova	-11.1	-21.7	-11.2	..	-	..
Russia	-8.0	-18.8	-4.2	..	-4.0	..
Ukraine	-4.8	-9.0	-2.6	..	-2.3	..
Ex-GDR Länder [b,f]	-33.2	-8.3	-37.8	-49.5	8.3	81.6

Sources: National statistics; OECD and DIW estimates.

Notes: The change in productivity (output per worker) is calculated by deflating the changes in output by the changes in employment. As the basis for employment and output statistics may vary slightly, these calculations should not be regarded more than orders of magnitude.

a   For 1992 changes, first quarter of 1992 as compared to the same period of 1991.
b   For 1992 changes, first three quarters of 1992 as compared to the same period of 1991.
c   1991 changes on 1990: enterprises with over 100 employees; 1992 changes on 1991: enterprises with over 25 employees.
d   For 1992 changes, first two quarters of 1992 as compared to the same period of 1991.
e   For 1991 changes, first three quarters of 1991 as compared to the same period of 1990.
f   For 1991 changes, the last two quarters of 1991 as compared to the same period of 1990.

all these countries is still low although rising quickly, but anecdotal reports suggest that registering as unemployed, at least in Russia, is difficult, because of the underdeveloped state of the network of labour offices. The official unemployment figures may therefore considerably understate the actual level of unemployment.[168]

Information on other aspects of labour market developments in these countries is limited. Vacancies appear to be falling but the number of unemployed in Russia only rose above the number of vacancies in the autumn of 1992, in sharp contrast to the experience of eastern Europe in 1990 and 1991, when registered vacancies plummeted to a fraction of the level of unemployment. Published statistics suggest that employment in Russia is falling only slowly, with falls in employment in state enterprises being partially offset by increases in private sector activities, but an ILO survey of Russian industrial firms reports considerable falls in employment (over 8 per cent) in the firms in their sample between September 1991 and June 1992.[169] The overall picture is therefore not clear but, in the light of the very sharp falls in output seen in the Baltic states and the CIS countries in 1992, larger rises in unemployment are to be expected.

The development of unemployment in the German *Länder of the former GDR* provides a sharp contrast to the other transition countries. Registered unemployment quickly rose to very high levels as loss-making enterprises were shut down and excess labour was reduced in enterprises which survived. Overall employment therefore fell drastically, particularly in the industrial sector and, despite large falls in output, productivity actually increased significantly. The rise in open unemployment would have been even quicker but for the existence of schemes supporting short-time work, of short-term agreements in certain industries limiting the extent of job losses (most of which ran out by the beginning of 1992) and the introduction on a massive scale of public employment programmes and retraining schemes.

The structure of unemployment

Some consistent features of the structure of unemployment are already emerging. In all countries, except for Hungary, the *unemployment rate for women* is substantially higher than that *for men* (table 3.2.6). The immediate factor behind higher female unemployment seems to be the relatively larger falls in female

---

[168] Official Russian unemployment statistics appear to be very narrow in their coverage by excluding first job seekers (for example school-leavers). The number of official unemployed in October was 367,500 whereas the number of registered nonemployed (which includes school-leavers and pensioners) amounted to 921,300. See S. Marni, "How prepared is Russia for mass unemployment?", *RFE RL Research Report*, 4 December 1992.

[169] See, for example, G. Standing, "Labour market dynamics in Russian industry in 1992", unpublished paper presented to an ILO conference on "Employment Restructuring in Russian Industry", St. Petersburg and Moscow, October 1992.

TABLE 3.2.4

European transition countries: Registered vacancies

(Thousands)

	December 1990	December 1991	March 1992	June 1992	September 1992	December 1992
Albania	..	..	..	..	..	..
Bulgaria	28.4	10.0	12.0	10.7	11.0	..
Czechoslovakia	72.2	56.6	90.8	98.2	106.6	..
Czech Republic	..	48.4	78.3	85.0	88.8	..
Slovak Republic	..	8.2	10.4	13.2	17.7	..
Hungary	16.8	11.5	15.1	25.3	25.6	..
Poland	54.1	29.1	27.2	31.7	40.5	..
Romania	..	..	..	..	..	..
Croatia	3.7	..	..	..	..	..
Slovenia	3.5	3.5	..	..	..	..
Yugoslavia (FR)	23.8	21.3	28.5	22.1	..	..
CIS states:						
Armenia	..	10.7	..	4.6	3.5	1.6
Azerbaijan	..	18.6	13.1	12.7	13.0	10.9
Belarus	..	63.8	33.4	31.4	24.2	18.2
Kazakhstan	..	58.1	45.3	55.0	50.4	41.2
Kyrgyzstan	..	14.8	3.3	4.6	6.2	4.6
Moldova	..	7.2	2.9	1.8	1.1	0.4
Russia	..	841.0	450.6	397.9	341.0	306.9
Tajikistan	..	0.8	3.1	3.0	1.4 [a]	0.4
Turkmenistan	..	3.0	..	..	..	..
Ukraine	..	273.9	..	..	..	129.1
Uzbekistan	..	50.6	24.5	26.7	26.2	19.8
Georgia	..	..	..	..	..	..
Estonia	..	..	..	..	..	..
Latvia	..	..	..	..	..	..
Lithuania	..	..	..	..	4.0 [a]	..
Ex-GDR *Länder*	17.5	35.4	33.3	31.8	31.3	32.9

Source: National statistics.

[a] August 1992.

employment than in male employment,[170] but the underlying causes of this tendency are not clear.

No explanation for the anomalous behaviour of Hungary has yet been put forward but it is clear that this feature is not a peculiarity of the monthly Hungarian statistics, as the Labour Force Survey data on unemployment also give a lower unemployment rate for women. In the fourth quarter of 1992, the female unemployment rate was 7.9 per cent compared to 11.3 per cent for men. One plausible reason can be that Hungarian female workers may be concentrated in sectors which are less recession-sensitive, such as health and education. In fact, male workers dominated the workforce in heavy industry, which was by far the most affected by employment losses. Furthermore, female workers may be less reluctant to take on low paid, low-skill jobs than their male counterparts. Similarly, unemployed females may tend to leave the labour market more rapidly and get jobs in the quickly growing underground economy. Unfortunately Hungarian employment statistics do not allow an empirical verification of these possible explanations.

The *regional patterns* of unemployment in eastern Europe, already coming into view in 1991, became more firmly established in 1992. Regional disparities in unemployment rates (and in the distribution of vacancies) generally increased between 1991 and 1992, and a general pattern emerged that capital cities and regions bordering Germany and Austria (western Czechoslovakia and Hungary) had lower than average unemployment rates and the more remote, rural regions, often with a high concentration of heavy industry, have the highest unemployment (table 3.2.7).

Although these regional divergences have emerged rather swiftly, it will not prove easy to reduce them. The usual mechanisms of regional adjustment through migration and, to a lesser extent, changes in relative wage rates between regions may prove slow to work in the transition economies as labour mobility is impeded by a shortage of housing and changes in relative wages may be delayed by controls on overall wages designed to keep inflation under control.

Less information is available on other characteristics of unemployment, such as *age* and *skill* patterns. However, unemployment appears to be concentrated

---

[170] For example between the end of 1990 and 1991, total female employment in Poland and Czechoslovakia fell by 3.8 per cent and 11.0 per cent respectively while male employment fell by only 3.6 per cent and 9.7 per cent. Note that the Czechoslovak figures are not comparable with those presented in table 3.2.9.

TABLE 3.2.5

European transition countries: Change in employment, 1990-1992
(Annual percentage change)

	1990	1991	1992
Albania	..	..	..
Bulgaria [a]	-5.8	-17.1	-18.5 [b]
Czechoslovakia [c]	-6.5	-13.1	-13.9 [d]
Czech Republic	..	..	..
Slovak Republic	..	..	..
Hungary [e]	-9.8	-16.7	-21.0 [d]
Poland	-8.1	-9.0	-8.6 [d]
Romania [a]	1.5	-11.0	-8.3 [b]
Croatia [a]	-3.6	-13.7	-12.4 [d]
Slovenia	-3.9	-7.3	-7.6 [b]
Yugoslavia (FR) [f]	-3.2	-7.3	-5.3 [b]
CIS states:			
Armenia	2.4	-2.5	..
Azerbaijan	-0.3	4.0	..
Belarus	-1.9	-2.5	..
Kazakhstan	1.3	-0.9	..
Kyrgyzstan	0.5	0.3	..
Moldova	-1.0	-	..
Russia	-1.0	-0.8	-2.1
Tajikistan	3.1	6.9	..
Turkmenistan	3.4	1.9	..
Ukraine	-0.6	-2.1	..
Uzbekistan	4.2	4.8	..
Georgia	..	..	..
Estonia	-2.0	..	..
Latvia	0.1	-0.3	-1.6 [g]
Lithuania	-3.7	0.4	..
Ex-GDR Länder [h]	-10.1	-19.1	-16.1 [d]

Sources: National statistics; IMF country reports; OECD and DIW estimates.

Notes: For countries where figures for 1992 are shown, all changes are based on averages of quarterly figures ("narrow measures" of employment) so that some comparability over time is maintained.

  a Socialist sector only.
  b First two quarters of 1992 as compared with the same period of 1991.
  c Changes based on raw data which cover all employees in state sector in 1989, all employees in enterprises with over 100 employees in 1990 and 1991 and all employees in enterprises with over 25 employees in 1992.
  d First three quarters of 1992 as compared with the same period in 1991.
  e Enterprises in the material sector with over 50 employees only.
  f Socialist sector and insured employees in the private sector only.
  g First two quarters of 1992 compared with figure for 1991 as a whole.
  h DIW estimates. These exclude those living in the area of the ex-GDR but working in the area of the former Federal Republic of Germany.

amongst the young and amongst prime age workers, which may be due to the fact that older workers have stronger links to their enterprise than do younger workers and so may be able to influence management's decisions on who to make redundant, similar to the "last in, first out" principle often adopted by western trade unions. Furthermore, those who are relatively uneducated and unskilled have above-average unemployment rates, at least in eastern Europe.[171] The average *duration* of unemployment is also increasing. For example, the average duration of benefit claims in Hungary increased from 149 days in June 1991 to 263 days in September 1992 and in Czechoslovakia the proportion of long-term unemployed increased from 3.4 per cent in December 1991 to 12.7 per cent in March 1992.

The structure of employment

The structure of employment is now showing the first signs of changing in response to the new environment. Though there are substantial variations between countries, employment in the traditional "productive" sectors of agriculture, industry and construction has fallen the most. Employment in health, education and other social services has generally fallen less than overall employment, but the only sector to show any growth in employment is finance and business services, as the commercial activities are now rapidly replacing the old institutions of central planning. Employment in business and finance grew particularly strongly in Czechoslovakia (though figures only refer to the end of 1991), Bulgaria and Poland (table 3.2.8). In Czechoslovakia, employment in transport and communications also increased, albeit at a much lower rate. This may to some extent be a spill-over effect of the rapid expansion of finance and business services, which usually increases the demand particularly for communications.

Measures to deal with unemployment and the social safety net

When reform programmes were initially being designed, the importance of a comprehensive social safety net to protect vulnerable groups of the population by providing them with a minimum level of income was stressed by all governments and their western advisers. In addition, it was proposed that programmes for retraining a significant fraction of the unemployed would be set up in order to ease the process of structural adjustment and reduce the risks of the emergence of long-term unemployment.

Now that the reform process is well under way, maintaining the planned type of safety net is proving increasingly difficult. Government budgets are coming under increasing strain and welfare entitlements have been increasingly restricted as governments seek to meet spending targets in the face of steadily rising unemployment.

As a result, the initial ambitious programmes for combining passive labour market policies (providing income support for the unemployed) with substantial active programmes (providing retraining for the unemployed to help them into new jobs) have had to be scaled down or abandoned as the budgets of official labour funds had difficulties in meeting even their obligations to pay unemployment benefits. Increasingly, entitlements to unemployment benefit have been scaled back as well.

Czechoslovakia, where unemployment has been falling in 1992, has been most able to fulfil the original programmes and, although only about 2 per cent of the unemployed (about 8,000) were involved in government training programmes in September 1992, a much higher number (nearly 130,000, equivalent to about 30 per cent of the number of unemployed) were employed in government-supported employment schemes. In

---

[171] The unemployed in Russia tend to be relatively highly educated. See S. Marnie, op.cit.

TABLE 3.2.6

Selected transition countries: Unemployment rates by sex
(Per cent of labour force)

Country	September 1991			September 1992		
	Total	Male	Female	Total	Male	Female
Bulgaria [a]	10.7	9.6	11.4	..	..	..
Czechoslovakia	5.6	5.0	6.3	5.2	4.3	6.3
Czech Republic	..	..	..	2.6	1.9	3.5
Slovak Republic	..	..	..	10.6	12.3	9.3
Hungary	6.1	6.6	5.5	11.4	12.9	9.8
Poland	10.7	9.4	12.2	13.6	11.7	15.9
Romania	2.4	1.8	3.2	..	..	..
Slovenia [b]	7.9	8.2	7.6	10.9	11.5	10.2
Yugoslavia (FR) [b]	15.7	12.8	19.8	16.4	13.0	21.2
Ex-GDR *Länder*	11.7	9.1	14.3	14.1	9.6	19.9

Sources: National statistics; and ECE secretariat estimates.

Note: Labour force by sex is estimated as follows: Bulgaria: Split into males and females assumed to be same as for employees on 1 August 1990, adjusted to be consistent with overall August reported rate. CSFR: reported rates for September 1991; for September 1992 calculated on end-1991 employment figures by sex and republic; January 1992 unemployment figures by sex and republic, normalized to accord with overall reported rate. GDR: reported rates. Hungary: ILO Yearbook 1991 for active population on 1 January 1990. Poland: sex split as of 31 December 1989 taken from 1990 Yearbook adjusted to be consistent with overall reported rate above. Romania: 1989 figures taken from ILO Yearbook 1991; made consistent with 1990 labour force figures in 1991 Yearbook. Yugoslavia: 1990 sex split for employees in social sector plus unemployed (1991 Yearbook), adjusted to be consistent with OECD Labour Force estimates.

[a] December 1991.
[b] June 1991 and June 1992.

TABLE 3.2.7

Selected transition countries: Regional dispersion of unemployment rates in September 1991 and September 1992
(Per cent of labour force)

Country	Number of areas	September 1991				September 1992			
		Average	Maximum	Minimum	Standard deviation	Average	Maximum	Minimum	Standard deviation
Bulgaria [a]	9	11.5	15.2	8.3	2.1	13.1	19.2	8.2	3.5
Czechoslovakia	12	5.6	10.6	1.5	3.0	5.2	12.5	0.3	4.1
Czech Republic	8	3.8	5.4	1.5	1.2	2.6	4.1	0.3	1.1
Slovak Republic	4	9.6	10.6	5.5	1.6	10.6	12.5	4.3	2.5
Hungary	20	6.1	12.6	1.8	3.1	11.4	20.5	5.6	4.3
Poland	49	10.5	17.3	3.8	3.9	13.6	23.5	5.8	4.6
Romania [b]	41	2.4	6.0	0.7	1.1	8.0	16.0	3.0	2.9
Ex-GDR *Länder*	6	11.7	14.0	10.4	1.1	14.1	15.4	13.0	0.9

Source: National statistics supplemented with secretariat estimates of regional labour forces. In some cases, regional unemployment rates are quoted using a different denominator from that normally used for presenting the national rates; there may therefore be some differences between the figures presented in this and the following table and those showing the development of unemployment rate over time.

Note: Standard deviations are weighted by the size of regional labour force.

[a] Figures refer to December 1991 and June 1992 respectively.
[b] Regional unemployment rates or labour force figures are not available for Romania. The estimates presented in the table assume that the regional distribution of the labour force is the same as that of the overall population; i.e., activity rate of the population is assumed to be the same for all regions.

fact, this is one of the major reasons for falling unemployment rates in Czechoslovakia during 1992. In most other countries, the proportion of the unemployed covered by such schemes is much lower: about 7 per cent of the unemployed in Hungary in 1992, a proportion which is set to fall further in 1993, less than 1 per cent of the unemployed in Poland (sharply down from 1991) and about 1.3 per cent of the unemployed in Russia (though this proportion is higher in some of the other CIS countries where unemployment is still very low).

Such figures are very different from those in the ex-GDR *Länder*, where a much faster rate of labour shedding and hence higher unemployment has been accompanied by massive expenditure on labour market programmes. At the end of September 1992, compared to a total unemployment figure of 1.1 million, over 560,000 people had taken early retirement, 250,000 were on short-time work, 490,000 were estimated to be on government retraining programmes and another 375,000 were employed on labour market programmes. In other words, in the absence of these active labour market schemes, numbers unemployed in the ex-GDR *Länder* would reach 1.8 million, or more than 20 per cent of the labour force, at the end of September 1992. The contrasts in the scope of labour market programmes between the ex-GDR territories and the other transition countries obviously reflect the huge differences in resources available for such programmes but, by extension, they imply differences also in the pace with which the structure of the economy can be changed. Mass unemployment will be far more difficult

TABLE 3.2.8

Selected transition countries: Change in employment by branch
(Percentage change at annual rate between 1990-QI and most recent available observation)

Latest observation:	Bulgaria 1992-QII	CSFR [a] end-1991	Hungary 1992-QIII	Poland 1992-QIII	Croatia 1992-QIII	Slovenia 1992-QII	Fed.Rep.of Yugoslavia 1992-QII	Ex-GDR Länder 1992-QIII
Agriculture	-23.6	-11.0	-24.6	-23.0	-4.3	-9.7	-3.9	-41.6
Industry	-17.9	-8.0	-13.5	-7.8	-12.0	-10.4	-5.4	-25.5 [b]
Construction	-27.5	-2.0	-26.3	-6.0	-19.3	-17.1	-9.8	-25.5 [b]
Trade	-25.8	-6.4	-17.5	-13.1	-10.8	-7.5	-7.0	-11.3 [c]
Transport and communications	-11.5	2.0	-9.8	-11.6	-11.7	-8.1	-5.3	-11.3 [c]
Finance, insurance	13.0	36.2	..	10.8	-13.2	3.2	-7.2	-8.9 [d]
Other	-8.1	-6.2	..	-6.7	-9.0	-5.4	-5.7	-8.9 [d]
Total	-17.5	-5.8	-16.9	-8.8	-11.3	-8.5	-6.0	-16.8

Sources: National statistics; DIW estimates.

Note: This table should be interpreted as giving some idea of relative employment changes by branch within each country. Because of the wide differences in statistical coverage between countries, it should not be used to compare the size of changes in employment between countries.

Country notes: Bulgaria: employees in state sector only and all changes are relative to average figures for 1990. CSFR: all forms of employment in all sizes of enterprises. Hungary: employees in enterprises in the material sector employing over 50 persons only. Poland: employees in enterprises employing five persons or more. Yugoslavia, Croatia and Slovenia: employees in social sector only. Ex-GDR: DIW estimates of total employment. The total figure includes those employed in ABM-special employment measures.

[a] As no figures are available for employment by branch in Czechoslovakia in 1992, the figures in the table give the change between the end of 1989 and the end of 1991 expressed at an annual rate.
[b] DIW do not give separate estimates for industry and construction.
[c] DIW do not give separate estimates for transport and distribution.
[d] DIW do not give separate estimates for finance and other branches.

TABLE 3.2.9

Czechoslovakia and Poland: Comparison of wide and narrow measures of employment
(Levels in thousands, rates in per cent)

	Narrow measure			Wide measure		
	Level end-year	Change Level	Change Rate	Level end-year	Change Level	Change Rate
Czechoslovakia						
1989	7 459	..	..	7 995	..	..
1990	7 208 [a]	-251 [a]	-3.4 [a]	7 686	-309	-3.9
1991	5 993	-1 215	-16.9	7 292	-394	-5.1
Poland						
1990	11 286	..	..	17 558	..	..
1991	10 167	-1 119	-9.9	16 474	-1 084	-6.2
1991	9 264	-903	-8.9	15 861	-613	-3.7

Source: National statistics.

Notes: The figures for the narrow measures of employment are those for the last quarter of the year; those for the wide measures describe the situation at the end of the year. For Czechoslovakia, the narrow measure covers employees in all enterprises with more than 100 employees (including agricultural cooperatives), the wide measure covers total employment in the whole economy (including women on maternity leave). For Poland, the narrow measure is the number of employees in enterprises employing five or more, expressed as full-time equivalents and excluding private agriculture; the wide measure covers all forms of employment in all sectors and is not on a full-time equivalent basis.

[a] Figures and changes are given for the first quarter of 1990 as figures for the final quarter of 1990 on a comparable basis are not available. If the changes between 1990-QI and 1990-QIV are adjusted to be on an annualized basis the fall in the level of employment becomes 335,000 or 4.5 per cent.

---

to tolerate in countries where the resources and expertise available to deal with it are more limited.

Carrying out the other main "active" function of labour administration, matching workers efficiently to the available jobs, is also proving difficult. Labour offices are generally understaffed and under-resourced relative to the level of unemployment and staff are not yet adequately trained to respond to the needs of the developing market economy.

As far as the provision of benefits is concerned, schemes were initially fairly generous and benefits amounted to a relatively high proportion of previous earnings. As unemployment increased, these benefit regimes were tightened to exclude those thought not to be genuinely unemployed. For example, in Poland the previously economically inactive could initially receive benefits by registering as unemployed; they were excluded from benefits as from July 1990. The maximum period of eligibility for unemployment benefit was lowered, generally to either six, nine or twelve months, levels of benefits fell in relation to average wages and more stringent conditions on claimants' job search activity were imposed.[172]

---

[172] Such conditions usually take the form of an obligation on the claimant to produce evidence that they have been looking for work or a formal commitment to accept any "suitable" job. In practice, such conditions are difficult to enforce when unemployment is rising rapidly and labour offices are understaffed.

While imposing time limits on the duration of unemployment benefit has often been proposed in market economies as a means of reducing unemployment by encouraging the unemployed to look harder for work, such proposals are usually combined with a government guarantee of retraining or employment once benefits run out and also the provision of "last chance" social assistance to those who would otherwise be destitute. In the transition economies, there is no prospect of such employment guarantees and social assistance programmes are extremely limited. There is therefore a risk that a substantial proportion of the unemployed will sink into poverty once their benefit entitlements run out, putting at risk the social consensus required for the transition to a market economy to be successful.

More generally the social safety net should not be interpreted narrowly to denote solely unemployment benefit programmes. If no other programmes are available, the labour market functions of unemployment benefit schemes will be overwhelmed by the need to provide income support for individuals who have much weaker links to the labour market (pensioners, the disabled, single parents etc.). Indeed, unemployment benefits should increasingly be distinguished from social assistance in both the level of payments and the method of financing if a reasonably efficient system of social security is to be set up.[173]

Income distribution and poverty

There have been widespread press reports that price liberalization has led to a widening of the income distribution and a sharp increase in the incidence of poverty. Unfortunately, given the large gaps in statistical coverage, a comprehensive analysis of this issue is not possible. However, some figures which illustrate the scope of *poverty* in Russia and Czechoslovakia are available.

In Russia, it is estimated that the proportion of the population with incomes below the "physiological" minimum (the level deemed necessary for physical subsistence) rose from 1 per cent in 1991 to a peak of 19 per cent in March 1992 and is estimated to have ended the year at just over 10 per cent. If the more generous "living" minimum is taken as the poverty level, over 40 per cent of the population were estimated to have incomes below the poverty line at the end of 1992.[174]

In contrast, it is estimated that about 3 per cent of Czechoslovak households at the beginning of 1992 received incomes below the "life minimum", the official poverty level used to allocate social benefits, while another 9 per cent had incomes which were only just above this level and below the more generous "social" poverty level. The number of households with "social" poverty level of real incomes were expected to decline by about a third by the end of 1992 due to the increase in real wages during the year.[175] The sharp difference in the incidence of poverty between the two countries appears not to be a statistical artifact and is probably due to a combination of the poor performance of the Russian economy in 1992, in particular, the very sharp rise in prices, and also because of the more developed social security system in Czechoslovakia.

As far as *income distribution* is concerned, it seems that the distribution of both income and wages is becoming more unequal, though again precise evidence is hard to obtain and the significance of these changes is hard to evaluate.

In Poland, evidence from household surveys indicates that the distribution of farmer's incomes actually became more compressed between 1989 and 1991 (the share of the top 20 per cent fell from 42.8 per cent of total farmers' incomes in 1989 to 38.2 per cent in 1991 while the share of the bottom 20 per cent increased from 6.4 per cent to 8.3 per cent) as average farm incomes fell sharply. Wage and salary earners on the other hand saw the income distribution widen slightly in the same period (the share of the top 20 per cent increasing from 34 per cent to 35 per cent of total income and that of the bottom 20 per cent decreasing slightly from 10.1 per cent to 9.9 per cent). Unfortunately aggregate figures for all households together are not available.[176]

The Bulgarian central bank reports that the wage distribution has recently widened with "the ratio between the highest wages and the lowest wages" rising from 1.5 in the first half of 1991 to 2.15 in the same period of 1992.[177] Finally, in Russia, the ratio between the top and bottom decile points of the income distribution is reported to have risen from 5.4 in 1991 to 6.5 in the second quarter of 1992.[178]

Note on the Hungarian Labour Force Survey

At the time of writing, Hungary is the only east European country to have published results of a household labour force survey. Such surveys measure unemployment according to the internationally accepted ILO definition and thus permit, in principle at least, international comparisons of the level and structure of unemployment to be made. Such comparisons cannot be made using the monthly count of the registered unemployed. While the results presented below

---

[173] For further details, see N. Barr, *Income transfers and the social safety net in Russia*, Studies of Economies in Transformation, No.4, World Bank, Washington, D.C., 1992; A.B. Atkinson, *The social safety net*, mimeo, London School of Economics, London, 1991.

[174] Centre for Economic Conjuncture and Prognosis, *Russia – 1992: Economic Conjuncture*, Moscow, 1992, p.63.

[175] *Prague Economic Papers*, April 1992, p.302.

[176] Central Statistical Office (GUS), *Rocznik Statystyczny 1992*, Warsaw, 1992, table 6, p.214.

[177] It is not clear what is the precise nature of this ratio. The figures are taken from *Bulgarian National Bank Report*, January-June 1992, Sofia, 1992.

[178] Russian Federation Goskomstat, op.cit., p.11.

TABLE 3.2.10

Comparison of registration- and survey-based measures of unemployment in Hungary, 1992
(Thousands and per cent)

	QI	QII	QIII
Number of unemployed (thousands)			
Monthly unemployment count [a]			
Total	458.4	523.8	601.5
Male	274.8	308.9	351.9
Female	183.6	214.9	249.5
Labour Force Survey			
Total	421.4	430.9	436.2
Male	251.9	256.1	255.9
Female	169.5	174.8	180.3
Unemployment rate (per cent)			
Monthly unemployment count			
Total	9.6	11.1	13.1
Male	11.2	12.7	14.9
Female	8.0	9.4	11.3
Labour Force Survey			
Total	8.9	9.1	9.5
Male	10.3	10.5	10.8
Female	7.4	7.7	8.2

Source: National statistics.

Note: The unemployment rates in this table have been calculated using the economically active population as measured by the corresponding 1992 household survey as the denominator. They will therefore differ from the figures presented in the other tables in this section, which use the figures for the 1991 economically active population.

a: Average of the figures for the three months in the quarter.

may be specific to Hungary and may be affected by their own statistical problems,[179] they do illustrate some of the problems which can arise in interpreting the regular monthly unemployment statistics, especially when these are produced by labour offices which have only been recently set up and may still be in the process of establishing their role in the labour market.

The Hungarian survey is carried out every quarter and the first published results cover the first three months of 1992. This note compares the regular monthly unemployment statistics with the survey-based measure of unemployment for the first three quarters of 1992. As can be seen from table 3.2.10, the monthly unemployment count is higher than the figures indicated by the survey, suggesting that, on a standardized basis, Hungarian unemployment is overestimated by the monthly unemployment count. Moreover, the difference between the two measures increased substantially over the first three quarters of 1992, with the survey-based measure indicating a much smaller increase in unemployment than the registration-based figures.

The reasons for this discrepancy are apparent from table 3.2.11, which breaks down those who are registered with the labour centres as unemployed into those who were already in employment as measured by the household survey, those who were not in the labour force (because they were either not looking for work or were unavailable for work and so are not classified as unemployed on the ILO definition) and those who were defined as unemployed according to the survey-based definition.

The table shows that the major reason for the increase in the registration-based measure of unemployment in the second and third quarters of 1992 was the sharp increase in the number of unemployed according to the ILO definition (line 3). By the third quarter of 1992 this process of increasing coverage of the unemployed by the labour offices was almost complete and only 5,600 ILO unemployed remained who had not registered with the labour offices (line 5). Thus, while changes in the monthly unemployment count did not reflect changes in the survey-based measure for most of 1992, it is now more likely that the two measures will move together in the future, albeit at different levels due to the exclusion of already employed (line 1) and economically inactive (line 2) from the survey-based measure. As a result, unless the pace of job losses accelerates markedly in the next few months or patterns of participation in the labour market change, it is to be expected that the monthly unemployment figures will show smaller rises in the next few months than they did in 1992.

The two measures of unemployment agree fairly closely on the structure of unemployment. Both measures show that the average duration of unemployment is increasing, that unemployment is higher among men than among women and much higher among the young than amongst older workers. Finally, the unemployment rate is much higher for those with no or

---

[179] Household surveys have their own statistical problems of which differential non-response (i.e., the response rate to the survey differs significantly between different groups so that the overall results are biased) is one of the most important. The Hungarian survey appears to suffer from this problem to some extent, as the Hungarian Central Statistical Office reports that the Labour Force Survey unemployment rate in the third quarter "corrected" for differences in response rates between the active and inactive populations would be 10 per cent, against an uncorrected figure of 9.5 per cent.

TABLE 3.2.11

Reconciliation of the two measures of unemployment in Hungary, 1992
(Thousands)

	QI	QII	QIII
Registered with the labour office (monthly unemployment count) [a]	458.4	523.8	601.5
of which:			
(1) Already employed	28.9	29.9	28.3
(2) Economically inactive	113.2	112.1	142.6
(3) ILO unemployed	316.3	381.9	430.6
(4) Total ILO unemployed (survey-based measure of unemployment)	421.4	430.9	436.2
(5) ILO unemployed but not registered with the labour offices (derived as residual (4)-(3))	105.1	49.0	5.6

Source: National statistics.

[a] Average of the figures for the three months in the quarter.

TABLE 3.2.12

Measures of employment change in Hungary, 1992

	QI	QII	QIII
Levels *(thousands)*			
(1) Total employment [a]	4 332.4	4 281.3	4 139.5
(2) Employment in material sector [b]	1 957.4	1 923.6	1 850.5
(3) Estimate of employment in small enterprises and non-material sector [c]	2 375.0	2 357.7	2 289.0
Change over previous quarter *(per cent)*			
(4) Total employment [a]	..	-1.2	-3.3
(5) Employment in material sector [b]	..	-1.7	-3.8
(6) Estimate of employment in small enterprises and non-material sector	..	-0.7	-2.9

Source: National statistics.

Note: For reasons given in the text, the estimate of employment in smaller firms and the non-material sector is only illustrative.

[a] Figures taken from the Labour Force Survey.
[b] Figures taken from quarterly statistics from enterprises in the material sector with more than 50 employees.
[c] Derived as residual, (1)-(2).

few educational qualifications than for those who are more highly qualified.

As far as the employment situation is concerned, the regular employment statistics produced in Hungary are based on reports from large enterprises and thus exclude most of the growing private sector. In addition, the quarterly Hungarian figures suffer from the additional disadvantage of only covering the material sector of the economy so that they exclude most of the service sector. These partial figures are therefore likely to be misleading indicators of the overall development of employment. On the other hand, figures derived from household surveys are far more likely to pick up the growth of small private enterprises and are not restricted in their sectoral coverage. In these circumstances, the labour survey provides the only comprehensive figures for the development of employment in 1992 which are yet available.

While the precise definitions of employment used by the two sources differ (for example, the enterprise-based figures for the material sector count the number of jobs; the household survey measures the number of people employed, so differences may emerge if some individuals have two jobs; the two surveys may have been carried out at different times and figures in the household survey may have been biased by non-response problems),[180] a comparison of the changes over time in the two figures for total employment may give some idea of the growth of the private and service sectors. Table 3.2.12 presents some illustrative calculations: as might be expected, the wider measure of employment shows a smaller fall in quarter on quarter fall in employment than does the narrow measure in both quarters, suggesting that employment in small enterprises is not falling as fast as it is in larger firms.

(ii) Prices and inflation

This section provides a brief summary of consumer price trends and developments in nominal and real incomes in the transition economies. Inflation generally abated in most east European countries, but none the less remained disturbingly high with at least double-digit annual rates; it was two orders of magnitude

---

[180] Figures for self-employment and employment in small enterprises may be particularly affected by non-response if respondents believe that statistical information from the survey will be made available to the tax authorities. Employees in large enterprises are likely to be less affected by such considerations.

TABLE 3.2.13

European transition countries: Rate of consumer price inflation, 1991-1993

(*Percentage change*)

	Period over period of preceding year					Month over preceding month						
	Jan.-Dec. 1991	Jan.-Mar. 1992	Jan.-June 1992	Jan.-Sept. 1992	Jan.-Dec. 1992	Aug. 1992	Sept. 1992	Oct. 1992	Nov. 1992	Dec. 1992	Jan. 1993	Feb. 1993
Albania	104.1 a	..	..	..	249.1 a	45.6	9	11	3.5	3.5	..	..
Bulgaria	338.5	..	..	..	79.3	1.2	3.4	6.2	6.7	4.6	6.9	4.7
Czechoslovakia	57.9	16.4	11.8	10.4	..	0.6	1.8	2.0	2.0	..	..	..
Czech Republic	52.0 a	15.7	11.5	10.6	11.1	0.6	1.9	1.9	..	..	8.1	..
Slovak Republic	58.3 a	17.7	12.5	10.3	10.0	0.7	1.6	2.1	1.7	1.0	..	..
Hungary	35.0	26.2	24.2	23.1	23.0	0.8	2.4	2.5	1.6	1.1	6.8	1.7
Poland	70.3	41.1	40.9	42.0	43.0	2.7	5.3	3.0	2.3	2.2	4.3	3.4
Romania	165.5	251.8	231.8	217.4	210.4	3.4	10.1	9.6	13.5	13.2	11.5	..
Yugoslavia (SFR)	118.0	466	..	..	..	..	..	..	..	..	..	..
of this:												
Croatia	124.2	..	384.3	467.8	634.0	21.5	28.9	38.2	29.0	25.3	..	..
Slovenia	117.7	284	..	257.1	201.3	1.4	2.7	3.4	2.8	1.1	..	..
Yugoslavia (FR)	121.8	475.3	1 397.2*	3 938.1	9 337.0	44.2	73.7	49.2	33.3	46.6	100.6	211.8
CIS	89.1	760	710	852	..	9.0	12.6	19.7	24.0	27.0	..	..
Armenia	95.6	557.0	584	633	729	9.2	9.3	22.2	34.0	18.5	..	..
Azerbaijan	88.6	717	662	836	..	10.0	2.4	16.0	22.0	94.5	..	..
Belarus	81.5	637	637	857	1 016	9.0	8.1	14.6	20.0	35.8	..	..
Kazakhstan	86.0	614	554	694	885	7.5	12.0	16.9	20.0	20.6	..	..
Kyrgyzstan	89.6	606	696	841	906	4.5	27.8	23.7	24.0	20.8	..	..
Moldova	100.7	809	817	791	941	4.5	11.1	19.2	32.0	21.8	..	..
Russia	91.8	807	775	901	1 110	8.6	11.5	22.9	26.1	24.9	27	23
Tajikistan	91.9	585	502	779	913	17.5	10.4	8.0	8.0	13.6	..	..
Turkmenistan	86.7	571	481	571	710	21.8	9.0	4.2	9.0	14.3	..	..
Ukraine	84.6	734	635	836	..	14.0	18.0	20.0	20.0	30.0	..	..
Uzbekistan	85.0	544	458	533	599	6.5	2.5	12.9	22.0	11.2	..	..
Georgia	..	..	..	..	..	..	..	..	..	..	..	..
Estonia	155.8	1 156	..	..	1 076	18.8	6.6	7.7	9.5	3.3	3.1	1.7
Latvia	172.2	567	546.7*	..	951.2	16.3	12.1	25.1	12.0	2.6	..	..
Lithuania	216.4	..	707.6	..	1 020.8	14.2	29.4	18.9	29.0	27.7	..	..
Ex-GDR *Länder*	17.3 b	15.6	14.9	..	..	..	..	..	..	..	..	..

Sources: National statistical publications and direct communications to ECE.

Notes: Albania, Bulgaria, Czechoslovakia, Hungary, Poland, Romania: consumer prices. Croatia, Slovenia: retail prices. Yugoslavia (FR): consumer prices. CIS countries: retail prices, goods only, except for the Russian Federation, where the monthly rates of change are taken from a new series of consumer prices (goods and services). Estonia: consumer prices. Latvia: consumer prices of goods and services. Lithuania: retail prices through May, consumer prices from June.

a Within-year (December from previous December).
b Second half of 1991, compared to the same period of 1990.

higher in most successor states of the former Soviet Union and some of the ex-Yugoslav republics. Nominal incomes outpaced price advances in some of the east European countries, but real incomes contracted sharply in the former Soviet Union.

Price developments

Inflation continued to be a significant concern in all of the transition countries in 1992, but the dimensions of the problem and its causes varied widely between countries. In those countries which had started their reforms earlier, inflation rates were gradually coming down, although the pace was in all cases considered excessive. In other countries, overall inflation remained high, although the rates fluctuated significantly in the course of the year (see table 3.2.13, second panel).

As measured by the *consumer price index* (CPI), the Federal Republic of Yugoslavia has been in full hyperinflation[181] since April 1992, and in Croatia the pace was only somewhat less devastating at some 30 per cent per month in the last quarter. On the other hand, in Czechoslovakia, Hungary, Poland and Slovenia, inflation ran 2-3 per cent a month in the same period, generally down from earlier in the year. Somewhat higher single-digit average monthly rates were registered towards the end of the year in Albania and Bulgaria, while in Romania the pace accelerated to some 12 per cent a month in the last quarter. In several countries, larger increases were noted in January-February 1993, but as these were generally associated with changes in fiscal regimes or other shifts in the environment, they may be one-on adjustments and not necessarily indicative of changes in the short-term trend. Except for Croatia and FR Yugoslavia, inflation thus generally ran at *annualized* rates of 20 to 100 per cent in the east European countries towards the end of the year.

---

[181] Conventionally defined as price advances of 50 per cent per month or more.

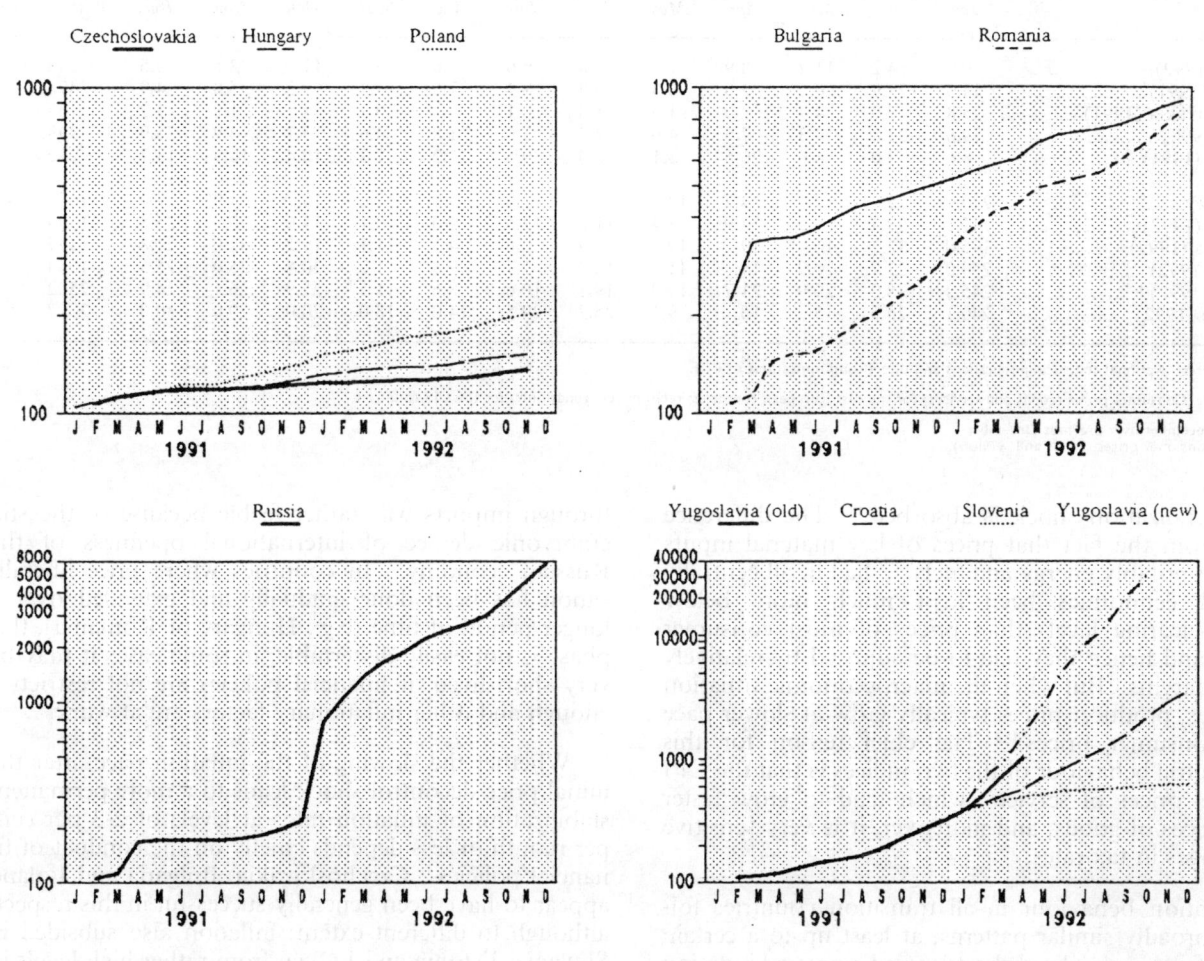

CHART 3.2.1
Rate of inflation, 1991-1992
(January 1991 = 100)

Source: National Statistics.

On the other hand, in most successor states of the *former Soviet Union*, consumer prices were rising at some 25 per cent a month in the last quarter of 1992 (1350 per cent annualized), with an accelerating pace. However, the Baltic states — except for Lithuania — had brought down inflation from double-digit monthly rates in the first semester of 1992 to some 3-5 per cent at the end of the year.

A survey of trends in *year-over-year* average price level changes, by contrast to the *current* pace of inflation, yields a somewhat different ranking (first panel of table 3.2.13). By comparison to the experience in 1991, a marked slow-down of year-over-year inflation can be observed in most east European countries in 1992, most strikingly in Czechoslovakia, but also in Bulgaria, Hungary and Poland. By contrast, inflation accelerated in Albania, Romania and Slovenia, and much more sharply in the other ex-Yugoslav republics, where it was chiefly fuelled by rapidly growing expenditures linked to war financing and the disruption of inter-republican trade links. Very much higher year-over-year price level changes than in eastern Europe and significant acceleration were registered in all the ex-Soviet states in 1992, primarily in reflection of the fact that for most of them this was the first year of substantial price liberalization. For various reasons, the initial price liberalization shock in these countries was much larger both than expected and than what had been experienced in eastern Europe at the same stage of reforms.

The consumer price index is not always the best measure of inflation, especially if the impact of inflation on investment decisions or on international competitiveness of a country is under review. For this purpose, a *producer price index* would be preferable. Available data suggest that producer prices tend to increase much faster immediately after price liberalization, but rise more slowly than consumer prices once the in-

TABLE 3.2.14

European transition countries: Rate of inflation after price liberalization shocks, 1990-1992, month over preceding month
(*Percentage change*)

	Jan.	Feb.	Mar.	Apr.	May	June	July	Aug.	Sept.	Oct.	Nov.	Dec.	Within-year [a]
Albania (1992)	9.9	14.2	11.2	8.9	7.8	5.8	6.6	45.6	9	11	3.5	3.5	248.7
Bulgaria (1991)	13.6	122.9	50.5	2.5	0.8	5.9	8.4	7.5	3.8	3.3	5.0	4.9	473.7
Czechoslovakia (1991)	25.8	7.0	4.7	2.0	1.9	1.8	-0.1	-	0.3	-	1.6	1.2	53.6
Poland (1990)	79.6	23.8	4.3	7.5	4.6	3.4	3.6	1.8	4.6	5.7	4.9	5.9	249.3
Romania (1991)	8.6	7.0	6.6	26.5	5.1	2.0	9.5	11.2	7.3	10.4	10.9	13.7	222.8
CIS (1992)	220	25	17	14.2	10.5	15.4	13.0	9.0	12.6	19.7	24.0	27.0	1680
Belarus (1992)	141.1	33.4	19.5	13.9	13.4	11.8	12.5	9.0	8.1	14.6	20.0	35.8	1949
Moldova (1992)	195.0	32.2	10	19	19	4.9	0.6	4.5	11.1	19.2	32.0	21.8	1377
Russia (1992)	250.0	23.7	20.9	14	11	13.4	7.7	7.8	11.2	20.8	25.0	25.5	1653
Russia (1992) [b]	245.3	38.0	29.9	21.7	11.9	19.1	10.6	8.6	11.5	22.9	26.1	24.9	2502
Ukraine (1992)	240.1	25.5	5	15	9.1	25.7	29.5	14.0	18.0	20.0	20.0	30.0	2077

Sources: National statistical publications and direct communications to ECE.

Note: Consumer prices (goods and services) for the east European countries; retail prices (goods only) for the CIS countries.

[a] December over previous December.
[b] Consumer prices (goods and services).

itial liberalization shock is absorbed.[182] The difference arises from the fact that prices of key material inputs — most notably energy and fuels — had in all centrally planned economies been kept much more below market-clearing levels than prices of most consumer goods, and hence they had to be adjusted by relatively larger margins. But when the macroeconomic situation stabilizes, producer prices typically grow at slower pace than consumer prices.[183] The chief reason for this asymmetric behaviour of the two indices is that the CPI includes prices of services, which tend to grow faster than prices of goods, and moreover, it is very sensitive to changes in tax rates.

Inflation behaviour in all transition countries followed broadly similar patterns, at least up to a certain point: from generalized shortage and repressed inflation under central planning in its final days, to an explosion of open inflation associated with price decontrol at the outset of the stabilization operation, and to gradual deceleration of inflation afterwards (see chart 3.2.1). The dynamics of the process and the duration of particular stages have, however, been different across countries, reflecting the varying degree of initial price controls, scope of subsequent price liberalization, and tightness of accompanying financial policies. This latter factor has also been responsible for dissimilarities in inflation behaviour after price liberalization. Typically, in all countries which went through a "shock" price liberalization, inflation dropped to single-digit monthly rates already after 2-3 months, as exemplified by the experience of Poland, Czechoslovakia and Bulgaria (see table 3.2.14). This slowing after the liberalization shock was, however, weaker in Russia and other CIS states, in part because the initial "corrective" inflation was much higher than in any of the east European countries while the moderating impact of foreign competition through imports was rather feeble because of the still embryonic degree of international openness of the Russian economy. In countries where price liberalization has been done gradually and extended over a longer period of time (e.g., Hungary or Romania), this phase is not easily discernible; in some cases it may be very short-lived, if financial policies are not restrictive enough and other inflationary factors are at work.

Whether the slowing of the inflation pace after the initial price decontrol shock leads to a more permanent stabilization of inflation rates at levels of 2-3 per cent per month or less depends chiefly on the conduct of financial policies. Czechoslovakia, Bulgaria and Poland appear to have been generally successful in this respect, although to different extent; inflation also subsided in Slovenia, Estonia and Latvia, from rather high levels in the course of 1992, curbed by financial restrictiveness following monetary reforms introduced in these countries. However, failure to persevere with financial restrictions results inevitably in renewed acceleration of inflation, as demonstrated by price developments in Russia, where the monthly rate first came down from 250 per cent in January to 21 per cent in March and less than 9 per cent in August, but then increased again, as noted, in November-December 1992.

Sources of inflation

The anti-inflationary policies of many transition countries in eastern Europe could claim success if not the disturbing and somewhat puzzling *persistence* of inflation after the initial adjustment shock. The phenomenon was first observed in Poland, where inflation dropped to 3-4 per cent per month at the end of the first quarter of 1990 but has only little declined since then. A similar situation developed in Bulgaria and in

---

[182] For instance, during the first month of the Polish stabilization programme, industrial prices increased by 110 per cent as compared with 79 per cent increase for consumer prices. A similar development can be observed in Russia, where the growth of industrial wholesale prices in January 1992 was faster that of consumer prices (380 per cent and 250 per cent, respectively).

[183] For instance, the industrial price index in Poland lagged behind the consumer price index both in 1991 and in 1992 (48 per cent versus 70 per cent in 1991, and 28 per cent versus 43 per cent in 1992, respectively). The corresponding figures for Hungary were 32 and 35 per cent in 1991 and 9 and 22 per cent in 1992.

Czechoslovakia, although in the latter case the inflation rate was significantly lower than in the two other countries.

The reasons why inflation remains at relatively high levels should probably be sought in a combination of factors ranging from government policies to wage and labour market rigidities, structural constraints, and exogenous shocks. These factors operate both from the demand and from the supply side, but their relative significance seems to have changed over the last three years.

Monetary overhang, which was notorious in centrally planned economies, is the cumulative result of long-term excess demand under controlled prices, and translates into distorted demand structure, increased velocity of money circulation ("excess" liquidity), falling demand for domestic currency, and "dollarization" of the money stock. Price liberalization results in onetime upwards adjustment of the general price level, which normally "eats up" most of the monetary overhang and thereby restores a "stock" equilibrium. If the stabilization operation is credible, money supply is not accommodating, and the growth of incomes is controlled, price expectations tend to reverse quickly, demand for money starts to grow, "excess" liquidity disappears gradually, and the inflationary pressure coming from the demand side alleviates. *Demand-side* inflationary pressures can be expected to continue after the price liberalization phase, however, if macroeconomic policy is unable to restrict growth of monetary incomes, and if expectations are not reversed. In this case, inflation is fuelled by a "flow" disequilibrium which now becomes the driving force. This type of inflation mechanism appears to be present in the transition economies in which money supply grows excessively either due to unrestricted credit emission for enterprises or due to large budget deficits financed by central banks (Romania, Russia, Ukraine, several states of former Yugoslavia − see also section 3.4 for a discussion of monetary and fiscal sources of inflationary tendencies).

On the other hand, however, many new inflationary factors are likely to emerge on the *supply side*. Drastic increases of energy prices, typically an important part of stabilization programmes, coupled with sharp devaluation of domestic currencies, produce a radical change in relative prices which amounts to a powerful shock affecting producers. This effect is sometimes exacerbated by an increase in interest rates on existing debts, by the removal of subsidies and, occasionally, by an increase in direct taxation. The result is a sharp rise of *production costs* which may or may not be fully compensated by the rise of selling price. In addition to that, under deepening recession, unit costs of production will increase because of falling output levels, which will exert pressure for further price hikes.

If monetary policy is lax, and incomes growth unrestricted, the slowing of inflation after absorption of the price liberalization shock will be only temporary, or may not materialize at all. In such cases, inflation may easily develop into hyperinflation. The slowing of inflation after the liberalization shock will also be weak if, under moderate monetary restraint, the government subsequently raises controlled prices of important production inputs (fuels and energy), or undertakes fiscal measures affecting the general price level, such as turnover (or VAT) tax hikes, increase of customs duties, or increase of depreciation allowances. Prices are also likely to grow if the domestic currency depreciates against major foreign currencies, but then the impact depends on the conduct of monetary policy.

An important supply-side shock may come from abroad, in form of higher import prices, as was the case for the east European countries in 1991, or for the Baltic states in 1992 − both groups of countries experienced sharp rises of prices of imported fuels from Russia. The external shock may also be produced by the disruption of traditionally established trade links, which may in turn result either from economic causes (e.g., a sharp fall of production in a supplier country, removal of trade preferences by a foreign country) or from political ones (e.g., the disintegration of federal states). This factor definitely added to inflationary pressures in the former Soviet Union, where the impact of the January 1992 price liberalization was compounded by the collapse of intra-CIS trade. Finally, a supply-side shock can also be produced by entirely exogenous factors, such as natural catastrophes (earthquakes, droughts), while fluctuations of the price level around a longer term trend can be exacerbated by seasonal factors.

The inflationary impact of supply-side (or cost) factors may be neutralized to some extent by restrictive financial policies, but its complete elimination is unlikely unless monetary restrictions are extremely drastic. When profit margins are already squeezed, production costs are growing rapidly, and production technologies cannot be changed in the short run, prices are likely to increase even if aggregate demand falls. But the fall of aggregate demand is usually accompanied by a shift in the demand structure. Non-priority sectors are likely to suffer deeper recession, as households would spend higher proportions of their falling real incomes on foodstuffs and other necessities. Price behaviour in non-priority sectors would then depend on whether they are monopolized or not. In the first case, prices for non-priority goods may go up even faster than prices of necessities, because monopolies would react to the fall of demand with higher prices and reduced output. As a result, inflation will continue unabated. In the second case, cost-driven price advances may be somewhat slower in non-priority sectors, but overall inflation will continue, fuelled not only by rising prices of necessities, but also because their share in total market sales would increase at the expense of non-priority goods. Most of transition countries have painfully experienced this combination of recession and inflation in 1992.

One method of combatting inflation stemming from increased production costs or monopolistic profits is to allow for increased import competition. But for this instrument to work effectively several conditions have to be met. First, the economy should be open to imports and foreign transactions should not be restricted; second, the nominal exchange rate should be relatively stable in order to "anchor" domestic prices, and third, sufficient international reserves are needed to support the operation. Unfortunately, these conditions can be expected to be fulfilled only in open economies with liberal foreign trade regimes and access to external reserves (Hungary, Czechoslovakia and, to a lesser extent, also Slovenia and Poland), but clearly they are absent in economies such as those of Russia, Ukraine, Bulgaria, or the Baltic states. It should also be remembered that if import protection is removed too abruptly and factor markets are not functioning properly, competition from abroad may effectively destroy a large part of the industrial base of the country (for further discussion on this issue see section 3.5(iv)).

All these supply-side factors seem to have been in operation in transition countries and have been behind the continuous inflation registered in these countries in 1991 and 1992. However, their structure and impact varied from case to case. In this context, the low inflation rate in Czechoslovakia can be explained in a large part by the stability of the exchange rate and lack of "government-induced" inflationary measures, such as tax increases or energy price adjustments. This changed somewhat in January 1993, when the introduction of the VAT resulted in an immediate jump of inflation from 1-2 per cent a month to 8 per cent.[184] But this one-time impact is likely to dissipate quickly. Considerably higher inflation in Poland has been caused by repeated corrections of energy prices and turnover tax rates, the imposition of the 6 per cent import surcharge (in December 1992), as well as the creeping devaluation of the Polish zloty. An additional inflationary impulse can be expected to come in Poland in July 1993 when the VAT is to be introduced.

The *devaluation* factor played some role in sustaining inflationary pressures also in Hungary, Romania, and in the ex-Soviet states, especially in the second half of 1992. In most transition countries, rapidly rising unit costs of production may also be responsible for a large part of observed inflation. This is especially true in Poland and Bulgaria, where the profitability of enterprises has dropped sharply in 1992 and the current output levels are well below installed capacity levels. In Russia, by contrast, average profit margins in the enterprise sector have remained relatively high in spite of contracting output. In this case, the monopolistic market structure which permits corresponding pricing behaviour of Russian enterprises appears to be one of the most important inflationary factors operating from the supply side. Other factors which were clearly at work included administrative increases of energy prices,[185] and the depreciation of the rouble.

Interesting enough, *wages* do not seem to have been a significant source of inflationary pressures in transition economies. In most cases, the growth of average nominal wages lagged behind price inflation, especially in Poland and the CIS states (see table 3.2.15).

Inflation in the *former Yugoslav republics*, with the exception of Slovenia, had its roots not only in the transition process, but more importantly in the violent disintegration of the former federal state and the consequent war-type economies. Swelling budget deficits, caused by high expenditures linked with defence, care for refugees and social protection, have been behind acceleration of inflation in Croatia and FR Yugoslavia. The international embargo on trade with the latter republic resulted in acute shortages of fuels and many other goods, and added to inflationary pressures.

Most of east European transition countries have also suffered from a severe *drought* in the summer of 1992, which led to a sharp increase in agricultural and food prices. Harvests in Poland, Lithuania, Latvia, Belarus and Czechoslovakia were lower by an average of 10-15 per cent, and the impact on prices may be expected to be felt also in 1993.

(iii) Nominal and real incomes

Nominal wages continued to grow fast in all transition countries, linked to general price level increases through a variety of formal and less formal indexation mechanisms. As compared with 1991, the *monthly average nominal wage* in 1992 on a *year-over-year* basis was 20-40 per cent higher in Czechoslovakia, Hungary and Poland, 120-160 per cent higher in Bulgaria and Romania, and 200 per cent higher in Slovenia (see table 3.2.15). Not surprisingly, wage increases were much faster in high-inflation countries, such as Croatia (315 per cent), Latvia (717 per cent), Russia (891 per cent), or FR Yugoslavia, where an increase by 29 times was recorded over January-September 1992.

The *monthly data* confirm the seasonal tendency for nominal wages to accelerate sharply towards the end of year (table 3.2.16). The wage increases in December 1992 were in most countries much higher than the average monthly increases for the rest of the year.[186] In most cases these wage hikes can to a large extent be attributed to various one-time extra payments which typically occur at the end of the budgetary year (bonuses, compensations, etc.), in some cases they may reflect also more fundamental acceleration of wage and price inflation (e.g., in Russia, FR Yugoslavia, and Croatia).

---

[184] A price reaction of this magnitude is rather large given a restrictive monetary policy and will probably recede soon. It has been estimated that the introduction of a broadly-based VAT accelerates inflation by about 0.7 percentage points for each percentage point of the VAT rate, under the assumption that the monetary authorities allow money supply to grow to accommodate the introduction of the VAT — which is normally the case (S. Cnossen, "Key questions in considering a value-added tax for central and eastern European countries", *IMF Staff Papers*, Vol.39 (1992), p.225).

[185] Between December 1991 and December 1992, oil prices in Russia were increased in three consecutive rounds (January, May and September) 84 times, and prices for natural gas and hard coal by 15 and 55 times, respectively (see table 3.4.9).

[186] For instance, in Bulgaria the average wage increased in December 1992 by 17.4 per cent as compared with the November level, while the average monthly increase for January-November 1992 was only 4.2 per cent. Similarly, the corresponding figures for Hungary are 17.4 per cent and 3.4 per cent, for Poland, 9.3 per cent and 3.2 per cent, and for Romania, 14.6 per cent and 7.4 per cent.

TABLE 3.2.15

European transition countries: Nominal and real wages, 1989-1992

(Thousand national currency units; percentage change)

Country or group	Average monthly wage (Thousand national currency units)				Estimated real wage change [a] (Per cent)			
	1989	1990	1991	1992	1989	1990	1991	1992
Albania	0.465	0.468	..	..	..	..	..	..
Bulgaria	0.274	0.361	0.933	2.050	3.0	6.3	-41.1	22.5
Czechoslovakia	3.123	3.285	3.784	4.540*	0.1	-4.7	-27.0	8.1
Czech Republic	3.138	3.247	3.790	..	..	..	..	..
Slovak Republic	3.090	3.217	3.748	..	..	..	..	..
Hungary	10.571	13.446	17.948	22.939*	0.7	-1.2	-1.1	3.9
Poland [b]	0.207	1.030	1.756	2.466	8.3	-24.4	-	-1.8
Romania	3.063	3.384	7.489	19.756	2.7	6.0	-16.6	-15.1
Croatia	0.823	4.786	8.045	33.354	24.2	-16.2	-15.1	-43.5
Slovenia	1.180	5.657	10.314	30.813	18.4	-26.5	-15.2	-0.8
Yugoslavia (FR)	..	..	..	385.728	..	..	..	..
Soviet Union	0.240	0.275	..	..	7.3	9.1	..	..
Armenia	0.220	0.241	0.355	..	..	..	..	..
Azerbaijan	0.179	0.195	0.324	..	..	..	..	..
Belarus	0.228	0.265	0.539	..	..	..	10.8	12.4	..
Kazakhstan	0.234	0.265	0.459	..	..	..	..	..
Kyrgyzstan	0.198	0.219	0.363	..	..	..	..	..
Moldova	0.201	0.233	0.423	..	..	..	..	..
Russia	0.259	0.297	0.579	5.738	..	8.7	1.7	-18.1
Tajikistan	0.188	0.207	0.346	..	..	..	..	..
Turkmenistan	0.221	0.244	0.415	..	..	..	..	..
Ukraine	0.218	0.248	0.492	..	6.9	9.9	8.5	..
Uzbekistan	0.194	0.215	0.362	..	..	..	..	..
Georgia	0.198	0.214	..	..	..	..	..	..
Estonia	0.270	0.341	0.586	..	..	..	..	..
Latvia	0.250	0.291	0.570	4.565*	4.6	5.0	-28.0	-23.8
Lithuania	0.244	0.283	0.737	..	..	..	..	..
Ex-GDR Länder	..	..	..	..	..	..	..	..

Source: ECE secretariat Common Data Base, based on national wage and price data.

a Nominal wage change deflated by reported consumer price index.
b Million zlotys.

Despite the large increases in nominal terms, in some transition countries wages have lagged behind consumer prices, and hence they declined in real terms. A survey of year-to-year data on nominal wages deflated by CPI shows that in 1992 the *average real wages* fell by 2 per cent in Poland, 15 per cent in Romania, 18 per cent in Russia (and probably by the same magnitude in other European CIS republics), and a staggering 43 per cent in Croatia. In some countries, these declines continued for the second (or the third) year in a row, bringing up the cumulative fall in average real wage over 1991-1992 to 29 per cent in Romania, 40-42 per cent in the Baltic states, and to more than 50 per cent in Croatia (table 3.2.17).

Other east European countries[187] registered quite substantial increases of real wages in 1992, ranging from 4 per cent in Hungary to 22 per cent in Bulgaria. This widely disparate wage behaviour in different transition countries can partly be explained by their being at different stages of stabilization and reform programmes, and consequently by different economic policies pursued in individual countries. Typically, all transition countries suffered a sharp decline in real wages during the first year of stabilization — an essentially inevitable outcome of price liberalization under partial incomes controls. The first year of stabilization resulted in a fall of real wages by 24 per cent in Poland in 1990, and by 27 per cent in Czechoslovakia and 41 per cent in Bulgaria in the next year. The stabilization-induced decline was also rather substantial in Russia, although it came after a period of apparently unsustainable wage increases prior to 1992, as illustrated by table 3.2.15). While these data may overestimate the fall because they have been derived from changes in official prices and wages, without proper adjustment for availability of goods at official prices, the substantial drop in real wages seems to be generally undisputed.

The large initial falls in real wages are typically followed by substantial increases in the later stage of stabilization programme. This seems to demonstrate that both moderation in wage demands and the grip of incomes policies tends to get much weaker as the programme enters its second or third consecutive year, and wage earners struggle to recuperate real wage losses incurred at its early stage. For instance, real wages in Poland increased by 1 per cent in 1991, and in Bulgaria

---

[187] Except Albania for which wage statistics for 1992 are not yet available.

TABLE 3.2.16

European transition countries: Average nominal monthly wage
(Thousand national currency units, unless otherwise noted)

	1991 Aug.	1991 Dec.	1992 Jan.	Feb.	Mar.	Apr.	May	Jun.	Jul.	Aug.	Sep.	Oct.	Nov.	Dec.
Albania	..	..	..	..	..	..	..	..	..	..	..	..	..	..
Bulgaria	0.963	1.685	1.517	1.543	1.800	1.804	1.792	2.117	2.083	2.080	2.462	2.325	2.388	2.803
Czechoslovakia [a]	4.022	4.700	4.181	3.875	4.626	4.442	4.558	4.626	4.729	4.110	4.408	4.991	..	..
Czech Republic	..	..	..	..	..	..	..	..	..	..	..	..	..	5.462 [b]
Slovak Republic [a]	3.898	4.548	..	..	4.561	4.125	..	4.403	4.582	4.153	4.183	4.899	5.276	5.340
Hungary [c]	17.948	21.500*	17.862	20.477	22.433	21.395	21.639	22.310	22.732	23.491	23.270	23.566	25.795	30.293
Poland (mill.) [d]	1.818	2.286	2.082	2.112	2.308	2.405	2.277	2.337	2.464	2.482	2.603	2.753	2.944	3.218
Romania [e]	7.489	11.824	13.005	12.717	15.287	15.677	17.709	19.426	19.989	19.804	23.306	24.080	28.456	32.612
Croatia	8.045	10.916	12.508	13.619	15.172	17.593	20.806	24.245	26.892	37.430	45.342	50.323	61.894	74.417
Slovenia	10.314	16.603	18.169	20.208	24.093	26.195	30.364	30.817	33.819	34.191	35.516	37.425	38.945	41.936
Yugoslavia (FR)	8.300	14.600	15.420	19.490	27.400	43.580	68.250	118.890	181.530	273.610	418.580	658.990	1064.440	1735.560
CIS	0.530	..	..	..	..	..	..	..	..	..	..	..	..	..
Armenia	0.355	..	..	..	..	0.794	0.853	1.206	1.382	1.586	1.747	2.068	2.321	..
Azerbaijan	0.324	..	..	..	..	..	1.009	1.767	2.510	2.758	3.467	3.377	4.340	..
Belarus	0.539	..	1.313	1.598	2.268	2.606	3.484	4.319	4.424	4.709	6.464	7.502	9.312	..
Kazakhstan	0.459	..	..	..	..	2.096	2.709	3.689	3.794	4.232	6.238	6.646	8.170	..
Kyrgyzstan	0.363	..	..	..	..	1.093	1.191	1.641	1.822	1.798	2.439	2.757	4.370	..
Moldova	0.423	..	0.989	1.322	1.586	1.696	2.362	2.642	2.716	2.859	3.509	4.120	6.065	..
Russia	0.579	..	1.438	2.004	2.726	3.052	3.675	5.067	5.452	5.870	7.379	8.853	10.576	13.000
Tajikistan	0.346	..	..	..	..	1.045	1.327	1.955	1.873	..	..	2.827	3.307	..
Turkmenistan	0.415	..	..	..	..	..	..	..	..	..	..	..	..	..
Ukraine	0.492	..	..	..	..	2.973	..	..	..	5.500	..	..	..	..
Uzbekistan	0.362	..	..	..	..	1.229	..	1.908	2.013	2.306	2.788	3.934	4.568	..
Georgia	..	..	..	..	..	..	..	..	..	..	..	..	..	..
Estonia [f]	0.586	..	-	0.246	-	-	0.430	-	-	0.698	-	..	0.730	0.796
Latvia	0.570*	1.132*	1.385	1.850	2.250	2.790	3.590	4.675	4.905*	5.000	5.660*	6.350*	7.520	8.850*
Lithuania	0.737	2.165	2.279	2.798	3.240	4.118	4.978	5.768	..	..	..	..	..	..

Source: National statistics.

a Industrial enterprises with 25 workers or more for 1992 (with 100 employees or more for 1991); estimated from index on 1989 average.
b Last quarter of 1993.
c Enterprises with 50 workers or more, excluding agriculture and the budgetary institutions.
d Net monthly earnings (wages and salaries in the enterprise sector).
e Excluding private sector.
f 1991 in roubles, 1992 quarterly averages in Estonian kroon.

and Czechoslovakia by 22 and 8 per cent, respectively in 1992, even though recession continued in these countries unabated. This "catch-up" effect, however, may not manifest itself so strongly in the economies that have not gone through a radical stabilization operation (Hungary, Romania), or where wages have been affected by other exogenous factors (former Yugoslavia).

This uneven behaviour of real wages in transition countries reveals much more similarities if analysed in a *medium-term* perspective. Taking 1989 as the base year, it may be observed that real wages in east European countries were lower in 1992 by a margin of 23-25 per cent, with the magnitude of real wage decline being broadly the same across countries. This is a rather striking result, because it would suggest that real wage behaviour during transition is largely independent from widely differing national reform strategies and local conditions. But more plausible is another interpretation, which derives from resistance of real wages to fall below what is regarded by the main social groups as a minimum subsistence level. A one-fourth loss in real wage may be about a maximum sacrifice which wage earners in transition countries are prepared to accept in the medium run, and further cuts in real wages are strongly resisted. The medium-term real wage resistance may also explain the "catch-up" effect, which could be observed in those countries where the initial fall in real wages was sharper.

TABLE 3.2.17

Transition economies: Percentage change in real wages, 1990-1992
(Percentage change over preceding year)

Country	1990	1991	1992
Bulgaria	7.3	-43.0	..
Czechoslovakia	-5.4	-25.2	10.1 [a]
Hungary	-3.7 [b]	-8.0	..
Poland	-24.4	-0.3	-3.6
Romania	6.0	-16.6	-15.1
Croatia	-16.6	-25.1	-38.0 [c]
Slovenia [d]	-25.8	-11.1	-2.8
Yugoslavia (FR)	-4.7	-5.8	-50.0
Belarus [d]	12.0	-2.2	-38.3 [c]
Russia [d]	9.0	-10.5	-46.5 [c]
Ukraine [d]	11.0	-3.7	-30.0* [c]

Source: National statistics.

a Czech Republic only.
b Workers and employees in the non-agricultural sector.
c January-November 1992.
d Real incomes per capita.

It is interesting in this context to note a distinctively different wage behaviour pattern in *Hungary*, where real wages displayed remarkable stability, in contrast to other transition countries. As shown in table 3.2.17, between 1989 and 1992 the average real wage increased

TABLE 3.2.18

Transition economies: Changes in real incomes of population, 1989-1992 [a]

(Percentage change over preceding year)

Incomes, by categories	1989	1990	1991	1992
Bulgaria				
Total incomes	-0.8	11.2	-41.6	17.5 [b]
Wages	2.9	2.8	-52.6	..
Social transfers	0.0	2.5	-27.5	..
Farms' incomes	-4.5	29.9	-19.6	..
Other incomes	-6.2	24.2	-45.3	..
Czechoslovakia				
Total incomes	1.3	-2.2	-27.3	7.7 [c]
Wages	0.3	-7.5	-34.2	-1.8 [c]
Social transfers	3.6	-5.1	-21.9	4.1 [c]
Farms' incomes	1.7	-8.4	-52.3	-16.5 [c]
Other incomes	2.2	27.1	-3.1	41.3 [c]
Hungary				
Total incomes	-1.6	0.8	-9.0	-1.1 [d]
Wages	-2.0	-3.6	-10.4	-8.0 [d]
Social transfers	3.4	0.9	-5.3	-0.2 [d]
Farms' incomes	-6.4	-6.2	-33.5	-11.7 [d]
Other incomes	..	9.5	16.7	-5.0 [d]
Poland				
Total incomes	6.0	-14.7	5.9	..
Wages	6.3	-32.3	-6.6	-12.4
Social transfers	8.6	-14.3	29.3	-3.1
Farms' incomes	13.5	-49.9	-18.7	..
Other incomes	5.3	19.2	16.5	..

Source: Compiled from national statistics.

a Changes in nominal personal incomes deflated by CPI.
b January-June 1992.
c January-October 1992.
d January-November 1992.

by 1.5 per cent, and the annual changes were kept within a rather narrow band between -2 and +4 per cent. This is something of a puzzle in the light of the rather sharp decline in Hungarian output in 1991-1992, broadly comparable to that in Poland and Czechoslovakia, and in view of rapidly growing unemployment. In any case, the absence of sharp fluctuations in real wage level may have been an important reason for the much praised political and social stability in Hungary during last several years.[188]

It should be remembered that the data on changes in real wages should be interpreted only with great caution. First, if nominal average wages are deflated by the CPI, the results obtained do not properly reflect changes in living standards of particular groups of wage earners, because differences in consumption structure are ignored. But this deficiency can be dealt away through allowing for differences in households' consumption baskets. Table 3.2.17 offers a set of *official data* on changes real wages for east European countries, and changes of real per capita incomes for selected CIS countries, as provided by national statistical authorities. In most cases the data have been obtained with a different methodology, as they are derived on the basis of cost-of-living indices rather than of CPI. While the new picture thus obtained displays some differences as compared with the former one — e.g., contrary to earlier results it suggests a small decline of real wages in Hungary — the overall message seems to be the same: all transition countries registered substantial falls in real wages in 1990-1991, and that the fall has continued in majority of them also in 1992. The collapse of real wages and per capita incomes has been particularly dramatic in CIS countries and in some ex-Yugoslav republics, where the decline over the last two years has been of the order of 40-50 per cent.

A more fundamental reservation about the correctness of data on real wages refers to the very concept of "statistical real wage", which is clearly problematic if applied in the economy with market shortages and controlled prices. In most transition countries majority of prices were under administrative controls until 1990, and in some countries even in 1991 (the ex-Soviet Union). Under such conditions the real wage index, which is derived on the distorted base of "statistical" real wage from 1989 or 1990, is bound to overstate the magnitude of fall in purchasing power of the average nominal wage. It is unlikely, however, that the difference would be large enough to change the overall declining trend of real wages.[189]

A survey of changes in *real incomes* provides additional confirmation of this tendency. In all transition countries real incomes of population fell sharply in the first year of stabilization, although the impact varied for different social groups. As illustrated by data in table 3.2.18, the main "victims" were in all cases wage-earners and individual farmers, while residual incomes encompassed under the "other incomes" category generally increased.[190] The falling share of wages in total incomes, which is implied in the data on relative annual changes, is not surprising in the light of shrinking levels of employment and the surge of earnings from private commercial activities. Also, the relatively strong position of recipients of social transfers can be explained by growing unemployment compensation payments and accelerated retirements; nevertheless, the substantial increase of social transfers in real terms in Poland in 1991 or in Czechoslovakia in 1992 is more difficult to understand.

In all countries, for which data are available, real incomes fell more than the average real wages, and the difference is even larger for total wage incomes. This discrepancy may be explained by the presence of indexation mechanisms aimed at protecting individual wages under conditions of a general slump in aggregate

---

[188] One indication of social stability in Hungary is the almost total absence of strikes and other major conflicts in industrial relations. By contrast, the number of strikes in Poland increased dramatically from 305 in 1991 to 6362 in 1992, with the number of working days lost increasing from 517,000 to 1.837 million, respectively.

[189] See a more detailed discussion of this issue in the previous edition of this publication (United nations Economic Commission for Europe, *Economic Survey of Europe in 1991-1992*, New York, 1992, pp.52).

[190] This category includes non-wage incomes such as interest payments and dividends, net credits for households, transfers from abroad, insurance premiums, and in some countries also incomes from private businesses.

demand. Poland is an exception in this respect, chiefly because private sector incomes, which have been included in the "total incomes" category until 1991, accounted for a much larger proportion of the total.

The absolute and relative fall of wage earnings and farming incomes in total real incomes suggests a strong structural change in *income distribution*. Between 1989 and 1992 the share of wage earners in total incomes of population fell from 55.9 to 44.1 in Bulgaria, from 59.8 to 46.6 per cent in Czechoslovakia, and from 51.6 to 45.6 per cent in Hungary. A similar tendency has been earlier observed in Poland, where the proportion of wages declined from 45.7 per cent in 1989 to 33.9 per cent in 1991, and more recently in the former Soviet republics. This specific pattern of income distribution can be regarded as perfectly natural for countries attempting to switch their economic systems from central planning to a market mechanism, but policy-makers should be aware of social and political implications of this process.[191]

---

[191] The pattern of income distribution during transition was discussed more extensively in the last edition of this publication (see United Nations Economic Commission for Europe, *Economic Survey of Europe in 1991-1992*, New York, 1992, p.51-55).

3.3 FOREIGN TRADE AND PAYMENTS

(i) Foreign trade of the transition countries

An overview

In 1992, developments in the foreign trade sector of the transition economies broadly continued along the lines observed in the preceding two years, but the balance of of forces was somewhat different. *Intra-eastern trade* continued to contract sharply. *Trade with the market economies* also continued to fall in the case of the successor states of the *former Soviet Union,* but expanded further with a pronounced acceleration in the countries of *eastern Europe* (table 3.3.1). *Trade balances* worsened for both subregions (table 3.3.2).[192]

In *eastern Europe,* the contraction of *aggregate exports* observed for the last several years was arrested in 1992 and *aggregate imports* turned up slightly. These aggregates for eastern Europe as a whole, however, are strongly affected by the sharp downturn of trade in Croatia and the new Federal Republic of Yugoslavia under the impact of the war in that part of the region and the UN sanctions against the latter country (table 3.3.3). In all other east European countries but Albania exports increased in 1992, in most cases after steep falls in the preceding two years, and in most countries imports also expanded strongly.[193] The region's aggregate trade deficit deepened by an estimated $1 billion (table 3.3.2).

This relative improvement in the trade performance of east European countries contrasts sharply with the changes in the foreign trade of the *former Soviet republics.*[194] The large fall in output, the disruption of traditional trade links between the republics, high inflation, and a monetary and payments crisis have all led to a steep fall in trade levels, which in turn has had a negative feedback on output and consumption. These developments resulted in a fall of exports and imports of the CIS countries by about one quarter in 1992.[195] It is important to note that these large falls were not limited to sales to ex-CMEA partners. Both in 1991 and 1992, exports to the market economies fell sharply. On the *import* side, the contraction in 1991 was more rapid yet, with a fall of 36 per cent. In 1992, imports continued to fall, although at a slower rate than in 1991. The *trade balance* of the CIS countries jointly deteriorated in 1992 by some $4 billion (table 3.3.4).

The foreign trade of the *Baltic republics* in the first half of 1992 — when these countries were still largely in the rouble zone — is generally reported in rouble terms both as regards trade with the CIS countries and with the rest of the world. Nominal trade values rose very rapidly, but in the face of the rapid and uneven pace of internal price inflation and external rouble depreciation, the meaning of these data is very difficult to interpret. A large (but rapidly falling) part of external trade of Estonia, Latvia and Lithuania continued to be with the CIS countries.[196] Because of their high dependence on inputs from the CIS countries, the non-CIS foreign trade of the Baltic republics probably contracted by similar proportions in 1992 as that of the CIS countries.

The disintegration of the economic space of two large countries (Soviet Union and Yugoslavia) and the disappearance of the east German market continue to have a negative impact on the successor states and has depressed trade in the region as a whole. The impact of systemic reforms *within* the individual countries is less uniform. In some countries, the failure of stabili-

---

[192] This section updates the more detailed treatment of trade and finance in the transition economies in United Nations Economic Commission for Europe, *Economic Bulletin for Europe*, Vol.44(1992), New York, 1993, chapter 2. The foreign trade data used in this chapter exclude the trade among states which became independent towards the end of 1991 and in early 1992 — i.e., intra-USSR and intra-CIS trade, trade among the Baltic countries and between them and the CIS countries, as well as trade among the republics of the former Socialist Federal Republic (SFR) Yugoslavia. On the one hand, data on these trade flows are still quite scarce, and on the other, those that are available are difficult to put into the common framework of world trade analysis because of the very different internal price regimes that governed them until recently. For a discussion of the issue of "new" foreign trade, see section 3.3(v) below. All data refer to changes in the US dollar values of trade, partly for the sake of easy comparability, and partly out of necessity, as some eastern countries have stopped publishing foreign trade data denominated in national currency units.

[193] East European export growth without Croatia and Yugoslavia can be estimated at some 5 per cent in value and that of imports at some 6 per cent or more. These estimates reflect the ECE secretariat's feeling that strong export growth and probably even faster import expansion will eventually be recorded in Poland, for which at the moment only preliminary data on payments flows are available (see discussion below).

[194] Data through 1991 generally cover the territory of the former Soviet Union as it was prior to the resumption of independence of the Baltic states in August of that year. Statements about 1992 generally refer to the very rudimentary aggregate data for the countries of the Commonwealth of Independent States provided by the CIS Statistical Service. At least notionally these should cover the 11 republics which initially joined in the formation of the CIS (i.e., all former union republics but the Baltic states and Georgia). Lately, Azerbaijan has refused to ratify membership in the CIS. However, in any case the CIS data are likely to involve rather rough estimates, especially for states — such as, most importantly, Ukraine — which in other fields have been reluctant to provide much information to the CIS data base. In many instances, the (only slightly) more ample information on rates of change in the foreign trade of the Russian Federation alone are juxtaposed with data for the former Soviet Union in prior years.

[195] Data refer to changes in the US dollar value of trade with countries outside of the former Soviet Union. Since early 1992, the CIS states generally report all trade values in dollar terms. Comparable data for the same period in 1991 are sometimes also reported, but very little is known about the methods used to convert the transferable rouble portion of trade — still relatively important in the first half of that year — into dollars. The rates of change of total trade and trade with the ex-CMEA countries in 1992 reported here depend on these somewhat obscure valuations.

[196] Some information on this is provided in the section below on "new" foreign trade (section 3.3(v) below).

TABLE 3.3.1

European transition countries: Foreign trade, by direction, 1990-1992
(Value in billion US dollars; growth rates in percentages) [a]

Country or country group [b]	Exports				Imports			
	Value	Growth rates			Value	Growth rates		
	1991	1990	1991	1992	1991	1990	1991	1992
Eastern Europe, *to or from:*								
World	58.6	-2.8	-6.9	0.1	62.1	4.6	-4.1	1.4
Transition economies	17.3	-15.6	-24.6	-23.5 [c]	17.2	-11.2	-19.8	-8.0 [c]
Soviet Union successor states	10.6	-16.1	-25.1	-35.9 [c]	11.4	-10.8	-9.3	-3.4 [c]
Eastern Europe [d]	4.8	-25.6	-20.1	-7.6 [c]	4.5	-17.3	-25.8	-11.0 [c]
Developed market economies	34.9	9.9	6.6	24.3 [c]	37.0	19.1	7.8	22.3 [c]
Developing countries	6.4	-12.6	-11.8	15.4 [c]	7.9	6.7	-9.2	-38.1 [c]
Soviet Union/Russia, [e] *to or from:*								
World	46.7	-5.2	-24.6	-25.2	45.4	-	-35.9	-21.3
Transition economies	14.0	-24.3	-35.0	-25.8	14.3	-10.6	-43.4	-42.8
Eastern Europe [d]	9.2	-26.9	-40.8	-32.7 [f]	8.8	-12.1	-51.6	-49.7 [f]
Developed market economies	26.4	12.3	-16.2	-20.3	26.4	5.6	-31.0	-13.0
Developing countries	6.3	-9.5	-29.0	-44.0	4.7	3.8	-35.8	-2.6

Source: Secretariat of the United Nations Economic Commission for Europe, based on national statistical publications and direct communications to the ECE secretariat from national statistical offices.

a Growth rates are calculated on values expressed in US dollars. Trade with "transition" and east European countries through 1990 was valued on the basis of an adjusted dollar measure reflecting consistent rouble dollar crossrates. For details of the revaluation, see the note to table 2.1.4 and the discussion in box 2.1.1 and section 2.1(iii) in United Nations Economic Commission for Europe, *Economic Bulletin for Europe*, Vol.43 (1991), New York, 1991. All trade values for 1991 and 1992 were either originally reported in dollars or were converted to dollars at the appropriate national conversion coefficient (usually the "commercial" rate quoted by national banks).
b "Eastern Europe" refers to Albania, Bulgaria, Czechoslovakia, Hungary, Poland, Romania and Yugoslavia and its successor states. The partner country grouping follows the practice until recently prevalent in the national statistical sources, which differs from the breakdown usually employed in United Nations publications. Thus, "transition economies", which covers the ex-socialist trade partners, in addition to the east European countries, the Soviet Union, and the Asian centrally planned economies, includes Yugoslavia and Cuba. "Developed market economies" differs from the aggregate used in section 3.3.2 below by the exclusion of Turkey and the inclusion of Australia, New Zealand and South Africa.
c Four countries only (Bulgaria, Czechoslovakia, Hungary and Romania), January-September 1992 from January-September 1991.
d Excluding Yugoslavia.
e 1992 data refer to the Russian Federation.
f Trade with all former CMEA members (i.e., including Cuba, Mongolia, Vietnam). It can be assumed that trade with the non-European CMEA members fell more steeply than that with eastern Europe.

zation policies has led to increased imports without a commensurate rise of exports. In cases where the net result of economic reforms has succeeded in compressing overall domestic demand and hence imports, exports have been the only demand component that has grown.

The analysis of foreign trade developments in 1992 encounters even more severe *data problems* than in preceding years. In the first instance, it is now quite clear that in the wake of the far-reaching systemic transformations in the transition countries and the entry of many new actors (state firms as well as foreign and domestic private enterprises) into external trade activities, governments have not only largely abandoned their former ability *to conduct* foreign trade as a state monopoly, but have also to a significant degree lost the capacity *to know* what is going on in that sphere. The adoption of international trade registration procedures, mainly to be implemented by customs services not trained for this task, poses difficulties which in 1992 appear to have become more severe. Data collection is also hampered by conflicts of interest between enterprises and the statistical requirements: as the enterprises themselves become more profit-oriented they are increasingly reluctant to tie up resources in administrative activities which do not serve their immediate interests. In some countries there is also evidence to suggest that companies are deliberately misleading the authorities about their foreign trade activities in order to escape taxation or to evade foreign exchange controls.

There are three aspects of the deterioration of eastern foreign trade statistics: (1) growing time lags in reporting; (2) the presentation of published data in large aggregates only; and (3) cases when different authorities of the same country release contradictory information on trade developments. Inadequate detail of trade commodity components, and lack of information on methods of trade data conversion to a reporting currency also hamper the interpretation. Without this information, the reported data can only with difficulty be put into a meaningful analytical context. Thus even the data presented here without severe caveats may be subject to substantial revisions.

Review by country

The impact of trade changes arising from the break-up of the CMEA and the accelerating reorientation towards western markets is summarized in charts 3.3.1 and 3.3.2. Modifications in the *geographical patterns of trade* in the past three years in most countries resulted in striking shifts of market shares from the formerly socialist countries (or CMEA partners) towards the rest of the world.

These movements need to be seen against the background of the direction of change in absolute values (table 3.3.3). In the east European countries, the declining market shares of the ex-socialist and developing countries reflect *falling* trade levels and the gains of the developed market economies *rising* trade levels; in the case of the former Soviet Union, the redistribution of trade shares occurred against the background of a *contraction* of *all* trade flows.

TABLE 3.3.2
European transition countries: Trade balances, 1989-1992
(Billion US dollars)

Country group	1989	1990	1991	1991 a	1992 a	1992
Eastern Europe a						
World	2.8	-1.9	-3.5	-2.0	-1.5	-4.6
Transition economies	2.1	0.6	0.1	-0.4	-1.5	..
Developed market economies	1.4	-0.9	-2.1	-0.8	-0.8	..
Developing countries	0.2	-1.4	-1.5	-0.8	0.8	..
Soviet Union/Russia b						
World	-2.7	-5.9	1.3	5.7	-2.2	3.1
Transition economies	-1.2	-3.8	-0.3	0.9	1.9	3.3
Eastern Europe c	-1.9	-3.9	0.5	0.4 d	1.3 d	..
Developed market economies	-6.5	-5.1	-	3.1	-3.6	0.4
Developing countries	5.0	3.0	1.6	1.8	-0.5	-0.6

Source and country groups: As for table 3.3.1. Conversion of national currency data to dollars as in table 3.3.1.

a Data for eastern Europe in January-September 1991 and 1992 cover Bulgaria, Czechoslovakia, Hungary and Romania only.
b 1992 data refer to the Russian Federation only.
c Includes the former GDR prior to 1990 but excludes Yugoslavia (SFR).
d Russian Federation balance with all former CMEA members (i.e., including Cuba, Mongolia, Vietnam).

In *eastern Europe*, all countries registered marked changes in *export* performance in 1992. In *Bulgaria* and *Romania*, where earlier falls had been extremely large, modest growth was registered.[197]

Exports continued to grow in *Czechoslovakia* and *Hungary* in 1992, although at a slightly reduced pace in both countries (below 5 per cent). In Czechoslovakia a 20 per cent increase was recorded in sales of intermediate manufactured goods (SITC 6), while in Hungary most of the growth appears to have come from stepped up exports of industrial consumer goods which represent about one quarter of exports.[198]

Exports probably grew rapidly also in *Poland* in 1992, but the precise course of Polish foreign trade is at the moment somewhat obscure owing to recent major data revisions, of which at the time of writing only the trade balance impact is known.[199] It may be assumed that the revisions reflect primarily a much stronger growth of imports than registered in the balance-of-payments data published hitherto. The 10 per cent growth of exports shown in table 3.3.3 is what was reported by the National Bank of Poland prior to the revision.[200]

Among the successor states of *former Yugoslavia*, Slovenia, the largest exporter, succeeded in increasing its foreign sales by 8 per cent, while its imports remained basically unchanged. In Croatia and the new Federal Republic Yugoslavia (Serbia and Montenegro) reported trade levels fell steeply, as already noted.

For eastern Europe as a whole, the picture is more diverse when changes of *imports* in 1992 are reviewed, with rates of change ranging from *minus* 1-3 per cent (Hungary and Romania) to *plus* 70 per cent (Albania). In *Albania*, the huge rise in imports reflects mainly western assistance deliveries. Judging from data pertaining to the January-September period, the almost 30 per cent rise in *Bulgarian* imports in 1992 as a whole probably reflects much larger increases in some key groups which are highly relevant for industrial restructuring. The growth in imports of machinery and transport equipment, as well as that of semi-processed goods such as textiles and base metals, was well above 100 per cent. At the same time, imports of mineral products, which account for more than one third of total imports, stagnated in value.

The expansion of imports into *Czechoslovakia* in the reported magnitude (almost 15 per cent) appears to reflect a substantial acceleration in the fourth quarter. The January-September figures still showed a more moderate rise (less than 8 per cent). However, it is also possible that significant revisions of the trade data for the earlier part of the year are captured in this figure.[201] Czechoslovak import growth was fuelled primarily by

---

[197] The 1 per cent rise of total exports in Romania may understate actual growth if part of the reported capital flight was effected through the evasion of customs controls on exports. In February 1993, Prime Minister N. Vacaroiu estimated that Romanian firms hold more than $600 million in illegal deposits in western banks (*Magyar Hirlap*, 13 February 1993). Much of this must have been accumulated over the past 12 months.

[198] According to the statistical office such exports rose by 28 per cent in 1992. However, this figure may reflect multiple counting of goods produced under transfrontier production cooperation agreements between Hungarian and western firms (*Beszélő*, 6 March 1993).

[199] Instead of a *surplus* of $0.5 billion claimed by the National Bank of Poland of until recently, the new figures published in early March show a *deficit* on customs basis of $2.6 billion and of $1.6 billion on payments basis (*Rzeczpospolita*, 5 March 1993).

[200] Customs-based trade figures have so far been published only for the first quarter of 1992; these showed an 8 per cent rise in imports, as against the 5 per cent fall registered in the payments data for the same period. Since the National Bank trade data do not capture credit transactions financed abroad smaller than $10 million nor local purchases of convertible currency in amounts smaller $10,000, they are likely to be a poor indicator of current trade trends, especially as regards imports. Western data for the first nine months show a rise of 14 per cent in Polish exports to and of 5 per cent in imports from the developed market economies (table 3.3.10).

[201] In the first half of 1992, the results of customs-based data collection were so unsatisfactory that foreign trade figures were based on largely on expert estimates

TABLE 3.3.3

European transition countries: Change in foreign trade values and trade balances by partner region, 1990-1992

(Growth rates in percentages; trade balances in billion US dollars)

	Growth rates						Trade balance, in billion US dollars		
	Exports			Imports					
Country and trade partner groups [a]	1990	1991	1992	1990	1991	1992	1990	1991	1992
Albania									
World	-24.5	-64.2	-33.3	0.1	-46.3	66.7	-0.2	-0.2	-0.4
Transition countries	-31.8	-82.2	..	-90.5	171.9	..	-0.5	-	..
Developed market economies	-12.4	-36.0	..	6.8	-18.2	..	-3.4	-0.1	..
Developing countries	9.8	-70.3	..	-5.3	-16.7	..	0.6	-	..
Bulgaria									
World	-21.3	-34.2	1.6	-23.7	-51.5	27.5	-0.4	0.7	-
Transition economies	-32.0	-27.8	-25.7	-23.8	-43.1	-4.3	0.2	0.6	0.2
Developed market economies	-11.1	-36.3	61.6	-25.9	-59.8	79.3	-0.8	-0.1	0.1
Developing countries	9.7	-47.6	14.2	-19.0	-54.4	26.7	0.2	0.2	0.1
Czechoslovakia									
World	-10.5	5.6	3.2	0.3	-7.2	14.6	-1.1	0.4	-0.9
Transition economies	-27.4	6.8	-33.0	-17.1	0.3	-10.2	-0.7	-	-1.0
Developed market economies	13.4	6.9	26.4	24.6	-13.7	39.6	-0.6	0.3	-0.4
Developing countries	-10.9	-6.0	27.4	-12.0	4.4	-20.4	0.2	0.1	0.5
Czech Republic	..	..	2.6	..	..	26.1	..	..	-0.8
Slovak Republic	..	..	6.5	..	..	-1.1	..	..	-0.1
Hungary									
World	0.6	5.1	4.1	-0.7	30.2	-3.2	0.9	-1.2	-0.4
Transition economies	-21.5	-26.8	3.2	-18.8	2.8	1.8	0.4	-0.3	-0.3
Developed market economies	27.9	21.4	9.1*	8.6	44.3	-3.4*	0.5	-0.9	-0.1*
Developing countries	-12.8	21.8	-27.4*	52.9	29.0	-16.6*	-	0.1	-0.1*
Poland									
World	24.7	-18.5	9.7 [b]	-2.5	24.3	6.1 [b]	5.7	-0.6	-2.6 [b]
Transition economies	14.9	-62.0	..	1.8	-42.8	..	1.6	-0.5	..
Developed market economies	40.0	13.7	..	-4.7	71.7	..	3.4	0.3	..
Developing countries	3.2	-15.5	..	-17.1	151.0	..	0.7	-0.5	..
Romania									
World	-43.4	-7.1	0.9	18.1	-17.6	-1.4	-2.3	-1.4	-0.8
Transition economies	-45.5	29.2	-17.6 [c]	-13.7	-8.9	-33.7 [c]	-0.9	-0.2	-0.1 [c]
Developed market economies	-38.4	-22.8	2.5 [c]	116.7	-9.4	42.6 [c]	0.2	-0.3	-0.6 [c]
Developing countries	-51.0	-11.9	43.8 [c]	10.1	-32.7	-46.2 [c]	-0.9	-0.9	0.1 [c]
Yugoslavia (SFR)									
World	7.1	-3.9	..	27.5	-17.1	..	-4.6	-1.4	..
Transition economies	-8.1	-24.9 [d]	..	3.1	-22.1 [d]	..	-0.5	-0.2 [d]	..
Developed market economies	20.7	19.2 [d]	..	42.5	-8.3 [d]	..	-3.4	-0.4 [d]	..
Developing countries	-5.4	-23.3 [d]	..	19.1	-33.4 [d]	..	0.6	-0.2 [d]	..
Croatia	3.7	13.0	-5.0	25.2	-13.5	-10.4	-1.5	-0.5	-0.3
Slovenia	20.8	-6.3	8.0	47.0	-12.5	0.1	-0.6	-0.3	-
Yugoslavia (FR)	2.0	-19.1	-46.0	27.7	-25.6	-30.4	-2.0	-0.8	-1.3
Soviet Union, Russia [e]									
World	-5.2	-24.6	-25.2	-	-35.9	-21.3	-5.9	1.3	3.1
Transition economies	-24.3	-35.1	-25.8	-10.6	-35.9	-42.8	-3.8	-0.3	3.3
Eastern Europe [f]	-26.9	-40.8	-32.7 [g]	-12.1	-51.6	-49.7 [g]	-3.9	0.5	2.4 [g]
Developed market economies	12.3	-16.2	-20.3	5.6	-31.0	-13.0	-5.1	-	0.4
Developing countries	-9.5	-29.0	-44.0	3.9	-35.8	-2.6	3.0	1.6	-0.6

Source: Secretariat of the United Nations Economic Commission for Europe, based on national foreign trade statistics.

Note: Growth rates and trade balances are based on trade values in terms of US dollars. As an approximation to a consistent dollar valuation of rouble-denominated intra-group trade flows here, and in the longer time series in Appendix tables C.4 and C.5, the pre-1991 national-currency data on trade with the market economies were revalued, in national currency terms, at a common rouble-dollar crossrate and re-aggregated with the data on trade with the then centrally planned economies to obtain new trade totals (see box 2.1.1 and the discussion in section 2.1(iii) in United Nations Economic Commission for Europe, *Economic Bulletin for Europe*, Vol.43 (1991), New York, 1991). 1991 and 1992 trade flows were either originally reported in dollars or were converted to dollars with the relevant national conversion coefficient. Both procedures must be considered approximations to the desired standard of consistent valuation because (i) pre-1991 intra-CMEA trade flows contained convertible-currency components (which in 1990 were significant), and (ii) 1991 and 1992 trade flows still comprise some rouble-denominated trade.

[a] The partner country grouping follows the past practice of the national statistical sources, which differed from the breakdown usually employed in United Nations publications. Thus, "transition economies" − the former "socialist countries" − includes Yugoslavia and Cuba, in addition to the east European countries, the Soviet Union, and the Asian centrally planned economies.
[b] Growth of exports and imports according to unrevised balance-of-payments data of the Polish National Bank; trade balance according to preliminary revised figure reported in March 1993 (see footnote (199) in text).
[c] January-September 1992 relative to the same period in 1991.
[d] Excluding Croatia.
[e] 1992 data refer to trade of the Russian Federation.
[f] Excluding Yugoslavia. The former German Democratic Republic is included in the data for 1990, but not in those for 1991.
[g] Russian Federation trade with all former CMEA members (i.e., including Cuba, Mongolia and Vietnam).

a large rise in imports of machinery and transportation equipment (39 per cent), but substantial increases were recorded also in chemicals, intermediate manufactured goods and miscellaneous manufactures (17, 26 and 40 per cent, respectively).

---

of the Federal Ministry of Foreign Trade and the Federal Statistical Office (State Bank of Czechoslovakia, *Report on Monetary Development for the Period from January to September*, 1992, p.8). By the third quarter, however, the new data collection system was reported to have become workable.

TABLE 3.3.4
Conflicting foreign trade data for the CIS and Russia, 1991-1992

Source and method Country Group	Value (billion dollars or roubles)						Volume 1992 (1991 = 100)	
	1991			1992				
	Exports	Imports	Balance	Exports	Imports	Balance	Exports	Imports
CIS Statistical Committee [a]								
1. Russia (billion dollars)	36.8	25.6	11.2	..	..	..	..	..
2. CIS (billion dollars)	45.6	40.8	4.8	47	46	1	..	..
Russian Federation Statistical Committee								
3. Russia (billion dollars)	50.9	44.5	6.4	38.1	35.0	3.1	74 [b]	78 [b]
4. Russia (billion roubles)	64.2	44.7	19.5	7 353	6 755	598	..	..
Implicit exchange rate (rouble/dollar)	1.3	1.0	..	193.0	193.0	..	..	..
Russian Federation Ministry of International Economic Relations								
5. Russia (billion dollars)	50.9	44.5	6.4	45	42	3	100	
Addendum:								
Trade with other republics of the former USSR								
6. Russia (billion roubles)	134.8	103.3	31.5	2 147	1 849	298	63 [c]	70 [c]
Total "external" trade (4 + 6)								
7. Russia (billion roubles)	199.0	148.0	51.0	9 500	8 604	896	71 [d]	76 [d]

Sources: CIS Statistical Committee, *Strany-chleny SNG, Statisticheskii ezhegodnik 1992*, p.76; CIS Statistical Committee, *Ekonomika Sodruzhestva nezavisimych gosudarstv v 1992 gody*, p.28; Russian Federation Goskomstat, *O razvitii ekonomicheskikh reform v Rossiiskoi Federatsii v 1992 godu*, p.39, and direct communications to the ECE secretariat; data of the Russian Federation Ministry of External Economic Relations published in Gsokomstat *Rossiiskie Vesti*, 11 February and 2 March 1993; *Finansovye Izvestiya*, 4-10 February 1993; *Ekonomika i zhizn'*, No.13, March 1993, pp.14-15.

[a] According to the CIS Statistical Committee's 1992 report, turnover in 1992 *decreased* by about a quarter. The comparison with the corresponding 1991 level, published in the CIS statistical yearbook, would imply an 8 per cent *increase*.
[b] Stated to be "at constant prices", but the rate of change is identical to that of the reported dollar values.
[c] Estimate based on shipments measured in physical units form a sample representing some 60 per cent of trade flows.
[d] Weighted average.

In *Hungary* the within-year course of imports in 1992 seems to have been similar to that observed in Czechoslovakia: in the first half of the year imports contracted by more than 10 per cent, while in the third and the fourth quarters foreign purchases rebounded somewhat as industrial production stabilized. In terms of commodity breakdown, the growth of imports was fairly broadly based.

An upswing of imports in the last quarter of the year is likely to have occurred also in *Poland* owing to the strengthening of economic activity, but also because a 6 per cent import surcharge came into force in late December and enterprises were building up stocks in anticipation. However, the actual strength of Polish import growth in 1992 must be considered an unknown after the recent revision of the official trade balance estimates.[202] Until then, growth had been thought to be running at some 6 per cent on the basis of the pre-revision current-accounts data of the National Bank of Poland.[203]

Trade balances improved distinctly in Hungary and less pronouncedly in Croatia, Romania and Slovenia, all of which registered smaller deficits. A significant swing into a small surplus was thought to have occurred in Poland, but the latest data instead indicate a large increase of the trade deficit (to $2.6 billion). In Albania, Bulgaria and Czechoslovakia, the rapid import expansion resulted in worsened external balances, as did the collapse of trade in FR Yugoslavia.

The improvement in the trade performance of most east European countries contrasts sharply with the continued decline in the foreign trade of the former Soviet republics. For the *CIS countries jointly,* the disruption of domestic activity after the dissolution of the union resulted in a contraction provisionally estimated — in current US dollar terms — at one quarter for trade turnover, with imports falling more steeply than exports.[204] The impact apparently varied widely between republics, but little detail is available for states other than the Russian Federation. The exports of Ukraine fell by 20 per cent, and those of Belarus by 36 per cent. Only Russia and Kazakhstan are reported to have an external surplus ($3 and $1 billion, respectively).

In the absence of information on the geographical pattern of trade for the CIS countries jointly, the geographical pattern of foreign trade of the Russian Federation in 1992 is juxtaposed to USSR data for earlier years in charts 3.3.1 and 3.3.2 and table 3.3.1.[205] In this comparison, the large increase of the share of the developed market economies is the most conspicuous feature of the post-1989 developments. On the export

---

202 See footnote (199) above.

203 Western data show exports to Poland from the industrial market economies growing at some 5 per cent in the first three quarters of 1992 (table 3.3.10 below).

204 The CIS Statistical Committee values total CIS export and import flows in 1992 at about $47 and $46 billion respectively, noting that these figures are 25 per cent *below* the corresponding 1991 figures (*Ekonomika Sodruzhestva nezavisimykh gosudarstv v 1992 godu*, pp.28-31). However, the annual CIS statistical yearbook had reported exports of $45.6 billion and imports of $40.6 billion for 1991 (*Strany-chleny SNG: statisticheskii ezhegodnik 1992*, p.76), which would imply increases of 3 and 13 per cent in exports and imports, respectively.

205 The weight of Russia in aggregate CIS trade was 80 per cent for exports and 63 per cent for imports in 1991 (*Strany-chleny SNG*, op.cit., p.76).

CHART 3.3.1

European transition countries: geographical structure of exports, 1989-1992
(Per cent)

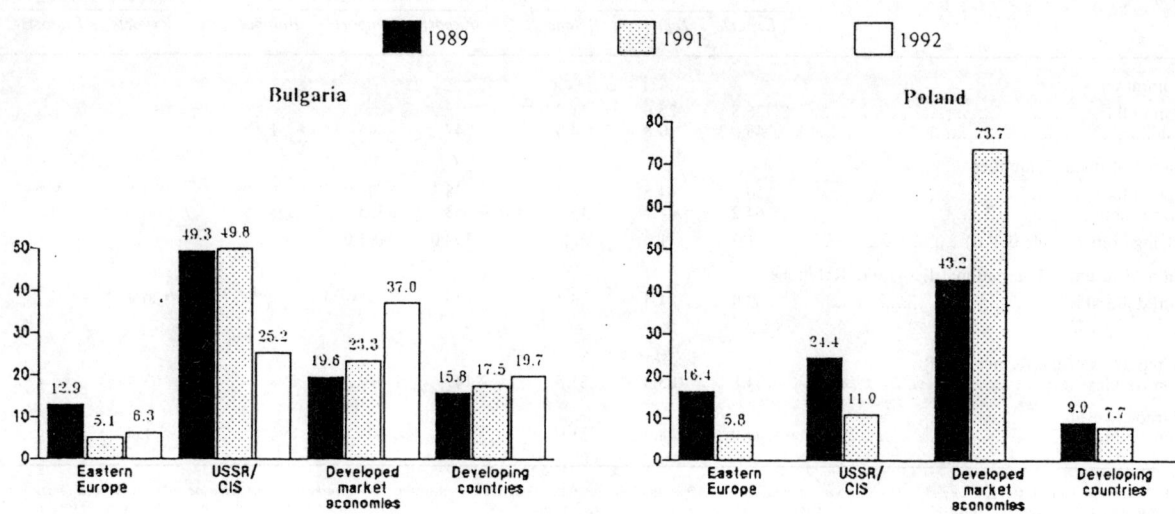
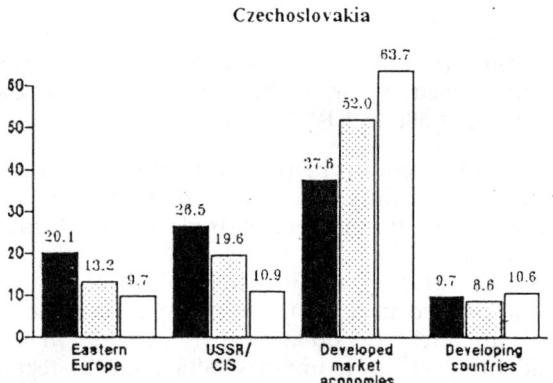
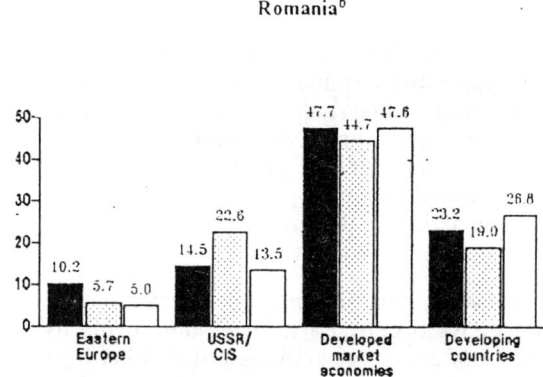
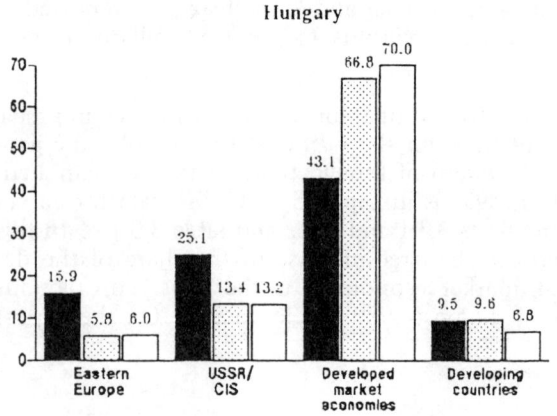
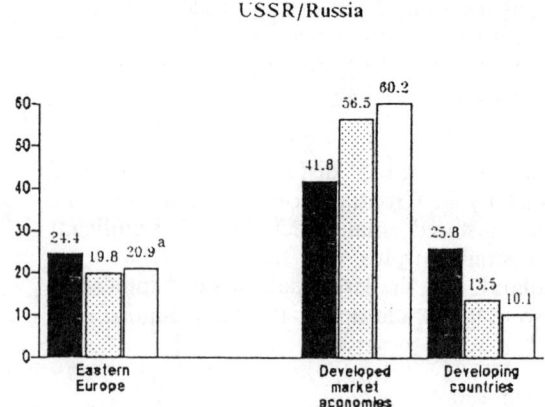

Source: ECE Common Data Base, national trade statistics.

Note: 1989 figures are calculated at standardized rouble-dollar crossrates (see United Nations Economic Commission for Europe, *Economic Bulletin for Europe*, vol.43 (1991), New York, 1991). The former GDR is included in the "eastern Europe" aggregates. Figures do not add up to 100 because trade with some smaller "socialist" partners (China, Cuba, etc.) is not shown.

[a] 1992 January-November.
[b] 1992 (QI-III).

side, the share of the developed market economies rose by more than 16 percentage points, while that of the developing countries fell by almost the same amount. On the import side, the shifts were somewhat smaller, but here the most striking loss in market share was that of the former socialist countries.

In 1992, the dollar value of exports of the *Russian Federation* fell by 25 per cent and that of Russian imports by 21 per cent, with "constant price" volume changes of the same magnitude, according to data of the Russian Statistical Committee (Goskomstat). However, these data are disputed by the statisticians of the Russian Ministry of External Economic Relations, which reports a 12 per cent fall in the value of exports (with stagnant, rather than falling, volume) and a 6 per cent contraction of imports.[206] In contrast to the divergent data claims, which are aired in public, the sources of the divergence appear so far not to have been publicly discussed. This leaves the outside observer in the unfortunate position of having to make a "blind" choice among conflicting data. The exposition below and in the tables to this section reflects the Goskomstat data, but should be read with this disclaimer in mind.

Russian *exports* to the countries of the former "socialist" group[207] and to the market economies contracted at approximately the same pace in 1992, by about one quarter, but exports to Russia's former CMEA partners[208] fell by about one third (table 3.3.3). In *imports* the bulk of the cutback fell on imports from the former socialist group (two thirds of the total contraction, or $6 billion), which declined by 43 per cent (50 per cent for the former CMEA countries); imports from the market economies fell by 11 per cent.

Russia's overall *trade surplus* halved from $6½ to $3 billion between 1991 and 1992, and shifted entirely the relationship with the former socialist countries: whereas in 1991 there had been a $5 billion surplus with the developed and developing market economies, trade with these countries closed with a small deficit in 1992. Since barter arrangements appear to have dominated in Russian trade with the former socialist countries, this surplus is probably of little importance for Russia's payments capacity.

Among main commodity groups, exports of *fuels and energy* brought about the same revenue as in 1991,

but accounted for 55 per cent of Russian export earnings in 1992, a 5 percentage point increase over 1991.[209] Exports of *oil and oil products* alone accounted for 35 per cent of earnings. In physical terms, crude oil exports increased by 17 per cent (to 66 million tons), but this was largely offset by a 23 per cent fall in the exports of oil products (to 27 million tons).[210] In view of the 14 per cent decline in the production of crude oil, maintaining the level of exports was possible only at the expense of domestic utilization and sales to the CIS member countries. *Natural gas* exports remained roughly unchanged in physical terms,[211] but a 10 per cent erosion of export prices resulted in a $1 billion decline in revenues. *Coal* exports fell by 24 per cent in volume and almost 30 per cent in value. No data are available on electricity exports. Exports of *machinery, equipment and means of transport* declined by 30 per cent in value and accounted for 9 per cent of total export earnings, or $3.5 billion − roughly equivalent to the combined investment goods exports of Czechoslovakia and Hungary. Out of this, more than $1 billion was earned from sales related to nuclear energy equipment and another $1.5 billion from the sales of road vehicles.[212]

On the import side, *machinery and equipment* imports fell by some 10 per cent in value, i.e., rather less than the overall fall of imports; this group now accounts for almost 40 per cent of Russian imports, as against 33 per cent in 1991. The imports of *grains* increased by 37 per cent in volume (to about 30 million tons), as did those of some other foodstuff items and raw materials (notably sugar, up 14 per cent). Most of the contraction appears to have been in industrial raw materials and intermediate products.

Over the past two years, foreign trade of Russian was shaped primarily by developments occurring within its borders. However, in addition to the domestic crisis, Russia's trade with the external world has been also negatively influenced by the chaotic environment prevailing around its borders with other former Soviet republics. It is reported that transit shipments were repeatedly temporarily blocked or threatened to be blocked, and that in some directions safe and speedy transportation is simply no longer possible. The year 1992 has brought a new development: Russian exporters have started to feel the competition of other

---

[206] Since both agencies appear to use the same dollar values for the base year of the comparison, 1991, and also similar figures for key commodity flows, the disagreement appears to turn mainly on the valuation of the trade flows, possibly on the manner in which soft-currency earnings are converted to dollars. See table 3.3.4 for a juxtaposition of the data.

[207] This includes China, People's Republic of Korea, Laos and former Yugoslavia in addition to the CMEA countries.

[208] Including Cuba, Mongolia and Vietnam in addition to the east European former CMEA members. Data on the east European countries alone are not yet available.

[209] The share of fuels in 1991 was estimated from Russian Federation Goskomstat, *Vneshneekonomicheskie svyazi Rossiiskoi Federatsii 1992*, Moscow, 1992, pp.11-15.

[210] Data of the Ministry of International Economic Relations (*Finansovye Izvestiya*, 4-10 February 1993). These data and the change compared to 1991 are broadly consistent with OECD trade partner statistics (IEA, *Monthly Oil Market Report*, Report for End-February 1993, Manuscript, pp.13-15).

[211] The Ministry of International Economic Relations reports exports of 90.3 billion cu m, a rise of 0.3 per cent, whereas Russian Goskomstat data show exports of 88.9 billion cu m, a drop of 0.8 per cent.

[212] *Finansovye Izvestiya*, 21-27 January 1993.

CHART 3.3.2
European transition countries: geographical structure of imports, 1989-1992
(Per cent)

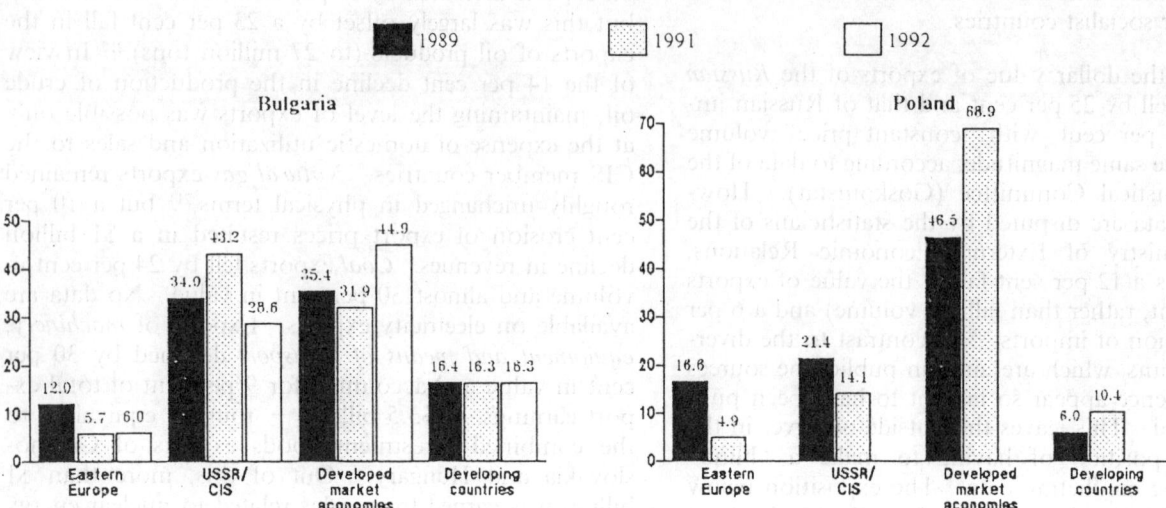
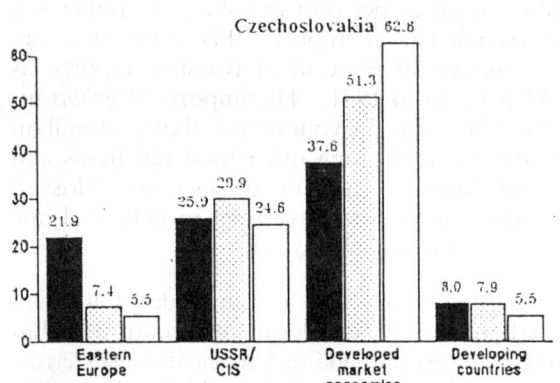
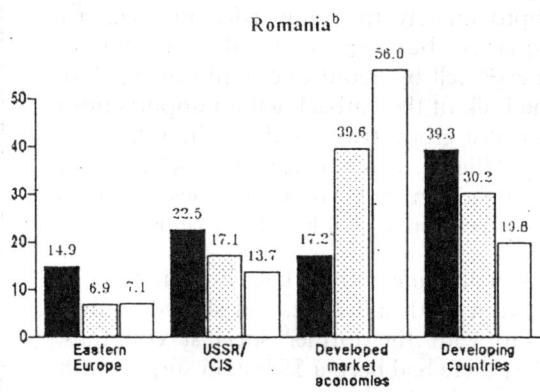
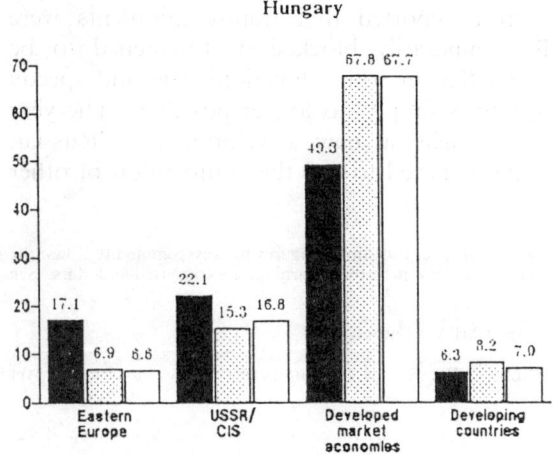
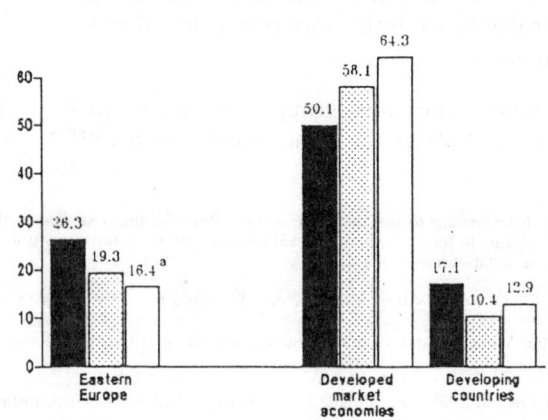

Source: ECE Common Data Base, national trade statistics.

Note: 1989 figures are calculated at standardized rouble dollar crossrates (see United Nations Economic Commission for Europe, *Economic Bulletin for Europe*, vol.43 (1991), New York, 1991). The former GDR is included in the "eastern Europe" aggregates. Figures do not add up to 100 because trade with some smaller "socialist" partners (China, Cuba, etc.) is not shown.

[a] 1992 January-November.
[b] 1992 (QI-III).

post-Soviet countries and complain about "dumping practices".[213]

While a rising proportion of the CIS countries' foreign trade is now formally conducted in convertible currency terms,[214] the share of barter trade appears to be rising even faster.[215] These linked export-import deals pose special valuation problems, and may provide the channel for capital flight.[216] The foreign trade data may therefore underestimate actual flows if traders intent on evading capital export prohibitions manage to by-pass frontier controls.[217]

The difficulties facing the CIS statistical authorities are certainly numerous. In the past, Soviet foreign trade was highly concentrated and conducted through specialized foreign trade organizations located primarily in Moscow. After the break-up of the Soviet Union, exporters and importers in the newly-emerging countries, and often regions within these countries as well, began to establish direct links with international markets. The sharp fall recorded in Russian foreign trade in 1992 may therefore in part reflect the rerouting of trade flows to the newly sovereign CIS countries. Customs borders in a physical sense still hardly exist in a large part of the post-Soviet economic space, and it is thus virtually impossible to distinguish intra-CIS from external trade flows — even if all parties were willing for this to happen. But economic interests may work in the opposite direction. As long as prices, exchange rates and trade regulations remain in their present state, large profits can be earned from re-exporting relatively cheap Russian raw materials obtained at artificially low, state-fixed prices into a country where prices are free or the regulations allow an advantageous rouble/dollar conversion.

(ii) **Current account, reserves and debt**

Current account

The overall improvement in the convertible-currency current accounts of the countries of *eastern Europe* continued in 1992 (tables 3.3.5. and 3.3.6).[218] In general the results were better than had been projected early in the year, in part because the expected upturns in economic activity did not materialize. In some countries, financing constraints caused the current-account deficits to turn out smaller than had been planned.

The convertible-currency current-account surpluses of Hungary and Slovenia increased while the deficits of Bulgaria and Romania declined. A decline is also indicated for Poland by the data shown in tables 3.3.5 and 3.3.6, but revisions of the Polish National Bank's trade balance estimates announced in March 1993 point instead to a deepening current-account deficit.[219] In most east European countries the improved trade balances contributed to similar developments in current accounts. Czechoslovakia, by contrast, experienced a considerable deterioration in its trade balance, but the current-account surplus diminished only marginally because of the offsetting improvement in the surplus on invisibles.

In Czechoslovakia, Hungary and Poland, current-account balances had improved through mid-year. Thereafter this tendency was reversed because imports strengthened. All three countries posted current-account deficits in the closing months of 1992. In Poland, imports rose in the second half of the year (after a contraction in the first half), presumably reflecting the upturn in economic activity. Also, the unusually severe drought in the region is estimated to have cut the country's trade surplus by $500 million.[220]

In Bulgaria and Romania the development of current-account balances in 1992 was partially shaped by shortfalls in expected external finance. As a result, the current-account deficits of these countries turned out to be smaller than foreseen in their economic pro-

---

213 Yu. Churakov, "Vhnesnaya torgovlya sokratilas', no balans ostalsya polozhitel'nym" (Foreign trade fell, but the balance remained in surplus), *Finansovye Izvestiya*, 21-27 February 1993.

214 With notable differences between republics: thus, convertible currency trade accounted for 65 per cent of Russian exports, but for only 40 per cent for those of Belarus, Turkmenistan and Uzbekistan.

215 The share of barter transactions increased from 4 per cent to 56 per cent in the exports and from 3 per cent to 41 per cent in the imports of Belarus between 1991 and 1992. Belarus may be the extreme case, but high barter components are noted in all CIS republics. For Russia, the share of barter transactions was estimated at about 25-30 per cent of trade turnover in 1992, but may be as high as 40 per cent if certain export-import transactions which are in reality "concealed" barter deals are included (*Ekonomika i zhizn'*, No.11, March 1993, p.1). About one third of Russian oil and natural gas exports (24 million tons) was reportedly disposed in the framework of such deals, including deliveries to former CMEA countries under the pipeline construction arrangements of the early 1980s (*Rossiiskie vesti*, 11 February 1993).

216 Russian Goskomstat re-estimated the declared value of flows in registered barter transactions in the first three quarters of 1992 at world market prices, concluding that there was a net valuation loss to Russia of $2.9 billion due "either to incompetence of the Russian side or to a deliberate understatement of the exchange proportions for mercenary purposes" (*Ekonomika i zhizn'*, No.11, March 1993, p.1).

217 Capital flight in the first three quarters of 1992 "through the channels of non-repatriation of export receipts and advance payments on fictitious import contracts alone" was put at $2-$2½ billion, or 7-9 per cent of export earnings during that period. Estimates of the research institute of the Russian Ministry of Foreign Economic Relations (*Ekonomika i zhizn'*, No.11, March 1993, p.1).

218 The analysis in this section is undertaken exclusively in terms of convertible currencies. Transactions in convertible currencies now account for the major part of these countries' external transaction. Financing issues are discussed in chapter 4.

219 See footnote (199) above.

220 GATT, Communication from Republic of Poland, L/7164, 11 January 1993.

TABLE 3.3.5
European transition countries: Convertible currency current-account balances, 1990-1993
(Million US dollars)

	1990	1991	1992	1993 Projection
Albania [a]	-95	-164	-200 [b]	-200 [b]
Excluding official transfers	..	-250	..	..
Bulgaria	-1 152	-887	-250 *	-1 500
Cash basis [c]	..	-77	451	..
Former Czechoslovakia	-1 104	356	226	..
Czech Republic	..	..	53	- [d]
Slovak Republic	..	..	172	-
Hungary	127	267	324	-200
Poland	716	-1 359	-269	-600
Romania	-1 650	-1 369	-944 [e]	-1 100 [a]
Former Yugoslavia	-2 664	-1 092	..	..
Slovenia [f]	680	315	793	705
Former Soviet Union [g]				
All currencies	-21 000	-3 300	-7 600 [h]	..
Convertible currencies	-4 800	-800	..	..
Russia [g]				
All currencies	-4 500	3 400	-3 200	..
Convertible currencies	2 000	5 500	..	..

Sources: National statistics; IMF, Russian Federation, *Economic Review*, April 1992; World Bank, *Russian Economic Reform*, September 1992; ECE estimates.

a All currencies.
b Trade balance only; figure for 1992 is a prelimary estimate.
c Includes only interest paid.
d Excluding ECU clearing transactions with Slovakia.
e January-November.
f Excluding transactions with former Yugoslavia.
g Including gold sales, but excluding gold swaps and grants. Excludes inter-republican trade.
h Projected.

grammes.[221] Increases in convertible-currency exports also contributed to the lower-than-planned current-account deficits.

The net *interest* burden of Czechoslovakia, Hungary, and Slovenia eased due to reduction of their external net debts and falling international interest rates (table 3.3.6). Romania's deficit on this item had been small, but has recently risen quickly as new debt built up. Bulgaria's interest obligations have continued to rise for the same reason, but only a small part of these have actually been paid (also see chapter 4).

Czechoslovakia and Hungary posted larger surpluses on the *tourism* item. Higher receipts were partially offset by the greater expenditures of their own citizens. Tourism earnings used to be a substantial source of hard currency for Bulgaria and Romania.[222] However, as a result of a fall in foreign visitors and increased travel abroad of the population these surpluses have vanished. The former Yugoslavia suffered an enormous loss in travel revenues in 1991 with the onset of hostilities and this persisted in 1992. Slovenia, out of the immediate area of conflict, has posted a large and growing surplus in tourism.

All east European countries recorded net inflows of *transfers* in 1992. Poland posted the largest net inflow, more than doubling the amount registered in 1991. This reflects the cancellation of part of the country's interest obligations under the agreement with the Paris Club (see chapter 4), which is shown as an inflow of official transfer payments in the accounts. A large surplus on private transfers enjoyed by the country until 1991 disappeared when domestic sales against hard currency payments were abolished.[223] Official transfers, chiefly humanitarian and commodity assistance from the Group of 24 financed a substantial share of the deficit in the other items in Albania's current account.[224]

---

[221] The Romanian economic programme envisaged a $1.3 billion deficit for 1992 (*Romania Economic Newsletter*, Vol.2, No.1, April-June 1992).

[222] In the mid-1980s Romania earned some $200 million annually from tourism.

[223] See United Nations Economic Commission for Europe, *Economic Bulletin for Europe*, Vol.43(1991), New York, 1991.

[224] EEC, IMF and IBRD, G-24 Meeting in Tirana, *Orientations for a G-24 Support Programme for the Restructuring of Albania's Economy*, presented to the G-24 meeting in Tirana, 17 July 1992. Albania's current-account deficit (exclusive of official transfers) came to some 25 per cent of GDP in 1991.

TABLE 3.3.6
European transition countries: Balance of payments in convertible currencies, 1990-1992
(Million US dollars)

	Bulgaria			Czechoslovakia			Hungary		
	1990	1991 [a]	1992 [a]	1990	1991	1992	1990	1991	1992
Current account balance	-1 152	-77	452	-1 104	356	226	127	267	324
Trade balance	-757	-32	485	-785	-447	-1 576	348	189	-49
Exports	2 615	3 737	5 093	5 994	8 341	11 280	6 346	9 258	10 028
Imports	3 372	3 769	4 609	6 779	8 788	12 856	5 998	9 069	10 076
Services balance	-503	-114	-76	-279	762	1 661	-948	-781	-486
of which:									
Transport	51	-14	-32	259	456	984	-164	-80	-116
Travel	78	-84	26	-71	445	660	345	560	590
Investment income	-688	-28	-76	-316	-65	9	-1 414	-1 331	-1 261
Transfers	108	69	43	-40	42	140	727	861	859
Capital account	-2 477	115	-32	326	47	41	-689	2 453	436
Direct investment	4	56	42	181	592	1 011	311	1 459	1 471
Other long-term capital (net)	-2 481	247	75	718	1 732	471	-107	1 611	-1 040
Assets	200	295	33	-40	174	129	-76	-57	-146
Liabilities	-2 681	-48	42	758	1 557	343	-31	1 668	-894
Short-term capital (net)	-	-188	-148	-573	-2 277	-1 443	-893	-617	5
Overall balance	-3 532	45	400	-778	413	266	-562	2 720	761
Reserves (net) [b]	888	-45	-400	1 102	-898	120	562	-2 720	-761
Special financing (net)	2 614	..	..	-	-	-	-	-	-

	Poland			Romania			Slovenia [c]		
	1990	1991	1992	1990	1991	1992 [d]	1990	1991	1992
Current account balance	716	-1 359	-269	-1 650	-1 369	-944	..	315	793
Trade balance	2 214	51	512	-1 743	-1 357	-841	..	18	50
Exports	10 863	12 760	13 997	3 364	3 533	3 366	..	1 972	4 184
Imports	8 649	12 709	13 485	5 107	4 890	4 207	..	1 954	4 135
Services balance	-3 479	-2 627	-3 710	93	-104	-143	..	253	658
of which:									
Transport	59	260	..	-208	-149	-113	..	34	97
Travel	-120	-20	..	-41	-1	-12	..	280	587
Investment income	-3 329	-2 863	-4 139	137	-5	-96	..	-184	-154
Transfers	1 981	1 217	2 929	-	92	40	..	44	86
Capital account	-6 893	-5 627	..	4	644	896	..	-89	-22
Direct investment	10	117	..	-18	37	73	..	42	109
Other long-term capital (net)	-4 163	-4 589	..	37	261	1 052	..	45	-
Assets	42	43	..	4	141	70	..	..	..
Liabilities	-4 205	-4 632	..	33	120	982	..	..	..
Short-term capital (net)	-2 740	-1 155	..	-15	346	-229	..	-175	-131
Overall balance	-6 177	-6 986	..	-1 644	-693	-122	..	227	771
Reserves (net) [b]	-1 938	1 188	..	1 644	693	122	..	-83	-603
Special financing (net)	7 755	6 569	..	-	-	-	..	-	-

Sources: As for table 3.3.5.

Note: Errors and omissions and valuation changes are not shown.

a Cash basis, reflecting only interest and principal repayments made.
b A negative sign indicates an increase in reserves.
c Excludes transactions with former Yugoslavia.
d January-November.

After a significant improvement in 1991 (table 3.3.5),[225] the current account of *Russia* swung into deficit in 1992, estimated at about $3.2 billion.[226] This was financed by new bilateral credits, arrears and the deferral of debt repayment (see chapter 4). A considerably larger deficit had been expected as the deficit recorded during January-September 1992 amounted to $6.4 billion. However, the trade balance improved dramatically in the final quarter of the year. An upturn in exports and a sharp decline in imports resulted in a $5

---

[225] This was the result of an increase in export earnings (including earnings from larger gold sales) and stagnant expenditures on imports. The authorities had aimed at generating a trade surplus to provide Vneshekonombank with the means to service its debt obligations, which it did until the end of 1991.

[226] This is a preliminary estimate of the Centre for Conjunctural Studies and Projections, *Ekonomicheskaya Konjunktura*, No.1, Moscow, February 1993.

billion trade surplus for that period yielding an estimated $2 billion trade surplus for the full year.[227]

It should be borne in mind that the trade and estimated balance of payments statistics[228] of the former Soviet Union and Russia are surrounded by considerable uncertainty.[229] In part this problem is shared by all countries in transition and stems from start-up difficulties in implementing a customs-based system of collecting trade statistics (a problem which may persist for years); the revaluation of trade originally conducted in roubles and through clearing accounts in a common currency, such as the US dollar; under-invoicing of exports to world markets, channelling exports through the other former Soviet republics in order to avoid various export restrictions and taxation, which results in the under-reporting of convertible-currency receipts; and the statistical problem of how to apportion the 1991 foreign trade of the former USSR among the individual republics in order to obtain a base-year figure for the calculation of growth rates for 1992.

National *projections for 1993* envisage some deterioration in the current account of most east European countries (table 3.3.5). These projections assume that economic activity will turn up or, in the case of Poland, strengthen. Most governments foresee an increase in output in 1993 (see section 3.1). In most countries, exchange rate policies will continue to be geared to curbing inflation, and any ensuing further appreciation of real exchange rates of domestic currencies can be expected to weaken trade balances. The widespread drought in 1992 will cut exports of agricultural products and in some countries necessitate additional food imports. Poland will be particularly affected with estimated $1 billion losses in net exports in both 1993 and 1994.[230] In a number of countries, the realization of the projected current-account deficits depends on the availability of external finance (see chapter 4).

Foreign currency reserves

In general, the *foreign currency reserves* of the countries of *eastern Europe* increased in 1992 (table 3.3.7). Those of Czechoslovakia, Hungary and Slovenia continued to rise, boosted by current-account surpluses, inflows of investment funds and in the case of the former two, some foreign borrowing. The reserves of Bulgaria rose sharply during the year, peaking at $1.1 billion in October. To achieve this level, the country forewent most debt servicing, borrowed $420 from the G-24/EC and IMF, and the national bank made net purchases in the foreign exchange market of about $310 million.[231] During the previous two years, Bulgaria's reserves had been very low, and the recent build-up was achieved at a very high opportunity cost in terms of lost income (see chapter 4). The $241 million in official reserves reported by Romania is grossly inadequate and falls short of the modest $500 million target for 1992 agreed with the IMF.[232] However, Romania's deposits at BIS reporting banks increased, possibly due to capital flight.[233] The position

TABLE 3.3.7

European transition countries: Foreign currency reserves and BIS deposits, 1989-1992
(Billion US dollars, end of period)

	1989	1990	1991	1992
Albania				
FCR	0.42	0.22	0.09	..
BIS	0.31	0.22	0.03	0.05 a
Bulgaria				
FCR b	..	-	0.31	0.86
BIS	1.18	0.58	1.03	1.43 a
Czechoslovakia				
FCR	2.16	1.10	3.05	4.18
FCR b	..	..	1.35	1.05
BIS	2.20	1.35	2.80	4.35 a
Czech Republic				
FCR	..	..	..	3.46
FCR b	..	..	..	0.73
Slovak Republic				
FCR	..	..	..	0.72
FCR b	..	..	..	0.32
Hungary				
FCR	1.25	1.07	3.94	4.35
BIS	1.18	1.78	3.74	4.02 a
Poland				
FCR	2.31	4.49	3.62	3.99
BIS	3.95	7.82	5.14	6.72 a
Romania				
FCR	1.76	0.37	0.37	0.22
BIS	1.84	0.58	0.58	0.75 a
Former Yugoslavia				
FCR	4.14	5.46	2.68	1.48 c
BIS	7.07	7.98	5.05	5.42 a
Slovenia				
FCR	..	0.27	0.37	1.16
Former Soviet Union				
FCR	..	..	-	2.00
BIS	14.70	8.63	8.87	12.18 a

Sources: BIS, *International Banking and Financial Developments*, Quarterly Reports, Basle, February 1993. IMF, *International Financial Statistics*, March 1993; national statistics.

Note: BIS: Deposits held with BIS reporting banks. FCR: Foreign currency reserves.

a September.
b Holdings of the national bank.
c June.

---

[227] Ibid.

[228] Official balance of payments statistics for Russia and the former Soviet Union are available for 1990-1991. These were reproduced in United Nations Economic Commission for Europe, *Economic Bulletin for Europe*, Vol.44(1992), New York, 1993, table 2.2.5.

[229] The uncertainty surrounding Russian trade statistics is partially reflected in the differences in the statistics produced by Goskomstat and the Ministry for Foreign Economic Relations (see below).

[230] GATT, L/7164, op.cit.

[231] BBC, *Summary of World Broadcasts*, EE/W0258 A/2, 26 November 1992.

[232] *Romania Economic Newsletter*, Vol.2, No.1 (April-June 1992).

[233] See footnote (197).

of Albania is even worse, the country having exhausted its small foreign currency holdings in 1990-1991 for badly needed imports.

The *Czech Republic's* foreign currency reserves are reported to have declined sharply towards the end of 1992. The main reason appears to be that the national bank increased sales of foreign currency to commercial banks whose positions had been weakening when domestic firms stepped up year-end imports, prior to the tax increase which took effect on 1 January 1993.[234] This decline appears to have been a consideration in the government's decision to seek a special IMF loan in February 1993 and undertake capital market operations (see chapter 4).

No official statistics concerning the foreign exchange reserves of the successor states to the *former Soviet Union* appear to have been released, but they are assumed to be very low. According to a recent statement, in 1992 the *Russian Federation's* currency reserves increased from zero to $2 billion (presumably at the end of 1992).[235] This includes drawings on the country's $1 billion IMF standby credit negotiated last year. Traditionally, deposits with BIS reporting banks had been used as a proxy for the former Soviet Union's foreign currency reserves.[236] However, it is now widely accepted that these assets are illiquid. They also seem to represent increasingly the deposits of citizens and enterprises of the successor states and thus are not easily accessible to the authorities.[237] In the third quarter of 1992, these assets rose by nearly $3 billion to $12.2 billion — the largest quarterly increase in more than 10 years — presumably because of capital flight.[238]

Russia's *gold reserves* are reported to have increased from 290 tons at the end of 1991 to 308 tons at the end of 1992.[239]

Indebtedness

The *gross* indebtedness of *eastern Europe* continued to rise in 1992 (table 3.3.8). Capitalization of unpaid interest and new official credits continued to boost Bulgaria's debt. Recourse to official financing also raised Romania's gross debt. It has increased rapidly since 1989 — the former government had virtually eliminated the debt by then — reaching more than $3 billion at the end of 1992. The gross indebtedness of Czechoslovakia rose, but since foreign currency reserves increased faster, net indebtedness declined. Hungary's net debt also declined because of reserve accumulation and also because of net repayments.

The gross debt of the former *Soviet Union* was reported to have reached $75.8 billion in mid-1992. This figure excludes lease-lend obligations to the United States (some $800 million) and covers only convertible-currency debt (i.e., it excludes debt to former socialist countries).[240] It exceeds the $70.7 billion debt reported last year by Vneshekonombank for end-May 1992.[241] A higher figure of $80 billion has been reported in the western press, but this may include some obligations in non-convertible currencies. The increase since the end of 1991 was due to new bilateral credits, rescheduling of some debt service obligations and the accumulation of arrears.

Repartition of debt of defunct states

In the wake of the dissolution of the former Czechoslovakia, USSR and Yugoslavia, a quick resolution of the issue of responsibility for these states' outstanding debt and servicing obligations became essential if the successor states were to assume normal relationships with creditors and obtain access to new credit facilities. The Czech Republic and the Slovak Republic agreed to divide the assets and liabilities of the former CSFR applying a ratio of two to one (excluding those *vis-à-vis* the IMF, see below).

The successor states of the *former Yugoslavia* have agreed on the repartition of debt, although no formal accord appears to have been forthcoming. The new republics are reported to have taken responsibility for the obligations run up by banks, enterprises, and official organs located on their respective territories. However, aside from IMF and World Bank loans it is uncertain what formula has been used to divide those loans obtained in common by the former federal government.

As regards the *former USSR,* the legal framework for assuring servicing and management of the inherited debt was set out in three inter-republican agreements.[242]

---

[234] M. Kucera, executive director of the Czech National Bank. *Wall Street Journal,* 15 January 1993.

[235] As reported by former Acting Prime Minister Ye. Gaidar, on the occasion of a visit to the United States. *Bulletin of the United States Mission (Geneva),* 5 March 1993. In the context the figure appears to pertain to the end of 1992.

[236] The IMF, for example, maintained this practice through the first part of this year. See IMF, *Economy of the Former USSR in 1991,* Washington, D.C., April 1992.

[237] During 1991 the BIS deposits of the former Soviet Union became increasingly illiquid since they were held as deposits with Vneshekonombank subsidiaries abroad, which themselves have sizeable convertible-currency liabilities to other depositors (IMF, *USSR,* op.cit., April 1992, p.23). Similarly, the BIS noted that the $1.7 billion rise in the deposits of the former Soviet Union with reporting banks in the second half of 1991 had occurred "presumably in part because of transfers to support Soviet banks operating in the reporting area" (BIS, *Maturity Distribution of International Bank Lending,* Basle, July 1992, p.3).

[238] BIS, *International Banking and Financial Market Developments,* Basle, February 1993.

[239] Gaidar, op.cit.

[240] *Rossiiskie Vesti,* 29 January 1993. This figure and definition of the debt was repeated by Minister of Foreign Economic Relations, S. Glazeva, although the Minister did not attach a date to it. *Izvestia,* 11 February 1993.

[241] World Bank, *Russian Economic Reform,* Washington, D.C., September 1992, table 4.2.

[242] There are "Memorandum of mutual understanding concerning the debt to foreign creditors of the Union of Soviet Socialist Republics and its successors in law", 28 October 1991 (*Izvestiya,* 29 October 1991); *Treaty on the legal succession regarding the external debt,* adopted 4 December 1991 (this treaty set up an

TABLE 3.3.8

Eastern Europe and former Soviet Union: External debt in convertible currencies, 1990-1992
(Billion US dollars, end-of-period)

	Gross debt			Net debt		
	1990	1991	1992	1990	1991	1992
Albania	0.3	0.5	0.6	0.1	0.4	0.6
Bulgaria	10.0	11.4	12.0*	10.0	11.1	11.1
Czechoslovakia	8.1	9.3	9.5	7.0	6.3	5.3
Czech Republic	..	..	6.9	..	..	3.4
Slovak Republic	..	..	2.6	..	..	1.9
Hungary	21.3	22.7	21.4	20.2	18.8	17.1
Poland	48.5	48.4	49.4	44.0	44.8	45.2
Romania	0.4	1.9*	3.5*	-	1.5	3.3
Former Yugoslavia	16.5	14.5	..	11.0	11.8	..
Yugoslavia, Fed.Rep.	..	..	9.0	..	..	..
Slovenia	2.0	1.9	1.7	1.1	1.5	0.5
Eastern Europe [a]	88.6	94.2	96.4	81.3	82.8	82.7
Former Soviet Union	61.1	65.3	75.8 [b]	..	..	..

Sources: National statistics; ECE secretariat estimates and table 3.3.7 for foreign exchange. Net debt equals gross debt less foreign currency reserves.

[a] Excluding Yugoslavia.
[b] Mid-1992, excluding US Lend Lease debt (about $800 million). A figure of $80 billion was cited by the western press in early 1993, but it may include some non-convertible currency debt.

Under these arrangements, the signatory republics agreed to take responsibility "jointly and severally" for the external debt contracted by the former Soviet Union. Vneshekonombank (or any legal successor) was entrusted with the tasks of managing the outstanding debt and negotiating with foreign partners on behalf of the republican signatories. The republics were allocated fixed percentage shares of the former Soviet Union's debt service obligations. Finally, in order that Vneshekonombank could make the scheduled payments on their behalf, the republics committed themselves to opening special accounts with the bank and to transfer the necessary funds in advance.

These arrangements have not worked out as intended. Estonia, Latvia, and Lithuania were not parties to any of the treaties and in March 1992 proclaimed that "the Baltic States are not legal successors of the former USSR and therefore cannot take responsibility for servicing its external debt".[243] Ukraine, the second most important republic in terms of allocated shares, initially insisted on settling its part of the obligations separately. Subsequently, Russia offered to assume the entire debt of the former Soviet Union on the condition that the other successor republics renounce all claims on the corresponding assets (the so-called zero option). This solution was accepted by all CIS states, including the Ukraine, in November 1992.[244] However, in late December 1992 Ukraine denounced this accord, citing

TABLE 3.3.9

IMF quotas and shares applied to the IMF assets and liabilities of the successor states to Czechoslovakia and Yugoslavia
(Per cent and million SDRs)

		Quotas [a]	
	Shares	Eighth review	Ninth review
Czechoslovakia	100.00	590.0	847.0
Czech Republic	69.61	..	589.6
Slovak Republic	30.39	..	257.4
Yugoslavia (SFR)	100.00	613.0	918.3
Bosnia-Herzegovina	13.20	76.9	121.2
Croatia	28.49	180.1	261.6
Macedonia	5.40	33.5	49.6
Slovenia	16.39	99.0	150.5
Serbia/Montenegro	36.52	223.5	335.4

Source: IMF.

[a] Quotas under the Eighth and Ninth General Review of Quotas.

the inability of the parties to agree on the value of Soviet property abroad.[245] In mid-January 1993, Russia and Ukraine concluded a new separate agreement on the division of ex-Soviet assets and Ukraine accepted responsibility for the 16.34 per cent of the debt of the former USSR (as originally provided for in the treaty of 4 December 1991).[246] However, negotiations undertaken by the two parties in mid-March again failed to

---

allocation key which distributed the debt of the former USSR as follows: Russian Federation, 61 per cent; Ukraine, 16 per cent; Belarus and Kazakhstan, about 4 per cent each; with smaller shares for the remaining republics); *Agreement on additions to the Treaty on the legal succession regarding the external debt*, adopted 13 March 1992. By the third agreement Ukraine gained one of the permanent chairmanships of the Interstate Council, together with the Russian Federation and a rotating third member, as well as a blocking vote on the Council (since 80 per cent of the voting power is required for a decision).

[243] The World Bank, *Financial Flows to Developing Countries*, June 1992, p.17.

[244] In November 1992, Ukraine had concluded a provisional agreement with Russia under which Russia assumed responsibility for the Ukraine's share of the Soviet debt and also the sole right to negotiate with western creditors. In return the Ukrainian government relinquished its claim on the debts owed to the former Soviet Union by developing country trade partners (reported to be some $124 million). The signatories envisaged further negotiations on a special agreement on sharing out remaining assets, reported to include Soviet embassies abroad and reserves of gold and diamonds (*Financial Times*, 24 November 1992).

[245] *Wall Street Journal*, 4 January 1993.

[246] The two parties undertook to divide the former Soviet Union's assets and debts and make separate arrangements for servicing their respective shares of the debt. This process is to be completed by 31 March 1993.

resolve the outstanding questions.[247] Final negotiations with the Paris Club on a debt rescheduling package have been suspended until these issues are resolved.[248]

The IMF has decided that the Czech Republic and the Slovak Republic shall be the successors to the assets and liabilities of the CSFR in the IMF and that Bosnia and Herzegovina, Croatia, the former Yugoslav Republic of Macedonia, Slovenia, and FR Yugoslavia (Serbia and Montenegro) are successors to those of the SFR Yugoslavia.[249] Their respective shares of such assets and liabilities, as determined by the IMF, are presented in table 3.3.9.[250] These shares are reflected in the quotas which have been set for the new republics. Similar issues do not arise with respect to the republics of the former Soviet Union because they joined the IMF already as sovereign states.

(iii) East-west trade developments

Eastern trade values, volumes and trade balances

The European transition countries' trade with the developed market economies[251] contracted in the first half of 1992, but then picked up somewhat in the third quarter. None the less, for the period January-September 1992 the volume of trade showed no rise from the same period in the previous year (table 3.3.10). East-west trade had stagnated already in 1991, after posting moderate-to-rapid growth in 1988-1990. In 1992, this sluggish overall performance reflected a further contraction of some 6-8 per cent in the western trade of the successor states of the *former Soviet Union*,[252] and an offsetting rise of about the same magnitude in that of the countries of *eastern Europe*. It should be noted, that the trade of eastern Europe with the west continued to gross markedly faster than overall western trade.

A new feature in 1992 was the emergence of import growth in Bulgaria and Romania. The western trade of Czechoslovakia, Hungary and Poland had been expanding strongly already in the preceding two years, as was also the case of Bulgaria's exports. However, the further contraction in the western trade of the former Soviet Union and the former Yugoslavia offset the positive tendencies in the trade of the other eastern countries.

This summary of recent developments in trade volume and the analysis below are based on trade value and price data recorded in the statistics of the western partner countries, because they are more complete in country coverage and commodity details. It should be noted that in the first nine months of 1992 pronounced discrepancies were noted between the *rates* of change registered in the statistics of the eastern countries and the aggregated statistics of their western partners (see box 3.3.1 for a juxtaposition of 1992 data, and table 3.3.10). Although in the course of the year these differences tended to narrow, in several cases the discrepancies remain very large.[253] As noted in box 3.3.1, there are serious reasons to doubt the reliability of recent eastern trade statistics.[254] The review below should be read with this in mind. It will proceed on the basis of the western data, while drawing attention to the most notable divergences observed (usually in footnotes).

The rate of decline in the volume of trade of the western countries with the successor states of the *former Soviet Union* slowed somewhat, from 8-14 per cent in 1991 to 6-8 per cent in the first nine months of 1992 (table 3.3.10).[255] Export revenues were diminished by the fall in international prices of fuels and metals.

The growth in the volume of *eastern Europe's exports* to the west slowed to a rate of 6 per cent in the first three quarters of 1992 (table 3.3.10). This overall expansion reflected a further rise in the sales of Bulgaria,[256] Czechoslovakia, Hungary and Poland. Czechoslovakia posted another strong surge in export

---

[247] *Financial Times*, 15 March 1993.

[248] According to O. Wethington, US Assistant Secretary of the Treasury for International Affairs, the G-7 were notified by Russia and Ukraine that they were not prepared to respond to the rescheduling offer. *Bulletin of the United States Mission (Geneva)*, 14 January 1993.

[249] IMF, *Press Releases*, Nos.92/92 and 92/96.

[250] Each successor state has been asked to notify the IMF within one month whether it agrees with its share. IMF quotas of all eastern countries are shown in United Nations Economic Commission for Europe, *Economic Bulletin for Europe*, Vol.44(1992), New York, 1993, table 3.3.1.

[251] The terms "developed market economies", "west" or "western countries" as used here refer to the countries of western Europe (including Turkey), North America and Japan. The term "east", "eastern countries", or "European transition countries" refers to eastern Europe and the former Soviet Union jointly. This terminology is also used in chapter 4.

[252] The western trade returns for 1992 on which this section is based generally do not yet report separately on trade with the successor states of the former Soviet Union and the former Yugoslavia, for which in any case base-year data would not be available. Trade with these countries is therefore reviewed in the form of country group aggregates.

[253] In the first three quarters of 1992, all rates of change carry the same *sign*. This was not the case in the first half of the year (see United Nations Economic Commission for Europe, *Economic Bulletin for Europe*, Vol.44(1992), New York, 1993, table 2.2.2 and box 2.2.1). Discrepancies in the directions of recorded change were earlier observed in in the case of Bulgarian exports (1990-1991) and imports (1991) and Czechoslovak imports (1991).

[254] Reliability questions of course also arise from the fact that the data for any given eastern country represent an aggregation of partner country data from 22 different sources.

[255] Statistics for the Russian Federation show a much stronger contraction in the value of exports than do western data for the CIS countries jointly.

[256] While western data show steady strong growth of Bulgarian exports since 1990, Bulgarian statistics register an increase in exports for the first time in 1992, with a rate of growth which is over three time higher than that recorded in western data.

BOX 3.3.1

European transition countries: Comparison of changes in the value of trade as reported in eastern national and western "mirror" statistics: January-September 1992 [a]

(Percentage changes in US dollar values)

	Bulgaria		Czechoslovakia		Hungary	
	Exports	Imports	Exports	Imports	Exports	Imports
Western data [b]	26	13	48	70	14	18
Eastern data [c]	83	94	30	35	20	1
Balance of payments [d]	45	35	38	37	18	9

	Poland		Romania		Russia	
	Exports	Imports	Exports	Imports	Exports	Imports
Western data [b]	14	5	-	18	-9 [e]	-1 [e]
Eastern data [c]	..	..	3	43	-35	-4
Balance of payments [d]	12	3	15	-	..	..

Sources: Western data: western trade statistics, as for table 3.3.10. Eastern data: trade statistics, as for table 3.3.1: trade statistics on a balance of payments basis from national sources.

   a Relative to the same period in 1991.
   b See source note in table 3.3.10 for the reporting countries covered.
   c Trade with developed market economies (west).
   d Trade in convertible currency.
   e Trade with countries formerly part of the Soviet Union.

The collection and interpretation of eastern foreign trade data have always been plagued by severe difficulties. Since 1989 the situation has clearly worsened in many respects, increasing the reasons to doubt the reliability of these statistics.

First, changing national borders and in some cases civil wars have made it more difficult for national authorites to monitor actual trade flows. The choice of exchange rates to value trade flows in a common currency continues as a major problem in interpreting the statistics of certain countries. A third difficiency is a consequence of the transition to a market-based foreign trade regime and the adoption of western statistical recording practices (i.e., collecting trade information at the customs border, adoption of the Harmonized System of classification (HS) and the Single Administrative Document (SAD), etc. These problems have been disussed in more detail in "Data availability and reliability: from bad to worse", **Economic Bulletin for Europe**, Vol.44(1992), New York, 1993, pp.54-55.

The detailed evaluation of the overall trade performance of the eastern countries must be based on national statistics. To some extent, a confrontation with partner ("mirror" data) can provide controls and help fill gaps. But here, too, the situation has deteriorated, especially in 1992 after the break-up of the Soviet Union and Yugoslavia. The majority of western countries have not yet adjusted their own country classifications to the new political realities and continue to report trade with the former Soviet Union and Yugoslavia in aggregate rather than by individual trade partner.

Nevertheless, a crude comparison of east-west data based on "mirror statistics" is possible and indeed essential. This comparision is provided above and in table 3.3.10. Although this crosscheck applies only to the western trade of the transition countries, the discrepancies in growth rates also bring into question the accuracy of their statistics for trade with their other partners and, of course, for their total trade as well. There is nothing that can easily be done to reconcile these differences in the data sources, except to sound a warning about the uncertainty surrounding any conclusions based on them.

growth,[257] while the expansion of Hungary's exports slowed sharply. The export volume growth of these four countries has now continued for a period of 4-5 years.[258] The multi-year slide in Romania's western exports was reversed in the third quarter of 1992. However, the average volume of exports in the first nine months was still somewhat lower than in the same period in 1991. Romania's western export's fell sharply in the early 1990s, and in 1992 amounted to only 60 per cent of the level recorded in the second half of the 1980s. The pace of Bulgaria's exports picked in the third quarter of the year.

In contrast, the exports of Albania and the successor states of former Yugoslavia jointly declined again, those of the latter at a quickening pace, reflecting the war in the region and probably the impact of the UN trade embargo on some of these states. However, the trade of Slovenia appears to have expanded.

Overall, eastern Europe's exports continued to expand faster than total western imports, which rose by less than 2 per cent in volume in 1991 and some 4½ per cent in the first three quarters of 1992 (see developed markets economies in chart 3.3.3).

---

[257] As in the past few years, western statistics register a considerably more dynamic development of the trade with Czechoslovakia in 1992 than is recorded in Czechoslovak statistics (see table 3.3.10).

[258] In the case of Bulgaria, this performance reflects the evidence of western trade data. Bulgarian data show sharp falls in exports to the west in 1990-1991. This discrepancy may be explained in part by the fact that Bulgarian trade data do not take into account exports by private companies and legal persons (Bulgarian News Agency, *Bulgarian Economic Outlook*, 17 February 1992, p.8).

TABLE 3.3.10
European transition countries: Trade with the west, 1990-1992
(Growth rates in percentage: trade balances in billion US dollars)

	Eastern exports			Eastern imports			Eastern trade balances				
	1990 a	1991 a	QI-III 1992 b	1990 a	1991 a	QI-III 1992 b	1990 a	1991 a	1991 b	QI-III 1991 b	QI-III 1992 b
Albania											
Value ......(W)	-10	-24	-21	20	36	95	-0.1	-0.2	-0.2	-0.1	-0.3
Volume ......(W)	-14	-18	-20	8	43	86	..	..	..	..	..
Bulgaria											
Value ......(E)	-11	-36	83	-26	-60	94	-0.8	-0.1	..	-0.1	-0.3
......(W)	25	26	26	-35	1	13	-0.6	-0.4	-0.5	-0.3	-0.2
Volume ......(W)	11	31	22	-42	4	8	..	..	..	..	..
Czechoslovakia											
Value ......(E)	13	7	30	25	-14	35	-0.6	0.3	..	0.4	0.3
......(W)	18	27	48	34	20	70	-0.1	0.2	0.1	0.1	-0.8
Volume ......(W)	5	32	42	19	24	63	..	..	..	..	..
Hungary											
Value ......(E)	21	28	20	4	54	1	0.5	-0.9	..	-1.0	-0.5
......(W)	27	15	14	17	17	18	0.1	-	-0.2	-0.2	-0.4
Volume ......(W)	12	19	9	3	21	13	..	..	..	..	..
Poland											
Value ......(E)	40	14	..	-5	72	..	3.4	0.3	..	0.6	..
......(W)	46	6	14	26	55	5	0.9	-2.7	-3.0	-2.1	-1.6
Volume ......(W)	32	11	10	13	61	2	..	..	..	..	..
Romania											
Value ......(E)	-38	-23	3	117	-9	43	0.2	-0.3	-	-	-0.6
......(W)	-29	-16	-	103	-9	18	0.3	0.1	-	-	-0.3
Volume ......(W)	-39	-12	-3	89	-5	16	..	..	..	..	..
Yugoslavia											
Value ......(E)	21	..	..	43	..	..	-3.4	..	..	..	..
......(W)	22	-7	-15	38	-25	-19	-2.1	-0.5	0.4	-0.3	0.6
Volume ......(W)	9	-4	-18	23	-23	-22	..	..	..	..	..
Eastern Europe											
Value ......(E)	10	7	24 c	19	8	22 c	2.5	-0.8	..	0.3	..
......(W)	20	5	10	28	7	11	-1.7	-2.5	-3.2	-2.5	-3.1
Volume ......(W)	7 c	9	6	14 c	11	7	..	..	..	..	..
Former Soviet Union											
Value ......(E)	12	-16	-35 d	6	-31	-4 d	-5.1	-	..	3.1 d	-3.6 d
......(W)	20	-10	-9	-5	-17	-1	2.3	3.8	0.5	3.5	1.6
Volume ......(W)	1	-8	-6	-13	-14	-8	..	..	..	..	..
Total East											
Value ......(W)	20	-2	1	12	-3	6	0.6	1.3	-2.7	1.1	-1.5
Volume ......(W)	7	1	-	14	-	1	..	..	..	..	..

Sources: For western data, 1990-1991: United Nations commodity trade data base (COMTRADE); January-September 1991 and 1992: OECD, *Statistics on Foreign Trade,* Series A, Paris; IMF, *Directions of Trade and International Financial Statistics,* Washington, D.C.; national statistics; and ECE secretariat estimates based upon western data. These data reflect the trade of 22 western reporting countries (see Appendix table C.8).

Note: (E) eastern data, see table 3.3.1; (W) western data as above; volume changes are calculated on western value data. For the methodology and derivation, see United Nations Economic Commission for Europe, *Economic Bulletin for Europe,* vol.31, No.1, New York, 1979.

a Data include only the trade of the Federal Republic of Germany (i.e., excluding the trade of the former German Democratic Republic with the eastern countries, prior to reunification).
b Data include the combined trade of the Federal Republic of Germany and the former German Democratic Republic.
c Four countries only (Bulgaria, Czechoslovakia, Hungary, Romania).
d Russian Federation only.

In the current policy discussions concerning the achievements of economic reforms in the transition economies, the sustained growth of exports to the west has been viewed as a key indicator of success. The positive performance registered here demonstrates that eastern products can compete in western markets, contrary to widely-held doubts about their quality and the scope of the market for such goods. The gains in revenue have been particularly important in the early stages of the reforms, since — together with inflows of private capital and international assistance — they have enabled most of these countries to boost imports and rebuild foreign currency reserves. In the face of contracting domestic demand and the demise of the former CMEA market, western markets have provided the only source of demand growth for the economies of Bulgaria,[259] Czechoslovakia, Hungary, Poland and, more recently, Romania. In Poland, exports to the west have been credited with fuelling the recent recovery in industrial output.

---

[259] In the case of Bulgaria, trends based on national data do not warrant this conclusion.

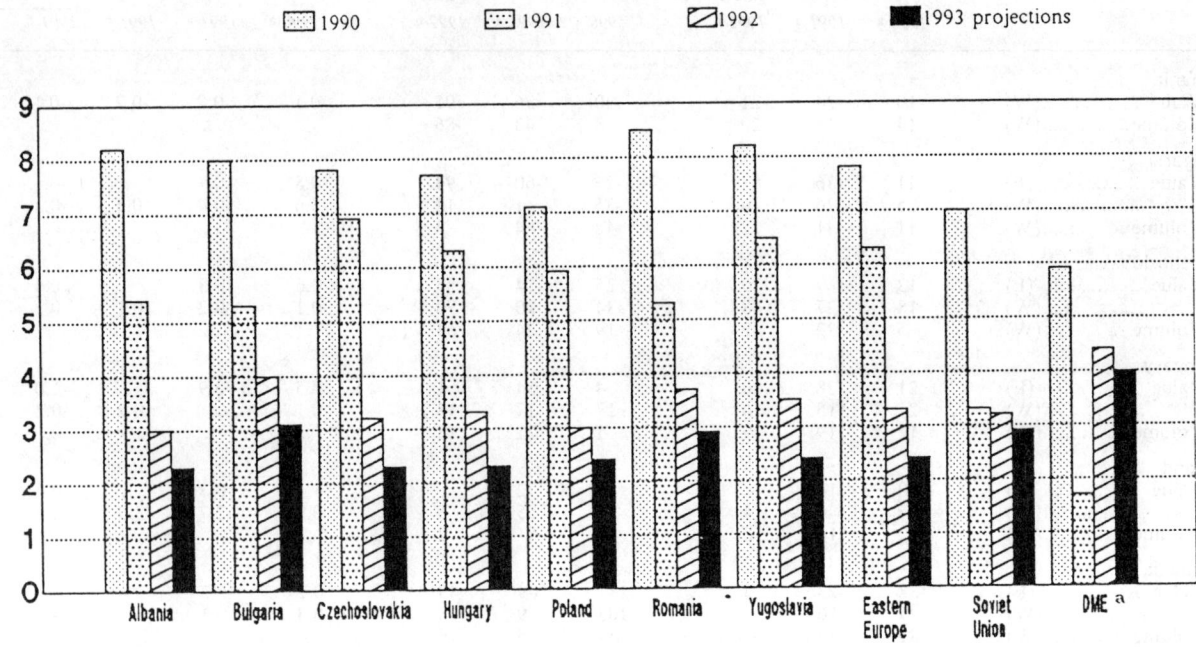

CHART 3.3.3

Specific western import demand facing eastern Europe and the former Soviet Union, 1990-1993
(Per cent annual change, volumes)

Source: ECE secretariat estimates.

Note: These are changes in the total import volumes of western countries weighted by the shares of those countries in the exports to the west of individual eastern countries. Imports include goods and services.

[a] Total import growth of developed market economies.

The volume of *imports* from the west into eastern Europe increased in the first three quarters of 1992 at a rate of 7 per cent — somewhat more slowly than in 1991 (table 3.3.10). This growth reflects chiefly developments in the third quarter of the year. In that period, Poland's western imports turned up (they had declined by 5 per cent in the first six months of the year), Czechoslovakia's expenditures quickened again (to a noteworthy 63 per cent in volume),[260] the growth of Bulgaria's imports strengthened,[261] and the imports of Romania turned up after having contracted in the previous year. In the last two countries growth strengthened as the year progressed.

Imports into Albania increased again, nearly doubling, financed by emergency food and commodity assistance provided by the Group of 24.[262] By contrast, the growth of Hungary's imports slowed, but remained relatively buoyant. The imports of the Yugoslav successor states jointly continued to contract rapidly.

Despite the recent increases, the western imports of Bulgaria and, especially, Romania remain far below 1989 levels. In this respect, their experiences contrast to those of Czechoslovakia, Hungary and Poland.[263] The imports of these three countries have risen significantly in recent years, which has undoubtedly facilitated the implementation of their reforms (see below and chapter 4).

The *trade balance* of the eastern region with the west swung into deficit in the first three quarters of 1992, due to the larger deficit of eastern Europe and the sharply lower surplus of the former Soviet Union (table 3.3.10).

Eastern trade by direction

Since 1989, numerous important changes have occurred in the geographical structure of the *east Euro-*

---

[260] Western statistics show a considerably more dynamic development of the imports of Czechoslovakia and Hungary than do national statistics (see table 3.3.10). The opposite is true for Romania.

[261] Attention is drawn to the discrepancies between the rates of change registered in western mirror data and Bulgarian statistics in 1991 and 1992. In the latter year, western data show a modest strengthening of imports, in comparison to near doubling of the US dollar value reported in the Bulgarian data.

[262] Already in 1991, nearly 40 per cent of Albania's imports had consisted of emergency food and commodity aid. EEC, IMF and IBRD, op.cit. However, the country's purchases of other goods, especially industrial inputs and energy, have fallen.

[263] This point was examined in more detail in United Nations Economic Commission for Europe, *Economic Bulletin for Europe*, Vol.44(1992), New York, 1993.

pean countries' western trade (table 3.3.11). In general the position of the EC strengthened, including in 1992. The EC now generally accounts for over 70 per cent, and in some cases over 80 per cent, of eastern Europe's trade with the west. The share of Germany, the largest individual partner, increased further. It should be noted that in the data for 1992 the share of Germany (and the EC) includes the trade of the new east German *Länder* while in the data for 1989 (i.e., prior to unification in 1990) the trade of the former German Democratic Republic with the eastern countries was not included in the German share. Thus the changed weight of Germany (and the EC) in 1992 reflects enlarged geographical coverage as well as the development of trade flows. In general, the shares accounted for by Italy, also traditionally one of eastern Europe's largest trade partners, and of Turkey also rose.

In several cases there was a pronounced strengthening of trade links between the eastern countries and neighbouring western countries. The shares of Turkey and Greece in Bulgaria's trade rose sharply, while that of Germany declined. Similarly, the shares of Germany and Austria in Czechoslovakia's trade rose as did that of Austria in Hungary's trade and that of Turkey in Romania's trade. The intensification of these regional trade ties was to be expected as the trade distortion stemming from intra-CMEA trade preferences disappeared.

The importance of the EC, Germany, Italy, and Turkey has also increased in the western trade of the *Soviet Union's successor states*. Neighbouring Turkey now accounts for 5-7 per cent of the western trade of these countries. The share of France has increased as well. By contrast, in the wake of the discontinuation of the long-standing bilateral trade agreement, Finland lost its position as the leading western trade partner.

Eastern trade, by commodity

East European exports of *manufactures* rose at above average rates in the first three quarters of 1992 (table 3.3.12).[264] As a result the share of manufactures in these countries' western exports has continued to rise, in contrast to its relative stability prior to the reforms. In most countries exports of *consumer* and *engineering goods* led the way. Exports of *semi-manufactures* rose at lower rates, except in the case of Czechoslovakia where the increase was only slightly lower than for the country's total western exports. Within this product grouping, sales of *iron and steel*, considered to be a sensitive sector, were mixed. Only the sales of Bulgaria, Poland and, especially, Czechoslovakia reflect growth in volume terms. In the latter case, however, the value of these exports rose by over one half.

As regards other sensitive items, east European exports of *clothing* increased rapidly. In general, *textile* exports rose at slower rates, and those of Hungary and Romania declined. In US dollar terms, several countries' sales of *food* rose modestly. However, since international prices of food products declined, the volume growth of these exports was larger. The drought in much of the eastern area may have already caused some exporters to reduce deliveries in the third quarter of 1992. It should be noted that the Hungarian authorities have reported significant increases in the sales of agricultural goods during 1992, apparently much more than is shown by the data in table 3.3.12.[265] Overall, these results suggest that while various restraints continue to affect eastern exports (see also chapter 4), particularly in the sensitive sectors, certain transition economies did manage to boost their sales.

The considerable increase in the exports of the former *Czechoslovakia* was broadly based, the sales of most sectors, particularly engineering goods, rising rapidly. In *Albania*, where exports had been declining for several years, the contraction in 1992 was concentrated in the primary goods sectors (especially food and ores), and to a lesser extent in consumer goods. Virtually all commodity groups registered falls in the exports of the successors of former *Yugoslavia*.

Due to the relatively favourable development of their financial situation, *Czechoslovakia, Hungary and Poland* have been relatively unconstrained in their capacities to increase *imports* during the past three years. In 1992, the growth of their western imports was broadly based. Their expenditures on *consumer goods*, and those of Czechoslovakia and Hungary on *food*, grew at above average rates. This was particularly true in the case of Czechoslovakia, where domestic consumption picked up. In Czechoslovakia and Poland, imports of *semi-manufactures* increased faster than their exports of these goods. In Poland this may reflect the upturn in the economy. In Czechoslovakia, however, industrial production continued to fall.

The development of these three countries' imports of *engineering goods* was mixed. Czechoslovakia's purchases rose by over 70 per cent. They accounted for over one half of the country's total incremental value of imports. This is remarkable in the light of the fact that domestic investment has only recently bottomed out after a steep decline (which continues in the other eastern countries). Only a small share of this increment was due to *road vehicles*, an import which has grown very rapidly in most eastern countries. This commodity group reflects largely passenger automobiles which are more often used as a consumption good than for busi-

---

[264] The commodity data presented in table 3.3.12 reflect the trade of a sample of 20 countries. The growth rates of total trade deriving from this sample differ slightly from those calculated from the returns of all 23 developed market economies in tables 3.3.10 and 3.3.11. These discrepancies result first of all from the absence in the sample of Italy and Turkey. Italy is a major participant in east-west trade, and both countries posted particularly dynamic trade results in 1992 (see table 3.3.11).

[265] For example, Hungarian exports of agricultural products to the EC were reported to have increased by 20-25 per cent in the first half of 1992, partly as a result of the entering into force of the association agreement with the EC (State Secretary at the Ministry of Finance Gyorgy Rasko, *Financial Times*, 9 September 1992). Agricultural exports, including food, rose by 16 per cent in value in the first three quarters of the year (Minister of International Economic Relations Bela Kádár, as quoted in BBC, *Summary of World Broadcasts*, EE/W0255 A/2, 5 November 1992).

TABLE 3.3.11

European transition countries: Trade with west, by major western partner [a]

(Levels in billion US dollars; shares in per cent; change in percentages)

	Eastern exports					Eastern imports			
	Level in 1991	Share in 1989	Share in 1992 [b]	Change in 1992 [c]		Level in 1991	Share in 1989	Share in 1992 [b]	Change in 1992 [c]
Albania: to/from									
DME	0.1	100.0	100.0	-21	DME	0.3	100.0	100.0	95
EEC.12	0.1	72.2	83.5	-9	EEC.12	0.2	83.8	85.9	124
EFTA.7	-	15.7	6.6	-61	NAM	-	3.9	5.6	87
NAM	-	1.9	3.2	-17	EFTA.7	-	9.7	3.8	-16
Italy	-	26.7	32.1	-18	Italy	0.1	32.9	28.8	97
Germany [a]	-	22.9	21.9	-25	France	-	4.4	25.7	364
Greece	-	4.8	16.0	108	Belgium and Lux.	-	1.5	13.8	1400
France	-	8.2	6.5	22	Greece	-	11.9	6.7	165
Japan	-	7.6	6.0	-51	Germany [a]	0.1	26.3	6.5	-49
United States	-	1.9	3.2	-16	United States	-	3.4	5.5	88
Belgium and Lux.	-	0.1	3.1	-40	Turkey	-	2.5	4.7	-15
Austria	-	4.8	3.0	-65	Switzerland	-	2.6	2.2	112
Netherlands	-	3.8	2.3	618	Netherlands	-	2.5	1.6	269
Bulgaria: to/from									
DME	1.3	100.0	100.0	26	DME	1.7	100.0	100.0	13
EEC.12	0.9	68.4	73.4	25	EEC.12	1.2	71.2	75.6	17
EFTA.7	0.1	22.2	7.3	22	EFTA.7	0.2	22.7	13.7	21
NAM	0.1	3.6	5.6	27	NAM	0.1	1.7	4.9	-25
Germany [a]	0.3	32.5	24.3	21	Germany [a]	0.5	40.5	28.5	7
Italy	0.2	9.0	15.8	43	France	0.2	6.4	14.6	75
Turkey	0.1	1.7	10.7	36	Italy	0.3	7.5	11.2	-27
Greece	0.2	2.3	9.0	-10	Greece	0.1	1.0	7.8	78
France	0.1	6.6	8.3	23	Austria	0.1	10.5	6.5	2
United Kingdom	0.1	6.2	5.4	34	United Kingdom	0.1	6.0	5.1	56
United States	0.1	2.1	4.9	22	United States	0.1	1.4	4.4	-31
Austria	0.1	12.2	4.2	27	Switzerland	-	6.0	3.7	59
Netherlands	-	4.4	3.3	41	Turkey	0.1	1.1	3.6	2
Former Czechoslovakia: to/from									
DME	6.6	100.0	100.0	48	DME	6.3	100.0	100.0	70
EEC.12	5.1	74.5	77.7	51	EEC.12	4.7	66.9	74.7	67
EFTA.7	1.0	9.8	15.6	45	EFTA.7	1.2	16.8	18.2	61
NAM	0.2	9.0	3.2	52	NAM	0.1	8.0	5.1	301
Germany [a]	3.1	22.3	49.6	60	Germany [a]	3.0	32.7	48.8	75
Austria	0.6	4.6	10.9	71	Austria	0.8	6.5	11.9	70
Italy	0.6	14.7	9.4	58	Italy	0.5	11.2	7.7	85
France	0.4	9.4	5.3	32	France	0.5	6.3	5.8	9
United Kingdom	0.2	7.1	3.5	41	United States	0.1	7.5	4.3	278
Netherlands	0.3	4.7	2.8	-2	Netherlands	0.2	2.8	3.6	68
United States	0.1	7.7	2.6	78	United Kingdom	0.2	5.9	3.5	72
Belgium and Lux.	0.1	2.2	2.0	42	Switzerland	0.2	6.4	3.1	30
Spain	0.1	5.0	2.0	79	Belgium and Lux.	0.1	2.7	2.3	68
Hungary: to/from									
DME	6.7	100.0	100.0	14	DME	6.6	100.0	100.0	18
EEC.12	4.6	64.1	70.7	18	EEC.12	4.3	70.5	66.6	21
EFTA.7	1.4	22.2	20.5	13	EFTA.7	1.8	23.8	25.6	13
NAM	0.4	8.2	5.2	-2	NAM	0.3	2.7	4.4	13
Germany [a]	2.6	32.2	39.4	18	Germany [a]	2.5	42.2	38.2	17
Austria	1.0	13.1	14.8	18	Austria	1.2	14.2	18.3	19
Italy	0.8	11.4	13.8	25	Italy	0.6	8.2	10.4	32
France	0.4	7.7	5.8	-	France	0.3	5.6	4.3	5
United States	0.4	7.3	4.7	-3	United States	0.3	2.6	4.1	9
United Kingdom	0.2	3.8	2.8	14	Netherlands	0.2	3.7	3.8	17
Belgium and Lux.	0.1	2.2	2.7	39	United Kingdom	0.2	4.2	3.7	24
Netherlands	0.2	2.9	2.5	6	Switzerland	0.3	5.1	3.5	-15
Sweden	0.2	2.9	2.1	5	Belgium and Lux.	0.2	4.0	3.4	45
Poland: to/from									
DME	9.9	100.0	100.0	14	DME	12.6	100.0	100.0	5
EEC.12	7.8	70.8	81.1	17	EEC.12	9.8	69.9	77.7	5
EFTA.7	1.3	17.4	12.5	6	EFTA.7	1.8	16.5	15.6	15
NAM	0.4	7.7	3.7	-	NAM	0.5	7.2	3.8	1
Germany [a]	4.4	31.7	46.6	21	Germany [a]	5.1	38.9	38.5	-2
Italy	0.6	8.0	8.0	41	Italy	0.9	7.3	8.5	27
France	0.6	6.7	6.1	7	Netherlands	0.9	4.7	7.5	8
United Kingdom	0.6	9.0	5.9	17	United Kingdom	0.6	5.2	7.4	71
Netherlands	0.5	5.9	4.3	-9	France	0.7	6.2	6.1	16
Austria	0.5	5.4	4.3	-2	Austria	0.6	6.5	5.1	7
Denmark	0.4	3.0	4.0	10	Denmark	0.7	2.5	3.9	-22
Sweden	0.4	5.1	3.9	17	Belgium and Lux.	0.5	3.4	3.8	-9
United States	0.4	6.4	3.3	1	United States	0.5	6.7	3.4	-1

TABLE 3.3.11 (concluded)

European transition countries: Trade with the west, by major western partner [a]

(Levels in billion US dollars; shares in per cent; change in percentages)

	Eastern exports					Eastern imports			
	Level in 1991	Share in 1989	Share in 1992 [b]	Change in 1992 [c]		Level in 1991	Share in 1989	Share in 1992 [b]	Change in 1992 [c]
Romania: to/from									
DME	2.3	100.0	100.0	-	DME	2.3	100.0	100.0	19
EEC.12	1.8	71.5	76.1	-3	EEC.12	1.6	61.5	74.1	25
EFTA.7	0.1	5.1	6.6	30	NAM	0.2	15.6	9.0	-9
NAM	0.1	11.5	5.1	25	EFTA.7	0.2	6.7	7.3	4
Germany [a]	0.7	21.5	33.1	6	Germany [a]	0.7	25.4	27.7	3
Italy	0.4	22.8	15.6	-5	Italy	0.3	7.4	16.0	73
France	0.3	12.2	11.9	-12	France	0.3	10.9	13.6	27
Turkey	0.2	6.1	8.4	9	United States	0.2	12.7	8.3	4
United Kingdom	0.1	5.0	4.6	4	Turkey	0.1	4.3	5.7	44
United States	0.1	9.2	4.0	26	Netherlands	0.1	5.5	4.7	76
Austria	0.1	1.7	3.9	56	Greece	0.1	3.5	4.2	44
Japan	0.1	5.0	3.6	-17	United Kingdom	0.1	5.1	3.8	14
Spain	0.1	2.9	3.1	-33	Austria	0.1	3.2	3.6	13
Netherlands	0.1	3.2	2.7	-15	Australia	-	7.8	2.7	50
Former Yugoslavia: to/from									
DME	11.3	..	100.0	-15	DME	10.5	..	100.0	-19
EEC.12	9.4	..	77.7	-21	EEC.12	8.5	..	77.7	-22
EFTA.7	0.9	..	10.1	10	EFTA.7	1.3	..	13.9	-8
NAM	0.7	..	5.3	-30	NAM	0.4	..	3.8	-25
Germany [a]	4.7	..	37.8	-22	Germany [a]	4.2	..	37.1	-24
Italy	2.7	..	21.9	-23	Italy	2.2	..	21.4	-18
France	0.8	..	8.8	6	Austria	0.8	..	9.3	1
Turkey	0.1	..	6.0	338	France	0.8	..	8.2	-8
United States	0.7	..	4.8	-31	Netherlands	0.4	..	3.8	-11
Austria	0.5	..	4.8	-8	United States	0.4	..	3.6	-21
Sweden	0.1	..	2.8	86	Turkey	0.1	..	3.5	267
United Kingdom	0.3	..	2.3	-17	United Kingdom	0.3	..	2.3	-46
Netherlands	0.3	..	2.2	-27	Switzerland	0.3	..	2.2	-35
Belgium and Lux	0.2	..	1.7	-5	Belgium and Lux	0.2	..	2.0	-30
Eastern Europe: to/from									
DME	38.2	100.0	100.0	10	DME	40.3	100.0	100.0	11
EEC.12	29.7	72.0	77.1	9	EEC.12	30.4	72.2	74.9	11
EFTA.7	4.9	14.0	13.4	17	EFTA.7	6.6	17.3	16.9	17
NAM	1.9	8.1	4.4	-5	NAM	1.7	5.5	4.6	15
Germany [a]	15.8	31.2	42.2	13	Germany [a]	16.1	38.5	39.0	9
Italy	5.3	16.6	13.5	3	Italy	4.8	12.1	12.0	11
Austria	2.7	6.7	7.7	22	Austria	3.7	8.3	9.7	22
France	2.7	7.6	7.0	7	France	2.8	6.8	7.2	14
United States	1.7	6.9	3.9	-5	Netherlands	1.9	3.8	4.8	15
United Kingdom	1.4	5.3	3.8	14	United Kingdom	1.6	4.6	4.5	31
Turkey	0.9	2.9	3.3	56	United States	1.6	5.0	4.1	15
Netherlands	1.4	3.7	3.0	-8	Belgium and Lux	1.2	2.8	2.9	8
Sweden	0.8	2.6	2.6	30	Switzerland	1.2	4.2	2.7	-4
Belgium and Lux	0.9	1.9	2.4	18	Sweden	0.9	3.1	2.2	10
Former USSR: to/from									
DME	32.0	100.0	100.0	-9	DME	28.7	100.0	100.0	-3
EEC.12	22.1	64.7	69.8	-8	EEC.12	17.8	47.6	64.5	3
EFTA.7	4.5	18.8	11.6	-26	EFTA.7	4.9	16.9	18.8	7
NAM	1.0	3.3	3.6	-	NAM	2.9	18.8	7.9	-24
Germany [a]	8.8	17.6	26.6	-14	Germany [a]	11.1	21.5	32.6	-18
Italy	4.7	13.8	14.2	-13	United States	3.6	14.9	13.5	11
France	3.1	10.5	11.6	11	Italy	2.5	9.0	11.0	38
Japan	3.3	10.6	8.2	-28	France	1.5	6.0	8.4	48
Turkey	1.1	2.5	6.8	86	Canada	1.3	2.0	5.3	-1
Finland	1.9	10.8	4.8	-23	Turkey	0.6	2.5	4.3	119
United Kingdom	1.6	5.5	4.0	-22	Japan	2.1	10.8	4.0	-51
Belgium and Lux	1.3	4.8	3.9	-13	United Kingdom	0.6	3.9	3.1	33
Netherlands	1.1	5.2	3.8	8	Netherlands	0.7	2.0	3.0	36
Spain	0.8	5.1	2.9	21	Austria	0.8	3.0	2.8	-1

Source: As for table 3.3.10.

Note: DME = developed market economies; NAM = North America. Shares of the EC and Germany in 1989 include only the trade of the Federal Republic of Germany. The levels (1991) and shares (1992) of the EC and Germany include the combined trade of the Federal Republic of Germany and former German Democratic Republic (see footnotes to table 3.3.10).

a Western partners are ranked by their shares in the western trade of the individual transition countries in 1992.
b January-September 1992.
c January-September 1992, relative to same period in 1991.

TABLE 3.3.12

European transition countries: Trade with the west, by commodity, 1992 [a]

(Per cent, million US dollars)

	Albania	Bulgaria	Czecho-slovakia	Hungary	Poland	Romania	Yugoslavia	Eastern Europe	Former USSR
Eastern exports (per cent)									
Total	-17	20	49	13	13	1	-18	10	-5
Primary goods	-25	24	42	8	14	-5	-25	9	5
Food	-21	-	16	7	-10	8	-26	-4	51
Fuels	118 400	-87	22	-12	-9	-43	-72	-11	-10
Manufactures	-31	26	52	17	16	5	-15	12	-
Semi-manufactures	-19	9	48	9	7	-6	-16	12	13
Iron and Steel	-36	15	57	2	17	4	-23	20	18
Chemical	10	-8	14	12	-8	15	-35	-3	16
Engineering goods	207	-12	60	16	17	9	1	19	-21
Machinery	86	10	38	-3	3	-2	-25	3	7
Consumer goods	-41	65	50	22	22	7	-23	8	5
Textiles	-55	64	29	-2	10	-17	-20	6	-8
Clothing	-39	78	55	24	28	29	-27	5	351
Eastern exports (increments, million US dollars)									
Total	-7	138	1 923	523	803	10	-1 141	2 248	-873
Primary goods	-6	62	233	84	240	-6	-186	421	168
Food	-1	-1	37	52	-90	5	-99	-97	176
Fuels	5	-18	52	-26	-72	-34	-68	-160	-1 045
Manufactures	-7	100	1 610	455	607	48	-850	1 963	13
Semi-manufactures	-1	12	540	52	92	-12	-160	524	229
Iron and Steel	-1	6	209	2	50	3	-47	221	59
Chemical	-	-7	64	39	-49	8	-120	-64	145
Engineering goods	1	-11	527	135	158	11	17	837	-232
Machinery	-	3	112	-9	7	-1	-75	38	8
Consumer goods	-6	100	542	268	357	49	-708	602	17
Textiles	-2	8	67	-3	13	-4	-37	42	-5
Clothing	-4	69	153	123	217	87	-467	178	93
Eastern imports (per cent)									
Total	106	22	69	16	4	11	-21	12	-
Primary goods	321	-15	43	12	-3	18	-22	6	28
Food	361	-25	85	27	-8	23	-22	9	30
Fuels	-59	-3	151	-22	31	-46	-26	16	61
Manufactures	-12	38	72	17	3	14	-22	12	-13
Semi-manufactures	-32	7	59	6	24	-14	-12	13	-20
Iron and Steel	-90	-12	70	-6	4	-47	1	4	-38
Chemical	39	19	51	2	18	-19	-13	10	-15
Engineering goods	-2	48	72	17	-8	22	-24	12	-15
Machinery	2	1	45	7	4	6	-23	10	-25
Consumer goods	4	62	94	32	6	37	-27	12	10
Textiles	3	65	78	23	28	47	-23	10	138
Clothing	-72	63	133	34	-4	25	-34	10	39
Eastern imports (increments, million US dollars)									
Total	107	203	2 673	666	303	144	-1 264	2 832	-68
Primary goods	117	-29	195	45	-32	64	-158	202	1 300
Food	118	-41	172	55	-83	64	-88	196	1 304
Fuels	-5	-2	63	-11	168	-46	-28	140	56
Manufactures	-6	242	2 390	623	172	124	-1 139	2 406	-1 478
Semi-manufactures	-6	14	460	69	345	-40	-170	671	-575
Iron and Steel	-5	-3	49	-5	7	-19	2	25	-287
Chemical	3	23	281	16	178	-34	-120	347	-264
Engineering goods	-	164	1 499	309	-262	83	-589	1 205	-1 025
Machinery	-	2	420	43	39	10	-151	363	-819
Consumer goods	1	63	432	245	89	81	-381	530	123
Textiles	-	27	109	64	149	57	-200	206	171
Clothing	-2	11	55	44	-5	11	-60	55	67

Source: United Nations Trade Data Base (COMTRADE). Components may not add to totals owing to gaps in reporting at the commodity level and exclusion of SITC 9. See text for coverage of western countries.

[a] January-September 1992 relative to same period in 1991.

ness.[266] Imports of road vehicles into Poland fell sharply in 1992 (after a huge increase in 1991),[267] presumably because of the higher tariffs imposed at the beginning of the year. Poland's purchases of *transport equipment* also fell steeply.

Over the past three years the generally improving financial situation of Czechoslovakia, Hungary and Poland has enabled these countries to raise considerably their imports of food, consumer goods, and passenger vehicles. The resulting greater availability of foreign goods is likely to have facilitated the introduction of painful economic reforms. At the same time, however, these countries' purchases of engineering and semi-manufactured goods also increased substantially (table 3.3.13). This runs counter to what would be expected on the basis of the steep decline in domestic industrial output and investment during this three-year period.

In *Bulgaria and Romania*, both constrained by shortages of convertible currency, imports turned up only in 1992.[268] In these countries, too, *consumer goods* imports increased rapidly, three times the rate of their total western imports (table 3.3.12). Due primarily to purchases of *road vehicles* and, in the case of Bulgaria, also of *transport equipment*, *engineering goods* imports grew relatively rapidly. Bulgaria's imports of *semi-manufactures* increased slowly, while those of Romania actually declined. Romania purchased more western *food*, perhaps to compensate for the effects of the drought. Overall, goods for current consumption appear to have benefited most from the upturn in these countries' imports.

Emergency food deliveries made available by the international community accounted for virtually all of the growth in *Albania's* imports. The country's imports of certain manufactures, including textiles, also rose, but the changes in US dollar value were small. The successor states of former *Yugoslavia* jointly posted another across-the-board decline of western imports.

Fuels continued to dominate the commodity structure of *exports* from the *former Soviet Union*. In the first three quarters of 1992, revenues from the sale of fuels declined by 10 per cent (table 3.3.12). However, since international prices of natural gas and petroleum and petroleum products fell by a similar amount, the volume of these exports is estimated to have remained roughly constant. Deliveries of both *natural gas* and *petroleum and petroleum products* to the west appear to have stagnated. This observation is consistent with OECD data covering the petroleum imports of member countries.[269] Deliveries of fuels to the west from the area of the former Soviet Union had further contracted during the first half of 1992 and turned up sharply in the

TABLE 3.3.13

Eastern Europe: Imports of western semi-manufactures and engineering goods, 1989-1992
(Billion US dollars)

	1989	1990	1991 a	1992 a
Bulgaria				
Semi-manufactures	0.6	0.4	0.3	0.4
Engineering	1.1	0.8	0.7	1.4
Total	1.7	1.2	1.0	1.7
Czechoslovakia				
Semi-manufactures	0.9	1.0	1.2	1.8
Engineering	1.7	2.6	3.3	5.4
Total	2.6	3.6	4.5	7.3
Hungary				
Semi-manufactures	1.4	1.5	1.6	1.7
Engineering	2.0	2.3	3.0	3.3
Total	3.4	3.8	4.6	5.1
Poland				
Semi-manufactures	1.4	1.3	2.1	2.7
Engineering	2.3	3.0	5.1	4.7
Total	3.7	4.3	7.2	7.4
Romania				
Semi-manufactures	0.3	0.6	0.4	0.4
Engineering	0.2	0.3	0.6	0.8
Total	0.5	0.9	1.0	1.2

Sources: As for table 3.3.12.

a Includes imports from the former German Democratic Republic, (see footnotes to table 3.3.10).

third quarter of the year. The earlier decline has been attributed to the widespread and growing problems in the oil industry of the successor countries.

Exports of *semi-manufactures* (including iron and steel, chemicals, other semi-manufactures) increased in volume, but incremental earnings of foreign currency were modest. Included in this grouping are *non-ferrous metals*, which the USSR and its successor states are reported to have exported in markedly higher quantities during the past two years. International prices of non-ferrous metals fell sharply in 1991, and again in 1992, although modestly. These price changes contributed to the decline in the value of exports of *ores*. However, *food* exports increased.

Purchases of western *food* dominate the changes in the total *imports* of the ex-Soviet countries in the first three quarters of 1992 (table 3.3.12). However, the growth of net imports of food was somewhat lower because food exports increased. The area's purchases of *consumer goods*, including clothing and textiles, rose, but the value increment was small. Within the *engineering goods* group, imports of *road vehicles* and *transport equipment* increased, but those of *machinery*

---

[266] Data for 1990-1991 are available in the United Nations Economic Commission for Europe, *Economic Bulletin for Europe*, Vol.44(1992), New York, 1993, Appendix table 13.

[267] Ibid.

[268] Romania's western imports increased temporarily in 1990, financed by the elimination of a large trade surplus and a sharp drawdown of foreign currency reserves.

[269] OECD data show little increase in the quantity of OECD members' petroleum imports during the first three quarters of 1992 (OECD, *Monthly Oil and Gas Statistics, October 1992*, Paris, January 1993). Russian data indicate that the Federation's total exports of oil fell by 16 per cent, natural gas by 2.7 per cent, and coal by 3.7 per cent in volume terms in the first seven months of 1992. The OECD data are of particular interest since they may more accurately record export flows than national statistics which, it is suspected, reflect under-reporting to hide the flight of capital.

fell. *Semi-manufactures* (including both chemicals and iron and steel) imports also declined. Overall, it appears that the bulk of recent import growth has been for current consumption. Purchases of western goods for production actually fell. This result and the finding that exports of semi-manufactures increased in the first three quarters of 1992 suggests that internal demand for production inputs was weak. However, certain products have been exported to earn convertible currency even as shortages have developed at home.

Factors determining east-west trade

On the whole, the western environment faced by the eastern countries deteriorated in 1992. Although the growth of *total imports* into the developed market economies increased from some 2 per cent in 1991 to 4½ per cent in 1992, the growth of *specific western import demand* facing the eastern countries *slowed* from over 6 per cent to some 3 per cent (chart 3.3.2).[270] These contrary tendencies are explained by the development of import demand in those western countries on which eastern exports are concentrated. In particular, the growth of total imports into Germany, which absorbs between one quarter to one half of the eastern countries' western exports, amounted to 10-12 per cent in 1990-1991, but slowed to only 2.8 per cent in 1992. Although the import growth of the other developed market economies tended to strengthen in 1992, this was not sufficient to offset the impact of the German deceleration. Overall, eastern exporters benefited from *improved access to western markets*, but certain new restrictions were added (see chapter 4).

It might be noted that the growth of "specific" western import demand facing the eastern countries is expected to slow further in 1993, due to a general slow-down in western, particularly German and Italian, import growth (chart 3.3.2).

Although the international trading environment has been important, at the present time *internal factors* and the *collapse of trade in the former CMEA* have had a greater role in shaping the development of eastern exports to the west. *Economic stabilization programmes* have tended to impact differentially on eastern exports and imports. In general, these policies have contributed to the drastic drop in domestic output and demand — by some 25-30 per cent — and to even larger falls in domestic investment (see section 3.1). This has tended to curtail import demand, particularly for fuels, industrial inputs and certain types of investment goods. At the same time, however, resources for export to other, particularly western, markets have been released.[271] While some standard products, above all many primary goods, are easily disposed of, eastern manufactured goods are generally less saleable. This problem has been exacerbated by the scale of unused export capacity resulting from the demise of trade among the former CMEA countries.[272]

The net impact of *structural reforms* on the development of eastern trade flows has probably been positive.[273] *Inter alia,* enterprises have been adapting to the acquisition of foreign trade rights and generally lower subsidies, and the role of the private sector in external trade has increased. The introduction of generally liberal *trade policy regimes* and *payments systems* (including internal convertibility for most current-account transactions) partially explains the considerable growth of imports into those eastern countries not constrained by serious hard currency shortages. In general, the importing operations of enterprises have been facilitated by the replacement of administrative controls with relatively low tariffs and some quotas (although several countries also adopted temporary restrictive measures, most often affecting consumer goods). Economic agreements concluded recently with the west by Czechoslovakia, Hungary and Poland have improved the access of western goods to certain eastern markets.[274]

Nominal exchange rates in Czechoslovakia and Hungary were relatively stable in 1992 (also see section 3.4(iv)).[275] In Poland the cumulative depreciation of the currency was larger, but still lagged behind the prevailing rate of inflation.[276] Under the impact of domestic inflation, this resulted in a further appreciation of the currency in real terms in all three countries. In Hungary, this will have contributed to the slowing of export

---

[270] The *specific* (market-weighted) western import growth facing Czechoslovakia, Hungary and Poland fell from 6-7 per cent in 1991 to around 4 per cent in the first half of 1992. For the other eastern countries, the slow-down was considerably smaller.

[271] Examples are presented in United Nations Economic Commission for Europe, *Economic Bulletin for Europe*, vol.44(1992), New York, 1993, p.68.

[272] The task of shifting exports from the former CMEA partners to the west has proved difficult even for certain Finnish enterprises used to operating in market conditions. See "Finland's Trade with the Soviet Union: Its Impact on the Finnish Economy", in United Nations Economic Commission for Europe, *Economic Bulletin for Europe*, vol.44(1992), New York, 1993, chapter 4.

[273] On these points also see United Nation Economic Commission for Europe, *Economic Bulletin for Europe*, vol.44(1992), New York, 1993.

[274] As of 1 March 1992, these three countries abolished tariffs on certain industrial products, reduced tariffs on others and in certain cases lifted quantitative restrictions. The obligations undertaken by the three countries are not uniform. See GATT, *European Agreements between the Czech and Slovak Federal Republic, Hungary, Poland and the European Communities*, Communication from the Parties to the Agreement, L/6992/Add.1, 7 April 1992.

[275] Early in 1992, the Hungarian authorities announced a policy of forint devaluation below the rate of inflation for the remainder of the year, with the aim of forcing exporters to increase efficiency to remain competitive (Peter Akos Bod, President of the National Bank of Hungary in *International Herald Tribune*, 18-19 April 1992). The forint was devalued by 1.9 per cent (March), 1.6 per cent (June), and by 1.9 per cent (November 1992). Since the rate of inflation was over 20 per cent during the year, the real exchange rate of the forint appreciated substantially. In consequence there has been increasing concern about the erosion of exporters' profitability (Bela Kádár, quoted in *Wall Street Journal*, 15 January 1993).

[276] In October 1991, Poland introduced a crawling peg exchange rate mechanism. Under the new system, the nominal value of the zloty has been devalued against a basket of five western currencies at the rate of about 1.8 per cent a month. However, a persistently higher pace of inflation caused the real exchange rate to appreciate in the closing months of 1991 and early 1992. In order to strengthen the incentive to export, the authorities devalued the currency toward the end of February 1992 by an additional 12 per cent. Since then the real exchange rate of the zloty has continued to appreciate, because the rate of crawl lagged behind inflation rate.

growth, but no such effect is observable in Czechoslovakia and Poland. On the import side, an appreciation of the real exchange rate is likely to have contributed to Czechoslovakia's import boom and, in the second half of the year, to the upturn in Poland's imports. The slow-down of import growth in Hungary in 1992, on the other hand, factors other than changes in the real exchange rate of the forint.

In Bulgaria and Romania, the development of exchange rates has been considerably more complex.[277] After a very substantial devaluation of the Bulgarian lev in early 1991,[278] the real exchange rate appreciated, including during most of 1992 when the nominal rate stayed in the range leva 20-22 per dollar. None the less, in 1992 the real rate remained lower than it had been prior to the March 1991 devaluation. The depreciated lev would have fostered the growth of Bulgarian exports observed in the western trade statistics in 1991-1992, or the boom registered with a lag in the national data in 1992. The appreciation of the real exchange rate in 1992 is consistent with the upturn noted in the country's imports.

The Romanian leu was devalued sharply in November 1991.[279] After a period of relative stability, the nominal and real rates of the leu depreciated rapidly in the second half of 1992. This may have contributed to the upturn in exports that the country posted in 1992, but imports turned up in spite of the depreciation.

(iv) **Intra-eastern trade continues to shrink**

Changes in trade levels and structure

The contraction of the transition economies' trade with each other, which had begun in 1989 and accelerated sharply in 1991, continued almost unabated in the first three quarters of 1992 (table 3.3.1). Exports of the east European countries to the successor states of the Soviet Union appear to have fallen by more than 35-50 per cent in value in 1992, while east European imports contracted by some 3 per cent on the strength of their own data or by well over 30 per cent if the Russian Federation data are taken as indicative.[280] Trade among the east European transition countries also continued to fall but at a slower pace — some 8-11 per cent.

These figures suggest that the period of steeply falling intra-regional trade connected with the break-up of the CMEA and the onset of radical reforms in eastern Europe has not yet run its course. In addition to economic reforms, which themselves tend to lower economic activity and trade in particular, mounting disorder in the post-Soviet economic space was a depressive factor, both directly and indirectly.

From the perspective of direct trade relations, the east European countries are affected by the limited import-absorption capacity of Russia, Ukraine and Belarus, their largest trade partners in the past. Apart from factors such as foreign currency shortages, disruptions in transportation and communications, the chaotic legal environment, etc., the recovery of these trade flows is also hampered by the declining relative competitiveness of east European exports. In many critical areas, such as food and pharmaceutical products, east European producers are losing markets to western suppliers who can supply financing and, in part, sell at subsidized prices or as a part of various assistance programmes.

The contraction of intra-eastern trade is beyond any doubt, but the actual decline may be less severe than shown by the official statistics (particularly in Russia) because an increasing proportion of this trade is conducted through third parties, often enterprises located in Austria and Switzerland. Although no quantified evidence is available to assess the scale of such triangular transactions for the region as a whole, the phenomenon can be illustrated for Hungary. Table 3.3.14 presents Hungarian trade transactions with five transition economies in two ways: by purchaser and seller country as well as by country of destination and origin. The discrepancy between the two sets of data gives a rough approximation of the relative importance of middleman trading. Generally the data show a larger volume of bilateral flows once these transactions are included, i.e., on the destination-origin basis. Thus, Hungary's exports to the countries of the former Yugoslavia, for example, appear to be conducted predominantly through third-country partners, perhaps in view of the war risks involved. In the case of trade with the former USSR, the share of middleman trade increased from 22 per cent to 35 per cent in both imports and exports as chaos grew in the post-Soviet economic space.

The indirect effects of the collapse of the eastern trading block and the Soviet economic space as its centre are also important. Only a few years ago, many east European countries saw themselves as potential bridges between the west and the large Soviet market. It was believed that western investors would bring substantial funds to eastern Europe to build bridgeheads

---

[277] Both countries experienced serious supply problems which are likely to have distorted trade responses to exchange rate movement. There are numerous reports of energy shortages constraining industrial activity, in part because the volume of oil imports fell by some 50 per cent in 1991. Imports of much needed fuel supplies for the winter of 1991-1992 are reported to have been diverted, by necessity, from industry to consumers. Economist Intelligence Unit, *Romania*, Country Report No.1, 1992.

[278] The Bulgarian currency fell from leva 2.88 per dollar at the end of January 1991 to leva 28.25 per dollar in the third week of February 1991, a depreciation of 90 per cent.

[279] On 6 November 1991, the leu was devalued from about lei 60 per dollar to lei 180 per dollar. After a period of stability in the first four months of 1992 it fell further to over lei 300 per dollar at end-June, then to some lei 400 in December.

[280] The lower figure in each case reflects trade value changes reported by the four east European countries for which data are available (Bulgaria, Czechoslovakia, Hungary and Romania) and the higher figure Russian Federation trade reports on trade with all former CMEA countries, of which eastern Europe constitutes the bulk. Similarly large discrepancies between east European and Soviet mirror data on mutual trade flows have been noted in the preceding two years (see table 3.3.1).

TABLE 3.3.14

Hungary's trade with the transition economies, 1991-1992
(Million US dollars, shares in per cent)

	Exports		Imports	
	1991	1992	1991	1992
CSFR				
Sales purchase	153	214	247	357
Destination origin	189	290	373	475
Share of middleman trade	19.0	26.2	33.8	24.8
Poland				
Sales purchase	139	115	94	107
Destination origin	200	143	213	173
Share of middleman trade	30.5	19.6	55.9	38.2
Romania				
Sales purchase	83	149	36	55
Destination origin	121	189	52	68
Share of middleman trade	31.4	21.2	30.8	19.1
Former Yugoslavia				
Sales purchase	190	158	135	121
Destination origin	374	378	150	140
Share of middleman trade	49.2	58.2	10.0	13.6
Former USSR				
Sales purchase	934	917	1 174	1 210
Destination origin	1 200	1 407	1 590	1 863
Share of middleman trade	22.2	34.8	26.2	35.1

Source: Hungarian Ministry of International Economic Relations.

Note: Non-dollar denominated transaction, which in 1991 still had a 5-10 per cent share in Hungary's trade with the eastern countries, is not included.

for their future expansion into a market of 280 million consumers. As the overall situation in the post-Soviet economic space has deteriorated, this motive for investing in eastern Europe has lost much force.

Another indirect effect of the break-up of the CMEA trading mechanism can be seen in the changes in the *commodity composition* of the remaining trade flows. In the course of the increasingly competitive process of market selection, the transition countries are losing their positions in the mutual trade in *manufactures*. Thus, in bilateral trade between Czechoslovakia and Hungary the share of engineering products in Czechoslovak exports fell from 19 to 11 per cent, and in Hungarian exports from 33 to 20 per cent between 1991 and 1992.[281] The shares of non-edible raw materials, mineral fuels and other semi-finished products increased. Similar structural shifts are observed in Czechoslovak exports to the successor states of the USSR. The shares of machinery and equipment, consumer goods and chemicals have declined, while that of food products (exports of which increased by 170 per cent) rose to 17 per cent in trade with the post-Soviet countries, more than twice the level of the share of these products in total CSFR exports.[282] Although it is the most industrialized of the transition countries, Czechoslovakia's role is thus shifting towards that of a purveyor of primary goods to the less developed economies of the group.

The much-discussed concerns regarding the vulnerability of eastern Europe to anticipated declines in Soviet *energy supplies* were not put to the test in 1992. These had been particularly prominent in Czechoslovakia and Hungary, both landlocked countries for which the re-routing of imports would have posed a very serious logistic problem.[283] In the event, Russia — and to a smaller extent Ukraine and Kazakhstan — were able to satisfy the energy needs of these two east European countries, as well as those of others which had enough hard currency to pay for such supplies. Czechoslovakia, for example, secured an estimated 9 million tons of crude oil in the first three quarters alone, which was actually more than the 7.5 million tons contracted for the year as a whole.[284] Poland had contracted for 5 million tons (about half of it's needs) and experienced no difficulties.[285] For Hungary detailed data are not available, but according to an official account Hungarian companies were able to obtain through barter and cash payments more oil and oil products than the country needed, permitting relatively large re-exports.[286]

Emerging new trade frameworks

On 21 December 1992, representatives of the Czech Republic, Slovak Republic,[287] Hungary and Poland initialled a number of documents in Cracow which are to establish a free-trade zone covering this market of 64 million people. The new *Central European Free Trade Agreement* (CEFTA) went into effect on an interim basis three months later, on 1 March 1993, pending full ratification by national parliaments in the course of 1993. This agreement is the first tangible fruit of efforts to establish post-communist cooperation among the countries of eastern Europe which had started at the Visegrád summit of the leaders of Czechoslovakia, Hungary and Poland in February 1991.

Negotiations between these countries had accelerated after the conclusion of association agreements between them and the EC at the end of 1991. Rapid action was needed because the tariff and other concessions granted to and obtained from the EC under these agreements implied discrimination against the trade with third parties, including the mutual trade of the three countries. An additional reason for accelerating negotiations within the group was that the EC wanted

---

[281] Data refer to January-September 1992 as compared to the same period in 1991. *CSFR in the International Economy*, vol.2 (1992), No.3, p.55.

[282] Ibid.

[283] About a decade ago, the two countries had jointly built an oil pipe line to the Adriatic port of Rijeka in preparation for such an emergency. This pipeline, however, was unusable in 1992 in consequence of the war in former Yugoslavia.

[284] *CSFR in the International Economy*, vol.2 (1992), No.3, p.71.

[285] *Rzeczpospolita*, 14-15 November 1992.

[286] Press conference of Minister for International Economic Relations Bela Kádár, quoted in *Figyelö*, 18 February 1993.

[287] The two republics had begun to participate in certain external negotiations already before the end of the year in anticipation of the impending formal dissolution of the Czechoslovak federation.

to treat the three countries as one customs area.[288] With the successful conclusion of trade negotiations with the EFTA countries in 1992-1993, the disparity between the rules of "east-west" trade on the one hand, and the rules applicable to trade among the three (soon to be four) east European countries on the other, had widened even further. These interrelated external developments created a need to counterbalance their impact by an agreement between the group's members to match the concessions already negotiated with their major western trading partners.

However, the degree of enthusiasm varied between the members of Visegrád group. This may reflect different perceptions of the risk that attempts to establish closer economic and political ties within the "triangle" could slow the process of catching up with the west. Also, the benefits from trade liberalization were considered relatively insignificant because of the current low level of mutual trade and different assessments of the chances of restoring it to earlier levels.[289]

The CEFTA participants agreed to eliminate tariffs and quantitative restrictions on *industrial goods* within eight years. For *agricultural products* full liberalization was not envisaged, but significant tariff reductions were agreed for the next five years. It was agreed that these liberalization measures would be implemented in three stages between 1 March 1993 and 1 January 2001. The precise timing is to be regulated by three bilateral agreements, in which the Czech and Slovak Republics agreed to submit a common list *vis-à-vis* the other two countries. The participants in the Cracow meeting also signed a letter of intent committing themselves to start negotiations on reducing the eight year implementation period to five years.[290]

As CEFTA negotiations attracted international attention, Slovenia and Ukraine expressed their interests in joining in, but no formal steps have been reported so far. On the other hand, the Czech Republic has recently made an official announcement of advanced negotiations with Slovenia on a fully-fledged free trade agreement, to be signed in May 1993.[291] If the Czech-Slovenian free trade agreement comes into force, it will be difficult to fit into the CEFTA framework, which excludes agriculture; but this might give an additional boost to the CEFTA countries to abolish these remaining restrictions.

The full texts of the Cracow agreements have not yet been published. None the less, sufficient detail has emerged in press reports and ministerial communiques to permit a preliminary assessment.[292] First it should be noted that the CEFTA agreement is based on symmetry of obligations and concessions, as opposed to the agreements with the EC and EFTA in which the western countries allowed their concessions to be implemented faster than those of the east European countries in order to give the latter breathing space to adjust. Second, the conditions agreed upon in the CEFTA documents appear to reflect strongly the lessons of 1991-1992. In those two years, the three countries had made significant steps towards opening their domestic markets to the rest of world, and all of them had been shocked by the intensity of import competition which resulted. Thus, when the time came to establish their mutual trade regimes for the next five to eight years, they all preferred to remain on the safe side and to make fewer concessions than they might have done two years earlier.

In February 1993, the Hungarian government launched a *Carpathian Euroregion* in February 1993 at a meeting in Debrecen, where the founding charter was initialled by the foreign ministers of Hungary, Poland and Ukraine and a representative of the Slovak Republic. The main objective of the new organization is to promote cross-border cooperation among the participating countries in a variety of domains, such as transport, environment, and trade links between small enterprises. The initiative enjoys the financial support of the Council of Europe and the American Ford Foundation.[293]

A potentially important component of a new trade and payments framework for the transition economies is a clearing system which was called into life by ten commercial banks operating in Bulgaria, the Czech Republic, Hungary, Poland and Russia. The *East European Clearing Union*, created in late February 1993 in Paris, is engaged in setting up a computerized transaction system in which member banks can mutually settle their convertible currency claims and liquidate the difference in ECU. The project, which unites not only small private but also a few large state-owned banks, such as the Hungarian Foreign Trade Bank and the Polish Bank Handlowy, enjoys the support of the European Bank for Reconstruction and Development (EBRD). The EBRD, in principle, is willing to play the role of lender of last resort, but the precise forms

---

[288] On 12 May 1992, representatives of the customs authorities of the three countries had signed an intergovernmental agreement in Warsaw on the application of their association pacts with the EC. Under this new agreement, which went into effect on 1 June 1992, goods originating in any of the three countries will be treated as if they were goods originating from one customs zone.

[289] Interestingly, these sentiments were aired in the form of public statements by leading politicians of the Visegrád countries a few weeks *after* the Cracow signing ceremony. A particularly strong echo was generated by an interview given by Czech Prime Minister V. Klaus to *Le Figaro* (8 January 1993) which was generally interpreted in the other two countries as an attempt to belittle the significance of the Visegrád cooperation.

[290] S. Richter and L.G. Tóth, "Szabadkereskedelem – avagy 'felzárkózás' Európához" (Free trade or "catching up" with Europe), paper prepared for the Vienna Institute for Comparative Economic Studies, as reported in *Figyelő*, 25 February 1993.

[291] Prime Minister V. Klaus, as quoted in *Magyar Hírlap*, 18 March 1993.

[292] See e.g., *Magyar Hírlap*, 27 February 1993.

[293] *Frankfurter Allgemeine Zeitung*, 16 February 1993.

and limits of this engagement still have to be worked out.[294]

In the *Baltic subregion*, disagreement among the governments of Estonia, Latvia and Lithuania has blocked the establishment of a free trade zone. The concept of a *Baltic Free Trade Area* began to be discussed even before the three countries regained their independence. So far, however, no progress has been made, while the development of Baltic mutual trade lags well behind that of total trade of the republics.[295] It seems that present economic conditions in the three states are too divergent and their recovery plans are based on different philosophies. Hence, common ground for agreement is difficult to find. It is likely that frequent changes in governments are also working against multilateral cooperation, as the results of one round of talks may become irrelevant in the next round. More recently, governments appear to be giving preference to bilateral negotiations, hoping that this will yield quick results on the most urgent issues for the countries concerned.[296]

Political efforts to create a new framework for trade relations among the *CIS republics* are witnessed by events simultaneously taking place in various forms and at various levels. In the bilateral framework, which is perhaps the most relevant for trade relations in the short term, some 160 trade agreements were reported to have been signed among the CIS countries in 1992 alone.[297] In the multilateral framework, negotiations have continued at all levels in 1992, including eight summit meetings of CIS head of states held in different capitals.[298] In late 1992 and early 1993, one of the key issues was the work on the *Charter of the Commonwealth of Independent States*. The Minsk Summit held in January 1993 agreed on a draft text which was signed by seven countries — Moldova, Turkmenistan and Ukraine being the exceptions. It is noteworthy, however, that all 10 participants signed a joint declaration allowing these three countries to adhere the charter within a year.[299] In the meantime, they will continue to enjoy the benefits of full membership.[300] Article 19 of the Charter concerns member countries' ideas on co-operation in the economic sphere, noting their desire to form "a common economic space on the basis of market relations and the free movement of goods, services, capital and labour."

Parallel to the protracted bargaining at the highest political level, some 250 CIS-wide agreements[301] were also signed at these summit meetings (30 per meetings on average), covering a vast range of details from military affairs to environment protection.[302]

One potentially important development which pertains both to bilateral and multilateral affairs of the successor states of the former USSR was the so-called *Surgut Agreement* of March 1993 on setting up an *Intergovernmental Council on Oil and Gas*. The significance of this agreement, which takes its name from the Tyumen Oblast city where the founding meeting was held, is underlined by the fact that all 15 successor states of the former USSR were invited and 13 actually participated, Turkmenistan and Latvia being the exceptions.[303]

Although details of mutual deliveries were not addressed in Surgut, the main result of the meeting appears to have been a consensus to promote trade through barter-type cooperation agreements which would link energy shipments with contributions to investment or other requirements of major energy projects or specific regions. It is not clear, however, to what extent this indicates a policy change to abandon or modify the intention of the main supplier, Russia, to move towards trade of oil and gas products against convertible currency at world market prices.

(v) The new foreign trade

One consequence of the political developments in 1991-1992 is the appearance of "new" foreign trade: the transformation of domestic trade conducted within established states into "foreign" trade across the new borders. After the break-up of the Soviet Union and Yugoslavia and, more recently, the dissolution of the Czechoslovak federation, 27 sovereign states now share the space where two years ago there were only eight.

The degree of "foreignness" of this new foreign trade varies substantially, depending upon the rapidity which the newly-independent states institutionalize the separateness of their economic space, independent (and different) economic policies, independent monetary

---

294 *Heti Világgazdaság*, 27 February 1993.

295 Estonia's total imports, for example, grew more than threefold in nominal Estonian kroon terms between the first and the fourth quarter of 1992, while imports from Latvia or Lithuania have not even doubled. Similar tendencies can be discerned on the export side, although here shipments toward Latvia appear to have fared relatively better (Statistical Office of Estonia, *Quarterly Statistical Bulletin*, 1992, No.4).

296 P. Morris, "Baltic free trade still in a deadlock", *The Baltic Independent*, 12-18 March 1993.

297 G.V. Gabunia, "Some Aspects of the Trade Policy of Russia", paper presented at an IIASA conference on *Economic Relations Among the Successor Republics of the USSR*, Vienna, 11-13 March 1993.

298 O. Bychkova, "CIS: Connections without consequences", *Moscow News*, No.5, January 1993.

299 For the text of the Charter, see *Rossiiskaya gazeta*, 12 February 1993.

300 A. Sheehy, "Seven States Sign Charter Strengthens CIS", *RFE RLE Research Report*, vol.2, No.9, 26 February 1993.

301 "Deklaratsiya o deklaratsii" (Declaration on the declaration), *Izvestiya*, 25 January 1993.

302 Issues directly relevant to the coordination of monetary policies and inter-bank cooperation are discussed in sections 3.4(vi)-(vii) and 3.7(iv)-(v) below.

303 Estonia was represented by an observer only. Interfax news agency, quoted by BBC, *Summary of World Broadcasts*, SU/W0271, 5 March 1993.

policy and separate currencies, the establishment of customs frontiers and trade controls. In some instances, as in the separation of Croatia and Slovenia from Yugoslavia, or — with less clear-cut modalities — that of the Baltic republics from the Soviet Union, these transformations were completed very rapidly once the new sovereignties were secured and the formerly internal trade becomes fully "foreign". In other cases, with more uncertainty about the form of future economic cooperation, as in the CIS states, a kind of "semi-foreign" trade developed — in Russia the term trade with the "near-abroad" (*blizhnee zarubezh'e*) has recently come into use — but this is unlikely to endure indefinitely.

The scale of the transformation of internal trade to "new" foreign trade is still largely unquantifiable. Generally, the immediate effect of establishing new state borders is that trade tends to decline sharply, but the evidence is very incomplete. Statistical services in the new countries are still being established; customs borders between the new states are in many cases rather porous; and controls over flows of goods and service are not very effective. For any comparison over time, measurement problems loom very large: where data that would permit a comparison of the *status quo ante* (the internal trade across the administrative lines which now have become customs frontiers) are available at all, they are valued at internal prices in the currency of the former state, whereas the new foreign trade will tend to be valued at world market prices and in the new or in international currencies.[304]

Given this scarcity of relevant information, in what follows it will only be possible to provide a very rough assessment of the development in 1992 of the "new" foreign trade in the CIS countries, the Baltic states and the successor states of former Yugoslavia.[305]

Intra-CIS trade

Inter-republican trade in the former Soviet Union had started contracting already in 1990, with a 5 per cent fall reported for that year. Comparable data are still not available for its development in 1991, but early assessments indicated that it declined by *more* than the 25-35 per cent fall registered in Soviet trade with third countries.[306] For 1992, the only available information concerns the trade of the Russian Federation with other CIS countries — trade flows which in 1990 had accounted for 40-45 per cent of the trade among the later CIS states.[307] According to Russian Goskomstat estimates, Russian exports fell by 37 per cent in volume in 1992, and Russian imports by 30 per cent.[308] These figures for the last three years suggest that intra-CIS trade flows in 1992 amount to some 40-45 per cent of their 1989 level. The fall thus was much steeper than the contraction of NMP or GDP of these countries over the same period. Given the almost equally stark decline in their third-country trade over the same period, the CIS economies clearly have suffered an external shock which was much larger than anything experienced by most countries of eastern Europe.

The rouble *value* of Russia's intra-CIS exports rose 16-fold in 1992, while a slightly larger (18-fold) increase was reported on the import side. The resulting change in the trade balance suggests that Russia made a determined attempt to reduce the imbalance experienced in 1991. In the last year of Soviet inter-republican trade, Russian "exports" had surpassed Russia's "imports" by 30 per cent; in 1992 this margin was halved to 16 per cent.[309]

Among Russian CIS exports, fuels deserve special attention: the fall in Russian deliveries of coal, crude oil and oil derivatives was estimated to be in the 20-50 per cent range. Inter-governmental contracts for 1992 deliveries, often already much below flows transacted in the year before, were frequently not implemented.[310]

"New" foreign trade of the Baltic states

The Baltic states were the first countries hit by the full force of conversion from intra-trade at rouble prices to "new" trade at world market prices after the intro-

---

[304] Owing to this lack of comparability over time (or the absence of 1991 data on the corresponding "internal" trade flows), the "new" foreign trade was also excluded from the foreign trade measures used in this chapter for those countries — Croatia and some of the Baltic states — which in their national data already add it to the "old" foreign trade (see footnote (192) above). Most transition countries, in any case, still report the "new" trade as a separate category.

[305] For a more extensive discussion of the issues raised by the phenomenon of "new" foreign trade, see United Nations Economic Commission for Europe, *Economic Bulletin for Europe*, Vol.44(1992), New York, 1993, section 2.4 (pp.83-82).

[306] C. Michalopoulos and D. Tarr, *Trade and Payments Arrangements for State of the Former Soviet Union*, World Bank Studies of Economies in Transition, No.2, Washington, D.C., 1992, p.3.

[307] The Russian Goskomstat data on trade with the "near-abroad" in fact often cover trade with all republics of the former Soviet Union — i.e., include trade with the three Baltic states and Georgia in addition to the 11 CIS states.

[308] Direct communication to ECE. Estimates are based on a sample, registered in physical units, of shipments to and from Russia covering 60 per cent of the value of all flows.

[309] In absolute terms the Russian Goskomstat figures show a R300 billion surplus in trade with the former republics of the USSR. This figure stands in striking contrast with the R3,000 billion trade surplus claimed by V. Mashchits, chairman of the State Committee for Cooperation with the CIS States, in a recent government meeting (*Izvestiya*, 19 February 1993). Prime Minister V. Chernomyrdin was quoted as saying at the same meeting that "in 1992 (Russia's) trade turnover with the 'near abroad' was twice as large as its trade with the 'far abroad'" (loc.cit.). The Goskomstat 1992 data reproduced in table 3.3.4 indicate precisely the opposite relationship.

[310] E.g., Russian coal deliveries to Azerbaijan, Tajikistan, Turkmenistan and Armenia materialized at one third of their contracted levels. As concerns natural gas, Belarus and Ukraine managed to secure 11 per cent more than their bilateral agreement with Russia envisaged, while Moldova and Kazakhstan got only a fraction of that — 6 and 16 per cent, respectively (CIS Statistical Committee, *Ekonomika Sodruzhestva nezavisimykh gosudarstv v 1992 godu*, Moscow, 29 January 1993, pp.26-27). From a Russian perspective, the underfulfilment of quotas pertaining to food imports was of special concern. Fragmentary data suggest that the shortfalls were rather large. The fulfilment rate varied within a broad range: sugar (2 per cent), milk and milk products (21 per cent), meat and meat products (35 per cent), vegetables (59 per cent). V.A. Maksimov, "Trade and Economic Relations of Russia with the Former Soviet Union Republics", paper presented at a IAASA Workshop on *Economic Relations Among the Successor Republics of the USSR*, Vienna, 11-13 March 1993.

duction of their new currencies in May-June 1992. In the past, 90-95 per cent of their cross-border merchandise transactions had been with other USSR republics rather than with the outside world, and it was obvious that a switchover to world market prices with unchanged trade flows would leave the three countries with extremely large trade deficits which their partners (chiefly Russia) would hardly tolerate. Thus a contraction of trade flows was inevitable.[311] How large this contraction actually was in 1992 is not clear. The statistical services of the Baltic countries are struggling with similar problems those of Russia, compounded by the measurement difficulties arising from the mid-year changes in currency. In a highly inflationary environment, rouble-denominated nominal flows are not comparable over time, and appropriate deflators or records of flows in physical terms are difficult to establish in the present transition stage. In the absence of such data, it may be useful to note the changes in the geographical patterns of trade.

The share of Russia in total imports of *Estonia* fell from 46 per cent in 1991 to 33 per cent in 1992, while on the export side the contraction was from 56 to 20 per cent (table 3.3.15). *Latvian* data show that the countries of the former Soviet Union still remain the most important trading partners, but their share in total exports (45 per cent) and imports (38 per cent) in 1992 are now much below the 97 and 87 per cent registered in 1991.[312] According to the assessment of the authorities in *Lithuania*, the share of CIS-related exports fell from 85 per cent to 71 per cent in 1992, with a very small shift recorded on the import side (from 83 to 82 per cent).

"New" foreign trade in former Yugoslavia

Trade among the former Yugoslav republics declined sharply already in 1991 owing to the violent disintegration of the country. Exact data on the volume of intra-Yugoslav trade before the split of the country are not available, nor a full set of data pertaining to 1992 trade levels among the five successor states. Thus, the preliminary assessment below is based on data from Slovenia and Croatia only. In fact, these data are also incomplete, as they give no indication on trade flows between these two republics on the one hand and the new FR Yugoslavia (Serbia and Montenegro) on the other. This is, of course, well explained by the war between these countries and the UN trade embargo on FR Yugoslavia.

Although in the past there was no need for systematic recording, in retrospect the assessment is that a quarter of all sales by the *Slovenian* enterprises was done in the market of the former Yugoslav partner republics. In that trade, Slovenia constantly registered a trade surplus.[313] In 1992, Slovenian exports to other republics of the former Yugoslavia came to some $1.4 billion, imports $1.1 billion. The share of inter-republican trade in Slovenia's total "external" trade (inter-republican plus foreign trade) has fallen from almost 60 per cent to about 25 per cent. In spite of the special payment agreements with the partner republics in the former Yugoslavia (except Serbia and Montenegro), the payment of the $300 million surplus is at a stalemate because of liquidity problem in the other republics. Croatia was the destination for 62 per cent of Slovenia's exports and the source of 68 per cent of its imports, with a surplus of $113 million.

In Croatia, the former Yugoslav republics were the destination for 32 per cent ($1.4 billion) of total exports and the source of 24 per cent ($1.0 billion) of total imports in 1992.[314] There was a trade surplus in this "new" foreign trade of $382 million. Slovenia was Croatia's main trading partner in the former Yugoslav region. According to the Croatian statistics, this country exported to Slovenia $1,014 million and imported $825 million worth of goods. The surplus in trade was $189 million. Since both countries reported a surplus in their mutual trade, it is obvious that there are discrepancies in the statistical coverage and reporting. Trade between the two states had a declining trend since the summer 1992 when Croatia introduced an import surcharge of 23 per cent and Slovenia reciprocated.[315] In trade with Bosnia and Herzegovina, Croatia had a surplus of $95 million. The future trade with Bosnia and Herzegovina, or one of its parts, is assisted by the abolishment of tariffs between Croatia and the Croat community in Bosnia and Herzegovina called Herzeg-Bosnia[316] which also uses the Croatian dinar as its own currency.[317]

The first months of "new" trade in former CSFR

On 1 January 1993, the 74-year history of the Czechoslovak federation ended with the assumption of full sovereignty by the Czech and Slovak Republics. The common currency of the two countries continued to function for five more weeks. As from 8 February 1993, trade between the two republics has been conducted in an ECU-based clearing system.

The economies of the two republics were closely integrated, but the relative dependence was much larger

---

311 See United Nations Economic Commission for Europe, *Economic Survey of Europe in 1991-1992*, New York, 1992, pp.163-166, for an *ex-ante* analysis of external trade relations of the Baltic states with the former Soviet Union and Russia.

312 State Committee for Statistics, *Latvia in Figures 1991*, 1992, p.17; *Bulletin of Latvian Economic Statistics 1992*, March 1993, p.20.

313 Centre for International Cooperation and Development (CICD), *Economic developments in Slovenia in 1992*, Ljubljana, CICD, January 1993, p.13.

314 Comparable data for 1991 are not available.

315 *Novi Vjesnik*, 26 November 1992, p.7a.

316 *Ekonomska politika*, 22 February 1993, p.17.

317 *Yugoslav Daily Survey*, 18 February 1993, p.5.

TABLE 3.3.15
Foreign trade of the Baltic countries, 1990-1992
(Percentages total trade = 100)

	Exports			Imports		
	1990	1991	1992	1990	1991	1992
Estonia: trade with						
USSR/CIS	94.1	82.2	34.3	82.0	71.7	40.1
Russia	55.0	56.5	20.5	51.8	45.9	32.7
Latvia	5.7	7.7	10.8	4.7	5.1	1.7
Lithuania	3.1	3.8	1.5	3.2	6.3	2.9
Latvia: trade with [a]						
USSR/CIS	95.2	96.8	45.2	74.5	87.2	37.9
Russia	50.2	49.8	26.3	41.9	44.5	28.1
Estonia	3.7	3.2	1.3	2.8	5.2	6.5
Lithuania	4.6	5.4	3.5	7.7	10.1	3.2
Lithuania: trade with						
USSR/CIS	94.1	85.1	71.1	80.1	82.7	81.7
Russia	52.7	56.5	..	43.6	49.6	..
Estonia	1.5	2.3	..	1.8	1.8	..
Latvia	3.4	6.7	..	4.2	4.7	..

Source: National statistics.

a In 1992, imports of fuel — 17 per cent of all imports from countries of the former USSR — are not identified by origin. If allowance were made for this, the shares of CIS and Russia would be probably considerably higher.

in the Slovak Republic;[318] hence the break-up is likely to have a more important effect on the economy of Slovakia. One early study estimated the loss of GDP in 1993 attributable to the separation at 2.1 per cent for the Czech Republic and at 5.7 per cent for Slovakia, presumably using the assumptions on the likely post-separation fall of trade then prevalent.[319] Initial estimates anticipated a rather small impact on mutual trade between the two republics — a fall in 1993 of some 10 per cent.[320] However, more recently this was raised to some 30 per cent.[321] Scattered data for the first months of 1993 suggest that the fall may be much steeper: trade turnover in January, the last month of the common currency regime, was reported to have been below Kcs 6 billion (Czech exports of Kcs 3.4 billion and imports of Kcs 2.3 billion).[322] The level of bilateral payments flows in the first month of trade conducted under the new clearing system (8 February to 4 March 1993) has been reported to be of the same order of magnitude.[323] This represents a 60 per cent decline from the average monthly turnover of some Kcs 15 billion in bilateral flows between the two parts of the dissolved federation in its last year.[324] If this fall is not contained within the next few months, the first-year costs of separation for the two economies are thus likely to be very much higher than the 2 and 6 per cent of GDP anticipated last year.

---

[318] For a review of these relations, see United Nations Economic Commission for Europe, *Economic Bulletin for Europe*, vol.44 (1992), New York, 1993, pp.88-91.

[319] I. Sujan, former Chairman of the Federal Statistical Office,in *Hospodarske noviny*, 13 August 1992, p.8, using the assumptions on trade then current.

[320] See the statement of the Czech Finance Minister I. Kocarnik made on Czech Television on 18 November 1992. BBC, *Summary of World Broadcasts*, EE/1545, 23 November 1992, p.B 5.

[321] J. Tosovsky, Governor of the Czech National Bank, speaking on Czech Television on 8 February 1993. BBC, *Summary of World Broadcasts*, EE W0269, 18 February 1993.

[322] *Rudé pravo*, 9 March 1993.

[323] According to a report prepared by the Prague-based correspondent of a Hungarian newspaper. J. Kárpáti, "Cseh-szlovák kereskedelem — Kinos az átjárás" (Czech-Slovak Trade: A Difficult Passage), *Heti Világgazdaság*, 20 March 1993.

[324] The volume of Czech deliveries to the Slovak Republic in 1992 amounted to Kcs 96.6 billion in 1992, and the counter-flow from Slovakia to the Czech lands to Kcs 81.8 billion (*Narodna obroda*, 24 February 1993, p.13).

3.4 REVIEW OF MACROECONOMIC POLICIES

(i) Overview

Macroeconomic policies in the European economies in transition in 1992 aimed generally at a relatively wide range of stabilization objectives, from combating inflation and restoring or maintaining financial balance on domestic markets to stopping recession and reducing unemployment. But priorities, instruments, specific actions and results differed among countries. In most general terms, anti-inflationary measures broadly dominated domestic policies in Bulgaria, Czechoslovakia, Hungary and Slovenia, while attempts to prevent output collapse and protect employment levels prevailed in the successor states to the former Soviet Union, Romania, Croatia and the new Federal Republic of Yugoslavia. In Poland and in the Baltic states, macroeconomic policies were guided by a mixture of priorities, reflecting mostly changes in the internal political situation and the external environment.

Only two or three years ago, the main policy objectives and instruments in all transition countries had been very similar — not only because all countries shared the common strategic goal of replacing their centrally planned economies by viable market systems, but also because these policies were initially formulated in a rather general, sometimes exhortatory manner. However, it soon became clear that the transformation would last longer and costs would be higher than originally anticipated. New challenges and specific problems required more detailed plans and more complex policy responses, which inevitably and to a varying extent reflected the interests of the main social groups. Political differences began to emerge within the pro-market coalitions in each of the transition countries. Unexpected difficulties in transforming old economic structures, as well as varying initial conditions, resulted in a wide differentiation of national macroeconomic policies.

The two most important factors which determined the economic policy mix in transition countries in 1992 were probably the progress already made in the market reform process and the internal balance of powers between the various political factions holding different attitudes towards the reform strategy. At the risk of some oversimplification, this can be interpreted as the balance between radical reformers and "gradualists". The pattern of policy evolution seems typically to range between short-term liberalization-*cum*-stabilization programmes and more versatile and better balanced policy packages laying greater emphasis on longer-term institutional and structural measures. Accordingly, in those countries which embarked on market reform programmes only recently, key policy approaches were directly determined by the large domestic financial and market imbalances. They typically included price liberalization and the fight against inflation, the elimination of market shortages, liberalization of foreign trade and the achievement of some measure of convertibility. The falls in output registered in these countries were generally deeper, but the social readiness to accept sacrifices was generally also greater. The emphasis on domestic stabilization measures, coupled with large-scale liberalization of economic activities, was probably most visible in 1992 in such countries as Albania, Bulgaria, and the Baltic republics.

In other countries, where market reforms were already more advanced and problems of price stabilization and goods-market equilibrium were largely solved or reduced to manageable proportions, the policy focus shifted to the removal of fiscal imbalance and industrial restructuring. Having completed price and foreign trade liberalization, and enjoying relatively balanced market situation, these countries have turned their attention to the institutional and social aspects of reforms, trying to move from recession to recovery while containing the costs of necessary adjustments and even resorting to some protectionist measures. This group of countries includes Hungary, Poland and, to a lesser extent, also Czechoslovakia and Slovenia.

Political considerations played a particularly important role in shaping national economic policies in Russia, Ukraine, Belarus, Romania, FR Yugoslavia and Croatia. In the CIS countries and the ex-Yugoslav republics, macroeconomic stabilization objectives were inseparably intermingled with political — and in many instances contradictory — goals of acquiring and strengthening national sovereignty while trying to minimize losses arising from the break-up of economic ties between the new entities emerging from the old federal states. In this context, Russia stands out: government efforts to liberalize and stabilize the economy, undertaken within an ambitious reform programme in 1992, were largely neutralized by simultaneous — and largely politically-motivated — attempts to maintain some degree of monetary unity in the space of the former Soviet economy, as well as by deep internal conflicts over the required course of market reforms.

As a result of the dissimilarities in starting conditions, political landscape and vulnerability to external impact, the transition countries of central and eastern Europe now follow a wide spectrum of macroeconomic policies. Liberal, strongly pro-market attitudes dominate the policy of the Czech government, while a much more interventionist approaches prevail in the policies of the Romanian and Slovak governments. Less pronounced, but also present, are differences in the policies followed by different CIS states, where the "shock" approach which dominated Russian policy in the first half of 1992 contrasted sharply with the more gradual course taken by Belarus and Ukraine.

Generally, macroeconomic policies were more successful in stabilizing national economies in eastern Europe than in the economies of the former Soviet Union. A partial explanation is that stabilization efforts in the post-Soviet states were weakened and sometimes compromised by the consequences of the breakup of the country into new sovereign entities and the many unsettled policy coordination issues this posed, given the close economic interdependence of the ex-Soviet states. But other reasons have also contributed, notably the generally much worse initial conditions, reflecting larger imbalances and structural distortions, the disastrous procrastination of reform prior to 1992 and the lack of adequate international assistance.

In most transition economies, the policy focus was primarily on various aspects of internal balance, with balance of payments and foreign debt issues being paid less attention. Except for Hungary and Czechoslovakia, none of the transition countries serviced their external debt obligations in full. Strong emphasis on domestic stabilization problems was typical of the initial stage of transformation, as demonstrated by Bulgaria, Poland and Russia. At a later stage, however, the mix of policy priorities became more balanced, and the issue of external balance can be expected to gain further in importance.

(ii) Fiscal policies: falling revenues and ballooning deficits

The role of fiscal and monetary policies considered as conventional *demand-management* mechanisms have as their principal purpose to ensure that potential output grows at the desired rate and that growth in actual demand is in line with growth in potential output under conditions of low inflation. Excess demand can be curbed by means of restrictive policies designed to restrain expenditures in the private and public sector. If, on the contrary, aggregate demand falls below the supply level corresponding to full utilization of production resources, fiscal policy instruments, including government spending, can be used to stimulate aggregate demand.

But fiscal policy is also an important instrument of *resource allocation*. By differentiating tax and subsidy rates, the government can change relative prices and costs and thereby influence production and consumption structure to achieve a more efficient allocation of resources. Finally, fiscal policy is also the principal instrument of *income and wealth redistribution*. All of these roles have important political and social implications.

The specific combination of fiscal policy tools and methods applied in any economy will differ depending on policy goals and established priorities. If inflation is the main concern, government expenditures may have to be cut or budget deficits reduced in order to lower aggregate demand. If unemployment is considered to be the key issue, an opposite policy may be pursued.

It seems that fiscal policies in most transition countries were dominated by social (income-redistribution) objectives, although in some cases stabilization objectives also played a role. The main (though not always very explicit) consideration was probably to protect large segments of households and firms from the impact of persistent recession through direct subsidies, indirect support and social safety nets. Another important consideration was to keep budget deficits within limits which would not trigger additional inflation. In no country was fiscal policy explicitly used to stimulate economic recovery or to reduce unemployment. Nevertheless, large deficits registered in some countries may have contributed to the expansion of aggregate domestic demand and may have had an anti-recessionary effect.

This specific *hierarchy of objectives* is to some extent understandable if one remembers that the initial stage of transformation is typically characterized by sharp contraction of output and an inflationary surge. But the preponderance of social protection and income distribution objectives may have also resulted from the long-established tradition of social security in formerly centrally planned economies. The policy of maintaining large state responsibilities in the area of incomes and employment has more than offset the more general tendency to reduce the degree of state involvement in the economy. Thus, if at first the share of budget expenditures in GDP in most transition countries declined in the initial stage of transformation, mostly because of sharp reductions in price subsidies, it subsequently rose again due to falling output and the requirements of social transfer spending. It currently remains at relatively high levels if compared with market economies. For instance, the share of state budget expenditures in GDP in Poland, after having fallen from 34-35 per cent in late 1980s to 29 per cent in 1991, increased again to 33 per cent in 1992. A similar pattern can be seen in Hungary, where the corresponding figure fell first to 31 per cent in 1990 and then increased to 37 per cent in 1992. Generally the structure of expenditures changed in favour of social transfers, with the investment share falling particularly sharply.

Fiscal balance in transition economies generally deteriorated in 1992, nearly all countries ending the year with higher budget deficits, varying from 1.8 per cent of GDP in the Czech and Slovak Federal Republic to 6-8 per cent of GDP in Poland and Hungary (table 3.4.1). In some CIS countries (notably in Ukraine) the fiscal deficit may have reached some 20-25 per cent of GDP.

Deficits chiefly resulted from budget revenues falling much below planned levels, the gap being particularly large in Hungary, Poland and the Baltic republics. But in some countries (FR Yugoslavia, Croatia, Romania, Ukraine, Moldova) the impact of revenue shortfalls was compounded by above-plan government expenditures. This outcome largely stemmed from overly optimistic assumptions concerning economic performance which as a rule underpinned the draft budgets in early 1992.

TABLE 3.4.1

Fiscal balance in transition economies, 1990-1992
(Per cent of GDP)

Country	1990	1991	1992
Bulgaria [a]	-4.9	-3.6	-3.1 [b]
Czechoslovakia	0.5	-1.8	-1.8
Hungary	-0.1	-4.9	-7.4
Poland	0.4	-3.8	-6.0
Romania	1.0	1.9	-2.0
Slovenia	-0.3	2.6	1.6
Croatia	..	..	..
Fed.Rep.of Yugoslavia [c]	-5.7	-9.6	..
Russia [d]	..	-10.0	-4.9
Estonia	..	..	..
Latvia	2.5	4.2	-3.6 [b]
Lithuania	-2.8	3.2	-1.5*

Source: Compiled from national statistics and ECE estimates.

[a] On a cash basis.
[b] January-June 1992.
[c] As percentage of gross social product.
[d] Consolidated budget deficit (federal budget deficit reached 8 per cent of GDP in 1992).

In Hungary, for instance, the government had forecast 0-2 per cent growth of GDP for 1992, and budget estimates of the number of unemployed for end-1992 were 520-580,000; the forecasts turned out to be wide of the mark: GDP in fact fell by 5 per cent and unemployment reached 660,000. The Hungarian budget deficit, originally planned at Ft 69.8 billion for full-year 1992, already exceeded this level in May, when it reached Ft 78.7 billion. The full-year outturn reached Ft 200 billion or 7 per cent of GDP. Similarly, in Russia the initial targets of 0-1 per cent deficit in the first quarter 1992 and a zero deficit for the second half of the year both proved to be completely unrealistic. The budget approved in Romania provided for a deficit of 89 billion lei for 1992, but the actual figure was well over 200 billion lei. In Poland, macroeconomic assumptions for the 1992 budget draft were more prudent as the government predicted zero GDP growth and a 5 per cent deficit for 1992. Nevertheless revenues were overestimated and the deficit ultimately reached 6 per cent of GDP.

Revenue shortfalls, observed in 1992 in most transition countries, were caused essentially by three main factors:

— continued recession, which reduced taxable incomes and sales;[325]

— an inadequate and ill-structured fiscal base (undue reliance on taxation of the state sector and on direct taxes);

— poor fiscal discipline (tax evasion, tax non-payment, inefficient bankruptcy laws).

Additionally, in countries such as Bulgaria and Hungary, budget revenues failed to meet targets because of institutional changes in existing tax systems, while in some federal states, such as Russia, the deterioration also resulted from the disruption of revenue transfers from local governments.

It is interesting to note that the fiscal crisis was most acute in two rather different groups of countries: those most advanced in the reform process, such as Hungary and Poland, and those which are still at the initial stage, such as Ukraine. While in the latter case the deficit has been generally caused by a flood of direct and indirect subsidies to avoid the closure of ailing state companies, in the former case a part of the deficit arose from revenue shortfalls due to the opposite policy. In *Hungary*, the introduction of new accounting and bankruptcy laws in January 1992 resulted in a drastic increase in the number of bankruptcies among state enterprises, which obviously affected the level of current tax revenues. In the first five months of 1992 alone the number of firms reported filing for bankruptcy jumped to 1,100 (some 5 per cent of the total number), and these firms were not meeting their tax obligations.[326]

Hungarian revenue shortfalls seem to be to a large extent side effects of the government's efforts to enforce tougher financial discipline on enterprises and to harden their budget constraint. In this sense, the 1992 fiscal crisis can be considered as the delayed consequence of many years of statistical and accounting juggling aimed at hiding financial losses of the enterprise sector. The revenue-reducing impact of the new bankruptcy law has been exacerbated by new banking regulations imposing higher loss provisions for commercial banks. The measure resulted in substantial cuts in enterprise credits, and this is likely to continue in 1993. On the other hand, budget expenditures proved very inflexible, especially those earmarked for social transfers, and are unlikely to be reduced in the next two or three years. Generally, the budget position will probably not improve until current restructuring is advanced enough and the Hungarian economy starts to grow.[327]

In Poland, a deficit of some 5 per cent of GDP was planned for 1992 at the beginning of the year, but, to the dismay of the IMF, which was involved in protracted negotiations with the Polish authorities on the reactivation of its assistance programme, this limit was revised upwards later in the year. The actual budget result for 1992 falls somewhere between the initial and the revised targets.

Budgetary problems in Russia, Romania, and especially Ukraine are of a different origin, resulting primarily from the general softness of fiscal policy. Not only

---

[325] An important recession-linked factor curbing tax revenues was the shift in the structure of consumer expenditures in favour of necessities, which are typically taxed less than consumer durables.

[326] K. Okolicsányi "The Hungarian Budget Deficit", *RFE RL Report*, Vol.1, No.29, 17 July 1992.

[327] The budget deficit seems to be the most troubling issue of Hungarian economic policy. In the recent blueprint of the three-year budget prepared by the Ministry of Finance, the deficit is planned to remain large: Ft 270 billion in 1994, Ft 220 billion in 1995, and Ft 222 billion in 1996, thus exceeding the 5 per cent GDP benchmark. The IMF has recorded strong objection to these projections. *Magyar Hirlap*, 10 February 1993.

were tax payments by enterprises forever delayed because of the generalized payments crisis, but the governments also took a rather lenient position towards debtors. Moreover, expenditures were boosted by large amounts of soft credits extended by central banks to the commercial banking sector at concessionary interest rates, the cost of which had to be covered by state budgets.

With a budget deficit of only 1.8 per cent of GDP (Kcs 16.7 billion, or some $600 million), *Czechoslovakia* demonstrated fiscal discipline under conditions of steeply falling output − an achievement not matched by any other transition country. This result is mainly due to drastic cuts in social security payments, which in real terms diminished sharply in 1992,[328] and to strict tax collection discipline. Performance in that country also reflects a benefit of rapid price stabilization: it may be easier to maintain fiscal balance when inflation is low, because pressures for income protection and transfers can be kept under control or are not transformed into self-sustaining indexation mechanisms. However, the fiscal deficit is expected to rise in 1993 as more and more firms come under financial stress.

The deterioration of fiscal balances has been addressed with different tools and with varying determination. Expenditure cuts, which was the method most resorted to in all countries and which was especially effective in Czechoslovakia, Poland, Bulgaria and the Baltic states, were in some cases accompanied by accommodating credit policies intended to improve enterprise liquidity (Russia, Romania, Croatia, Federal Republic of Yugoslavia, other CIS states), in part with the aim of enabling enterprises to pay overdue taxes. However, this technique backfired in most cases, because fiscal revenue gains were soon offset by the costs of concessionary lending which had to be financed from the budget. Nevertheless, under high inflation and generally negative interest rates, this option offered some limited short-term relief to the budget. The traditional method of containing budget deficits by raising taxes during the fiscal year was generally not employed, although some countries decided to introduce new taxes and duties in the course of 1992 (Poland, the Baltic republics).

Fiscal deficits were financed by a combination of measures, ranging from traditional borrowing from central banks, borrowing from commercial banks, selling government bonds (or bills) to the non-bank public and borrowing abroad. With external sources of funds very limited because of the low creditworthiness of most of transition countries and their generally high levels of external indebtedness, the choice among domestic sources of finance depends largely on the degree of development of financial markets. If financial markets are rudimentary or non-existent, as they are in many transition countries, the only method feasible on a larger scale is direct borrowing from the central bank. This was practically the only way of financing deficits in Romania, Russia and other CIS states. Primary emission was also behind the uncontrolled growth of spending in FR Yugoslavia and Croatia. In Czechoslovakia and Slovenia, government borrowing from central banks was occasionally supplemented by resort to domestic financial markets or borrowing from abroad.

Government borrowing from the central bank tends to "monetize" the budget deficit, because the government account held in the central bank is credited with new money which directly increases the existing money supply. A high degree of "monetization" of deficits may be expected to result in accelerated inflation in all countries where the domestic market situation is not stabilized. This was what happened in the CIS and Baltic states and also in Romania.

On the other hand, high reliance on borrowing from commercial banks requires a developed two-tier banking system and established markets for government securities. This method results typically in higher interest rates and may under certain circumstances preempt the access of non-budgetary sectors, i.e., enterprises and households, to financial resources for investment and other purposes, and thus strengthen recessionary forces. This method of covering fiscal deficits was used chiefly in Bulgaria, Hungary and Poland. Commercial banks in these countries continued purchasing government securities because of their relatively high yields, high degree of liquidity and almost zero risk. Nevertheless, some inflationary pressures may appear, depending on the structure of government expenditures.

Shifts in fiscal policies initiated in 1990-1991 continued in 1992. Generally, the real levels of budget revenues and expenditures declined, although their share in GDP in many cases increased or remained unchanged, because of falling output levels. The share of state budget expenditures in GDP varied from 32-33 per cent in Poland and 35 per cent in Hungary to 40 per cent in Czechoslovakia and Albania and 42-44 per cent in Russia.

Generally, fiscal targets for 1992 were as a rule not met in the transition countries,[329] − sometimes by large margins as in Hungary or Russia − and policy stances were of a passive, accommodating character. In those countries where governments managed to keep fiscal imbalances within relatively narrow limits, this was chiefly because of loose monetary and credit policy which helped state enterprises to meet their tax obligations. This was the case of Romania, the Baltic states and most of the CIS republics. This is not, however, the case of Czechoslovakia, the only transition country which succeeded in maintaining low inflation with a

---

[328] See the interview with the Czech Minister of Economy K. Dyba in *Rzeczpospolita*, 9 February 1993.

[329] While in some countries the actual budget deficit in 1992 was broadly in line with initial targets, this cannot in all cases be regarded as a symptom of healthy fiscal balance. For instance, in Bulgaria the moderate deficit was achieved only because the country suspended service on its external debt (if the state budget deficit is recalculated on a "commitment" basis instead of a "cash" basis, the deficit would rise from 3 to 12 per cent). In Poland, the government's last-minute attempts to raise additional financing towards the end of 1992 were unsuccessful because commercial banks were unable or unwilling to purchase additional amounts of government securities. (See the interview with the Polish Minister of Finance J. Osiatynski, *Zycie Gospodarcze*, No.5, 1993).

TABLE 3.4.2

Changes in money supply in transition countries in 1990-1992
(Per cent, end-of-year over end-of-year)

Country	M0 [a]			M2		
	1990	1991	1992	1990	1991	1992
Bulgaria	..	66.9	-2.3 [b]	..	120.4	35.1 [b]
Czechoslovakia	-7.2	28.8	12.1 [c]	0.1	27.3	6.9 [c]
Hungary	16.2	24.0	9.9 [c]	28.7	28.3	15.5 [c]
Poland	298.2	43.3	37.7	157.9	47.4	56.6
Romania	..	..	..	..	101.0	49.1 [d]
Slovenia	..	..	163.5	..	..	131.6
Croatia	..	..	..	..	..	393.1 *
Fed.Rep.of Yugoslavia	142.9	243.3	5 643.6	36.8	127.6	14 570.0 *
Russia	..	..	864.0	..	..	660.0
Latvia	5.7	54.9	..	6.8	98.8	..
Lithuania	135.7	25.0	..	55.4	143.0	..

Source: National statistics and IMF data (for the Baltic states).

[a] Cash only, except for the Federal Republic of Yugoslavia, where changes in M1 are given.
[b] January-October 1992.
[c] January-September 1992.
[d] January-November 1992.

TABLE 3.4.3

Foreign currency deposits held with the banking system in transition countries
(Million US dollars, end of period)

Month, year	Bulgaria	Czechoslovakia	Hungary	Poland
December 1990	1 929	1 150	1 823	6 297
June 1991	1 735	1 279	..	5 327
September 1991	1 865	1 502	..	5 576
December 1991	1 958	1 625	2 608	5 824
March 1992	2 027	1 963	..	5 809
June 1992	1 909	2 315	..	5 922
September 1992	2 055	2 826	2 850	6 452
December 1992	2 544 *	..	..	6 491

Source: Compiled from national central banks data.

balanced budget. In other countries, where monetary policy tended to be more restrictive, fiscal deficits were large (Poland, Bulgaria), and in some cases ran out of control (Russia).[330]

(iii) Monetary policies: restrictive but inefficient

Like fiscal policy, monetary policy can be considered essentially as a *demand-management* instrument. In this role, monetary policy determines the level of aggregate demand by regulating the money stock in the economy to keep demand in line with supply changes.[331] But it has also a direct influence on the supply side through changes in interest rates and conditions of credit. Monetary policy is also regarded as the key instrument for protecting the value of the domestic currency.

The transition countries were at various stages of macroeconomic stabilization in 1992, with inflation and recession being the features common to all of them. Under such conditions of "slumpflation" (prolonged recession and high inflation), monetary policy is confronted with conflicting concerns: restoring price stability and facilitating recovery. In most transition countries monetary policies were intended to be restrictive, which suggests that the goal of fighting inflation was considered more important than the goal of fighting recession. It may also reflect beliefs — probably excessive but none the less commonly shared by policy-makers in transition countries — that monetary policies are efficient in fighting inflation, no matter what are the primary causes of inflationary pressures. On the other hand, declarations concerning the fundamental importance of monetary restraint were not always followed by corresponding policy actions, and in some countries monetary restraint, though badly needed, has not yet materialized.

---

[330] Large divergences between targets and results may also be an outcome of deliberately over-optimistic assumptions underlying proposed budget drafts. To put some gloss on planned figures may be a good political tactic for governments facing strong parliamentary opposition; it may also be helpful in convincing foreign creditors, in particular the IMF, that the budget policy will not become a source of financial instability.

[331] It should, however, be remembered that fiscal and monetary policy tools work through different channels, at different speeds and with different time lags.

TABLE 3.4.4

Cumulative change in real money stock and in GDP over 1990-1992 [a]

(Per cent)

Country	M1	M2	GDP
Bulgaria	-75 [b]	-60 [b]	-35
Czechoslovakia	-31 [c]	-27 [c]	-22
Hungary	-25	-10	-19
Poland	-53	-64	-18
Romania	..	-57 *	-33 *

Source: Compiled from national statistics.

[a] Nominal money stock deflated by CPI.
[b] 1991-1992.
[c] 1990-1991.

The actual degree of restrictiveness of monetary policy can be assessed using two broad indicators: money supply and credit conditions. Unfortunately, analysis of monetary developments in many transition countries faces difficulties, partly because monetary surveys are not published for all of them, and partly because changes in classification and reporting arising from the disintegration of former federal states have disrupted comparable and reliable time series. Monetary data are also blurred by the existence of an important foreign currency component in the money supply of many transition countries and uncertainty about the techniques used to incorporate this in the data.

Nevertheless, available evidence suggests that, while the *nominal* growth of money was substantial, the *real money supply* generally diminished in most transition economies over the last three years (table 3.4.2). Using consumer price changes as the deflator, the *real money supply* contracted over 1990-1991 in almost all countries, and this tendency continued in 1992. The monetary deflation was particularly strong during the initial stages of stabilization programmes – as in Poland in 1990, Bulgaria in 1991-1992, and also Russia in 1992. The fall in the real money stock seems to have been also rather strong in Romania, although complete data are not available.[332] The only country where the broad money stock (M2) increased in real terms in 1992 was Poland, but the apparent relaxation has been moderate and followed a two-year period of sharp contraction.

The decline in the real money supply was most dramatic in Bulgaria and Poland (until 1991), the two countries where initial imbalances were probably largest. The fall was more moderate in Czechoslovakia and Hungary. At the other extreme, credit emission was high in Croatia and Russia, and virtually unrestricted in FR Yugoslavia.

Reported changes in the nominal money supply may not in all cases reflect monetary policy decisions because they are also influenced by changes in stocks of *foreign currency deposits*, which in most countries are included in the total money stock. Their share in the money supply has varied not only across countries but also in individual countries as a result of the substitution of dollars for domestic money, or vice versa, and also as a result of changes in national official dollar exchange rates.

Foreign currency deposits held by firms and individuals with the banking system constituted a large but declining proportion of the nominal money stock in transition economies. Among those countries for which relevant statistics are available, the highest share was observed in Poland (where it fell from 72 per cent to 25 per cent between end-1989 and end-1992) and in Bulgaria (where it fell from 40 per cent to 30 per cent between mid-1991 and end-1992). But these figures clearly overstate the actual magnitude of the change, as they have resulted mainly from the appreciation of domestic currencies following initial devaluation. The lowest degree of "dollarization" is still observed in Czechoslovakia, in spite of an increase of this share from 9 per cent to 17 per cent of the total money stock over the last two years, and in Hungary where it remained at a constant 17-18 per cent of the total money stock.[333] In principle, the declining dollarization of the money stock can be taken as a symptom of gradually strengthening domestic currencies and the restoration of market balance. A contrary trend probably prevailed in Russia and the other ex-Soviet states, where dollarization most probably increased substantially in 1992 because of the rapid loss of confidence in the rouble and the lifting of many legal restrictions on convertible-currency deposits with the banking system.

In regulating the money supply, monetary policy is considered to be the most important tool for combating inflation. But the relationship between changes in the money supply and price levels depends largely on the dynamics of prices and expectations. As is known from the experience of hyperinflation in some developing countries, rising velocity of circulation may offset constrained money supply growth. By contrast, if inflation is low and declining, the real money supply may increase because of the replenishment of cash balances by households and economic entities (declining velocity). The first type of situation has been observed in Russia and other CIS countries and FR Yugoslavia, while the second was present in Bulgaria, Poland and probably in Czechoslovakia.

In order to assess the impact of changes in the real money stock (nominal stock deflated by the CPI) on output, monetary data have to be compared with changes in real GDP for the same periods. Because GDP has fallen in all transition countries, the actual size of monetary deflation is smaller but nevertheless substantial. The magnitude of the monetary shock in selected transition economies is illustrated in table 3.4.4.

---

[332] The stock of broad money increased in Romania from the *monthly average* of leu 625 billion in 1991 to leu 1,322 billion in August 1992 and to leu 1,446 billion in October 1992; similarly, cash in circulation increased from leu 120 billion to leu 269 billion and to leu 332 billion, respectively. Unfortunately, lack of end-year data does not permit calculation of annual changes; nevertheless, if the monthly average in 1991 is taken as a reference, the nominal money growth through October 1992 was 131 per cent for M2 and 177 per cent for cash money. Compared with consumer price changes over the same period (an increase of roughly 280 per cent), these changes suggest real monetary deflation of a similar magnitude as in Poland or Bulgaria.

[333] However, in absolute terms the *level* of convertible currency deposits has been increasing steadily over the last two years (table 3.4.3).

TABLE 3.4.5

Changes in net credit emission in transition economies, 1990-1992
(Per cent)

Country	Total credit			Credit for non-government sector		
	1990	1991	1992	1990	1991	1992
Bulgaria	..	148.1	44.8 a	..	112.7	33.1
Czechoslovakia	9.7	16.2	10.0 b	1.4	24.6	14.3 b
Hungary	11.1	7.1	6.6 b	20.2	3.4	-6.1 b
Poland	249.4	39.9	48.2	283.9	69.7	25.0
Romania	..	..	..	..	101.1	26.8 c
Croatia	-	125.8	259.7*	-	..	..
Slovenia	-	-	84.4	-	..	90.7
Federal Republic of Yugoslavia	429.0	165.6	9 995.2	41.4 d	120.3 d	13 212.0 d
Russia	..	..	1 230.0	..	..	..
Latvia	-17.6	95.7	..	6.3	131.7	..
Lithuania	11.1	57.4	..	20.5	128.2	..

Source: National statistics and IMF data (for the Baltics).

a January-October 1992.
b January-September 1992.
c January-November.
d Credit for enterprises only.

It is not certain whether the contraction of real money supply reflected simply the elimination of the inflationary overhang, or whether it also contributed to the fall of output. Nevertheless, it seems unlikely that monetary contraction on such a large scale could have left output levels unaffected. This appears to have been especially likely in the case of Poland and Bulgaria, where the falls in the real money supply were particularly large in the initial periods of stabilization.

The real monetary contraction has probably also contributed to the fall of output in Russia, where the CPI between end-1991 and end-1992 rose by more than 26 times, while M2 increased only 7.4 times, cash emission 9.6 times, and total credits 11.6 times. A similar situation has developed in the Baltic republics. It may be assumed that limited cash emission in the first half of 1992, officially attributed to technical problems with printing banknotes, was to some extent a deliberate policy by the Russian central bank aimed at containing demand and expenditures through reducing cash in circulation. The policy was only partly successful because other CIS states immediately resorted to the issue of surrogate monies, which inflated the supply of cash money in the whole rouble area.[334]

The second principal indicator of the restrictiveness of monetary policy is the availability of, and general conditions for, granting *credit*. Expansion of domestic credit was the dominant source of monetary expansion in all transition economies, but its structure differed between them. Credits for the government were more freely available than those for the enterprise sector in Poland and Hungary. In Czechoslovakia, by contrast, it was the other way round. Credits for households remained negligible in all transition countries. In Russia, another important source of money supply growth were the interstate credits extended to other CIS states through correspondent accounts in the Central Bank of Russia (CBR).

The growth of nominal credit emission (table 3.4.5) was generally slower than the growth of money supply in the east European transition countries. This may indicate that foreign currency assets have grown fast, accounting for the difference between the growth of the money supply and of domestic credit. By contrast, in Russia and probably also in other CIS states, the growth of domestic credit exceeded the growth of the money stock, suggesting some depletion of foreign currency assets. However, the difference may arise from failure to include the rouble value of foreign currency holdings with Russian banks in total money supply data.

On the other hand, credit emission for the non-government sector (enterprises and households) was generally more restrictive in 1992 than credit for the government in all east European countries for which data are available except Czechoslovakia. This is not only a distinct change as compared with 1990-1991, when credit expansion for the economy was generally faster than for the government (except in Bulgaria), but it is also another confirmation of the deteriorating fiscal position of these countries in 1992. In fact, due to that deterioration, the non-government sector has been receiving a declining proportion of all credits. For example, the share of credits for enterprises in total credit emission in Poland declined from 43 per cent in 1991 to 30 per cent in 1992, and in Bulgaria from 58 per cent in 1991 to less than 50 per cent in 1992. In Hungary, out of the total credit increase of Ft 154 billion in 1991, the enterprise sector (excluding small entrepreneurs) received Ft 106 billion, or 69 per cent; but in January-September of 1992 total credit emission grew only by Ft 20 billion, and the net credit position of enter-

---

[334] The deferral of wage payments which resulted was indeed to some extent anti-inflationary, because the arrears were not indexed and hence their real value declined. But this impact was partly offset in Russia by the issuance of privatization "cheques" which were freely transferrable; their sale is likely to have transferred purchasing power from strata with a lower to strata with a higher propensity to spend.

TABLE 3.4.6

Monthly nominal and real interest rates in Bulgaria, Czechoslovakia and Poland, 1991-1992

(Per cent)

Year, month	Bulgaria [a]		Czechoslovakia [b]		Poland [c]	
	Nominal	Real	Nominal	Real	Nominal	Real
1991						
January	1.2	-12.4	0.80	-25.0	3.7	-9.0
February	2.8	-121.1	0.80	-6.2	4.6	-2.1
March	3.2	-47.3	0.80	-3.9	4.6	0.1
April	3.2	0.7	0.80	-1.2	4.6	1.9
May	3.2	2.4	0.80	-1.1	3.9	1.2
June	3.2	-2.7	0.80	-1.0	3.9	-1.0
July	3.2	-5.2	0.80	0.9	3.5	3.4
August	3.7	-3.8	0.80	0.8	3.1	2.5
September	3.7	-0.1	0.77	0.5	3.0	-1.3
October	3.7	0.4	0.76	0.9	2.8	-0.4
November	3.7	-1.3	0.76	-0.8	2.8	-0.4
December	3.7	-1.2	0.76	-0.4	2.8	-0.3
1992						
January	3.7	-1.1	0.76	-0.2	2.8	-4.7
February	3.7	-2.1	0.76	0.3	2.8	1.0
March	3.7	-0.2	0.76	0.4	2.8	0.8
April	3.7	0.5	0.73	0.2	2.8	-0.9
May	3.7	-8.2	0.73	0.3	2.8	-1.2
June	3.7	-2.1	0.73	0.4	2.8	1.2
July	3.3	0.5	0.73	-0.2	2.7	1.3
August	3.2	2.0	0.73	0.1	2.7	-
September	3.0	-0.3	0.73	-1.1	2.7	-2.6
October	2.9	-3.3	0.73	-1.3	2.7	-0.3
November	2.9	-3.8	0.73	-1.3	2.7	0.4
December	2.9	-1.7	..	..	2.7	0.5

Source: Compiled from national statistics.

Note: Real interest rates obtained by deflating nominal rates by CPI.

[a] BNB refinacing rate.
[b] C-SSB discount rate.
[c] NBP refinancing rate.

prises declined by Ft 10 billion. A similar tendency has been observed in household borrowing. The share of households in the total credit stock declined between 1990 and 1992 from 10.7 per cent to 4.3 per cent in Bulgaria, from 7.8 per cent to 6.5 per cent in Czechoslovakia, from 21.7 per cent to 13.2 per cent in Hungary (including small entrepreneurs), and from 5.1 per cent to 2.9 per cent in Poland.

All these figures reveal a clear shift in the allocation of credit in favour of the government sector. Growing budgetary deficits are probably the main factor behind the shrinking pool of funds left for economic entities in east European countries. Among the many implications of this shift is growing pressure on interest rates, tougher conditions for credits and the general slowdown of economic activity.

Emission of new credits can be regulated through direct measures, such as bank-specific credit ceilings and credit rationing, or through more indirect measures such as reserve requirement regulations, interest rate policy, or "moral suasion". *Interest rates* remained key instruments of monetary policies in the east European countries. Most central banks declared their intention to keep interest rates positive in *real* terms. In practice, however, this principle has not always been observed. While remaining relatively high in nominal terms, central bank interest rates in most countries were negative in real terms (nominal rates deflated with the consumer price index). In countries with a relatively stable macroeconomic situation − Czechoslovakia, Hungary, Poland, and also Bulgaria − the basic interest rates of central banks (typically a refinancing rate or a discount rate) fluctuated, in real terms, between -3 and +1 per cent monthly. But in Romania, the CIS states and in some ex-Yugoslav republics, interest rates were continuously negative in real terms. The Baltic states gradually shifted to a policy of high nominal and positive real interest rates once they had introduced their own currencies and decoupled from the rouble zone.[335]

Variations in inflation resulted, among other things, in high volatility of real interest rates (table 3.4.6). In Bulgaria, for instance, jumps in the inflation rate in 1992 led to wide fluctuations in the real interest rate from -60 to +30 per cent. In Poland and especially in Czechoslovakia the fluctuations were smaller − from -32 to +17 per cent and -16 to +7 per cent, respectively.

Lending rates charged on credits by commercial banks were normally much higher and widely differentiated between banks and their clients. In Hungary, the average lending rate for all commercial banks declined gradually from 36 per cent at end-1991 to 29-30 per

---

[335] The picture might be different if producer price indices were used as deflators, but the relevant data are not available for all countries.

BOX 3.4.1

Inter-enterprise credit: Accounting exercise of economic problem?

When a firm does not pay for goods or services supplied by another firm in due time, the latter may extend a short-term credit to the former. Typically, the delay in payments results from a liquidity squeeze, which in turn may be of a temporary or a more fundamental character. In the first case, when the repayment is likely to be made soon, a supplier credit can be extended automatically; in other cases the repayment and credit conditions may be subject to negotiation. This inter-enterprise (I-E) credit is a normal phenomenon in market economies with well developed and flexible financial sectors and strict legal regulations, where it is likely to remain within limits established by accounting and financial practice. However, it gains more importance when general credit restriction are coupled with lax financial discipline on enterprises. This is precisely the case of the east European countries in transition.

The sharp increase of I-E credit in Hungary, Czechoslovakia, Poland, Bulgaria, and more recently also in Romania and Russia, raises doubts about the effectiveness of macroeconomic financial policies in these countries and is sometimes seen as an important impediment to industrial restructuring.[1] On the other hand, many economists argue that the economic impact of the I-E credit is not as harmful mainly because it is an illiquid monetary asset and as such supposedly has no effect on money supply. Besides, if all enterprises are considered, the I-E credits simply cancel out, thus leaving monetary aggregates unaffected.

However, these statements overlook some important and economically dangerous implications of excessive inter-enterprise credit. Consider three firms: A, B, and C. Firm B supplies to A materials worth, say, $5,000, but firm A does not pay for the delivery on time because of liquidity problems. To cover the resulting cash-flow gap and to pay its monthly wage bill, firm B has essentially three main options: first, it may draw on its reserves or sell some of its assets; second, it may borrow from a commercial bank; and third, it may stop paying its own supplier, C.

In the first case, assets which are relatively illiquid (bank deposits, securities, inventories) are converted into highly liquid assets (cash) which are paid out as wages. Total money supply does not change, but the stock of "narrow" money (M1) increases at the cost of relatively less liquid monetary assets. The substitution of less for more liquid assets leads to an increase in the velocity of money circulation and translates into higher demand on markets for goods. This is the "substitution" case.

In the second case, a commercial bank extends credit to firm B, thus enabling it to pay the monthly wage bill. In this way the I-E credit gets "monetized". The money supply will increase by that amount times the multiplier; moreover, the increase will mostly raise the stock of highly liquid assets (M1). Of course, the "monetization" of the I-E credits can take place only if the bank's credit activities are not restricted by binding reserve requirements or credit ceilings. This is the "monetization" case.

In the third case, the cash flow problem is passed over to firm C, which in turn is faced again by the same three possibilities. If the third option is again selected, e.g., firm C does not pay firm A, the circle closes and all three companies end up with higher level of receivables and liabilities, which arithmetically cancel out. This is the "circular" case, probably the most typical once I-E credits assume large proportions and, consequently, exert macroeconomic effects.

It is sometimes argued that unless the "circular" credit is monetized by the banking system, its effect is of a purely accounting character, and restricted to a worsening of the balance sheets of the enterprises involved. But in practice, the effects of "circular" I-E credits spill over beyond simple book-keeping operations. First, it is unlikely that a monetization (at least partial) can be avoided in a process which involves many firms and many banks. Some banks may be more liquid than others, and they may be expected to inject some new credit into the enterprise sector, thus increasing the money supply. They may be willing to do so for a number of reasons, both economic ones (to help sound enterprises to overcome temporary difficulties) as well as non-economic ones (to respond to political pressure to rescue some big state enterprises).

If banks are not identical, neither are firms. A redistribution of (il)liquidity is likely to take place from firms enjoying a comfortable liquidity position to firms faced by liquidity shortage. Thus, good firms will support, largely unintentionally, employment levels in loss-making firms. On the macroeconomic level, this will result in a shift in the internal structure of the monetary stock in favour of more liquid assets (mostly cash), and will be followed by a corresponding increase in wage inflation.

Even in the best case, when the money supply is not increased and no substitution between assets of various degree of liquidity takes place, excessive I-E credits entail other important costs. First, the redistribution of financial funds from good to bad firms distorts microeconomic adjustment, because the I-E credit allocation is not based on any consistent economic considerations and typically occurs across the board. Second, overall financial discipline weakens, as no interest is normally charged on I-E credits and no decisive sanctions are imposed on insolvent firms. Third, the fiscal base shrinks because of lower profits and inevitable tax credits and reliefs.

When I-E credits are partially or fully monetized, regular credits are "crowded-out". While current production levels may thereby be protected, in many cases this means a continuation of loss-making operations. Moreover, new investments are adversely affected and the proportion of non-performing loans increases. In the case of "substitution", the inefficient reallocation of credit which takes place in the economy also contributes to a general credit crunch, the accumulation of bad loans by banks and excessive enterprises debts. The impact of restrictive monetary policy at the macroeconomic level is thus effectively neutralized by allowing loss-making firms to circumvent credit limits.

Excessive levels of I-E credits require an urgent policy response. It should probably start with some multilateral compensation scheme, but this may be technically difficult and applicable only to a limited number of firms. A more general solution would be to convert the overdue liabilities into short-term commercial paper convertible into equity shares issued by indebted firms for sale at a discount to commercial banks. The banks can then either impose stricter financial discipline on firms to speed up the necessary adjustment (e.g., through acquiring some equity stake in the troubled enterprise), or initiate bankruptcy. This would, however, worsen the already overloaded balance-sheets of commercial banks. The recapitalization of banks may provide some relief, but it would require additional funds from the budget. They could nevertheless, be provided if fiscal revenues are expected to increase as a result of the operation. A third option would be to write off some of the debts. An inevitable "moral hazard" problem could be minimized by making a credible policy commitment to restrict the book-cleansing to a one-time operation. Whatever method is selected, it must be accompanied by measures to prevent the I-E credit from re-emerging. This can be ensured by "commercialization" of banks, improved legal regulation of financial flows including "trigger" provisions for initiating bankruptcy procedures and fast-track rules for closing down inefficient firms.

---

[1] United Nations Economic Commission for Europe, *Economic Survey of Europe in 1991-1992*, New York, 1992, pp.100-101.

cent at end-1992. But for more risky investments, bank lending rates in the latter year went up to 42-44 per cent.[336] In Poland the spread between the refinancing rate and the average lending rate was narrower, varying by around 6-8 percentage points above the central bank rate. In Czechoslovakia, the ceiling of 14 percentage points above the central bank rate imposed on the lending rates of commercial banks in early 1991 was abolished in mid-1992, but the spread remained relatively low (not exceeding 4-5 points). In the Baltic republics, current rates on commercial loans vary between 120 per cent per annum in Estonia to 220-300 per cent in Lithuania.[337] Record high levels were reached by interest rates in FR Yugoslavia, where rampant inflation, relative flexibility of the banking system and uncertainties originating in the civil war and sanctions sent annual domestic currency lending rates from 260 per cent in April 1992 up to 5,000 per cent in August, to 20,000 per cent in November 1992, and to a dizzying 50,000 per cent in February 1993.

As a result of these very wide differentials, interest rates on commercial credits in the east European transition countries were generally positive in real terms on an annual basis, with the real cost of credit typically increasing after the initial stage of stabilization. By contrast, in Russia, and probably most other states of the former Soviet Union, commercial interest rates remained strongly negative in real terms in spite of several increases in the base rate by the Central Bank of Russia — from 6-9 per cent in December 1991 to 50 per cent in April and to 80 per cent in May.[338]

In the course of 1992 there was some cautious relaxation in central bank interest rate policies in some transition countries as central banks reacted to gradually falling inflation rates and growing pressures to ease monetary policy in the recessionary environment. In Bulgaria, the refinancing rate was reduced from 54 to 49 per cent in June 1992 and to 41 per cent in September, while in Czechoslovakia the rediscount rate dropped from 8 per cent in the second half of the year. In Poland, the refinancing rate was cut from 40 to 38 per cent in July 1992, and then to 35 per cent in February 1993. A general decline in interest rates has also been observed in Hungary and Romania, where the central bank rate was reduced from 80 to 70 per cent.

Interest rates remained the principal tools for regulating credit emission and the money supply in the east European countries, but in the rouble zone the weakness of interest rates forced the monetary authorities to resort to other, mostly administrative instruments of credit allocation, such as bank-specific *credit limits*. *Credit ceilings* were used also by the central banks of Bulgaria, Czechoslovakia, and Poland, although their role declined as they were gradually replaced by instruments such as interest rates and changes in mandatory reserve requirements. Credit ceilings will be discontinued and replaced by open market operations in 1993 in Poland and in the Czech Republic. In Russia also, treasury bills are to be introduced from February 1993, and are intended to establish the foundations of a money market and to increase the flexibility of monetary policy. To this end, the CBR will allow trade in treasury bills at its discretion. Other instruments such as central bank surveillance and persuasion have also been extensively used to implement monetary policy in all transition countries.

As in the case of interest rates, credit allocation policies also differed between eastern Europe and the CIS countries. In eastern Europe, credits for the enterprise sector, especially for new businesses, and for households were generally expensive and difficult to obtain. Credit conditions typically included high interest rates, short maturities, high collaterals,[339] the presentation of elaborate and detailed business plans, etc. Also, the accumulation of bad loans from the past reduced current lending possibilities. In all cases for which data are available, as already noted, a high and growing proportion of total credits granted went to the financing of the government, which may suggest a "crowding out" phenomenon. In the CIS states, after a temporary tightening of monetary policies in the first quarter of 1992, credit allocation became largely automatic as the CBR changed its priorities from fighting inflation to protecting enterprises. Commercial banks in the CIS countries are more dependent on their big corporate clients, and the problem of bad debts has been alleviated by high inflation.

If monetary policy remained rather restrictive in most transition countries, its efficiency was low, especially in countries at an early stage of transformation and where financial market institutions are underdeveloped. The case of Russia and other CIS states is illustrative. In the absence of bankruptcy regulations and effective enforcement of contracts and payments obligations, the tightening of the real money supply in the first half of 1992 resulted in an *interenterprise arrears crisis*. Interenterprise arrears occurred in all transition economies, but they reached truly astronomical proportions in the CIS countries (see section 3.4(vi)). As discussed in box 3.4.1, excessive interenterprise credits are inflationary and at the same time hamper the efficient allocation of resources.

Attempts to solve the arrears crises included clearing operations (as in Russia and Romania), write-offs of old debts (privatization deals), or debt consolidation schemes (as in Poland), but none of these proved very successful. Possible solutions include: write-offs of old debts from before the reform (their nominal value being

---

[336] National Bank of Hungary, *Monthly Report*, October 1992.

[337] An interesting phenomenon has been observed in Lithuania, where loans in domestic currency ("talonas") carry lower interest than loans in dollars, which may cost up to 10 per cent per month, or 314 per cent per annum (*The Baltic Independent*, 12-18 February, 1993). This can be explained by the opportunities offered by some lucrative import transactions. A similar situation has been observed in FR Yugoslavia.

[338] Commercial bank lending rates were higher (they increased from 30 per cent in January 1992 to 95-100 per cent in the summer and to 130-135 per cent in December 1992), but nevertheless remained strongly negative in real terms.

[339] For instance, collaterals amounting to 3-5 times the loan value are not uncommon in Poland.

typically eaten up by inflation anyway); reductions in the scope of permitted cash settlement; and division of arrears into two categories: regular liabilities to be settled within, say, 60-90 days, and other liabilities to be converted into bonds and consolidated within a single state agency which would manage them within a long-term restructuring programme. In principle, an injection of additional funds from the government into the banking system would be necessary to strengthen banks financially, and to allow them to support long-term restructuring programmes.

In view of the specific conditions of the transition economies, monetary policy *stricto sensu* has been a crude tool. Given its primary goal of fighting inflation, it had to be much more heavy-handed than in market economies which respond more flexibly to macroeconomic measures. Thus the impact of monetary policy in transition countries was restrictive enough to reduce investments and current output, sometimes by considerable amounts, but it contributed little to the elimination of inefficient enterprises and the reallocation of resources, partly because the relevant monetary variables (especially interest rates) were often brought to bear only with restraint. Neither did it effectively stop inflation. The problems were of course rooted primarily in institutional impediments and structural rigidities inherited from central planning; monetary restraint alone is of little help in this context.

(iv) Foreign exchange policies

Foreign exchange policy is concerned with two broad categories of issue: determination of the level and of readjustment to the exchange rate of domestic currency, and the conditions for the convertibility of domestic currency into foreign currencies. The first group of problems is settled within the framework of current *exchange rate policy*, while the second group is addressed by measures determining the institutions of the country's *foreign exchange regime*. Exchange rate policy, in turn, typically addresses two distinct stabilization objectives. Firstly, it helps to maintain the competitiveness of domestic *vis-à-vis* foreign producers. In that role it is also the key instrument for restoring or maintaining external balance (by switching expenditure between domestic and foreign goods, as well as between tradables and non-tradables). Secondly, because of the impact of changes in the exchange rate on the domestic price of tradable goods, exchange rate policy is an important determinant of the level of domestic inflation.

The two goals of maintaining external and internal (price) balance may lead to conflicting policy prescriptions with respect to the level of and modifications to the exchange rate. If a country suffers from a current account deficit *and* domestic inflation at the same time, restoring external balance requires the devaluation of domestic currency — which, however, would result in a further acceleration of inflation. The decision thus depends on individual country conditions and policy judgements as to which imbalance should be curbed first. Most transition countries are currently faced simultaneously by current account deficits and inflation ranging from relatively low to very high. The policies adopted to tackle the twin problems of external and internal imbalances follow broadly a two-stage pattern, common for almost all countries.

(a) Two-stage adjustment: "shock" devaluation and gradual real appreciation

First, all the transition countries underwent deep, sometimes very drastic, devaluations of domestic currencies, usually at the initial stage of stabilization programmes. They were intended to eliminate one of the most pervasive distortions existing under central planning (that of wide differences between domestic and international prices) and to establish new relative prices more in line with international prices. In addition, devaluations were intended to prepare the ground for introducing some measure of domestic currency convertibility. In the second stage, after the initial devaluation was effected, exchange rate policy was used mostly as an anti-inflationary device, thus concerned more with maintaining internal balance.

This pattern has been followed by all transition countries, even though the modalities and timetable of policy decisions differed. Sharp initial devaluations increased the official exchange rate against the US dollar by a factor of 3 to 9, and in some countries even by far more (as in Russia in January 1992). The magnitude of "devaluation shocks" can be illustrated by data shown in table 3.4.7. But after the initial exchange rate realignment, further adjustments generally lagged behind domestic inflation. As a result, in all transition countries domestic currencies appreciated strongly in real terms in 1992.

The initial devaluations were accomplished either in one stroke, as in Bulgaria or Russia, or as a series of several consecutive devaluation decisions concentrated within relatively short period of time, as in Czechoslovakia or Poland. The choice of the method was typically a reflection of the general reform strategy ("shock therapy" versus gradualism), but it was also dictated by the scope of initial market imbalances and by availability of external financial support. On the other hand, the magnitude of devaluation was subject to policy decisions only where a fixed rate regime was adopted; in other cases it was simply the outcome of the removal of restrictions on foreign exchange markets and the ensuing more or less free float of the domestic currency. This was so in Bulgaria, where the dollar rate jumped eightfold following liberalization, and also in Russia and Romania.

The initial exchange rate realignment was in many countries coupled with the introduction of a considerable degree of convertibility of the domestic currency. In most countries, convertibility was generally applied in trade transactions for goods and most services, but some restrictions remained in tourism and foreign transfers (Czechoslovakia, Hungary, Bulgaria), and in factor services (Poland until July 1991, Romania). Capital account transactions remained under strict control everywhere.

TABLE 3.4.7

Initial foreign exchange rate adjustments in transition economies

Country	Period	Nominal exchange rate [a]		Rate of change (per cent)
		Before	After	
Albania	September 1991-January 1992	10.0	50.0	400.0
Bulgaria	February 1991	2.88	23.61	719.8
Czechoslovakia	September 1990-January 1991	15.71	28.00	78.2
Hungary	January 1991	60.95	68.59	12.5
Poland	August 1989-January 1990	988	9 500	861.5
Romania [b]	November 1991	59.53	201.74	238.9
Yugoslavia	January 1992	20.31	64.87	219.4
Russia [c]	January 1992	1.76	110.0	62 times

Source: Compiled from national statistics.

[a] In units of national currency per US dollar.
[b] The 11 November 1991 devaluation reported here marks the introduction of internal convertibility, but this decision was preceded by two earlier devaluations in November 1990 (by 77 per cent) and in April 1991 (by 66 per cent).
[c] The devaluation to R110 $ on 1 January 1992 referred to non-government transactions only.

There are strong arguments for early convertibility. It allows the "import" of undistorted price relatives and competition from abroad. It also facilitates trade and reduces the transaction costs of acquiring foreign exchange. The convertible currency system is the most transparent system of allocating foreign exchange because it works through the market. On the other hand, two arguments are frequently raised against early convertibility. First, it involves a danger of running down international reserves, which as a rule are very limited. Second, it may cause large transitional unemployment.

Both cautions assume implicitly that the supply side response will be sluggish, and that therefore a devaluation will not improve the trade balance unless it is very large. But a very large devaluation tends to be reduce real wages substantially, which may be politically intolerable. This is why west European countries decided in the late 1940s to achieve convertibility only gradually.

Reservations about the capacity of transition economies to absorb large devaluations and to maintain convertibility ("elasticity pessimism") proved unfounded; countries which introduced internal convertibility have generally improved their external balance through export expansion and/or import contraction. The analogy with western Europe is misplaced insofar as early convertibility in transition countries is also sought for other important benefits, such as realignment of price relatives and the "import" of competition from abroad. International reserves will be depleted only if the exchange rate is overvalued, but if wages are not allowed to rise too rapidly (and their level in post-socialist economies is relatively low anyway) and expenditures are kept under control, the risks of early convertibility can be contained.

After the initial devaluations, exchange rate policy was used mainly to counter inflationary tendencies resulting partly from the preceding devaluations. In all countries except Romania, exchange rates appreciated in real terms in 1992, even in those where floating rate systems were in force (table 3.4.8).

Strong real appreciation of the domestic currency took place in Czechoslovakia, Poland and Slovenia, and to a lesser extent in Bulgaria. In Romania, the currency depreciated further following the introduction of limited convertibility in November 1991 and foreign exchange auctions in June 1992, while in FR Yugoslavia the currency appreciated strongly at the official exchange rate because of restrictions imposed on domestic convertibility in late 1991.

A similar real appreciation can be observed in the former Soviet republics. The sharp devaluation in January 1992 set the official rouble exchange rate at R110/$, a rate which was broadly maintained until June; but it collapsed in the second half of the year, falling from R125/$ in July to R440/$ in December 1992. Meanwhile consumer prices in Russia increased 9 times during the first semester, and 26 times during the whole year, thus largely offsetting the impact of the initial devaluation. In the Baltic republics, the real appreciation was strongest in Estonia, where the exchange rate remained unchanged after June 1992 while domestic prices increased by 80 per cent between June and December.

A real appreciation need not necessarily lead to a deterioration of trade performance. In some cases, it may simply be a "mirror" reaction to the initial "overshooting" and can be regarded as another adjustment in search of a "correct" or equilibrium exchange rate. It should also be noted that the consumer price index is not the best deflator for gauging changes in international competitiveness. Producer price or nominal wage cost indices would offer better deflators, but are not available for most transition economies.[340] If the wage index is used for deflation, the extent of real appreciation is much smaller, and for some countries a moderate real depreciation may even be observed instead (Hungary). In any case, the tendency to restrict

---

[340] It should also be remembered that the exchange rates reported in table 3.4.8 are dollar rates, and hence may be deceptive in assessing the trade competitiveness of transition economies vis-à-vis European market economies. The fall of the dollar in mid-1992 and its subsequent rise towards the end of the year can be clearly seen in east European dollar-rate fluctuations.

TABLE 3.4.8

Changes in nominal and real foreign exchange rates in transition countries, 1991-1992 [a]

(December 1990 = 100)

Year, month	Bulgaria	Czechoslovakia	Hungary	Poland	Romania	Fed.Rep.of Yugoslavia	Slovenia [b]
Nominal devaluation							
March 1991	563.8	118.8	119.2	100.0	103.2	137.3	-
June 1991	640.8	127.7	126.3	119.9	175.9	220.2	-
September 1991	648.2	124.1	123.9	117.4	174.6	209.9	-
December 1991	779.8	118.0	126.0	116.5	534.8	193.1	114.9
March 1992	836.9	120.0	130.7	141.5	570.1	..	154.7
June 1992	819.5	117.5	128.7	144.2	752.0	1 093.2	159.7
September 1991	795.4	112.4	127.0	145.7	1 163.0	1 901.1	176.8
December 1992	879.4	..	136.0	162.6	..	7 129.3	204.6
Real devaluation [c]							
March 1991	147.9	83.7	102.0	79.6	83.3	115.4	-
June 1991	153.8	85.6	101.0	86.3	104.7	144.2	-
September 1991	128.5	82.9	96.6	80.4	79.5	105.2	-
December 1991	135.9	76.3	94.2	72.7	175.1	57.9	69.5
March 1992	126.6	76.2	90.5	79.1	126.2	..	65.5
June 1992	101.4	73.8	86.2	73.5	136.1	19.4	57.1
September 1992	91.6	68.3	82.2	67.8	179.1	8.7	59.5
December 1992	85.4	..	83.6	70.3	..	..	64.9

Source: Compiled from national statistics.

a Average official (market) dollar rates for respective months.
b October 1991 = 100.
c Deflated by CPI.

adjustments in the exchange rate may in the medium and longer run result in a loss of competitiveness and deepening recession.

In fact, foreign trade balances deteriorated in many transition countries in 1992 (see data in section 3.3). The emergence of large deficits in Bulgaria and Czechoslovakia, and in Poland in the second half of 1992, may be symptoms of diminishing competitiveness linked to appreciating exchange rates.

Are east European currencies overvalued or undervalued? No answer to this question can be given because knowledge of the working of the "fundamentals" determining long-term exchange rate levels under the highly volatile economic conditions in the reforming countries is very incomplete. Most observers tend to regard east European currencies as still substantially undervalued, especially if purchasing power parities (PPP) are taken into consideration. According to one estimate, the ratio of the official (market) rates to PPP rates exceeded unity in all east European countries in mid-1992, varying from 1.6-1.7 in Hungary and Poland to 2.4-2.6 in Bulgaria, Czechoslovakia and Romania.[341] On the other hand, trade balances are very fragile in all countries and international reserve levels are generally low (although higher than in previous years). On balance, while in some countries the exchange rates do not seem to be coming under immediate pressure, as in the Czech Republic, in some other countries devaluation pressure is gradually accumulating. This is most easily discernible in Hungary and Poland.[342]

The anti-inflationary thrust of exchange rate policies in 1992 appears to have had most impact under the fixed rate regimes. Part of the success of stabilization in Czechoslovakia may certainly be attributed to the fixed exchange rate (even though its initial "overshooting" may have resulted in stronger corrective inflation than would otherwise have been necessary). The "anchoring" effect of the fixed rate can also be seen in Estonia, where inflation was brought down from monthly rates of 20 per cent when national currency convertibility was introduced in June 1992, to 3-4 per cent by the end of 1992 (albeit the adjustment was more sluggish than in Poland and Czechoslovakia). Floating rates do not seem to have exerted any mitigating effect on domestic inflation in Romania, Lithuania or in the CIS states. In Bulgaria and Latvia, the central banks managed to maintain the float around some "reference" rate, despite high inflation. In Bulgaria, the exchange rate fluctuated around lev 23/$ for most of the year, and in Latvia the rate stabilized in the last quarter of 1992 at 165-175 "roublis" per dollar.

Most of the east European transition countries did not substantially change their exchange rate regimes during 1992.[343] In *Czechoslovakia*, the system of internal convertibility coupled with a fixed rate of the *koruna* against a basket of major convertible currencies has continued in operation. Technically, internal convertibility is secured by the interbank exchange market, operating daily with the participation of major commercial banks and the central bank. The role of the central

---

341 "The Role of Exchange Rates in Western Investment Decisions in Eastern Europe", *PlanEcon*, Washington, D.C., May 1992.

342 Another indicator of the strength of expectations for further devaluation is the steady growth of foreign currency deposits held by domestic residents, even in countries where large interest rate differentials existed in favour of domestic currency holdings (see table 3.4.3).

343 United Nations Economic Commission for Europe, *Economic Bulletin for Europe*, vol.44(1992), New York, 1993, pp.58-62.

bank is to equate demand and supply of currencies through occasional interventions, guided by the so-called "rules of liquidity" requiring commercial banks to keep the ratio of foreign assets and foreign liabilities within a limited range of 0.85-1.05.[344] The *koruna* is pegged to a basket which is to be changed periodically to reflect shifts in the structure of international payments.[345]

In *Hungary*, similar restrictions on internal convertibility in 1992 mostly pertained to trade transactions in certain monitored categories, which also limited the access of physical persons to the foreign exchange market. A major development was the introduction of an interbank foreign exchange market in July 1992: its daily quotations are used by the National Bank of Hungary to fix the spot exchange rates. Consistent with the logic of the adjustable peg regime, there were three relatively minor devaluations in 1992 of 1.6-1.9 per cent each (in March, June and November), followed by another 1.9 per cent devaluation in February 1993. As in the case of most other east European currencies, the forint rate is also based on a basket of currencies, which since the end of 1991 consists of the ECU and the US dollar in equal proportions.[346]

Poland has continued its crawling peg exchange rate policy in 1992. The established rule since end-October 1991 has been to devalue the *zloty* in daily micro-steps of Zl 9, amounting to an average monthly rate of devaluation of some 1.8 per cent. This rate is supposed to reflect the combined effect of two policy targets: to maintain a competitive real rate and to avoid major changes in exchange rates which would encourage speculation. The policy has been generally successful on both scores, although it had to be supported by an additional one-time realignment of the dollar rate by some 12 per cent at the end of February 1992. The Polish currency is also pegged to a basket of currencies, with the US dollar and the deutsche mark accounting jointly for 80 per cent of the total.[347]

Interbank foreign exchange market systems have also been in operation in countries with floating exchange rate regimes: *Bulgaria* (since February 1991) and *Romania* (since June 1992). A relatively well developed interbank foreign exchange market also exists in Slovenia, Croatia and FR Yugoslavia. Interbank currency markets, although taking the less sophisticated forms of periodic currency auctions, have been established in Russia in 1991, then in the Baltic republics in 1992, and subsequently in other CIS states.[348]

(b) The choice of an exchange rate regime: some general remarks

The methods and tools of exchange rate policy used in pursuance of the objectives of internal and external balance depend chiefly on the mechanism of foreign exchange rate determination adopted. Essentially, three different regimes have been in operation in the transition economies in 1990-1992: the *fixed rate regime*, based on the concept of "nominal anchoring" (Czechoslovakia, Poland from January 1990 to October 1991, Estonia), the *managed float regime* (Bulgaria, Slovenia after October 1991, Romania after November 1991, Latvia, Lithuania, Russia after June 1992), and the *adjustable (or crawling) peg regime* (Hungary, Poland after October 1991, Russia between January and June 1992, FR Yugoslavia).

The *fixed rate regime* has a number of advantages which makes it attractive for small open economies. First, the fixed rate helps to keep inflation under control and may serve as a "nominal anchor" (for a time). Second, a real commitment to the fixed rate imposes a measure of discipline on macroeconomic policies, especially on monetary policy. Excessive credit expansion would lead to an excess demand for foreign exchange — and, in the absence of continuous external financing, the country's international reserves would run down. Third, if the commitment is credible, it contributes to lowering inflationary expectations and speculation. Fourth, the fixed exchange rate, if it is credible, contributes to the stability of the macroeconomic environment and reduces uncertainties.

Fixing the exchange rate means pegging the domestic currency to an international currency or to a basket of currencies. The advantages of the fixed exchange rate regime will be greater if the domestic currency is pegged to a foreign currency which is considered stable and which is used for settling a major proportion of the foreign trade transactions of the country. The last argument rests on the theory of "optimum currency areas".[349]

The fixed rate regime, however, has also some disadvantages. The most commonly cited drawback is that it requires relatively large stock of international reserves in order to defend the established exchange rate. Another shortcoming stems from the loss of economic "sovereignty": the fixed rate regime means that the economic policy is to a considerable extent dictated by

---

[344] State Bank of Czechoslovakia, *Reform of the banking system in Czechoslovakia*, 1992.

[345] In 1992, the structure of the basket was as follows (in per cent): $49.07, DM 36.15, S 8.07, Sw F 3.79, Ff 2.92.

[346] See: National Bank of Hungary, *Annual Report 1991*, p.106.

[347] The composition of the basket reflects the average structure of current account payments, and since May 1991 has been as follows (in per cent of the total): $45, DM 35, £10, Sw F 5, and Ff 5.

[348] Most recently, 43 Ukrainian commercial banks having a licence to conduct currency operations established the Ukrainian Interbank Currency Exchange, with a charter fund of 500 million karbovanets (*Radio Ukraine World Service*, 5 January 1993, reported in BBC, *Summary of World Broadcasts*, SU W0264, 15 January 1993).

[349] The theory states that the internal gains from exchange rate predictability grow, though at a decreasing rate, as a region of mutually fixed rates is extended geographically. On the other hand, the likelihood of macroeconomic adjustment problems also grows if the region is extended to include other economies which are subject to domestic disturbances. An optimum occurs when the marginal gain from greater exchange rate predictability equals the marginal loss due to reduced economic stability.

developments in the external sector: one important policy instrument — the exchange rate — is given up, so that the burden of any necessary adjustment falls entirely on monetary and fiscal policies. There is also a risk of picking the wrong rate level; if it cannot be changed, the necessary adjustment will have to be done through a change in real wages. This shortcoming is important in transition countries, where large price adjustments are taking place, because the exchange rate has to be fixed in anticipation of future price and wage increases. This normally requires a larger initial devaluation, which may result in "overshooting".

The advantages of the *floating rate regime* are in most cases the disadvantages of the fixed rate regime. Much smaller foreign reserves are required to operate a floating system and demand-supply balance is assured at every moment of time. But its disadvantages are serious. They are connected mainly with the inherent volatility of exchange rates: under the floating system, there is a tendency for exchange rates to overshoot in response to any change in monetary policies. The recent experience of Bulgaria, Romania and Russia also seems to indicate that floating rates are more inflationary than fixed rates.

To avoid fluctuations of the exchange rate and thus of costs and prices, a predictable monetary and fiscal policy is needed. It is, however, unlikely that such a policy can be pursued during the initial stage of transformation because uncertainties and risks of policy mistakes are simply too high. This raises the danger of exchange rate speculations. Under such conditions the exchange rate regime becomes a source of considerable instability. The method used by some transition countries to reduce the dangers of speculation and overshooting was to intervene on foreign exchange markets. Central banks were responsible for keeping exchange rate fluctuations within certain limits through sales (or purchases) of convertible currencies. The capacity for central intervention, however, depends on the level of reserves available: for instance, the lack reserves in Bulgaria during the initial months left only very little possibility of "managing" the float and a similar problem was faced in Romania in the first half of 1992.

A compromise solution is to adopt an *adjustable rate regime*, in order to minimize the shortcomings of the two "pure" systems. If the exchange rate is adjusted frequently, the *crawling peg* applies, which may follow a predetermined path or be adjusted in a discretionary manner. In both cases speculation may become dangerous and adjustments should thus neither be too substantial nor postponed for too long.

The choice of the best solution again depends on initial conditions. In view of the price adjustments necessary during transition, the fixed rate regime carries the risk of rapid appreciation — as was the case in Poland — while the floating rate adds to inflation. If internal stabilization is the principal immediate objective, then the domestic currency should initially be devalued and a fixed rate maintained for a period of several months (probably 3 to 6) thereafter. This can be followed by a switch to a crawling peg once the initial corrective inflation is brought under control. An additional argument in favour of this solution exists if the exchange rate is an important determinant of inflationary expectations (as in Poland).

The crawling peg looks like the preferable solution for the period immediately after the initial price liberalization shock if domestic prices cannot be fully stabilized. The currency should be allowed to depreciate sufficiently, because domestic inflation is likely to persist for some time, and relative prices continue to change. The domestic currency could be pegged either to a trade-weighted basket of main foreign currencies, or simply to the ECU. Pegging to the deutsche mark makes economic sense, but may be politically problematic in some countries. Pegging to the dollar could be justified in some countries (Poland, Russia) for psychological reasons, but should perhaps be avoided in view of the volatility of that currency even in the short term. Since more than 50 per cent of east European foreign trade is with the EC countries, the dollar peg would introduce an element of unnecessary volatility.

If a fixed exchange or crawling peg rate is selected, policy-makers have to decide on the initial *level of the exchange rate*. Even in a floating rate system, it is not advisable to leave that issue entirely to the market. In the shift from non-convertibility to convertibility there is a tendency for the exchange rate to depreciate excessively, i.e., to "overshoot" — as was the case in Bulgaria in 1991. Some authors link this to the short-run rigidity of foreign exchange supply,[350] but a shift to convertibility in fact should substantially increase the supply of convertible currencies as exporters are then allowed to sell earnings on the market rather than surrender them to the government. The uncertainty factor, however, may play a greater role, because exporters may initially refrain from selling foreign currencies as they wait for the market to stabilize.[351]

Whatever regime is selected, it is important to establish a reasonable initial exchange rate. It will obviously depend on the level of international reserves and the volume of external assistance. In this context, support for convertibility by a stabilization fund is useful to help to absorb the initial surge of demand for foreign currencies. Both the reserves and the stabilization fund are of much greater importance in the case of a fixed exchange rate regime, but they are also necessary even under the floating rate regime precisely to avoid an initial "overshoot".

How should the initial exchange rate be determined? One way would be to examine the range of values between the official and the black market rate and to start with an average weighted according to the shares of

---

[350] E.g., W.M. Corden, "Trade Policy and Exchange Rate Issues in the Former Soviet Union", The World Bank, *Working Papers in Economic Policy*, WPS 915, May 1992.

[351] When the lev was allowed to float in Bulgaria in February 1991, the exchange rate jumped initially from lev 3/$ to lev 23/$, before falling to lev 15/$ several weeks later.

both these market segments in the total foreign exchange market. This "equilibrium" rate would then have to be adjusted to take into account corrective inflation or price liberalization effects. But there are strong reasons to believe that the selected rate will not be the real equilibrium rate, because both the demand and supply of convertible currencies tend to adjust during the process.[352]

The experience of transition countries demonstrates that initial overshooting is very likely. One of the reasons behind this is probably the conviction that it is better to devalue too much than too little. Both Poland and Czechoslovakia probably devalued too much, and paid for that with higher inflation rates. But too little devaluation might have been even more dangerous, as it could lead to the destruction of tradable goods industries.[353]

(v) Price and incomes policies

Rapid *price liberalization* is commonly considered as one of the cornerstones of radical pro-market reforms. Most transition countries went through a one-stroke large-scale price decontrol operation in the early stages of their stabilization programmes during 1990-1992. Poland, Czechoslovakia, Bulgaria, the former Yugoslav and Soviet republics liberalized between 70-90 per cent of prices at the outset of their respective reform programmes.[354]

Typically, prices for three categories of goods were excluded from these sweeping decontrols. The first were consumer necessities such as basic foodstuffs, utilities, housing rents, and transport fares; second, the prices of fuel and energy, both for households and for enterprises; and third are procurement prices for most important agricultural products. Prices for all three categories of goods and services have remained under some form of centralized control in all transition countries, even though the degree of government intervention in the price formation mechanism, as well as

TABLE 3.4.9

Changes in fuel prices in Russia, 1991-1992
(Roubles per unit)

Month, year	Crude oil (tons)	Hard coal (tons)	Natural gas (1000 cu m)
December 1991	75	20	14
March 1992	354	140	78
June 1992	1 940	865	201
December 1992	6 350	1 091	207
Index (Dec.1991 = 100)	8 466.7	5 455.0	1 478.6

Source: Russian Federation Gosskomstat Report, 1992, p.18. *Ekonomika i Zhizn'*, various issues.

the scope of the controls, varies considerably between countries.[355]

Public housing rents as well as utility prices and public transport fares are under state control in all transition countries. But in some countries a limited range of consumer goods are also subject to government monitoring or direct controls, e.g., in the CIS republics[356] or Bulgaria. A much longer list of goods is covered by controls in Albania and Romania, where approximately half of consumer goods prices are still restricted in some way.[357]

Except for electricity, heating and in some cases natural gas, prices for coal, oil and oil products have to all intents and purposes been liberalized in Poland, Hungary, Czechoslovakia, Slovenia and the Baltic states. The governments can, however, influence these prices indirectly to some extent through changing export tax and excise duty rates. Energy prices as well as transport and telecommunication fees are adjusted occasionally, the frequency of changes varying between once a year in Czechoslovakia to 3-4 times a year in other countries, depending on the inflation rate. Even though the motive for these increases — keeping relative prices close to international standards — is justified, they are nevertheless an important source of domestic inflation.

---

[352] One possible explanation takes into account two effects of the stabilization operation in transition countries: the fall of real incomes and change in inflationary expectations. If the demand for convertible currencies is assumed to be a function of real incomes (y p), and of the difference between expected depreciation of the domestic currency vis-à-vis foreign currencies and the domestic interest rate, then the "overshoot" will result from the downward shift of the demand schedule due to falling of real incomes and altered expectations. The result is that the equilibrium exchange rate will first rise and then fall, and the initial devaluation may be too large. However, if it is not large enough, it would fail to reverse expectations. Probably, a "correct" rate of the initial devaluation may be found after all, but the mechanics of the interaction between expectations, real income changes and demand for foreign exchange still need to be examined more thoroughly. See D. Rosati, "Foreign Trade Liberalization in the Transition to the Market Economy", WIIW, *Forschungsberichte*, Vienna, 1992.

According to another hypothesis, overshooting is inevitable because the supply schedule is vertical in the short run (supply is inelastic). Under this assumption, the initial equilibrium rate will be established at the intersection of the demand and short-term supply schedules and will only later gradually move down along the demand schedule to the point of intersection with the long-run supply schedule (W.M. Corden, op.cit.). But the assumption of short-run rigidity of supply is problematic under conditions of large-scale liberalization.

[353] An extreme example might be east Germany after the substantial *appreciation* of the GDR mark in 1990-1991.

[354] For instance, Poland liberalized some 90 per cent of non-agricultural prices in January 1990 (agricultural prices had already been decontrolled in August 1989). Price changes were even more sweeping in Czechoslovakia, where the share of centrally controlled-prices dropped from 86 per cent prior to the January 1991 liberalization to some 13-16 per cent thereafter, as measured by GDP value.

[355] In some countries, however, legal provisions have been maintained or reintroduced which allow governments to impose a wider range of price controls. In Bulgaria, for instance, Decree No.56 on Business Activity, passed in July 1992, empowers the Council of Ministers to set projected and minimum prices, maximum commercial mark-ups, threshold prices, etc. (BBC, *Summary of World Broadcasts*, EE W0241, 30 July 1992).

[356] In Russia, prices for basic foodstuffs were permitted to increase only by predetermined margins during January and February 1992; afterwards, decisions on imposing price controls on food products and other consumer goods were decentralized to local government and municipal levels.

[357] Romania avoided one-time liberalization of prices, opting for a gradual lifting of controls coupled with gradual elimination of subsidies and a build-up of social protection measures (BBC, *Summary of World Broadcasts*, EE 1371, 4 May 1992).

In the CIS republics, *fuel prices* still remain under government control. This currently takes the form of mandatory ceilings imposed on profit margins for oil, gas and coal producers. These controls have not prevented fuel prices from increasing sharply during 1992; in fact, they grew much faster then most other prices.

Russian fuel prices have been raised three times during the last year: first, a five- to sevenfold increase formed part of the January 1992 price liberalization package; second, as a result of a government decision on the regulation of prices of 18 May 1992; and a third time as a result of a Presidential decree of 17 September 1992 (table 3.4.9). In sum, oil prices increased 85 times over the twelve-month period between end-1991 and end-1992, i.e., much faster than consumer prices (26 times) or wholesale prices (over 30 times). Thus internal relative prices definitely improved, but fuel is still very cheap in terms of international prices in spite of these differentiated increases: one ton of oil costs the equivalent of some $12-15 at the current exchange rate. This obviously strongly distorts domestic-to-foreign price ratios and provides the wrong signals to domestic energy users.[358]

It should be noted that price liberalization has not in all cases involved the elimination of subsidies. While the process of curbing subsidies has progressed in all transition countries, it has nevertheless been slower and less comprehensive than price liberalization. Subsidies to domestic prices remain particularly widespread in Romania and in the CIS states.[359]

Incomes policies continued to be used in east European transition countries in pursuance of the twin objectives of fighting inflation and protecting living standards of large strata of the population. The principal instruments of these policies include minimum wage legislation, various indexation mechanisms and penalty taxes for excess wages.

In *Bulgaria*, the growth of nominal incomes was kept under strict control from the beginning of the stabilization programme in February 1991. In the first half of the year, targets for the nominal wage bill, nominal average wage and the minimum wage were set through a negotiation process within a trilateral commission including representatives of the government, trade unions and employers' organizations. In the second half of 1991, targets for the average wage were dropped to allow enterprises to negotiate individual wages more freely within the predetermined wage bills. The modification was intended to encourage enterprises to rationalize employment levels through shedding redundant labour. Enterprises which exceeded established wage limits were penalized through taxation at highly progressive rates, varying between 50 and 400 per cent if the wage bill exceeded the limit by more than 4 per cent.

The policy was rather successful in imposing strict wage discipline on enterprises in the first year. The aggregate wage bill remained well below the target, while revenues from the excess wage tax amounted to less than 1 per cent of all tax receipts. As a result, real wages (statistically measured) declined in 1991 by 42 per cent, the steepest fall in all transition countries. This sharp deterioration of living standards led to much tougher income and wage negotiations in 1992. Because of the largely divergent positions of the three parties involved, an economy-wide agreement could not be reached before July, when the National Council for Social Partnership (the tripartite body) agreed that the state sector wage growth limit for the whole of 1992 would be 26 per cent.[360] In fact, this ceiling was in many cases surpassed, and despite generally restrictive monetary policies, real wages increased in 1992 as compared with the end of 1991, although output and labour productivity contracted significantly.

Incomes policy in *Czechoslovakia* was also based on trilateral negotiations and economy-wide wage growth targets. In 1991 the limits were established on a quarterly basis, but they were subsequently revised upwards because of much higher inflation than anticipated. The limits were increased from 5-6 per cent in the first quarter to 32 per cent in the fourth quarter. As in Bulgaria, restrictive monetary policy and high penalty taxes prevented enterprises from paying wages above the limit, and the overall growth of nominal average wage in the whole 1991 was only 23 per cent — well below the inflation rate for the year (57.9 per cent).

In 1992, the minimum wage was increased by 10 per cent (to Kcs 2,200 per month), while the target for the average wage increase was set at 10-15 per cent. This was tantamount to a substantial relaxation of wage discipline because the annual inflation rate was expected to fall below 10 per cent. The growth of wages was in fact moderate throughout most of 1992, but accelerated sharply in the last quarter.[361]

By contrast with wages, pensions were not indexed (except in the first half of 1991 when the coefficient was 9-11 per cent). Instead they were periodically increased by a predetermined absolute amount (Kcs 110 in the second half of 1991 and Kcs 30-290 in 1992).

---

358 It certainly contributes to the excessive (and still rising) energy-intensity of production in Russia.

359 In Romania, the government's efforts to control inflation and to alleviate the impact of price liberalization on the living standards of the population resulted in continuous use of price subsidies. In the first half of 1992, price distortions were very large: the producer price of 1 kWh of electricity was 12.7 lei, but consumers paid only 0.65 lei. The corresponding prices for fuel bricks were 14,090 lei and 930 lei per ton; for butter — 125 and 25 lei per kg; for sugar — 96 and 32 lei per kg; and for bread 24 and 8 lei per 0.5 kg. The differences had to be covered by hefty subsidies coming from the strained state budget (BBC, *Summary of World Broadcasts*, EE 1371, 4 May 1992). In mid-1992 the Romanian government announced a programme for the gradual removal of these subsidies over the three-year period up to 1994 but, even with the large compensation grants planned for most of population, the programme was strongly opposed by labour unions. BBC, *Summary of World Broadcasts*, EE 1470, 27 August 1992.

360 *Demokratsiya*, 4 July 1992.

361 Preliminary estimates indicate that the average nominal wage in the Czech Republic in 1992 was about 20 per cent higher than in 1991, mostly a result of the surge in the last quarter of 1992 (*Rudé pravo*, 20 February 1993).

In *Hungary*, the system of tripartite wage negotiations was modelled on much earlier Austrian experience. According to the agreement reached for 1992 punitive taxes would be activated only if (net) nominal wages rose more than 23 per cent for the enterprise sector as a whole. The threshold was not surpassed in that year as the increase in wages was only 21-23 per cent. This outcome probably led the Ministry of Finance to suggest to scrap the central wage control system altogether in 1993.[362]

Poland had been the first transition country to apply wage-tax based incomes policies in 1990. But the system of setting wage limits was quite different from that applied later in Czechoslovakia, Bulgaria or Romania. No formal negotiations with labour unions were required; instead, the government linked the indexation coefficient to observed *ex-post* consumer price inflation and determined the coefficient values in a rather arbitrary manner. Low values during the initial period of the stabilization programme, which was characterized by very high corrective inflation, resulted in a sharp decline of real wages in the first quarter of 1990.[363] A catch-up increase in wages towards the end of 1990 allowed the decline to be reduced to some 25 per cent. Drastic sanctions for unauthorized wage hikes were imposed on enterprises in the form of punitively progressive excess-wage taxes to be paid out of net profits.

Modifications introduced in 1991 were intended to promote privatization and restructuring in the enterprise sector: private companies were exempted from the excess-wage tax, and those state enterprises which transformed themselves into joint-stock companies could obtain preferential tax rates. At the same time the wage growth target was changed from the total wage bill to the average wage for an individual enterprise – a move intended to work against much feared mass lay-offs.

The impact of incomes policy in 1991 was disappointing, however. Pressure from powerful workers' councils and labour unions resulted in wage increases much beyond the established limits (the indexation coefficient was fixed at 0.6 of the current inflation rate). In turn, high nominal wage increases not only led to a moderate growth in the real average wage in 1991 (by some 3 per cent), but also to a sharp deterioration in the financial position of enterprises because of high payments of excess-wage taxes. This experience appears to confirm the widely shared opinion that state enterprises tend to attach much higher value to marginal wage increases than to profits.[364]

Incomes policy rules were further relaxed in 1992, when excess-wage tax rates were reduced and exemptions extended for other categories of state enterprises. But the growth of wages was generally kept below the inflation rate throughout the year – chiefly because of the very strained financial position of enterprises. On the other hand, salaries in the budgetary sector and pensions – which were periodically adjusted to maintain a predetermined "parity" with wage increases – were held back in 1992 because of the mounting budget deficit.

Incomes policy in *Romania* in 1991-1992 rather closely resembled the system applied in Bulgaria and Czechoslovakia. The two main differences were that the results of annual government negotiations with the three groups of the nation-wide labour unions had to be translated into specific wage agreements at the enterprise level, and that indexation coefficients were based on an inflation indicator including both the past period and the four-month forecast. The indexation refers to the total wage fund for a firm, with individual wages to be negotiated separately. Penalty taxes have to be paid on wage increases exceeding established levels. The general policy orientation for 1992 was to protect 80 per cent of real wage levels compared with the previous year.

Some relaxation of incomes policy in Romania was observed in the second half of 1992, when some exceptions were permitted to allow additional wage increases. Specifically, the limit on the total wage bill could be extended if the company increased the number of employees or improved "efficiency", measured by profitability.[365]

No incomes policy of the type being used in the east European countries has been applied in the *former Soviet republics*. When the stabilization programme was under preparation in Russia in late 1991, various wage controls schemes were contemplated, but none was ever implemented.[366] In the opinion of Russian government officials, such policies would be neither feasible nor effective under the specific conditions of Russian industrial relations. Centralized limits on wage increases were considered impossible to implement because of the lack of social and political support from both the government and trade union organizations.[367]

---

[362] *Figyelő*, 17 December 1992, and *Heti Világgazdaság*, 22 December 1992.

[363] Indexation coefficients were fixed at 0.3 for January, and 0.2 for February-March 1990; but since the inflation rate in the first quarter was more than 130 per cent, it can easily be seen that real wages dropped by 44 per cent within three months.

[364] The rationale commonly given for the excess-wage tax was that in the absence of genuine owners of state enterprises no economic actor is interested in maximizing profits – hence punitive taxes on excess wages are needed to keep wage growth within limits. But if profits are indeed "irrelevant" to enterprise decision makers, the excess wage tax – which is paid out of profits after all – cannot be an effective deterrent. There is thus a fundamental fallacy underlying the scheme, as Polish experience confirms.

[365] These reliefs were legalized by the government resolution adopted in December 1992, pending trade union approval. BBC, *Summary of World Broadcasts*, EE/1571, 23 December 1992.

[366] In its "Memorandum on Economic Policy" published in March 1992, the Gaidar government announced that tax-based wage policies would be applied if wage increases were deemed excessive and were thus fuelling further inflation and contributing to unemployment ("Memorandum ob ekonomicheskoi politike Rossiiskoi Federatsii", *Ekonomika i zhizn'*, No.10, March 1992). But these measures were never invoked.

[367] Some western observers believe that the Russian government was politically too weak at that time to impose and enforce effective limits on wage increases in enterprises. M. Ellman, "Pochemu shokovaya terapiya w Rossii poterpela proval", *EKO*, No.9, 1992.

(vi) The failure of the Russian stabilization programme in 1992

(a) The Russian economy in 1992: from bad to worse

The failed coup in August 1991 triggered off a series of radical social and political changes in the then Soviet Union, two of which had important implications for future economic developments. First, the federal state itself disintegrated within a few months, giving birth to 15 independent states, with political ties between them substantially reduced. Second, the process of economic reform strongly accelerated after the fall of the conservative government and the collapse of communist party structures. The need for fundamental economic changes was indeed very urgent, as production levels were falling, inflation was on the rise, and market shortages were widespread.

Precipitate transfer of state powers from the all-union to republican governments was most visible in Russia, where a number of decisions were taken in late 1991 which prepared the ground for the introduction of an ambitious economic reform programme in 1992. The then new republican government, headed by Russian President B. Yeltsin and dominated by reform-minded, liberal economists led by Deputy Premier Ye. Gaidar, was determined to follow a radical variant of market reforms, with shock-type financial stabilization being its most conspicuous and most immediate component. In January 1992, after several weeks of intensive preparations, the Gaidar team launched a programme which at that time was considered to be the broadest and perhaps the boldest attempt ever undertaken aimed at economy-wide reconstruction.

Initially, the programme seemed to have been working: following the steep January rise in prices of some 250 per cent due to massive liberalization, inflation slowed down to 22 per cent in April, credit and money supply remained under strict control, and the budget deficit diminished sharply. The market situation improved and shortages began to disappear. One important symptom of the gradual regaining the macroeconomic control was the rapid eclipse of barter deals between enterprises which had spread in the second half of 1991. However, supply responses remained weak and restructuring in the enterprise sector was not forthcoming; instead, enterprises started to build up debts in trying to protect employment levels and wages. Under strong pressures from Parliament and powerful industrial lobbies, the Gaidar government had to agree to a softening of macroeconomic policies in early summer. Credit started to rise again, the rouble fell steeply against convertible currencies, and inflation accelerated again. Efforts to reimpose financial discipline in the autumn failed, the government's political position weakened further and macroeconomic control was eventually lost.

At the beginning of 1993 the Russian economy seems to be in much worse condition than a year earlier when the programme started. Output fell by a further 20 per cent in 1992, the level of consumer prices rose 26 times, and trade and payments within the rouble zone went into disarray. Sharp economic decline was accompanied by growing social and ethnic tensions, fed by rapidly deteriorating living standards and reduced job security. Political conflicts between the government, the Parliament and the President led eventually to the fall of Gaidar's government, which also appears to mark the end of his unsuccessful "shock" therapy.

Numerous reasons can be identified for the failure of the Gaidar strategy, but three of them seem to be of particular importance. First, given the specific conditions of the Russian economy, the initial reform programme was not conceptually adequate and was too narrowly focused on stabilization goals alone. Second, the government lacked the political support necessary to persevere with drastic and unpopular stabilization measures. Third, the programme has not been supported by an appropriate external assistance package. Other factors included the lack of coordination of the macroeconomic policies of Russia and the other former Soviet republics, and the lack of both institutions and management skills to carry out macroeconomic policies in a country of that size in profoundly changed conditions.

(b) The stabilization programme: targets, instruments and first results

The Gaidar programme assumed final and finished form only in mid-1992, long after its implementation had actually started. In fact, it was rather a series of plans which were constantly reformulated and updated as somewhat hasty responses to current developments, immediate needs, and the results of continuing negotiations with international financial institutions. The first draft, published in November 1991,[368] clearly stated that the main objective of transformation was to achieve financial stabilization and to establish a market economy. Next, key policy targets and measures were further elaborated and refined in collaboration with the IMF and independent international experts, leading to a more detailed and somewhat more realistic version of the programme published in February 1992.[369] This version also served as the basis for prolonged negotiations with the IMF and the G-7 group of industrialized countries concerning the forms, conditions and magnitude of a financial assistance package. In the course of these negotiations, and also in response to generally worse-than-anticipated performance of the Russian economy in the first half of 1992, the programme underwent further modifications before it was finalized and published in July 1992.[370]

---

[368] "Stabilizatsiya i vykhod iz krizisa. Osnovnye napravleniya ekonomicheskoi politiki pravitel'stva RSFSR", Moscow, November 1991. See also an interview with Deputy Prime Minister Ye. Gaidar in *Izvestiya*, 2 January 1992, p.2.

[369] See: "Memorandum ob ekonomicheskoi politike Rossiiskoi Federatsii", *Ekonomika i zhizn'*, No.10, March 1992.

[370] See *Rossiiskie vesti*, 11 July 1992.

The Gaidar government was clearly aware that without financial stabilization the market transformation cannot succeed. Hence, any serious attempt to stabilize the Russian economy had to aim at combating inflation as its top policy priority. On the other hand, price liberalization was regarded as the cornerstone of the whole reform – even though it was expected to result in a massive increase of the overall price level. Thus the government's intention was to carry out price liberalization right at the beginning of the programme, in a massive, shock-type operation. According to the plan, prices for 80 per cent of producer goods and 90 per cent of consumer goods were to be liberalized, with government price controls remaining only on a few primary commodities such as energy and fuels and on 14 basic consumer products.[371] While maintaining controls on prices for energy and fuels, the government nevertheless raised these prices sharply (5-7 times) on 1 January 1992 in an attempt to reduce substantially one of the most pervasive distortions characteristic of centrally planned systems.[372] This adjustment, as well as the general price liberalization, was expected to result in a wave of corrective inflation. Consequently, the key policy objective for 1992 was to reduce corrective inflation to single-digit monthly rates as early as possible, to keep it within limits of 350-580 per cent for the whole year and to stabilize prices thereafter. At the same time, the government planned to introduce limited convertibility of the rouble in order to increase competitive pressure from abroad on domestic producers and to accelerate the realignment of relative prices.

These stabilization goals were to be achieved through a combination of massive fiscal adjustment and restrictive monetary and credit policy. Budget expenditures were to be cut substantially, in particular those connected with subsidies, military spending and centralized investments. Budget revenues were expected to hold level – mainly due to the strengthening and broadening of the tax base through the imposition of VAT and taxes on energy exports. As a result, the consolidated budget deficit was supposed to come down to zero as early as the first quarter of 1992.[373] The stabilization plan was to be buttressed by a series of important institutional measures, including substantial liberalization of foreign trade operations, the start of the privatization of state property,[374] and deregulation of procurement and distribution systems.

With the benefit of hindsight, it can be observed that the implementation of the strategy went through three distinct stages in the course of 1992. In the first, the government broadly succeeded in imposing a restrictive policy. While the consumer price index, triggered by liberalization of some 80 per cent of all prices, jumped by 655 per cent between end-December 1991 and end-April 1992, total credit emission increased much less – from R439 billion to R1,015 billion, i.e., by only 131 per cent, and cash issue by only 91 per cent (from R157 billion to R299 billion). The drastic fall in the real supply of credit was obtained initially through overall credit limits (a 15 per cent ceiling on the increase of credit stock in the first quarter) and later through increases in the basic interest rate – from 2-9 per cent to 20 per cent in January, to 50 per cent in March, and to 80 per cent in May. Sharp cuts in government expenditures allowed the budget deficit to be reduced to R50 billion or some 3.5 per cent of GDP by the end of the first quarter. Although higher than planned, that deficit indicated an enormous improvement over the yawning gap registered in 1991.

In a normal market economy, a financial squeeze of that magnitude would immediately have driven many firms out of business and provoked large rises in unemployment. This was probably also the outcome expected by Gaidar and his team. In Russia, however, it quickly turned out that the bankruptcy laws and the disciplines of contract enforcement left much to be desired. In fact, they were weak or simply non-existent – hardly surprising for anyone acquainted with the results of the earlier Polish stabilization experiment of 1990-1991. Accordingly, the reaction of enterprises was profoundly different from what had been planned: liquidity shortage resulted in massive non-payments and rapidly increasing indebtedness of enterprises *vis-à-vis* each other as well as to the budget and commercial banks. Interenterprise arrears increased from R40 billion at the end of 1991 to R751 billion at the end of March 1992 – equivalent to more than twice the amount of credit emission by the banking system. Also, the restrictive money emission resulted in arrears in cash wage payments for employees (of R21.7 billion by the end of April – 8 per cent of total monthly incomes of the population).

The major miscalculation in the first stage of stabilization was that corrective inflation following price

---

[371] See: *Ekonomika i zhizn'*, No.18, May 1992 and *Izvestiya*, 25 November 1992. The magnitude of the price shock can be illustrated by the following average price increases for main groups of goods which occurred within January 1992 only: 4 times for food, 2.5 times for non-food products, 3.5 for all retail prices and 5 times for wholesale prices.

[372] The price of crude oil was increased from R75 to R354 per ton, the price of natural gas from R20 to R140 per 1,000 m3, and the price of hard coal from R14 to R78 per ton.

[373] In the February version of the programme, the consolidated budget deficit was increased to 1 per cent of GDP in the first quarter, or to 4.4 per cent if the net balance of external financing is included. It should be noted that available data on the Russian budget give rise to some confusion, both because of the 1991 settlements with the discontinued all-union budget and the different methodologies used for budget deficit calculations. For instance, the Russian Goskomstat methodology excludes the exchange rate differentials on foreign credits extended to the Russian government by G-7 countries on purchase of grain and some other foodstuffs at the end of 1991, which were transferred to Russian importing enterprises at the preferential exchange rate of R1.75 $. The exchange rate differential between the preferential rate and the special commercial rate of R55/$ may be considered as an implicit subsidy to imports financed by the federal budget. If this subsidy is taken into account, as is done by the IMF, the consolidated budget deficit in 1991 rises to 19 per cent of GDP.

[374] Privatization procedures were initiated by the presidential decree No.341 of 29 December 1991. Initial plans for 1992 envisaged privatization of 50 per cent of the enterprises and organizations in the construction materials sector, wholesale and public catering, 60 per cent in the food industry, agriculture and retail trade enterprises, and 70 per cent of enterprises in light industry, construction, motor vehicle transport and maintenance work (see "Memorandum ob ekonomicheskoi politike ...", op.cit.). According to preliminary reports, these ambitious goals for 1992 were not reached, although in some areas, such as wholesale trade, progress has been substantial (see Russian Federation Goskomstat, *O razvitii ekonomicheskikh reform v Rossikoi Federatsii*, Moscow, 1993, p.20-21).

decontrol had been seriously underestimated: in January 1992 alone the CPI rose by 245 per cent as compared with December 1991, and the planned upper limit for annual inflation (500 per cent) was already surpassed in March (618 per cent).[375] Two factors seem to have been primarily responsible for this price explosion, which was much more violent than in other transition countries: the high degree of specialization and monopolization of production in the Russian economy and the relative weakness of import competition under restricted convertibility and a depreciated market exchange rate. Faced by a credit squeeze and supply problems, Russian enterprises behaved like typical monopolies: they raised prices and reduced output. The only possible source of short-term substitution of domestic supplies was imports from abroad; but limited access to convertible currencies, unstable rules on foreign trade licensing and import protection, as well as the sharp currency depreciation at the outset of the programme, made this possibility rather illusory, at least for most consumer goods and services. The depreciation gave effective protection for domestic monopolies and was at the same time a powerful inflationary factor, as domestic prices of tradables were aligned with foreign prices converted into roubles at the market rate. A similar type of monopolistic price behaviour was observed earlier in Poland and in Czechoslovakia, but the import reaction had been nevertheless much faster and stronger than in Russia, mainly because the two central European economies are smaller and were much more open to international markets.[376]

High inflation was probably to some extent also fuelled by the Russian surplus in trade with other CIS republics. Lack of regulations to limit technical credits in bilateral trade, coupled with practically unrestricted possibilities for other republics' central banks to issue credit to local banks and enterprises, resulted in large Russian trade surpluses, which were financed through an accumulation of claims on the correspondent accounts held by other republics' central banks in the Russian central bank. The share of interstate payments in the monetary base of the CBR increased from nil at the end of 1991 to 13.6 per cent by the end of the first quarter and to 29.2 per cent by the end of the first half of 1992 (table 3.4.11). This permitted additional credits to be extended to commercial banks, thus enabling them to finance Russian exporters.

The higher-than-expected jump in prices led to a much sharper decline in the real credit and money supply, and liquidity problems became widespread. If only a few firms come under financial strains they can be closed down without a major impact on the rest of the economy; also, the threat of bankruptcy will appear to be real and evoke genuine restructuring efforts. But if non-payments accrue on a massive scale and affect hundreds of state enterprises, immediate bankruptcies become socially and economically problematic. The possibility of closing down one fourth or one third of the state sector may not be taken very seriously by enterprises. This is probably what happened in Russia. The government's failure early in 1992 to eliminate loss-making firms quickly taught them that non-payments do not trigger immediate bankruptcy procedures, and this permitted the rapid spread of interenterprise credits.[377] A problem which first appears to be microeconomic was suddenly transformed because of its magnitude into a macroeconomic one. The result was a fatal loss of credibility of the whole programme.

(c) Second stage: from restrictions to financial accommodation in mid-1992

The massive scale of payments arrears and cash shortages translated into political pressure on the government to soften macroeconomic policy. The shortage of cash roubles — a phenomenon which can only partly be explained by "technical" problems in printing new banknotes[378] — was particularly damaging for the government's political position. Suspension of the payment of wages in many enterprises — in a situation where wages were anyway falling by 30 per cent in real terms — was commonly interpreted as a symptom of the indolence and inability of the authorities, government and central bank alike. This provoked strong protests by workers' organizations and allowed the stepping up of attacks against government policy. The representatives of both enterprises and various political forces criticized the Gaidar programme on the grounds that "shock" therapy under Russian conditions would inevitably lead to massive bankruptcies and mass unemployment on a socially unacceptable scale. The government, which never commanded a majority in Parliament and lacked its own strong political base, was quickly forced into an uneasy dilemma: either to comply or to resign. The room for manoeuvre became even more limited after the Central Bank of Russia — a close government ally in the initial stage — came under strong pressure to loosen monetary policy in May.

The relaxation took two general directions, although it does not seem to have been the result of any deliberate plan. First, the fiscal authorities began to tolerate increasing arrears in tax payments — a move which effectively amounted to granting subsidies. Revenues from VAT and from income tax had been in-

---

[375] There is ample evidence suggesting that the programme's authors had reckoned with a much lower price reaction. In an interview in 1991, Ye. Gaidar had predicted a doubling of prices following liberalization (*Moskovskye Novosti*, 10 November 1991). Equally unrealistic were the hopes of a rapid restoration of market equilibrium expressed by Russian government officials (e.g., K. Kagalovski expected market shortages "to disappear within a week", see *Moskovskie Novosti*, 5 January 1992).

[376] As a result, market shortages disappeared very quickly in Poland after the implementation of the programme, while they proved much more pervasive in Russia. Even in mid-1992, the market situation in many regions in Russia (except Moscow) was still characterized by endemic shortages in spite of price liberalization (see L. Szamuely, "Makroökonómiai stabilizációs kísérlet Oroszországban, avagy a sokkterápia újabb kalandja", *Külgazdaság*, No.10, 1992, p.14).

[377] The bankruptcy law was passed in November 1992, but with effect from 1 March 1993 only.

[378] Many experts interpret the cash shortage as an informal instrument used by the CBR to curb the inflationary pressures on the domestic market. See KOPINT-DATORG, *Economic Trends in Eastern Europe*, Vol.1, No.3, 1992.

TABLE 3.4.10

Russian state budget revenues in January-September 1992
(Billion roubles)

Revenues	QI	QII	QIII
Revenue			
Direct taxes [a]	130.6	354.8	509.4
Profit tax	98.0	289.3	399.8
Household income tax	32.4	61.1	109.6
Indirect taxes	155.7	256.4	621.3
Excise and turnover taxes	23.0	28.3	47.6
VAT sales taxes	106.6	192.3	454.0
Foreign activity taxes	26.0	35.8	119.6
Non-tax revenues	78.6	64.9	134.1
Tax on natural resources	6.6	8.3	27.8
Privatization revenues	0.7	2.6	15.6
Others	71.5	58.4	90.8
Total revenues	365.0	676.1	1 264.8
Expenditure			
National economy	105.8	281.1	552.4
Administration	6.3	14.9	28.7
Education, health, etc.	90.9	222.6	382.6
Defence	56.5	158.8	238.0
Foreign trade activity	15.9	10.5	89.7
Other	69.7	224.1	303.1
Total expenditure	345.0	912.0	1 594.5
Balance (excluding transfers)	19.9	-235.9	-329.8
GDP	1 402.0	2 235.9	4 612.1
Balance as per cent of GDP	1.42	-10.55	-7.15

Source: Russian Economic Trends, Vol.1, No.3, Table A16, p.80.

Note: Table includes all consolidated central and local expenditures and revenues. The numbers may not sum exactly because of rounding.

[a] Including property taxes.

creasing between January and April, but declined in May, while other revenues remained broadly unchanged despite continued inflation. A comprehensive picture of government finances in 1992 cannot be presented through lack of the necessary data, but available evidence indicates a significant deterioration in the second quarter. The faltering discipline of tax collection may to some extent be illustrated by the slow growth of state budget revenues in the first half of 1992 (table 3.4.10). On the other hand, expenditures were rising steadily. As a result, the federal budget deficit increased from R50 billion in 1991 to R301 billion, i.e., to 7.5 per cent of GDP during the second quarter alone.

The second area where the policy changed was credit emission and money supply. The relaxation of credit came first in response to the widening budget deficit in May, but it was soon seconded and strongly exacerbated by the expansion of central bank credits to commercial banks. This was directly connected with the uncontrolled increase of interenterprise indebtedness, which grew fourfold, to R3,100 billion, in the second quarter of 1992. In turn, this triggered considerably smaller, but politically much more explosive arrears in wage payments, which grew from R21.7 billion to R65.1 billion between 1 May and 1 July, i.e., to roughly one fifth of all incomes of the population in June.

The mechanism of accelerated monetary expansion in the first half of 1992 can best be illustrated by changes in the assets of the Russian central bank (table 3.4.11).[379] This shows how the significance of the main components of the monetary base changed over time. In the first quarter of 1992 the main source of credit emission was lending to commercial banks, and also, albeit to a lesser extent, credits to other republics and revaluation of international reserves following the sharp devaluation of the rouble in January 1992. But in the second quarter, credit to the government linked with the growing federal budget deficit became the key source of monetary expansion. The relative slow-down in commercial credit expansion in April and May was probably linked with the increase in the interest rate on centralized credits by the Central Bank of Russia from 20 to 50 per cent on 1 April.

The more fundamental shift in the monetary policy came, however, in June, and was a direct reaction to important policy decisions taken in May. First, fuel and energy prices were sharply increased by presidential decree in an attempt to correct relative prices, which had again become deeply distorted after the January-April inflation wave.[380] Second, the CBR decided at the end of May to raise the basic interest rate from 50 to 80 per cent in an effort to curb the growth of credit. Both decisions had a knock-on effect on enterprises' financial position. The increase of energy prices raised their need for working capital and short-term credits, but at the same time higher interest rates made this credit more expensive even though they were still negative in real terms.

The coincidence of these two decisions, probably unintentional, spurred a new wave of criticism against government policy and — this time — also against the central bank. In response, the CBR chairman, Mr. Matyukhin, announced his resignation in an open protest against mounting pressures from Parliament for a relaxation of financial policies, and specifically for revoking the CBR's earlier decision to increase the interest rate.[381] Initially, the resignation was rejected by the Parliament, but in view of the strong opposition to the Gaidar programme, a shift in favour of more interventionist policies was nevertheless considered desirable. Mr. Matyukhin stepped down several weeks later and was replaced by Mr. Gerashchenko, who had been known as an advocate of much more active lending policies for indebted enterprises. The change was accompanied by a distinct, albeit implicit, shift in the central bank's policy priorities from protection of the domestic currency to the protection of output and employment in the enterprise sector.

---

[379] The consolidated balance sheets of the Russian banking system was not available at the time of writing.

[380] The May adjustment in energy prices was of a similar magnitude as in January: oil prices went up to R1,940 per ton (5.5 times), natural gas prices up to R865 per 1,000 cu m (6.2 times) and coal prices up to R201 per ton (2.6 times).

[381] See *Rossiya*, Moscow, No.22, 27 May-2 June 1992.

TABLE 3.4.11

Changes in the monetary base in Russia in January-June 1992
(Per cent)

Asset	Structure [a]				Monthly rate of increase [b]		
	December 1991	March 1992	June 1992	September 1992	QI	QII	QIII
Credits to the government	65.5	48.5	41.4	35.8	6.3	23.6	44.1
Loans to commercial banks	29.0	41.8	35.5	32.1	32.8	23.3	30.6
Interstate credits	-	5.2	18.3	29.9	132.0	56.2	55.4
Foreign currency and gold	-	1.3	2.5	..	3 000	74.3	..

Source: Compiled from *The National Economy of Russia in January-June 1992*, The State Committee on Statistics of Russia, Moscow, 1992.

a End of period.
b In nominal terms.

TABLE 3.4.12

Monetary survey of the Russian Federation, selected items, 1992
(Billion roubles, end of period, unless otherwise stated)

	Stock of credit	Emission of cash money [a]	Stock of cash	Broad money (M2) [b]	Foreign currency deposits [c]
December 1991	439	-	167.0	936	203
January 1992	517	16.9	183.9	1 054	241
February 1992	697	23.4	207.3	1 179	279
March 1992	918	37.6	244.9	1 345	319
April 1992	1 025	64.3	309.2	1 481	376
May 1992	1 054	45.3	354.5	1 617	369
June 1992	1 393	87.2	441.7	2 069	-
July 1992	2 300	184.7	626.4	2 569	-
August 1992	2 100	182.0	808.4	3 422	-
September 1992	2 800	164.9	973.3	4 515	-
October 1992	3 700	191.8	1 165.1	5 722	-
November 1992	4 490	228.7	1 393.8	..	-
December 1992 [d]	5 100	286.0	1 679.8	..	-

Source: Compiled from *World Bank Country Survey: Russia*, The World Bank, Washington, D.C., 1992, and Russian Federation Goskomstat, *O razvitii ekonomicheskikh reform v Rossiyskoi Federacyi*, Moscow, 1993.

a Monthly emission.
b Excluding foreign currency deposits.
c Converted into roubles at the current CBR rate.
d Preliminary estimates.

Already in June 1992, CBR loans to commercial banks increased sharply, resulting in a jump in the total credits to the economy from R1,054 billion at end-May to R1,393 billion at end-June, i.e., by 32 per cent. But the real monetary explosion came in July (table 3.4.12). The overall credit stock increased by 65 per cent within one month, while cash emission increased by 42 per cent.[382]

The July monetary expansion provided some relief to the enterprise sector. Payments arrears fell by more than half, to R1.020 billion, over July-August, as enterprises used new bank credits to pay off outstanding liabilities.[383] Arrears in wage payments also dropped by two thirds to R21.5 billion by end-September. Inflation seemed to be under control, falling to its lowest monthly level of 9 per cent in August.

But these gains proved to be very short-term, especially as monetary policy also softened in other domains. Along with credits extended to enterprises to repay their debts, the central bank issued large amounts of concessionary credits to the agricultural sector (R750 billion) and to finance deliveries of goods to the northernmost territories (raiony Krainego Severa – R234 billion). The total amount of concessionary credits pumped into the economy during the third quarter reached R1.5,000 billion. This, in turn, provoked a sharp deterioration in the budget, which was responsible for financing the cost of interest rate differentials for concessionary credits. By the end of August, the consolidated budget deficit had increased to R666 billion (10.7 per cent of GDP) and the federal budget deficit to R820 billion (13.2 per cent of GDP).[384]

The relaxation of financial policies did not have any positive impact on production either: the fall of indus-

---

[382] At the same time, the CBR acting chairman, V. Gerashchenko, declared that the credit limit of R700 billion agreed upon with the IMF for the second half of 1992 was unrealistic and needed to be renegotiated.

[383] *Moscow News*, No.39, 1992.

[384] Russian Federation Goskomstat, *O razvitii...*, 1993, op.cit., p.11.

trial output in fact accelerated in January-September to 17.6 per cent relative to the corresponding period of 1991 (as against 13.5 per cent in January-June). The increase in the money supply did not translate into an expansion of economic activity; rather, enterprises used the newly-available funds partly to repay debts and partly to hedge against inflation. The latter motive was reflected in a strong fall in the auction rate of the rouble from R130/$ in mid-July to R205/$ at the end of August. Clearly, the most obvious effect of the widely-publicized change in credit policy of the CBR was to strengthen inflationary expectations and encourage speculative behaviour by enterprises and banks.

The depreciation of the rouble in August was also an outcome of the new foreign exchange rate policy. In pursuance of the earlier-stated goal of unification of exchange rates, the government and the CBR announced a shift to a single rate on 1 July. Simultaneously — though not announced at the time — the CBR reduced its intervention in foreign exchange auctions.[385] Cutting the supply of dollars and at the same time inflating the supply of roubles could have only one effect: a precipitate fall of the rouble.[386]

Depreciation of the rouble and credit expansion could not leave inflation rates unaffected. Inflation accelerated in September to 12 per cent per month, and then nearly doubled in October (23 per cent), reflecting an increase of food and agricultural prices and — more importantly — a new round of increases in fuel and energy prices in September. That new acceleration of inflation brought about another wave of enterprise indebtedness and non-payments and subsequently, a new wave of bank credit expansion. Arrears increased again to R2.1 at the end of October and the stock of credit jumped from R2,100 to R3,700 thousand billion, i.e., by 76 per cent during September-October alone. The attempt to solve the payments crisis through large-scale credit expansion clearly had failed: the only tangible result was runaway inflation, free fall of the rouble and continuous recession in the production sector.

(d) Third stage: loss of macroeconomic controls

Since the last quarter of 1992 the Russian economy has been in state of flux. A victim of continuous and intensified political struggles, the government had less and less control over macroeconomic policy. There were some attempts in the autumn to restore financial discipline, reflected in increased efficiency of tax collection, centralization of budget revenues, and further cuts in expenditures. As a result, the budget situation improved to some extent in October-November; but none the less, according to most recent estimates, the deficit of federal budget for the whole 1992 reached R958 billion, i.e., 6.4 per cent of GDP.[387]

Much more worrying has been the unceasing emission of CBR credit to banks, which has assumed a largely automatic character: Russian banks have been unable to screen their clients in order to allocate credits in a selective way to assure more efficient and less risky use of funds. The lion's share of all credits have been tied up in financing the working capital requirements of enterprises. The share of short-term credits in total credit stabilized in the second half of 1992 at some 95 per cent. Such an unusually high ratio can be explained not only by the reluctance of banks to provide funds for long-term project financing, which is quite natural under high inflation and strongly negative real interest rates, but may also indicate that Russian banks are unable to resist their clients when they come with requests for emergency and ad hoc financing.[388]

The lack of control over credit emission was reinforced by the policy of fixed and negative real interest rates, which failed to respond adequately to the rise in inflation. The CBR "central" rate remained at 80 per cent since end-May, but the lending rate of major commercial banks varied around 90-120 per cent — much below the inflation rate which in the last quarter was running at 900-1,000 per cent annually — while the rate on sight deposits remained at 20 per cent. Negative real interest rates helped debtors to survive, but penalized savers and destroyed financial discipline. The result has been loss of control over inflation, which in January 1993 was already running at a monthly rate of 27 per cent, and another steep fall of the rouble exchange rate from R410-415/$ in December 1992 to R650/$ at the beginning of March 1993.

The fall of the Gaidar administration in early December, despite strong support from the President, also marks the end of "shock" therapy in Russia, at least in its initial, perhaps "romantic" form. The new line of macroeconomic strategy which seems to have been emerging after the December session of the Congress of People's Deputies puts much more emphasis on gradualism and interventionism.

---

[385] As announced recently by the head of international operations in the CBR, A. Potemkin, the central bank's intervention on the currency market accounted for approximately 30 per cent of the supply in the second half of 1992, down from some 60 per cent during the first half (*Interfax News Agency*, Moscow, 4 February 1993, quoted from BBC, *Summary of World Broadcasts*, SU/1606, 6 February 1993).

[386] *Moscow News*, No.39, 1992. The monthly volume of currency sold at Moscow Inter-Bank Currency Auction (MIBCE) fell from $309 million in June to $253 million in July and $261 in August, before rising again to $453 million in September. The total volume of sales in 1992 was $2,862 million, of which $1,055 million, or 37 per cent, came from direct interventions of the CBR (*Izvestiya*, 6 February, 1993).

[387] Russia's GDP for 1992 is estimated at R14.9,000 billion (Russian Federation Goskomstat, *O razvitii* ..., 1993, loc.cit. and *Neue Zürcher Zeitung*, 12 February, 1993. But some estimates put the deficit much higher. According to A. Khandurov, Deputy Chairman of CBR, the actual federal deficit was R1,167 billion, i.e., 7.8 per cent of GDP; but if all credits extended to the Ministry of Finance are included, the deficit would rise to R1,971 billion, or 17 per cent of GDP. *Rossiiskie vesti*, 2 March 1993, p.3.

[388] The weaknesses bank policies are particularly visible when it comes to wage payments: in December alone, the average wage grew by 23 per cent, to more than R13,000, for the first time in 1992 almost matching the rate of price increase (see Russian Federation Goskomstat, *O razvitii* ... , 1993, op.cit., p.45).

(e) Lessons from the Russian experience: what went wrong?

At first glance, the derailed Gaidar programme may be regarded as yet another example of an unsuccessful attempt to stabilize a high-inflation economy characterized by structural rigidities. But the Russian transformation cannot be compared with any other reform attempt in post-war history: the size of the country, the heritage of the past, the scope of change and the role played by social and political developments are unprecedented. With the success of stabilization policies depending on so many different factors, most of them remaining largely beyond government control, the decision by the Gaidar team to opt for a radical approach was courageous, but it was also very risky, if not altogether unrealistic.

As mentioned already, the three principal reasons for the failure of Russian stabilization policy are the deficiencies of the programme itself, inadequate political support and lack of external assistance. As regards the first issue, the key miscalculation was the assumption that standard macroeconomic policies would work efficiently in conditions of fundamental structural imbalances in the post-communist Russian economy. In fact, the Gaidar programme was probably more "orthodox" than any other stabilization programme applied in eastern Europe, while the Russian economy was definitely the least suitable of all the transition economies for such a programme.

Its underlying idea was to curb inflation through a combination of restrictive monetary and fiscal policies. It was decided not to use "heterodox" elements such as incomes policy or extensive price controls. While the decision on price liberalization is fully understandable (as there is no other possibility of restoring market balance in conditions of such deeply distorted prices), the rationale for rejecting some form of centralized wage control is not clear. The argument which was commonly given at the time was that such controls would be neither economically effective nor politically acceptable under Russian conditions.[389] While it is probably true that it might have been difficult to impose in practice, this difficulty does not justify the lack of any attempt by the government to establish some measure of wage control.[390]

It should be noted, however, that neither incomes nor wages in fact grew excessively in 1992; the two accelerations which took place coincided with the relaxation of credit policies in June-July and in November-December. For the year as a whole, wage increases lagged behind price increases and led to a substantial fall in real average wages and also in real incomes of some 40 per cent. Thus, it seems that in the specific Russian environment the lack of an incomes policy was not of primary importance for stabilizing the economy, though it could have served as a useful additional instrument,[391] especially if it limited wage increases in loss-making enterprises.

The second deficiency, which appears to be more important, is that the Russian stabilization started with a floating exchange rate and with rather limited scope for rouble convertibility. In doing so, the government opted for a different road to convertibility than in Poland or in Czechoslovakia. It was taken for granted that a more general convertibility, combined with a fixed-rate regime in highly uncertain conditions and with very modest international reserves at the outset of the programme, would be a risky solution which could end in the depletion of reserves and suspension of convertibility. That judgement was probably wrong. Indeed, although a fixed exchange rate may be difficult to defend under current-account convertibility, the Polish experience suggests that it is viable if the rate is depreciated enough and the programme is credible. In any case, once selected, the option for a floating rate regime should have been accompanied by a much larger degree of internal convertibility. Import competition was the only immediate possibility of imposing a check on corrective inflation under monopolized structures; and this possibility was not exploited because of the very limited access to foreign exchange for potential importers and the resulting very high dollar rate.[392]

The limited convertibility concept was based on convertible currency auctions organized first once, and, in the second half of 1992, twice a week in Moscow for commercial banks. The number of participating banks was strictly limited through a licensing procedure, and the amount of hard currency sold was very modest; as a result, the currency auction operated at the margin of the otherwise restricted foreign exchange market, with all its characteristic distortions.[393] The amount of dollars sold in individual sessions increased gradually from an average of $5-10 million in January-February 1992 to $30-60 million in the last quarter; nevertheless

---

[389] See the interview with Ye. Gaidar in *Rossiiskaya gazeta*, 11 January 1992, p.3, and M. Ellman, "Shock Therapy in Russia: Failure or Partial Success?", *RFE/RL Research Report*, Vol.1, No.34, 28 August 1992.

[390] In the memorandum on economic policy presented to the IMF in February 1992, the government admitted the possibility of imposing strict wage controls in the form of punitive taxes on wage growth exceeding predetermined levels (see: "Memorandum ob ekonomicheskoi politike Rossiiskoi Federatsii", *Ekonomika i zhizn'*, No.10, March 1992). But these plans remained paper only.

[391] In view of weaknesses in the mechanism of centralized wage negotiations, it seems that it would not be advisable in Russia to link the growth of wages to the current inflation rate, a solution applied in Poland, but rather to link wages with the effective profits of enterprises.

[392] In addition, imports in January were hampered by lack of regulations on access to foreign trade transactions. Only on 29 January did presidential decree No.65 establish general freedom for such activities unless they are specifically forbidden by law. Had this liberalization come earlier, the price jump in January could have been substantially lower. For a critical assessment of Russian exchange rate policy in the first stage of stabilization, see D.M. Nuti, J. Pisani-Ferry, "Post-Soviet Issues: Stabilization, Trade and Money", Commission of the European Communities, *Economic Papers*, No.93, May 1993.

[393] One of the implications of segmentation of the foreign exchange market was the multiplicity of exchange rates, which varied in the first half of 1992 from R5.4/$ for centralized imports, R55/$ for the compulsory surrender of 40 per cent of export revenues and R230/$ for decentralized imports financed through currency auctions. The exchange rates were only unified in July 1992. See L.S. Goldberg, "Foreign Exchange Markets in Russia: Understanding the Reforms", IMF Paper on Policy Analysis and Assessment, International Monetary Fund, January 1993.

the sales covered less than 10 per cent of the total value of Russian imports in 1992 (table 3.4.13).

The limited amounts of currency sold at auctions inevitably resulted in high dollar rates, much above purchasing power parities. Currency was used for lucrative imports of a selected range of up-market consumer goods such as electronics or cars, or simply for hoarding against inflation, but the auction rate was commonly used as reference for economy-wide pricing decisions. Thus the adopted system not only effectively protected large segments of domestic production against foreign competition but it was also highly inflationary because of the impact of the auction rate on production costs.

The third deficiency of the Russian programme was that, in contrast to the Polish stabilization package, interest rates in Russia have not been raised to levels which would reverse inflationary expectations and encourage financial assets to be kept in domestic currency deposits rather than converting them into goods and foreign exchange. Only at the end of May was the central bank rate raised to 80 per cent per annum. But that move was not only insufficient to match expected inflation, but it was also hopelessly delayed and politically not credible. Moreover, many reductions were granted on so-called preferential credits. In sum, the May increase does not seem to have significantly lowered enterprise demand for new credit.

Lack of wage controls and a floating rate deprived the Russian programme of *the nominal anchors* which, as demonstrated by the Polish and Czechoslovak cases, are of crucial importance in the initial stage of stabilization. In such a situation, the main burden of stopping inflation fell on monetary policy. But that alone cannot bring about stabilization even in economies much less affected by structural imbalances than Russia — especially when interest rates are low and fiscal policy is lax. Hence, as at the very beginning, restrictive monetary policies subsequently had to be softened if not completely abandoned.

The reaction of Russian enterprises also confirms and amplifies one of the conclusions obtained from the earlier Polish experience: a drastic cut in credit emission is a necessary but not a sufficient condition for the "hard" budget constraint to work. It is equally important to restrict access to other possible sources of short-term financing, such as non-payments of taxes and debts, and to impose efficient and strict rules and procedures of "exit" for loss-making enterprises. Unfortunately, the Russian monetary shock was not accompanied by these collaterals, and the hard budget constraint thus did not materialize. In fact, the financial position of Russian enterprises has only marginally deteriorated during 1992, and generally remained much better than in other countries.[394] The proportion of loss-making firms increased from 14 per cent in end April to some 20 per cent by end-December. Meanwhile the profitability ratio in industry fell from 65 per cent in the first quarter to 47 per cent in the second quarter and to 34 per cent in the third quarter — figures which suggest only a moderate financial crunch.[395]

It is sometimes argued that the Russian programme should not be called "shock" therapy because such crucial elements as wage controls, real positive interest rates and internal convertibility were missing. In this context it is often critically compared with the Polish "shock" therapy of 1990.[396] While it is true that the success of Polish stabilization hinged crucially on convertibility and high interest rates, it cannot be denied that in some areas — such as the money and credit supply — the adjustment in Russia was indeed of a "shock" character. But because the monetary shock was not accompanied by other measures necessary for tightening the effective budget constraint for enterprises, its effect degenerated into payments arrears and monopolistic pricing.

The absence of nominal anchors has been another important reason for the low credibility of the programme. Many observers have pointed out that the Gaidar programme had only a slim chance of success in the absence of incomes controls, and that corrective inflation was likely to set in motion strong indexation mechanisms which would generate a further inflation-

TABLE 3.4.13

Results of convertible currency auctions in Russia in 1992 [a]

	Rate (Rouble/dollar)		Volume of currency sold
	Market [b]	Central Bank	(Million dollars)
January	230	110	18.4
February	186	102	33.3
March	152	92	98.9
April	154	100	46.9
May	122	94	100.4
June	128	89	308.9
July	143	142	253.0
August	168	168	261.3
September	220	225	453.1
October	354	353	407.9
November	426	427	353.3
December	415	415	452.7
Total for 1992	–	–	2 788.1
Memorandum item:			
Value of imports in 1992			35 000.0

Source: Ekonomika i zhizn', various issues, and national statistics.

[a] Monthly results of sales at the Moscow Inter-Bank Currency Exchange. If sales on the Saint Petersburg Currency Exchange are added, the total amount of currency sold increases to $2,868 million.
[b] Weighted average rate.

---

[394] For instance, average gross profitability of enterprises in Poland (measured by the ratio of aggregate before-tax profits to total sales revenues) fell from 23.1 per cent in 1990 to 6.9 per cent in 1991 and to 3.1 per cent in 1992 (January-November). The share of loss-making units in Poland increased from 23 per cent in 1990 to 43 per cent in 1992.

[395] Russian Federation Goskomstat, *O razvitii ...*, 1993, loc.cit., pp.12-13.

[396] See e.g., M. Dabrowski, "Shokovoi terapii v Rossii ne bylo i net", *Finansovye Izvestiya*, 17 December 1992.

ary spiral.[397] But perhaps the most important factor which vitiated the credibility of the programme was the unstable political position of the reformist government. The group of young economists around Gaidar did not have any strong political party rallying behind them; on the contrary, the stance of the main political forces represented in the Russian Parliament towards "shock" therapy varied from skeptical criticism to total opposition.

The criticisms of the Gaidar programme generally took three directions. Probably the most outspoken were concerns about the social costs of the radical shock therapy, mostly voiced by leftist and post-communist groups. Representatives of the powerful industrial complex pointed to the risks of collapse of entire sectors of manufacturing suddenly exposed to macroeconomic restrictions and new relative prices. Some even suspected that one of the purposes of the programme was to destroy the high-technology sectors of the Russian economy and to force unilateral disarmament of Russia's military forces.[398] Finally, a limited group of experts and policy-makers expressed concern over the internal inconsistency of the programme and the lack of any strong institutional support for macroeconomic stabilization measures.

In sum, unlike in other transition countries, where stabilization programmes were implemented under conditions of general political support and even popular enthusiasm, in Russia such a climate was clearly missing from the very beginning. The Gaidar team came under constant attack from both conservative and moderate political factions, and the lack of a workable majority coalition in Parliament undermined the position of the government and cast serious doubts on the sustainability of the programme (and of the government itself) in the medium term.

One important implication of the lack of political consensus was the disparate policy courses followed by the government and the central bank since early summer 1992. While the government was making efforts to impose financial discipline on enterprises in order to force them into restructuring, the central bank offered them more credits. Clearly, not only the policy objectives of the two institutions were contradictory, but also their roles seem to have been curiously misplaced: the government pursued the goal of financial stabilization, while the central bank was concerned with output and employment stabilization.

Probably the last real chance for saving the Gaidar programme was lost in the late spring 1992, when the Russian authorities and the IMF failed to agree on the terms and conditions of foreign financial support. The negotiations with the IMF were well under way already in late 1991, but many unsettled issues, mainly pertaining to the disintegration of the Soviet Union and monetary and payments arrangements among the newly-emerged independent states, but also to differences about the design of the stabilization programme, prevented the conclusion of an agreement by the two sides. Consequently, the negotiations with G-7 group on the rescheduling of Russian foreign debt and the extension of new credits were also allowed to run down. In fact, as pointed out by observers, the Russian government decided to go ahead with the programme in January 1992 without any clear commitment from the IMF to provide external support.[399]

This was one of the key considerations behind the decision by the Russian government to introduce only limited convertibility and to start with a floating rouble rate − a transitory solution planned initially to last for a short period only. It was expected that the rapid arrival of western assistance would allow a more stable system of convertibility. This, however, has never materialized. Dramatic appeals by Russian leaders for financial assistance[400] were met with sympathy and friendly declarations, but funds were not forthcoming. Alarmed by rapidly deteriorating economic performance of the Russian economy after several months of implementation of the programme, the IMF decided in June to release the first tranche ($1 billion) of what was planned to be a multi-billion assistance package even though no formal agreement had been concluded. But that was far below the needs of the strained Russian balance of payments; besides, even this limited amount has not been utilized in view of uncertainties surrounding the economic situation in Russia in mid-1992. While it is risky to speculate, it seems that if the programme had been supported from the very beginning by a currency stabilization fund of some $3-4 billion, it would have then been possible to introduce a much larger degree of convertibility − possibly based on a 100 per cent surrender requirement for exporters, and to maintain a less depreciated and more stable rouble rate for a longer time. This move − if combined with realistic interest rates on rouble deposits and credits − could have greatly helped to weaken inflationary expectations and restore relative price stability in the summer of 1992.

The stabilization fund − if not to be wasted − would of course have had to be supplemented by other indispensable institutional and policy measures missing in the initial programme. Among the most important ones are a balanced budget, incomes policies to secure a systematic lag between the growth of incomes and inflation, and an enterprise reform aimed at the effective restructuring of loss-making units. These measures will have to be undertaken in 1993 if the new Russian government is to succeed in its fight against inflation and economic chaos.

---

[397] See e.g., M. Dabrowski, "Pierwsze polrocze przeobrazen w Rosji", *Rzeczpospolita*, 30 June 1992.

[398] See e.g., *EEM Moscow Bulletin*, 24 July 1992.

[399] M. Ellman, "Shock Therapy in Russia...", op.cit.

[400] See e.g., Ye. Gaidar, "Russia needs three kinds of economic aid − and quickly", *Financial Times*, 22 January 1992; G. Burbulis, "Help Russia to Reform, and Come to Make and Sell", *The International Herald Tribune*, 23 January 1992; Ye. Gaidar, "If the west wants to help Russia, now is the time", *Financial Times*, 4 March 1992.

(vii) The break-up of the rouble zone

Immediately after declaring their independence, most of the ex-Soviet republics announced plans to introduce their own national currencies. The economic motives were clear: leaving the rouble zone was considered a necessary step to bringing down inflation, imposing some measure of monetary control and restoring macroeconomic stability. Political considerations were no less important, national currency being seen as one of the most conspicuous symbols of national sovereignty. However, the monetary reforms were postponed, and by the end of 1992 only the Baltic states and Ukraine had actually completed the process of separating their monetary systems from the rouble zone. Most other republics had introduced parallel currencies on their territories but maintained some form of limited monetary union with Russia. By the end of 1992, the rouble remained in unrestricted circulation only in Russia and some Asian republics.

A number of motives slowed the full conversion to national currencies. In the absence of domestic macrostabilization, it was almost certain that the new currencies would quickly lose value as a result of inflation – and their subsequent depreciation would undermine the political credibility of the new governments. Trade concerns also were important. New currencies would complicate payments in intra-CIS trade, which in turn could have disastrous effects on output levels in view of the high dependence of all former republics on mutual trade. Because the large role of rouble zone trade, the domestic currencies would have to be linked to the rouble through some kind of fixed rate regime if major price fluctuations were to be avoided.[401] Technically, it was therefore easier to stick to roubles in settling trade payments. Moreover, introducing a national currency requires a certain minimum level of international reserves to defend the exchange rate, but these were virtually non-existent in most of the new independent states.

There was probably also a "free rider" motive: in the absence of a central monetary authority, the temptation must have been strong for the 15 independent central banks and sovereign governments of the ex-Soviet republics to pursue expansionary fiscal and credit policies in the hope that the inflationary impact would be dispersed over the whole rouble zone.

Relying exclusively on the Russian currency, however, also created problems. The Russian central bank, in an attempt to control inflation, did not provide enough cash roubles to the other republics, thus creating a severe cash crisis in the rouble zone in the early part of 1992.

The cash shortage and the tightening of monetary policies by Russia in the first half of 1992 prompted other republics to introduce parallel currencies. Ukraine, Belarus, Lithuania and Latvia, one after the other, issued their own money surrogates which first co-existed with the rouble, representing a growing proportion of the cash stock, and eventually became interim currencies. Restrictions imposed by republican authorities on the use of roubles in certain transactions led to a gradual outflow of cash roubles to Russia which accelerated when the rouble lost its currency status in some republics. One consequence was an an uncontrolled increase of cash holdings in Russia. The uncoordinated and at times chaotic introduction of quasi-currencies in individual successor states thus had undesirable macroeconomic implications for the rouble zone as a whole.

Estonia took the lead in making a full break with the rouble when it introduced its national currency, the *kroon*, on 20 June 1992. The monetary reform was driven mostly by immediate stabilization goals. The kroon was made internally convertible and subjected to the working principles of a currency board: kroon bank notes and reserve deposits of commercial banks are fully backed by gold and foreign exchange, and convertible into deutsche mark at a fixed exchange rate. The return of pre-war gold reserves to Estonia by Sweden allowed for the initial capitalization of the currency board. The principle of full capitalization means that the amount of kroons in circulation (plus bank reserves) can never exceed the level of foreign exchange reserves, converted at the official rate.

The rouble was withdrawn from circulation in Estonia the day the kroon was introduced. The conversion of roubles into kroons involved an element of confiscation: individuals were allowed to convert cash holdings of roubles at the base rate of 10 roubles/kroon only up to an amount of 1,500 roubles; for amounts in excess of that sum – probably the equivalent of two weeks' wages at the time – the rate was 50 roubles/kroon. However, this limit did not apply to banked assets.

Latvia was the second country to depart from the rouble zone. In an attempt to mitigate the impact of the cash shortage on its economy, the Latvian government decided first to introduce a parallel currency, the Latvian rouble, and then to substitute the Latvian rouble for the Russian rouble. The Latvian rouble entered into circulation on 7 May 1992, its rate fixed at parity with the Russian rouble. Wages and salaries were to be paid in Latvian roubles only, bank accounts denominated in Russian roubles were initially frozen and later exchanged at very depreciated rate (1 to 10). In addition, sanctions were imposed on those who refused to accept the new currency.

The move was strongly criticized by monetary experts (some called it a "suicidal step")[402] as opening the way to uncontrolled inflation and to precipitating a further collapse of trade with Russia because the non-convertible Latvian rouble could not be used for imports from the CIS countries. These gloomy

---

[401] As emphasized by some authors, continued membership in the rouble zone also gave access to low-priced energy and raw materials, at least in the short run (D.M. Nuti, J. Pisani-Ferry, "Post-Soviet Issues: ...", op.cit., p.17).

[402] See *Financial Times*, 15 May 1992. The commentator of *Izvestiya* also predicted a "catastrophe" (*Izvestiya*, 16 May 1992).

predictions have not materialized. The inflation rate continued to decline and the impact on trade with Russia was limited; the main obstacles instead being supply-side constraints on Russian raw-material exports.

The Russian rouble was entirely withdrawn from circulation on 20 July,[403] to be exchanged at a rate fixed by the Bank of Latvia at 10 Russian for 1 Latvian rouble. This resulted in a drastic fall in the real cash balances of rouble account holders, and thus amounted to a more "confiscatory" monetary reform than in Estonia. The introduction of the permanent national currency, the *lat*, began in March 1993.

Lithuania soon followed Latvia's example. Immediately after having declared its independence in 1990, the republic announced a plan to introduce its own national currency, the *litas*. However, the monetary reform was postponed for months, and no precise date for introducing the litas was fixed. Instead, the Lithuanian government decided to remain for some time within the rouble zone and to use ad hoc measures to protect the country from the adverse effects of the general monetary disorder in the CIS.[404] In an effort to overcome the problem of the cash rouble shortage, Lithuania introduced quasi-money in the form of special coupons (*talonas*) for a transition period in April 1992, fixing its rate at parity with the Russian rouble. After several weeks, 60-70 per cent of all cash transactions were realized in talonas.[405] On 1 October 1992 Lithuania withdrew from the rouble zone and the talonas became the only legal tender in the country. The operation raised many technical problems because roubles were to be exchanged for talonas within one week, and conversion at the rate 1 to 1 was allowed only for rouble-denominated bank accounts. Cash in roubles, if it were not to be lost, had therefore to be either spent or deposited in banks.

Preliminary assessment of the Baltic republics' withdrawal from the rouble zone suggests that the monetary reforms contributed to domestic stabilization, although the impact differed between countries. Latvia and Lithuania followed a different reform pattern from Estonia. In Estonia, monetary conditions will be determined primarily by developments in the country's external accounts: a surplus (or a deficit) in the overall balance of payments would increase (or decrease) the supply of high-powered money (cash plus bank reserves). In Latvia and Lithuania, the principles for regulating the money supply have not been clearly established.[406] The rules for trade settlements with the CIS countries remained in disarray until late fall, when after difficult negotiations a system of convertible-currency clearing arrangements was put in place.

Some of the transition problems are gradually fading away. The unwanted inflow of Russian roubles into the banking system of the Baltic states was effectively stopped when Estonian and Latvian central banks decided to suspend purchases of roubles.[407] On the other hand, both the Latvian and the Lithuanian interim currencies are printed on low quality paper and can easily be counterfeited.[408] Nevertheless, while both currencies started at parity with the Russian rouble, they have so far proved stronger and more stable.

The increasing strength of the new currencies in the Baltic republics can be illustrated not only by gradually falling inflation rates but also by the slow-down of their depreciation against the dollar. As can be seen from table 3.4.14, the Estonian kroon has remained remarkably stable since its inception (being pegged to the deutsche mark, its dollar rate fluctuations reflected changes in the dollar rate *vis-à-vis* the deutsche mark). The Latvian roublis stabilized after some initial adjustment. Only the Lithuanian talonas is still falling against convertible currencies but it has, nevertheless, appreciated strongly against the rouble. The experience of the Baltic countries shows that an early separation from the rouble zone may be helpful in restoring macroeconomic stabilization. But the price the three countries paid was a much deeper fall of output, connected with the severance of their trade ties with the rouble zone and especially with Russia.

Although *Ukraine* was the first CIS country to issue a parallel currency in 1992, it progressed more slowly with a definitive withdrawal from the rouble zone. "Coupons" were introduced partly as an interim measure in preparation for the introduction of a national currency independent of the rouble and partly to stem the purchase of food and other goods by residents of other CIS states. State stores were instructed to accept only coupons and 25 per cent of wages were paid in them rather than in roubles. This share was raised to 50 per cent on 1 February 1992. On 1 April the rouble was formally abolished for current transactions in favour of the coupon, although it still retained a role (at 1:1 parity) as an accounting unit.

Initially, the cash transaction parity of the coupon against the rouble was fixed at 1:1 and was set through the exchange rate of the coupon against the dollar (10 roubles and 10 coupons to the dollar), yielding a rouble-dollar cross rate roughly the same as on the

---

[403] *Izvestiya*, 21 July, 1992.

[404] In November 1991 the Lithuanian parliament adopted the law introducing the litas. But the government stressed on several occasions that Lithuania did not intend to quickly replace the Russian rouble with its own currency, mainly because of the high share of rouble trade in the total foreign trade of the country (BBC, *Summary of World Broadcasts*, SU/W0199, 4 October 1991; SU/1223, 7 November 1991; *Izvestiya*, 6 November 1992).

[405] *Ekonomika i zhizn'*, No.22, May 1992.

[406] The Latvian rouble is formally allowed to float but, in practice, the central bank's policy has been to peg it to the deutsche mark.

[407] The roubles were coming from those CIS countries which had not yet introduced their own currencies but became centres of excessive monetary emission (e.g., Ukraine, Belarus, Azerbaijan).

[408] Numerous reports confirm that more and more counterfeit Lithuanian talonas and Latvian roublis are being detected in circulation. BBC, *Summary of World Broadcasts*, SU/W0260, 11 December 1992; BBC, *Summary of World Broadcasts*, SU/W0265, 22 January 1993.

TABLE 3.4.14

How did the new currencies fare? Exchange rates in the post-Soviet states
(Average central banks' rates in units per US dollar)

Date	Estonian (kroon)	Latvian (roublis)	Lithuanian (talonas)	Ukrainian (karbovanets)	Russian (rouble)
1992					
15 July	11.87	145	145	-	130
5 August	11.83	142	160	-	161
9 September	11.16	186	220	-	208
23 September	11.76	184	224	-	241
7 October	11.37	180	247	-	342
18 November	12.86	174	264	608	448
16 December	12.58	169	313	635	418
1993					
20 January	12.94	171	392	651	474
3 February	13.19	170	408	931	572
24 February	13.03	157	443	972	576
10 March	13.39	152	472	1 233	650

Source: National statistics.

Moscow interbank exchange. The black market rate of the coupon had by July 1992 none the less fallen to 60 kopeks and remained at that level for the rest of the year — a reaction to massive monetary emission and to the generally bigger market imbalances in Ukraine.

On 13 November, Ukraine decided to leave the rouble zone altogether and extended the use of the temporary "coupon" currency to non-cash transactions — a long delayed decision, originally scheduled for March and then for mid-1992. The "coupon", which has been renamed the *karbovanets*, is to remain in circulation until the macroeconomic situation stabilizes and the introduction of the permanent national currency, the *hryvnia*, becomes possible.

Despite long procrastination, the monetary reform in Ukraine does not seem to have been well prepared. No attempt was made to support it by establishing some disciplinary mechanism of linking the karbovanets with foreign currencies. Nor was it accompanied by cuts in budgetary spending and credit emission. As a result, inflation accelerated and the exchange rate fell steeply — not only *vis-à-vis* convertible currencies, but even with respect to the rouble.[409]

Among the remaining CIS states, only *Azerbaijan* decided to separate its monetary system from the rouble zone, although its national currency, the *manat*, has so far been introduced for cash transactions only; the non-cash circuit is served with "national" roubles, still partly linked to the Russian rouble. In *Belarus*, the original plan to introduce its own currency (first known as the *taler*, and later renamed the *rubel*) before April 1992 has also been postponed to a later date. A temporary parallel currency, the "settlement note", was introduced in May 1992 to alleviate the cash shortage. But the rouble is still used for all commercial transactions and the country does not have immediate plans to leave the rouble zone.[410] A similar position has been taken by *Armenia* and *Moldova*.

As a result of changing priorities, conceptual differences and the varying speed of implementing monetary reforms, the former Soviet republics currently remain at various stages of separation from the rouble zone. The monetary union existing until the end of 1991 was replaced de facto by a set of separate though interdependent national currency systems. A summary of the monetary systems in operation in these countries at the beginning of 1993 is shown in table 3.4.15.

The spontaneous disintegration of the post-Soviet monetary system resulted not only in the emergence of national cash surrogates, but also led to the somewhat unexpected appearance of "national roubles" used in non-cash settlements. This was caused chiefly by large imbalances in trade between the republics. The central banks of Russia and the Baltic states were increasingly reluctant to accept non-cash roubles in payment for exports to deficit countries such as Ukraine. Towards the end of 1992, Latvia introduced differentiated exchange rates for non-cash roubles originating in different CIS states, and this practice was quickly adopted by the central banks of some other republics (Russia, Belarus).[411] These differentials have also been reflected in national official and market dollar exchange rates. The multitude of non-cash roubles and cash surrogates with varying degree of convertibility has been an im-

---

[409] Immediately after the monetary reform, the exchange rates established by the National Bank of Ukraine were 647 karbovanets per dollar and 1.45-1.55 karbovanets per rouble (*Izvestiya*, 27 November 1992). But in less than three months the dollar went up to 970 karbovanets (end-February 1993). The rouble rate was slightly higher at 1.65 karbovanets.

[410] After September 1992, the Russian rouble could not be used for cash transactions in Belarus, but it was restored as legal tender, along with the Belarus "talon", at the beginning of March 1993 (*Rossiiskie Vesti*, 27 February 1993).

[411] For instance, the exchange rate of the Latvian roublis against the "national roubles" of CIS states, fixed weekly by the Bank of Latvia, varied in the period 15-22 February 1993 between 0.205-0.225 for roubles coming from Kazakhstan, Georgia, Uzbekistan and Turkmenistan to 0.257-0.266 for roubles from Moldova and Armenia, and 0.290-0.296 for roubles from Belarus and Russia. The Ukrainian currency was quoted at 0.112 Latvian roublis per karbovanets (*Rzeczpospolita*, 19 February 1993).

TABLE 3.4.15

Monetary systems in the former Soviet republics, as of February 1993

Monetary system	Convertibility		Country
	Foreign currencies	Roubles	
1. National currency	Yes	No	Estonia
2. National currency	Partly	-	Russia
3. Temporary currency	Partly	Partly	Latvia, Lithuania Ukraine
4. Mixed system: Cash circuit - only national surrogates; non-cash circuit - national roubles	No	Partly	Azerbaijan
5. Mixed system: Cash circuit - roubles and national surrogates; non-cash circuit - national roubles	No	Yes	All other CIS states

Sources: Compiled from: "The break-up of the rouble zone", *MBES Information Bulletin*, No.6, November-December 1992, M. Dabrowski, "Obumieranie strefy rublowej", *Rzeczpospolita*, 23 November 1992, and national data.

portant factor behind the sharp decline in intra-CIS trade in 1992.

Chaos in the monetary and payments relations between the CIS countries continued practically unabated until the Russian authorities took more decisive steps to protect the rouble and curb intra-CIS trade imbalances in the fall of 1992. The other republics were offered the alternative of either leaving the rouble zone entirely and shifting to national currencies, or remaining in the rouble zone (accepting the rouble as national currency, or pegging the national currency to the rouble within a stable and credible mechanism) and agreeing to close coordination of monetary and fiscal policies. Provided its dominant position is preserved, it seems that Russia remains interested in keeping the rouble zone in some form and is prepared to grant substantial trade and payments preferences to those republics which decide to remain within it.

This policy produced first results at the CIS summit meeting in Bishkek in October 1992, when an agreement to establish a "single monetary system" and to coordinate macroeconomic financial policies was signed by eight republics.[412] The agreement, however, was mainly a declaration of intent rather than a plan of action for the immediate future. For instance, the decision to create a CIS-wide central bank was accepted in principle, but was not elaborated to the point permitting concrete steps towards its establishment. Moreover, some countries specifically stated that they did not exclude introducing their own national currencies while remaining in the rouble zone.[413] Nevertheless, the Bishkek agreement may have marked a turning point in the spontaneous unravelling of the monetary system of the post-Soviet community of states.

However, the next step in strengthening the foundations of the rouble zone, which was to have been taken at the Minsk summit in January 1993, produced only meagre results. While the eight republics agreed to establish a common bank with a 50 per cent Russian participation (but a two-thirds majority requirement for decisions on key issues), they did not reach agreement on the principles of a common monetary policy. The activities of the new Inter-State Bank will, for the time being, be limited to settling payments in mutual trade transactions conducted under intergovernment trade agreements.[414] Had this agreement not been signed, Russia would probably have introduced its own "rouble" to protect itself against uncontrolled credit emission from other republics.[415] The road to a new, viable rouble zone may therefore be long and difficult, and success is by no means certain.[416]

---

[412] Armenia, Belarus, Kazakhstan, Kyrgyzstan, Moldova, Russia, Tajikistan and Uzbekistan. Georgia and Turkmenistan decided to remain outside the new system.

[413] Belarus, Armenia, Kyrgyzstan and Russia. See "The break-up of the rouble zone", *MBES Information Bulletin*, No.6, November-December 1992.

[414] *Financial Times*, 22 January 1993.

[415] A very sceptical assessment of the summit's decisions by the Deputy Chairman of the CBR, A. Solovov, may indicate that Russia does not exclude some unilateral decisions if CIS countries fail to agree on key points of monetary coordination. *Izvestiya*, 25 January 1993.

[416] The United Nations Economic Commission for Europe secretariat strongly emphasized these difficulties in the previous edition of this publication, concluding that "...the best chance of achieving macroeconomic stabilization within the present rouble area lies not in attempts to restore and strengthen a single monetary and fiscal centre, but in moves to create separate successor state currencies". United Nations Economic Commission for Europe, *Economic Survey of Europe in 1991-1992*, New York, 1992, p.132. This sceptical opinion about the feasibility of maintaining a monetary union has recently come to be shared by some other international organizations. In its annual report for 1992, the EBRD strongly criticized the attempts by the Russian authorities and by some international organizations to keep the rouble zone together. The report argued that the zone is economically ineffective, inflationary and damaging to intra-CIS trade, and that CIS countries should rather introduce their own currencies as soon as possible. The Bank sees this as the only possible way to impose prudent financial policies and secure trade in the CIS republics. See EBRD, *Annual Economic Review 1992*, London, 1993, pp.18-25.

3.5 MARKET REFORMS

(i) Markets and competition

The juridical framework for freedom of economic activities has been established in all the transition countries since the early stages of transformation. General liberalization was combined with the establishment of *de jure* equality between the public and private sectors and, in some countries, the latter now even enjoys substantial legal and economic preferences in many areas. Liberalization took the form of lifting bureaucratic restrictions on starting and expanding business activities, streamlining administrative procedures and reducing the scope of activities kept under the control of the state.

Internal competition has been greatly enhanced by two broad types of measure. First, early liberalization of foreign trade exposed domestic producers and consumers to international markets. Second, antimonopoly legislation was passed in most countries to protect domestic consumers against possible abuse by the powerful monopolistic organizations inherited from the central planning era.

Even though the many radical reforms effected since 1989 have revamped the institutional set-up of the transition economies, it would be premature to assume that a fully-fledged market mechanism is already in operation in these economies. Liberalization is only the first step on the road to building up a new system. To remove restrictions is a much easier task than establishing new, sound rules for commercial activities – including enforcement of contracts, effective tax collection, bankruptcy, debt repayment discipline, etc. Thus, though most transition countries have gone through a stage of sweeping liberalization, building up new institutions on what remains of the defunct socialist system has proved to be a very time-consuming, socially sensitive and politically loaded process. Moreover, the high costs of transformation have quite dramatically changed the overall mood in the transition countries and part of the initial enthusiasm for free markets and competition may have gone for ever. Symptoms of all kinds of protectionism – from calls for high trade barriers to the restoration of unconditional job security – are emerging and gaining support. However, these temptations have to be resisted if the market reform is to be carried out successfully.

Of course, markets are economically effective – i.e., they allocate resources optimally and enhance competition – if they are allowed to operate in an unrestricted and undistorted manner. The initial liberalization has achieved a lot in this direction, but the market mechanism is still very imperfect, even in those countries which are most advanced in the reform process. Existing impediments include very high uncertainty caused by frequent changes in the legal framework, structural rigidities rooted in existing physical and human capital stock, old attitudes and inadequate regulations and laws. Greatest progress has probably been made in the *goods markets*, where the expansion of private trade, made possible by demonopolization of traditional retail and wholesale trade networks, currency convertibility and foreign trade liberalization have assured a considerable degree of competition, at least in the east European countries. The situation also improved in the former Soviet Union where market imbalances diminished significantly, even though the high incident of monopoly has resulted in soaring inflation and a fall in sales volumes. Nevertheless, the emergence and rapid development of many commodity exchanges (*birzhy*) in the ex-Soviet republics has dramatically changed supply conditions, which allows the break-up of traditional monopolies in distribution networks and removes many notorious bottlenecks.

All transition countries have put in place anti-monopoly legislation, in some cases as early as 1989-1990 (Hungary, Poland). More recently, anti-monopoly laws were passed in Bulgaria, Czechoslovakia, Romania, Russia and other CIS states.[417] Even though the transition economies remain highly monopolized, relatively few cases of monopolistic practices appear to have been registered (but information is scarce). The persistence of price inflation in many countries suggests that monopolies continue to dominate, especially in non-tradable sectors and in services.

The *Russian economy*, where prices are to a considerable extent dictated by huge, state-owned enterprises, provides the most obvious case of a strongly monopolized goods market. It is symptomatic that in the first stage of the stabilization programme in 1992, enterprise profits as a proportion of GDP increased considerably despite drastic cuts in the real money supply.[418] This jump can to a large extent be attributed to monopolistic pricing, which was not contained by existing legislation and institutions. The reasons for the apparent weakness of anti-monopolist policies in Russia vary from inadequate reporting practices and inefficient proce-

---

[417] In Bulgaria, antimonopoly policy is based on the Competition Protection Act, passed on 2 May 1991, while in Czechoslovakia the key legislation on monopolies and competition is included in the new Commercial Code which entered into force on 1 January 1992.

[418] As reported by some sources, the proportion of total profits to GDP was 33 per cent during January-April 1992, up from 25 per cent for 1991, while hard currency holding of enterprises increased from $1.8 billion to $4 billion (*Institutional Investor*, September 1992, p.70).

dures to the general reluctance of the Russian authorities to take unpopular decisions on the breaking-up or closure of existing large enterprises, chiefly for sociopolitical reasons.[419]

Markets for *factors of production* are less developed and less competitive. The working of the *labour market* is in most countries hampered not only by the absence of strong employers' organizations, but also by administrative and structural rigidities in wage determination and labour mobility. *Capital markets* are underdeveloped, and have been established formally only in some countries. Hungary has operated the Budapest stock exchange since 1989, but the volume of transactions and the impact of stock trading on reallocation of capital resources in the economy as a whole is negligible.[420] In Poland, where the Warsaw stock exchange opened in April 1991, the number of companies listed has increased from 5 to 16, and the volume of weekly transactions fluctuates around $10-15 million. Stock exchanges of a similar size also operate in Belgrade (FR Yugoslavia)[421] and in Ljubljana (Slovenia). A formal stock exchange was also opened in Sofia (Bulgaria) in January 1992, and one is planned to begin operations in Prague (the Czech Republic) in 1993.

In *Russia*, more than 100 "wild" stock exchanges, or stock-trading departments, mostly attached to existing commodity exchanges, have emerged during recent months.[422] So far, they trade only in shares issued by some new commercial banks and insurance companies. The flow of shares of enterprises may be expected to increase rapidly in 1993, when the voucher privatization programme enters its second stage. But the rules of trading have not been determined, and the long-awaited law on securities exchange is not likely to be adopted by the legislature before the end of 1993.

In contrast, *foreign exchange markets* have generally been an area of unquestioned success. Internal convertibility, rightly considered to be an essential condition for competitive markets, has been introduced successfully in most transition countries, although restrictions remain on access to foreign currencies for some categories of transactions. In most east European countries the foreign exchange market was effectively stabilized within weeks through a combination of liberalization measures and the devaluation of domestic currencies.

Freedom of entry and exit is essential for a proper working of market competition. The *"entry"* for new businesses is formally unrestricted in almost all activities, except for those typically falling under state control (arms, nuclear energy and weapons, drugs, some utilities, natural monopolies). Nevertheless, starting economic activities may be a difficult task because of lengthy and unclear bureaucratic procedures, high capital costs, or gaps in legal systems. These freedoms were in principle established in most countries in the relatively early stages of transition.[423]

On the other hand, the *"exit"* procedure is in many cases more problematic, especially for enterprises which are deeply indebted, insolvent or de facto bankrupt. Contrary to many other institutional changes, bankruptcy laws have not yet been passed in many countries — they are still before parliament in Bulgaria and some CIS states. Even in cases where explicit bankruptcy provisions are formally in force, their implementation is plagued by a number of obstacles, mostly sociopolitical, but also procedural. In some countries, such as Czechoslovakia or Russia, new laws on bankruptcy have been promulgated, but their application is in suspense for technical reasons.[424] The only relatively effective bankruptcy legislation appears to be that of Hungary: the law, passed in November 1991, has been in force since January 1992 and several thousand bankruptcy cases have been brought before the courts.

One of the key features of the new bankruptcy law in *Hungary* is that it obliges an enterprise to file for financial restructuring within eight days if its payments arrears exceed 90 days; but if the enterprise had undergone restructuring already, it must file for liquidation. Another characteristics is that for the initiation of restructuring the agreement of all creditors is required. Moreover, once the enterprise files for restructuring or liquidation, its assets are protected and no payments can be effected before the court's ruling, which resembles protection given to insolvent firms under Chapter 11 of the US bankruptcy legislation.

---

419 The Russian Federation State Committee for Antimonopoly Policy and the Promotion of New Economic Structures, created in the autumn of 1990, has been understaffed and initially busy with other issues not directly concerned with monopolistic practices (e.g., the creation of free economic zones). On the other hand, the Law on Competition and the Restriction of Monopoly Activities in Commodity Markets was passed only in March 1991 and, according to some observers, many of its provisions are either not applicable or ineffective in the specific conditions of Russia. (V. Capelik, "The Development of Antimonopoly Policy in Russia", *RFE RL Research Report*, Vol.1, No.34, 28 August 1992.)

420 In January-October 1992, only two new companies went public, thus raising the number of traded stocks from 19 to 21.

421 Money market operations also take place on the Belgrade stock exchange, including transactions in Treasury bonds.

422 *Financial Times*, 23 February 1993.

423 In some cases, the general rule that "everything is allowed unless it is specifically prohibited by law" has been incorporated in new commercial codes (as in Czechoslovakia), or in special laws on undertaking business activities (Bulgaria, Poland).

424 For instance, the new bankruptcy law passed in Czechoslovakia in July 1991 does not give precise criteria for triggering liquidation. The law stipulates that a debtor firm can be considered bankrupt if it is unable to meet payments for a "protracted period of time" or if its liabilities are excessive. The new law provided a grace period of one year until October 1992, and recently Parliament voted an extension of the grace period for another six months. See *European Bank for Reconstruction and Development*, Annual Report 1992, p.10. Besides, some amendments aiming at better protection of companies facing serious financial difficulties are currently planned in the Czech Republic, apparently reflecting growing concern about the possibility of "chain bankruptcies" caused by the high indebtedness of state companies. *Financial Times*, 17 February 1993.

In *Russia*, the "Law on the Insolvency of Enterprises" was adopted by the Parliament on 19 November 1992, but will be implemented only on 1 March 1993. The reason for the delay is that appropriate accounting procedures, including rules for the valuation of enterprise assets and indexation have not yet been introduced. See *Rossiiskaya Gazeta*, 30 December 1992.

This particular solution was expected to result in a large wave of bankruptcies immediately after the grace period elapsed; and indeed, the number of bankruptcy cases brought to the courts jumped to some 1,100 in May 1992, and to almost 12,500 or nearly 25 per cent of all enterprises at the end of September.[425] One important by-product of this surge of filings was a sharp drop in tax revenues. Abuses have been noted in that many firms apply for liquidation in order to obtain a legal cancellation of all their debts so as to then re-establish their business activities with a clean balance sheet. It seems likely that in view of these unwanted repercussions the bankruptcy law in Hungary will be subject to further changes and amendments.

A rather different policy is followed in *Poland*, where no new bankruptcy law has been passed and existing legislation is based on the 1934 Commercial Code provisions and on Article 19 of the 1981 Law on State Enterprises. The two legal enactments are not very consistent; moreover, they display a clear bias towards liquidation as opposed to restructuring. This may not be the best framework for a transition economy, where massive restructuring is both desirable and necessary because of the transformation from plan to market. Contrary to Hungarian legislation, in Poland it is the creditors who take the initiative in filing for the bankruptcy of insolvent debtors. But an obvious weakness of this procedure is that there are many incentives not to do so. Creditors in Poland, but also in other countries, have so far displayed astonishing passivity in bringing insolvent clients to courts and forcing them into bankruptcy, and the number of registered bankruptcies is still negligible. As observed by one expert, if all cases of insolvency were brought to the courts, some 45 per cent of Polish state enterprises would have been declared bankrupt in 1992.[426]

Reasons for the passivity of creditors, and especially the commercial banks, include their high dependency on their troubled clients, reluctance to acknowledge non-performing loans and losses, hopes (sometimes justified) that the government will bail out debtors in one way or another and, finally, political and social motives.[427] Many of these problems can be solved only through a more fundamental reform of the financial sector and more resolute privatization.

(ii) Reform of the banking system

Until very recently, the reform of the banking system was not among the top priority tasks on the reform agenda of transition countries. When they started their race towards a market system in 1989 (or even earlier in Hungary and Poland), banking and financial sector reforms were planned and in some cases even started, but their significance appeared to be far less than that of stabilization, liberalization and privatization.[428] Nevertheless, the experience of three years of transformation with all its successes and failures has clearly demonstrated that fundamental revamping of the banking sector is not only one of the crucial conditions of successful reforms, but it is also the area where these reforms have to be most comprehensive and intrinsic.

Banks in centrally planned economies had very little in common with modern banking systems in market economies. The passive role of money, which was subservient to central planning in terms of real flows, and the prevalence of state ownership led to the development of a highly concentrated and monolithic banking structure embodying the central bank and its various branches – dubbed "monobanks" in western literature – highly dependent on the political authorities, the government, and on the Ministry of Finance in particular. Its main functions were to tap savings, to allocate low-interest credit to enterprises via plan-based rationing, and to finance the state budget, while commercial activities were reduced to a minimum. Regular lending, servicing the non-cash circuit of the economy, the role of "lender of last resort" and money supply management were all mixed together and performed by one, highly-centralized institution.[429]

Transition countries thus faced the difficult task of reforming their banking sector from the very beginning of transformation with a model which was clearly incompatible with the logic of the market system. It is worth noting that some changes in this direction had been made already: Yugoslavia (in the mid-1960s) and Hungary and Poland (in the 1980s) were among the early starters. However, the reforms were cautious and gradual, and only after the 1989 breakthrough did they gain a new dimension and gather speed. The process goes typically through two distinct stages. In the first, the break-up of the "monobank" structure takes place and a net of commercial banks emerges following general liberalization. In the second stage, consolidation and restructuring of commercial banks is accompanied by the strengthening of the central bank's primary functions and the development of money and financial markets.

---

[425] As reported by *Magyar Hirlap*, 2 November 1992. The process of ruling, however, advances very slowly, as Hungarian courts seem to have been completely unprepared for such large number of cases. See also K. Mizsei, "Bankruptcy Laws in the Post-Communist Economies", *International Economic Insights*, January/February 1993.

[426] K. Mizsei, op.cit., p.44.

[427] European Bank for Reconstruction and Development, *Annual Economic Report 1992*, pp.4-7; J. Mitchell, "Creditor Passivity and Bankruptcy: Implications for Economic Reform", paper presented at the CEPR-BBV conference, San Sebastian, 27-28 March, 1992.

[428] As one prominent government adviser noted: "...The experience (of Poland) suggests that the reform of the banking system, although important, can wait until a later stage of the reform process...", S. Gomulka, "Poland" in P. Marer, S. Zecchini (eds.), *The Transition to a Market Economy*, OECD, Paris, 1991, Vol.I, p.67.

[429] See D.M. Nuti, "Socialist Banking" in *New Palgrave Dictionary on Money and Finance*, J. Eatwell, M. Milgate, P. Newman (eds.), The Macmillan Press, New York, 1992.

TABLE 3.5.1

Banking reform in transition countries

Country	"Two-tier" banking since	Number of commercial banks				Foreign banks banks operating [a]
		1989	1990	1991	1992	
Albania	1992	..	..	..	..	..
Bulgaria	1990	4	67 [b]	73 [c]	80 *	Yes
Czechoslovakia	1989	3	14	38	53 [d]	Yes
Hungary [e]	1987	..	30	36	..	Yes
Poland [e]	1988	14	..	92	104 [d]	Yes
Romania	1991	..	..	..	15 [f]	Yes
Belarus	1991	..	..	..	35	No
Russia	1991	..	..	..	1 700 *	Yes
Ukraine	1991	4 [g]	18 [b]	68	..	No
Estonia	1991	..	..	54 [d]	43	No
Latvia	1992	5 [g]	..	14	40	No
Lithuania	1992	5 [g]	..	..	20 *	No

Source: Compiled from national sources.

[a] As of 1 February 1993.
[b] January 1990.
[c] March 1991.
[d] June 1992.
[e] Excluding cooperative banks and their branches.
[f] September 1992.
[g] Branches of Soviet specialized banks.

(a) Two-tier banking and the development of commercial banks

Separation of central bank functions and commercial banking functions is typically one of the first steps of banking reform. The rationale behind this separation has been clearly recognized for a long time:[430] while the central bank's main concern is stability of the currency and prices, commercial banks serve as "go-betweens" between savers, capital owners and investors and should be subject to market and competition rules.

Normally, the separation of central banking and commercial banking functions in transition countries is achieved through administrative decisions: territorial and sectoral branches of the central bank are given wide autonomy or full independence, the commercial functions of the central bank are limited or discontinued altogether, the central bank itself is given a new charter. As a result, several — most typically between five and ten — big state banks are established in this manner and assume the commercial functions of the former "monobank". This operation, which takes commercial activities away from the central bank, leads to the establishment of a "two-tier" banking system (as opposed to a "one-tier" or "monobank" set-up).[431] Table 3.5.1 shows that all transition countries either have already completed, or are at the final stages of completion of this step of reform.

The rapid increase in the number of commercial banks following decentralization of the banking system is indeed remarkable in some countries (Bulgaria, Estonia, Poland). While many of the new banks are direct ramifications of the break-up of the old "monobank", there is a growing number of banks, state or private, which have been established from first principles. Also, the presence of foreign banks, which evolves typically from representative offices to joint-venture participation and the establishment of local branches, is becoming more visible. Nevertheless, concentration in the banking sector remains very high; typically, several large state banks dominate the sector, accounting for more than 50 per cent of all assets (Poland, Hungary, Czechoslovakia, Romania).[432]

The impressive expansion of a commercial banking network is certainly a reassuring phenomenon, as it responds to the fast-growing needs for financial intermediaries in the transition economies. It is also a symbol of financial liberalization and of the market orientation of reforms. A larger number of banks also assures some degree of competition among them, to the benefit of their clients. But many new problems are also emerging as a result of this dynamic process, and some of them have become serious obstacles to the further progress of market reforms.

First, decentralization of the central bank's functions has been carried out in an administrative way, rather than on the grounds of more specifically economic considerations. Generally the new banks have been created on the basis of the regional branches of central banks, without any restructuring of their balance

---

[430] At least since the Robert Peel's Act in England in 1844, which marked the abandonment of the principles of the banking school in favour of those of the currency school.

[431] For instance, in Poland the process resulted in the establishment of nine big, "universal" banks and six "specialized" banks in 1988, each with total assets of $2-4 billion. In Romania, the largest commercial bank (Romanian Commercial Bank) was emancipated from the National Bank of Romania in 1991, with assets of $4 billion. See *Central European*, October 1992, p.45-48. Similar changes took place in Bulgaria, Russia and the Baltic republics.

[432] For instance in Hungary the five largest banks (14 per cent of the total number) control 75.4 per cent of all the assets of the banking system. *National Bank of Hungary*, Annual Report, 1991, p.52.

TABLE 3.5.2

Biggest commercial banks in Russia: Assets, capital and financial ratios
(Ranking by assets, as of 1 January 1993, in billion roubles, unless otherwise stated)

Category of banks	Total assets	Own capital	Solvency ratio [a]	Profits (in 1992)	Profit ratio [b]
First "ten"	5 245.1	52.7	1.01	114.6	217.4
First "twenty"	6 541.7	100.1	1.53	158.9	158.7
First "thirty"	7 334.1	109.2	1.49	181.1	165.8
First "forty"	7 969.5	116.9	1.47	199.9	162.4
First "fifty"	8 445.7	122.6	1.45	215.7	175.9

Source: *Finansovye Izvestia*, 4-10 February 1993.

a Ratio of own capital to total assets, unweighted, in per cent.
b Ratio of profits to own capital, in per cent.

sheets, and particularly of their assets portfolios. This has naturally resulted in a high degree of sectoral and regional specialization. Thus, decentralization has not established competitive banking structures, but rather banks with largely separate areas of interest and groups of clients. The two characteristic features of the new set-up are relatively high dependence of the banks on their principal clients and a heavy burden of past loans which bear no relation to the new bank's tasks and strategies.

This has led to the emergence of the so-called "bad loans" problem. In no transition country have the banks been able to prevent the accumulation of unpaid debts by financially pressed state enterprises, and in many cases the process went so far as to undermine the financial viability of the creditor banks. The characteristic phenomenon, which is called "too big to go bust", explains the passivity of banks *vis-à-vis* their notoriously insolvent clients. Because of the high proportion of credits extended to one or two big companies relative to other assets, the banks are unwilling to initiate formal bankruptcy procedures for these companies, and even conceal the latters' financial problems, since their close-down could be equally fatal for the banks as for the companies.

Another problem is the low capitalization of new banks. This is not only observable for the new state-owned banks, for which the size of own capital had never been essential under central planning because the state guaranteed all deposits anyway, but it is also true for the new private and cooperative banks. Since financial criteria for establishing new banks have not been very demanding and the policy approach was rather liberal, the result is a real boom in the banking sector, manifested as demonstrated by the mushrooming of hundreds of very active but small, poorly-endowed and inexperienced local banks. For instance, out of some 60 Polish private banks, only four had total assets exceeding $100 million and only two had equity capital of more than $10 million. The privately-owned BIG Bank, which is the largest in Poland, ranks only fifteenth among all Polish banks, and with $300 million worth of assets cannot seriously compete with even medium-size foreign banks.[433]

The process of asset growth outpacing the growth of own capital, which has been observed in all transition countries, can perhaps be best illustrated by data on the financial structure of the top 50 commercial banks in Russia shown in table 3.5.2. The capital/assets ratio is extremely low in Russian banks, at least two to three times lower than normal by international standards.[434] On the other hand, profit ratios are high, reflecting low capitalization and galloping inflation. Of all the 1,700 banks in operation in Russia, the capitalization of some 82 per cent of them is below R 50 million (currently the equivalent of some $100,000).

Low capitalization, inherited debts, shortage of professionals in banking and financial services, lack of experience and traditions of prudential lending have all substantially increased the risks inherent in banking activity in transition countries. The banks are generally highly vulnerable to financial losses as their portfolios are heavily exposed to "bad loans" and new credits carry high risks. As a result, the banks' credit policy, while in most cases excessively risk-averting, is not necessarily prudent. Common banking practice in eastern Europe is to ask for high loan collaterals in the form of physical or financial assets; also, banks tend insist on short loan maturities in order to reduce uncertainty. For instance, 95 per cent of all credits in Russia are short-term credits. In Romania, between 70 and 80 per cent of all loans is extended for one year or less, and out of these more than three quarters have maturities of three months or less.[435] On the other hand, the tendency to grant loans only for projects with very high expected returns may in fact increase the overall risk of banks' portfolios, because project screening procedures for lending purposes are very imperfect and do not prevent projects with seriously understated investment costs or overstated sales revenues being accepted. Finally, banks tend to charge very high interest

---

[433] According to the ranking of world's largest banks published recently by *Institutional Investor*, August 1993, the two-hundredth largest bank, the First Commercial Bank (Taiwan), has total assets of $29,955 million, i.e., 100 times more than the BIG.

[434] For instance, the average capital/assets ratio for the world's ten largest banks is 3.7 per cent, and for the largest 30 banks it is 3.8 per cent. *Institutional Investor*, August 1992, p.96.

[435] *Central European*, October 1992, p.45-48.

> **BOX 3.5.1**
>
> *The Estonian banking crisis*
>
> On 17 November 1992, the Bank of Estonia (BoE) suspended the operations of the three largest state-owned commercial banks in Estonia: Union Baltic Bank, Tartu Commercial Bank, and North Estonian Shareholders' Bank. The three banks' combined liabilities amounted to half of all deposits held by Estonian enterprises. The primary reason for the BoE's decision was the liquidity crisis provoked by the freezing of the commercial banks' deposits held at their accounts in the former Vneshekonombank, the ex-Soviet banking institution in charge of servicing foreign trade and payments transactions.
>
> But the roots of the crisis seem to lie elsewhere. More important was probably the inadequate handling of the liquidity crisis by the management of the banks. It was not recognized in time that the assets frozen at Vneshekonombank (totalling some $78 million, or some 40 per cent of all assets of the Estonian banking system) were completely inaccessible for the foreseeable future and hence in practice lost. The banks in fact continued to lend imprudently, especially to finance Estonian exports to Russia and other CIS states — even though these receivables were at best dubious. The shortage of funds was not properly addressed and no reserve provisions had been made against possible losses. Instead, when faced by the BoE's decision, the troubled banks suggested other solutions, such as a massive "rescue" lending operation by the central bank or a devaluation of the Estonian currency in order to increase the central bank's reserves. They also warned about the risk of chain bankruptcies in the Estonian banking system.
>
> But the Estonian authorities decided to take a tough line. Requests by the Tartu Bank and Union Baltic for refinancing credits were rejected by the Bank of Estonia, and the decision won full support from the Estonian government. On 18 December the board of BoE decided to close down the Tartu Bank — the first closure of a bank in a former Soviet Republic since the Second World War. Among the reasons given for the closure were imprudent lending to the ex-USSR, inability to adjust the bank's operations to the new monetary system after the introduction of the Estonian national currency in June 1992, and, last but apparently not least, the extravagant habits of some senior staff.[1]
>
> It was decided that the other two banks would be merged under the name of Allied Nordic Bank, and account holders would probably be able to recover most of their deposits. To finance the merger, the government and the BoE put together a restructuring plan based on a joint issue of BoE bonds for the amount of kroon 300 million ($25 million).[2] A near-full compensation within 12 months was subsequently promised to depositors of the North Estonian, a state bank owned directly by the BoE. The preferential treatment extended to the North Estonian was strongly objected to by the shareholders of the Tartu Bank, which was the least indebted of the three banks. None the less, its request to be rescued was turned down by the central bank. The depositors of the Union Baltic can hope to recover no more than 70 per cent of their money, depending on what proportion of the doubtful assets can be recovered. As to the shareholders of the Tartu and Union Baltic, they lost all their assets, because the proceeds from the liquidation of the two banks will in the first place be used to compensate the losses incurred by depositors. By February 1993, the Tartu Bank had ceased to exist and the new Allied Nordic Bank began regular operations.
>
> The recent banking crisis seems to have taught the Estonian authorities a lesson; it also has a number of important implications. First, it demonstrated the vulnerability of new commercial banks in a volatile and uncertain economic environment. The vulnerability is enhanced by the fact that the banks are owned by large enterprises which are in a position to shape the banks' policy according to their needs. Second, it underlined the importance of legal guarantees for bank deposits, which have not been promulgated in Estonia. Third, it also showed that it may be very embarrassing for a central bank to have substantial stakes in a commercial bank which goes bankrupt because of imprudent lending and lack of controls. More important, the crisis accelerated some structural measures aimed at strengthening the banking system. The BoE tightened controls over commercial banks, renewed all licences issued earlier, and increased the minimum limit for the bank's own capital to kroon 6 million (some $0.5 million). As a result of these measures, by end-January 1993 only 12 out of 43 registered banks were legally permitted to operate. The operations of two other banks (North-East Estonian and Revalia Bank) were suspended for insufficient liquidity, while the licences of 7 other banks were withdrawn because of their failure to meet the minimum capital threshold requirement. Also, 10 small regional banks serving mostly the rural sector have merged to form the Estonian Union Bank.[3] In mid-February 1993, only 25 banks were formally operating in Estonia, down from 43 three months earlier.
>
> ---
>
> [1] *The Baltic Independent*, 24 December 1992-7 January 1993.
>
> [2] *Estonian Radio*, Tallinn, 19 January 1993, quoted in BBC, *Summary of World Broadcasts*, SU/W0266, 29 January 1993.
>
> [3] *The Baltic Independent*, 29 January-4 February 1993.

rates on loans — very much above average deposit rates to try to compensate losses on some assets with high yields on performing loans.[436] But this policy is also dangerous as it put excessive burdens on debtors' cash flows and may undermine their financial position.

All these techniques may have been helpful in heading off some of the more immediate turbulences, but they could neither cure structural problems nor protect banks against the impact of recession. As a result, the banking sector in all transition countries has come under financial strains which have in some cases assumed crisis proportions. The first cases of bank bankruptcies have already been registered in Hungary, Poland and Estonia in 1992. The Estonian banking crisis is particularly illuminating, as it demonstrates very clearly all the deficiencies and dangers characteristic of the early stage of transition from socialist banking to market-based banking (see box 3.5.1).

---

[436] In Bulgaria, average lending rates in 1992 were 5.0-5.2 per cent per month, while average deposit rates for time deposits were 4.1-4.3 per cent and on sight deposits 2.72 per cent per month. These differences correspond to a spread of 20 percentage points between the annual lending and deposit rates. See Bulgarian National Bank, *Report*, January-June 1992, p.94. In Hungary, the differentials in 1992 were even larger in relative terms: while annual rates for time deposits of less than one year fell from 31.9 per cent to 18.4 per cent between January and October, lending rates for short-term credits declined only from 36.0 to 29.6 per cent, with the spread therefore more than doubling. See National Bank of Hungary, *Monthly Report*, September 1992, p.31. Very high interest rate differentials have been also observed in Poland and especially in Russia, where the annual lending rates charged by the commercial banks increased to 120-130 per cent towards the end of 1992 while deposit rates remained at the exceedingly low level of 20 per cent. See *Finansovye Izvestiya*, 24-29 December 1992.

(b) The second stage of reforms: consolidation and restructuring

Growing difficulties in loan management, the maintenance of balanced assets portfolios and sufficient liquidity in the banking sector evoked some countermeasure, both by central banks and by commercial banks. The main thrust of structural reforms has been to strengthen commercial banks financially in order to diminish their dependence on ailing state companies and to increase their flexibility and ability to channel funds into investment projects. In this context, the measures undertaken have typically led to two general directions: the strengthening of the capital base of banks, and the alleviation of the problem of "bad loans".

In most countries, minimum equity capital requirements for banks have been substantially raised, in order to restore banks' solvency ratios to correspond with western standards. In *Bulgaria*, most of some 80-plus commercial banks existing in 1992, which had been formed on the basis of BNB branch offices, were small and poorly endowed with resources. In April 1992, the Bulgarian National Bank increased the minimum own capital level from lev 50 million to lev 200 million (i.e., from some $2.1 million to some $8.7 million), with a grace period extended until May 1993. At the same time, the BNB created the Bank Consolidation Company, a specialized institution charged with the task of carrying out a series of mergers between the existing 69 state banks. The objective was to create six to eight large banks with equity capital around lev 1 billion each (some $40 million).[437] This limit will also be required to obtain a general licence for banking activities. However, banks with private and foreign participation will not be covered by the compulsory consolidation process.

In *Poland*, the minimum limit for bank's own capital for domestic entrepreneurs was raised twice: from Zl 20 to 35 billion in 1990, and next to Zl 70 billion in April 1992 (equivalent to some $5 million). The limit for non-residents was maintained at $6 million. At the same time, the capital adequacy coefficient of 8 per cent of risk-weighted assets has been adopted, in accordance with BIS recommendations.[438] Other prudential lending rules introduced in 1992 include stricter limits on credit allocation to one borrower and in any single asset: currently not more than 10 per cent of own capital can be engaged in one single credit transaction, while not more than 15 per cent can be lent to one single borrower.

Similar disciplinary measures were taken in *Czechoslovakia*, where the minimum capital requirement for "universal" banks was raised from Kcs 50 million to 300 million in late 1991 (some $10-11 million), and in *Hungary*, where the minimum liquidity requirement ratio has been increased from 5 to 10 per cent of liabilities.

The problem of *bad debts* has probably been one of the most debated and most controversial of all the economic and financial challenges confronted in 1992. Most banks in transition countries were faced with the issue, which suddenly gained critical importance in the context of rapidly falling profitability and uncertain prospects for economic recovery. The magnitude of the problem varies between countries but it assumes substantial proportions practically everywhere. The share of bad loans in total bank assets cannot be determined with sufficient precision because of inadequate and largely differing techniques of measurement, but available estimates suggest that it may be around 20 per cent in Bulgaria and Czechoslovakia, 30-40 per cent in Poland, and probably even more in Hungary. As can be seen from table 3.5.3, which gives some data on the proportion of bad loans in the total assets of the 12 largest Polish banks, the reserves in most banks are far too small to cover potential losses.

In the most general terms, the problem of "bad loans" can be approached according to three distinct policy variants: leaving them outside the government's policy altogether and thus enforcing mass bankruptcies, bailing out debtors (either via direct subsidies or additional credits for banks or by the state take-over of the debts) or by mass write-offs of debt. There may also be a combination of these policies. It is interesting to observe that, although the problem itself is essentially of the same character in all countries, it has been addressed by different policy measures in individual countries. No country decided to opt for either of the two extreme solutions, i.e., for mass bankruptcies or for mass write-offs. But while in Hungary the government directly involved itself in taking over bad debts from the banks, in Poland it was decided to help banks cope with the problem through the infusion of additional capital. In Romania and Russia, the policy adopted was to increase credits from the central bank to enterprises via commercial banks.

In *Hungary*, the problem was addressed within the framework of the new bankruptcy law and new accountancy legislation adopted in December 1991. The banks can sell the debts which are qualified as bad to the Hungarian Investment and Development Company for 50 per cent of the book value if the debts accrued before the end of 1991, or for 80 per cent in case of other debts, plus a flat rate of Ft 100 for each Ft 10 million of debt.[439] By the end of 1992, 14 banks and 112

---

[437] The first results of this process can already be observed. In September 1992, the Bulgarian United Bank SA was created through a merger of 22 smaller banks, with equity capital of lev 679.5 million, 90 per cent of which belongs to the Bulgarian Consolidation Company. See *MBES Informatsionnyi Byuleten'*, No.6, November-December 1992.

[438] The standard rule recommended by the Bank for International Settlements is that commercial banks' equity capital should not be lower than 8 per cent of total assets, their value weighted by risk coefficients reflecting collectability. According to the BIS rule, half of the 8 per cent should be "core" capital (common equity, preference shares and retained earnings), while the second half can include subordinate debt instruments, capital gains provisions and reserves.

[439] Debts qualified as "bad" before 31 December 1991 could be sold to the Hungarian Investment and Development Company or the Hungarian State for half of the book value plus Ft 100 per each Ft 10 million. Debts qualified as "bad" in 1992 would be saleable for 80 per cent of the book value plus Ft 100 for each Ft 10 million of debt. According to the new accountancy law, assets with an irregular status must be classified by banks into one of the three categories:

TABLE 3.5.3

"Bad loans" in the largest Polish banks, as of 30 June 1992

Bank	Total credits outstanding (Billion zloty)	"Bad loans" [a] (Per cent)	Reserves (Billion zloty)
Bank PKO BP	40 537	9.2	84
Bank for Food Economy (BGZ)	30 865	17.9	773
Industrial-Commercial Bank (BP-H)	11 562	6.4	470
General Credit Bank (PBK)	10 608	35.7	600
Silesian Bank (BS)	10 164	9.7	827
General Commercial Bank (PBG)	9 885	6.8	220
Pomeranian Credit Bank (PBK)	7 089	19.0	719
Gdansk Bank (BG)	6 619	23.8	886
Wielkopolski BK	6 163	17.9	330
Western Bank (BZ)	6 058	16.2	442
Deposit-Credit Bank (BD-K)	4 694	23.5	71
Export Development Bank (BRE)	2 964	31.0	553

Source: Gazeta Bankowa, No.32, 1992.

a Outstanding loans in "irregular" status.

savings cooperatives had offered Ft 151 billion worth of bad debts for sale (book value), which implies an expenditure of Ft 118 billion by the government. But these offers do not have to be accepted: both sides could withdraw from the bargain until 10 March 1993, if they deem that the sale conditions are not attractive enough.

In *Poland*, bad loans are classified into two main categories — "doubtful" and "loss" — and the reserve provisions are 50 per cent and 100 per cent, respectively. The draft plan for solving the problem was limited initially to the nine largest state-owned banks, where the share of bad loans varies between 20 and 50 per cent.[440] The plan calls for the creation of a special department within each bank, staffed by experienced personnel, to which all bad debts would be transferred. Debtor enterprises are given a period of time (usually four months) to come up with restructuring proposals; if they fail to submit a programme, or if the proposals are not accepted, the bank forecloses on the assets of the enterprise, or initiates forced liquidation pursuant to bankruptcy. The legislation leaves open the possibility of entering into a "conciliation" procedure with creditors. For the funding of the plan it is proposed to rely on external sources, partly originating with the "dormant" stabilization fund of $1 billion, and partly from new loans obtained from the World Bank. These resources are to be used for re-capitalizing banks in order to strengthen their motivation to initiate bankruptcy procedures and accept irrecoverable losses.

The scheme adopted in *Romania* was aimed not only at the "bad" debt problem, but also at eliminating interenterprise arrears, which by the end of 1991 had grown to more than 40 per cent of enterprise turnover. The authorities decided to use a one-time credit injection to debtors to pay off enterprise arrears. The National Bank of Romania released leu 426.4 billion worth of "compensation" credits in the first quarter of 1992, secured on the assets of debtors, carrying maturities of up to six months and charged at market interest rates.[441] The "global compensation scheme" allowed arrears to be reduced to one quarter of the original gross total, but did little to change the structural tendency of Romanian companies and banks to accumulate bad loans and inter-enterprise debts.[442] As a result, the arrears problem re-emerged, and by mid-1992 enterprise arrears had again reached 25 per cent of turnover. Moreover, the inevitable increase in money supply resulted in higher inflation.

An essentially similar approach was applied in *Russia*, where the central bank, faced by the spontaneous and uncontrollable growth in inter-enterprise arrears amounting to R3,200 billion at the end of June 1992, decided to extend "soft" credits to settle the unpaid debts. The arrears indeed fell to R1,200 billion in September 1992, but later the scenario was identical to that in Romania: arrears increased again and inflation accelerated sharply.

The conclusion which may be drawn from this review of policies in transition countries is that the problem of bad debts is essentially of a structural character, connected with the heavily distorted structure of incentives facing enterprises and banks in post-communist economies. "Fast-relief" operations cannot cure it permanently, although they can become an important element of a broader package of structural measures.

(a) "sub-standard" if they involve considerable "branch risk", (b) "doubtful" if the borrower made losses in the preceding two years or his arrears exceed 60 days, and (c) "bad" if the borrower is under bankruptcy proceeding. This legislation further requires that provisions be made against 20 per cent, 50 per cent and 100 per cent, respectively. The World Bank, *Transition*, Vol.3, No.11, December 1992-January 1993, p.15; Government Spokesman's Office, *Hungarian Review*, No.1, 1993, p.5.

440 Republic of Poland, Ministry of Finance, *"Selected Aspects of the Financial Sector in Poland"*, Warsaw, November 1992.

441 *Romanian Economic Newsletter*, Vol.2, No.1, April 1992.

442 *Transition*, The World Bank, Vol.3, No.10, November 1992, p.3.

(iii) Fiscal reform

Budgetary problems experienced in all transition economies have demonstrated serious deficiencies in their traditional fiscal regimes and accentuated the need for sweeping reforms of the *budget and tax systems*. The first step of budget adjustment is typically a drastic cut in subsidies, coupled with price liberalization. Also reduced are budget-financed investments and expenditures for "collective" consumption and national defence. The second step, which normally requires much longer preparation and extensive legislative work, is to overhaul the tax structure, to strengthen the fiscal revenue base through more efficient and widespread taxation. The central component of this reform is the shift to the VAT and the introduction of personal income taxation.

This reform pattern is only partly determined by the general logic of market transformation; in practice, it also reflects attempts to tackle the fiscal crisis which tends to emerge once the initial stage of transformation is complete (see section 3.4(ii)).

Most European transition countries have either initiated comprehensive fiscal reform or are in the advanced stages of its implementation.

Hungary, which was first to introduce a VAT in 1988 together with individual income tax, now enjoys the most efficient and probably best-structured fiscal system among the transition countries. In *Czechoslovakia*, a series of laws passed in 1992 have radically changed the country's tax system. Form 1 January 1993 the following taxes came into force: (a) value-added tax, (b) excise tax, (c) tax on income of legal persons, (d) tax on income of physical persons, (e) tax on real estate, (f) tax on transfer of real estate, (g) tax on inheritance and gifts, (h) tax for environmental protection, and (j) road tax. These taxes replace the large variety of taxes existing prior to 1993. Among these were separate taxes on salaries, taxes on the incomes of artists, taxes on agricultural income, etc.[443] The most important change was the substitution of the turnover taxes by a VAT and the unification of income taxes.

Tax reforms in other countries were less comprehensive, but nevertheless substantial progress has been made. *Bulgaria* replaced a highly differentiated profit tax rate (which varied from 15 to 95 per cent) by a uniform tax rate of 40 per cent in 1991, and the number of turnover tax rates has been reduced to three (zero, 10 and 22 per cent) – an important preparatory step for the introduction of a VAT. In *Poland*, the personal income tax entered into force in January 1992, but the introduction of the VAT has been repeatedly postponed, apparently for technical reasons. In *Romania*, the three most important fiscal changes included: (a) replacement of the previous, highly progressive profits tax by two flat rate levies, which were, respectively, 30 per cent on profits of less than 1 million lei and 45 per cent for higher profits, (b) a new dividend tax (10 per cent), and (c) a new tax on earnings from the sale of assets (20 per cent).

A somewhat different sequence has been followed in *Russia* and other *newly independent states of the CIS*. VAT had already been introduced in early 1992, accompanied by taxes on production and exports of mineral resources. It was chiefly thanks to the latter that the Russian budget did not sink into a disastrous deficit in 1992. VAT rates in the CIS countries have been fixed at relatively high levels: 18 and 28 per cent in Russia, 28 per cent in Ukraine, and even 50 per cent in Belarus. But due to hasty preparations and general economic disorder, the revenues from VAT as well as from personal income tax fell much below targeted for 1992.[444]

Legal changes in the fiscal systems have in all countries been accompanied by changes in the structure of revenues and expenditures. On the *expenditure side*, drastic cuts in subsidies and in central investments have been effected in Bulgaria, Czechoslovakia and Poland, especially during the initial periods of the stabilization programmes. For instance, the share of *subsidies* in total budget expenditures in Poland diminished sharply from 32.7 per cent in 1989 to 9.1 per cent in 1991, and in Bulgaria from 20.3 per cent in 1990 to 9.0 per cent in 1991. Similarly, the share of *investment* expenditures declined from 9.5 per cent to 6.5 per cent in Poland and from 5.2 per cent to 4.4 per cent in Bulgaria. This tendency coincided with a relative expansion of *social transfers*. The share of social security and welfare payments in Polish budget expenditures increased from 5.8 per cent in 1989 to 20.4 per cent in 1991, and in Bulgaria from 20.3 per cent to 29.2 per cent between 1990 and 1991.

In principle, this should be seen as a positive development, as it demonstrates a general tendency of transferring the social protection function away from enterprises to the budget and the transfer of investments to enterprises, both of which are consistent with the logic of market reforms. But there are two imminent dangers in this process. First, while productive investments can indeed be carried out in a more efficient way by private sector companies than by government agencies, cuts in centrally financed infrastructural investments (roads, telecommunications networks, public utility systems, public transport, etc.) may hamper recovery and discourage private and foreign entrepreneurs from starting or expanding economic activities. Even so, it appears that drastic austerity measures applied in transition countries have also been extended into these infrastructural areas, even though the technical infrastructure is commonly considered to be seriously underdeveloped. Secondly, there seems to be a systematic tendency to construct social safety nets which later turn out to be both economically inefficient and financially very costly. The generous unemployment compensation schemes, relatively high pensions indexed

---

[443] See Act No. 212/1992, *Czechoslovak Foreign Trade*, No.7, 1992, p.10.

[444] For more on this issue, see e.g., D.M. Nuti, J. Pisani-Ferry, "Post-Soviet Issues: Stabilization, Trade and Money", *Commission of the European Communities*, Economic Papers, No.93, May 1992.

TABLE 3.5.4

Selected items of budget revenues in transition countries, 1990-1992
(Per cent of total revenues)

Country	Corporate profit tax			Turnover (or VAT) taxes			Customs duties and import levies			Individual income tax		
	1990	1991	1992	1990	1991	1992	1990	1991	1992	1990	1991	1992
Bulgaria	28.4	31.2	13.8	17.1	18.0	..	..	..	..	8.0	8.7	..
Czechoslovakia	49.9	40.9	25.1 [a]	30.5	26.4	25.8 [a]	..	..	3.1 [a]	..	14.4	35.3 [a]
Hungary	14.7	10.8	8.1	39.8	40.1	43.2	7.9	8.6	11.9	8.2	17.4	18.0
Poland	45.1	25.5	16.5	20.2	29.9	33.0	1.9	8.1	8.8	..	..	22.9
Romania	..	..	23.7 [a]	..	..	31.7 [a]	..	..	5.3 [a]	..	..	32.4 [a]
Slovenia	16.8	2.0	1.9	17.0	22.4	..	..	..	..	..	..	..
Russia	..	..	34.1 [b]	..	..	36.9 [b]	..	..	7.9 [b,c]	..	..	8.8 [b]
Latvia	28.7	20.0	..	42.9	31.0	..	-	0.1	..	9.3	9.9	..
Lithuania	25.8	16.2	9.9	36.5	30.8	32.5	0.9	0.2	..	11.9	12.3	6.7
Estonia	..	27.2	23.5	..	32.0 [d]	52.1	..	0.2	3.6	..	..	..

Source: National statistics.

[a] January-October 1992.
[b] January-September.
[c] Including export taxes.
[d] Including excise duties.

to wages, and voluntary retirement plans introduced in Poland, Czechoslovakia or Bulgaria distort the working of labour markets, cause deficits, and are unsustainable in the longer run.

The unexpected depth of the recession and growing social tensions connected with high levels of unemployment led to some revisions of preliminary plans for fiscal reform. After the initial resolute departure from the policy of massive subsidies and unsound income equalization, symptoms of some retreat have appeared. In Poland, Romania, Russia and other CIS countries, redistribution of funds from profitable to loss-making firms is likely to increase again, largely for political and social reasons. The change has been visible in Poland, where several "structural" programmes have been prepared by the government to help whole sectors and individual enterprises particularly hard hit by the recession and collapse of export markets in former CMEA countries. Such programmes have already been announced for the coal mining sector, the aircraft industry, selected shipyards and also for some regions plagued with high unemployment. In view of the well-known weaknesses of the capital market in Poland, these programmes are probably necessary to avoid major social clashes and the economic losses which would arise from mass closure of state enterprises. But as long as the details of these programmes are not published, and without a credible timetable for restructuring, there is a danger that they may take the least efficient form of permanent protection of vulnerable sectors from outside competition. This possibility is looming large in Russia, where the new government recently authorized the issue of large amounts of cheap credits for the fuel and energy sector in order to improve its liquidity position.[445]

In some cases there has been a return to more differentiated tax rates, a move which obviously contradicts the fundamental principle of neutrality of taxes. For instance, the Russian government decided to differentiate the share of participation of particular regions in total VAT revenues. The share, which had been initially set at a standard 20 per cent with no exceptions, has been recently diversified as between 5 and 50 per cent, depending on the revenue generation potential of the region.[446]

Important changes have also occurred on the *revenue side* of government budgets. A general tendency observed in almost all transition countries has been the shift from direct to indirect taxes and duties. The heavy reliance on net income (profit) taxes, primarily concentrated in the state enterprise sector, was the key factor behind the sharp fall of budget revenues when recession squeezed companies' profits and destroyed a large part of the tax revenue base. Change has been most dramatic in Poland, where the share of corporate profit tax in total budget revenues dropped from 45.1 per cent in 1990 to 16.5 per cent in 1992, while the share of turnover tax increased from 20.2 to 33.0 per cent. A similar trend has been visible in Hungary, where the share of corporate profit tax fell from 14.7 per cent in 1990 to 8.1 per cent in 1992, while the share of turnover taxes (VAT plus excise tax) increased from 39.8 to 43.2 per cent (table 3.5.4).[447]

A second tendency throughout the region was replacement of the obsolete system of itemized turnover

---

[445] The decision by V. Chernomyrdin, the new Russian Prime Minister, to extend R.200 billion "soft" credits to the energy and fuel complex was aptly criticized in *Finansovye Izvestiya*, 14-20 January 1993.

[446] *Izvestiya*, 5 January 1993.

[447] This tendency is probably of a more permanent character. As pointed out by many observers, the share of profit taxes in GDP in centrally planned economies typically varied between 15 and 20 per cent, compared to a 3 per cent average for OECD countries (see e.g., V. Tanzi, "The Tax Reform in the Economies in Transition: A Brief Introduction to the Main Issues", *IMF Working Paper*, WP/91/23, March 1991). The transition process may be expected to entail a substantial decline in the proportion of corporate profit taxes to levels corresponding with those observed in western market economies.

TABLE 3.5.5

Tax regimes in transition economies, as of 1 January 1993
(Tax rates in per cent)

Country	Profit tax rate	VAT Since (date)	VAT Rate	Turnover tax rates	Individual income tax	Social security tax	Land tax
Albania	..	-	-	..	..	..	..
Bulgaria	35 a	-	-	0, 10, 22	0-40	20-42	..
Czechoslovakia	45 b	January 1993	0, 5, 23	-	5-47	50	Yes
Hungary	40	January 1988	0, 6, 25 c	-	0-40 d	51 + 10 e	Yes
Poland	40	July 1993	0, 7, 22	1-35 f	20-40	45	Yes
Romania	30, 45	-	-	3-15	6-45	..	..
Croatia	40	-	-	5-50	..	11 + 10 e	..
Slovenia	30	-	-	5-20	..	22.7 e	..
Yugoslavia FR	30-40	-	-	..	0-30	..	..
Belarus	30	January 1992	28	-	0-50	..	..
Russia	32	January 1992	10, 20 g	-	0-40	38	..
Ukraine	0-21	January 1992	10, 28	-	0-50	..	..
Estonia	35	January 1992	0, 18	-	16-50	20	..
Latvia	35 h	January 1992	6, 12	-	15-35	37 + 1 e	Yes
Lithuania	29 i	January 1992	0, 18	-	18-35	30 + 1 e	Yes

Sources: *Business Eastern Europe*, 15 February 1993, p.7, *Central European*, May 1992, pp.40-43, and national sources.

a But 30 per cent tax rate for joint-venture companies with 49-plus per cent participation of foreign capital, and 40 per cent tax rate for joint ventures with foreign capital participation less than 49 per cent.
b 20-55 per cent until end-1992.
c Subject to changes in 1993.
d The upper limit was reduced from 60 per cent in 1989 and from 50 per cent in 1990-1991.
e The second figure is the contribution paid by employees.
f Turnover taxes will be replaced by the VAT in July 1993.
g Until 31 December 1992 the VAT rates were 15 and 28 per cent.
h 45 per cent for service companies since October 1992.
i 10 per cent in preferred sectors such as agriculture and transport.

taxes by a uniform VAT system. Hungary was the first country in eastern Europe to introduce a modern VAT system based on three basic rates (zero, 6, and 25 per cent) as early as in 1988, but it was followed by all former Soviet republics in 1992 and by the Czech and Slovak Republics in 1993. Other changes in the tax structure included greater reliance on individual income tax, which in the past had played a purely symbolic role. Also, there has been a tendency to exclude social payments from regular budgets and to finance them entirely from special contributions by enterprises and individuals.

Because fiscal reforms in transition countries have proceeded at various speeds, current systems not only differ across countries but also cannot be considered as very stable (table 3.5.5 summarizes existing tax regimes in transition countries). Further changes can be expected in the course of 1993, such as the introduction of VAT in Poland scheduled for the beginning of July 1993.

The most characteristic feature of fiscal reform, common to all transition countries, is probably the *massive shift to VAT* and away from traditional turnover taxes. This uniformity should not be surprising for a number of reasons. To begin with, VAT enables all products (on which the same rate is levied), to be treated equally – no matter how many times they have been traded before they reach the final consumer or what production techniques have been used. In addition, unlike the cascade-type of turnover tax, VAT is neutral with respect to foreign trade, i.e., it allows for precise identification and rebate of the tax on exports, which can then leave the country free of tax. Imports can be taxed at exactly the same rates as domestic production.

Another reason is that the VAT is a very stable and flexible source of government revenue and opportunities for tax avoidance and tax evasion are more limited than under income tax.[448] Thus VAT reform permits the stabilization of the fiscal revenue base, the streamlining of collecting procedures, and enforces the operation of an accurate and precise accounting documentation which thus discourages the "shadow" economy.

These arguments have been important to the transition countries, not only because of the critical state of their budget accounts, but also because of their intention to integrate their economies with those of other industrialized countries (VAT is currently in use in 21 out of 24 OECD countries). Before 1989, VAT was used only in Hungary; but by the end of 1993 it can be expected to have been introduced in all transition countries.[449]

---

[448] See S. Cnossen, "Key Questions in Considering a Value Added Tax for Central and Eastern European Countries", *IMF Staff Papers*, Vol.39, No.2, June 1992.

[449] Poland and Romania are at various stages of preparation but both have the necessary legislation in place and intend to shift to VAT in 1993. These plans are typically part of a broader tax reform, involving a reduction in the number of the various taxes, compression of tax brackets and rates and strengthening of the

(iv) The dilemma of trade liberalization

(a) The concept of trade liberalization

The trade regime or the trade system as understood hereafter comprises the set of rules determining the way decisions on prices and quantities traded are taken and also includes trade policy and foreign exchange rate policy. In the most general terms, trade liberalization means that the government's role in the decision process is diminished, i.e., not only are decision-making powers shifted to autonomous (private and public) enterprises, but also the economic environment in which they operate is less regulated by direct government actions.[450] Thus, a liberal trade regime may be considered as one based on market forces and with little government intervention.

More specifically, the *concept* of trade liberalization covers three important aspects: the level of trade restrictions; the form of trade restrictions; and the structure of trade restrictions. Correspondingly, the evaluation of a country's trade regime depends on three factors: the degree of international openness; the degree of transparency; and the degree of neutrality.[451]

Traditionally, international *openness* has been measured by the ratio of exports and imports to GDP or to national income.[452] But this approach does not allow for distinctions between large and small economies. More importantly, it lacks a straightforward relationship with economic policy. Specifically, it may be argued that high shares of exports and imports in a country's GDP do not necessarily imply that the country is indeed "open" to international trade in the sense that the domestic market is easily accessible to foreign exporters or, vice versa, that domestic firms can freely engage in international transactions. An alternative approach, based on the concept of so-called "institutional" openness, is more policy-oriented and seems to be more useful.[453] According to this approach, a country is considered more open to international trade the fewer the restrictions faced by domestic agents on entry to international markets and by foreign agents on entry to the domestic markets (including not only trade barriers but convertibility restrictions as well).[454]

The trade system is considered *transparent*, when it is regulated predominantly by well-established, clear-cut and widely-publicized rules rather than by actions left to the discretion of the government and/or political authorities. Substitution of tariffs for quotas is one example of a move towards a more transparent regime. Another example is the replacement of administrative allocation of quotas by a public auction of import rights.

The degree of transparency is analytically more difficult to measure, but some insights can nevertheless be obtained through estimating the proportion of tariff restrictions in total trade restrictions (i.e., including both the tariff restrictions and non-tariff barriers such as quotas, currency controls, licensing, etc).

The *structure of protection* is of vital importance for the optimal allocation of resources and, therefore, for the growth of a country's welfare. As domestic producers are always confronted with the possibility of choosing between the domestic and the foreign market, the level of protection offered to the domestic market compared with the level of protection (support) offered to exporters determines the resource allocation proportions between import-substituting and export-oriented activities.

All three aspects of trade liberalization are important in the reform process in transition economies, although progress on each front has not been identical. It may be argued that it should not be attempted to establish all the components of trade liberalization instantaneously, and that largely diversified initial conditions may require a country-specific sequence of liberalization steps.

(b) Benefits from trade liberalization

A liberal trade regime is not an aim in itself but rather an important tool to promote economic development. But should it be taken for granted that trade liberalization is always and everywhere a good thing? There seems to be a consensus among the economic profession that higher and more equitable economic growth tends to be associated with higher growth of exports, and that a more outward-oriented trade policy gives better trade performance in terms of export expansion and external balance.[455] There is also wide agreement that transparent measures are generally more effective than administrative, discretionary policies. But there are controversies on whether neutrality should be achieved through import liberalization, i.e., reduction

---

tax authorities. Bulgaria also announced plans to introduce VAT in 1993, although its legislation is less advanced (see *Bulgaria: An Economic Assessment*, OECD, Paris, 1992, p.19). Only Albania has set no specific date for the tax reform.

450 O. Havrylyshyn and D. Tarr succinctly define trade liberalization as a "... movement away from too much and wrong government intervention" ("Trade Liberalization and the Transition to a Market Economy", *Working Papers*, WPS 700, The World Bank, July 1991).

451 See D.K. Rosati, "Institutional and policy framework for foreign economic relations in Poland", in *Reforms in Foreign Economic Relations of Eastern Europe and the Soviet Union*, Economic Commission for Europe, United Nations, New York, 1991, pp.21-31. Although not mentioned explicitly, the three aspects are nevertheless implicitly recognized in most of the World Bank's recent studies on trade liberalization.

452 See e.g., B. Balassa, "Export Incentives and Export Performance in Developing Economies", *Weltwirtschaftliches Archiv*, (1140), 1978.

453 D.K. Rosati, K. Mizsei, "Adjustment through opening of socialist countries", *WIDER*, Working Papers, No.52, Helsinki, January 1989.

454 Analytically, the degree of institutional openness towards international markets can be measured by the ratio of the amount of indirect taxes and subsidies to trade, plus the tariff equivalent of quantitative restrictions, foreign exchange restrictions and licensing, to the total value of tradables at international prices.

455 See A. Krueger, *Foreign Trade Regimes and Economic Development: Liberalization Attempts and Consequences*, Balinger, published for NBER, Cambridge, MA., 1978; S. Edwards, "Openness, Outward Orientation, Trade Liberalization, and Economic Performance in Developing Countries", The World Bank Working Paper, June 1989, mimeo.

of protection measures, or via compensatory export-promotion measures. Similarly, issues concerning the optimum sequencing of trade liberalization are not entirely clear. While it is relatively simple to establish theoretical links between liberal trade regimes and improved economic performance, it is more difficult to support these links by consistent empirical evidence which would demonstrate beyond doubt the causal relationship existing between trade liberalization and economic growth in individual cases.[456]

Let us first look into the effects of increased *openness*. The benefits stemming from the removal of import restrictions can best be illustrated within the familiar partial equilibrium framework. A reduction of a tariff (quota) reduces the domestic price of imported goods and thus increases net welfare (measured as the difference between the consumer surplus and producer loss plus government revenue loss). Lower domestic price also means increased competitive pressure on domestic producers. But the increased openness towards international markets also allows the "importation" of international relative prices, which affect the product mix in the country. The welfare benefits stemming from improved relative prices can in turn be illustrated within the general equilibrium framework. Domestic production shifts towards exportables, both exports and imports increase and the level of welfare increases as a result of more efficient use of domestic resources. These results are valid almost universally, except in the case of the optimal tariff for a large country, an exception which does not apply to most transition countries (with the possible exception of Russia). A wider range of goods available on the domestic market not only extends consumer choice, but also allows imports of better and more appropriate technology, thus directly contributing to economic growth.

The benefits stemming from a more *transparent* system of trade fall into two main categories. First, tariffs are superior to quotas because they do not eliminate external competition (unless they are prohibitively high) and in most cases provide a more reliable and stable basis for government revenues. Second, significant social costs are avoided because profit-seeking and sheer corruption, so notorious under administrative regimes, is diminished under transparent systems.

The benefits connected with a *neutral* structure of incentives can again be illustrated within the general equilibrium framework. A structure of protection biased towards the domestic market results in excessive import-substitution in production and excessive consumption of exportables. A balanced structure of protection removes these distortions and moves production, consumption and trade towards Pareto-optimality, thus increasing the level of welfare.

While the benefits of free trade have strong theoretical underpinnings, *empirical evidence* is less sanguine.

Extensive studies, made mainly under the aegis of the World Bank, confirm many of the theoretical conclusions. Yet many countries which started liberal reforms failed to improve their economic performance and eventually retreated from the reform track.[457] This may suggest that, while trade liberalization as a general policy orientation is most probably a good thing, the ways and means of its implementation are not as universal and simple as considered by many and should differ according to the specific conditions of individual countries.

The recognition of the significance of different local conditions also means, however, that the theoretical validity of the conclusions obtained may be limited. It is not the logic of the free trade argument which is called into question, but rather the standard *assumptions* underlying the theoretical analysis of gains from trade — and especially the general equilibrium model which has been the main driving force of normative thinking in the area of economic policy. Three assumptions seem to be of special importance in this regard, because they are very unlikely to be fulfilled in the context of transition economies. First, the assumption of *competitive markets with profit-maximizing agents and perfect information* is clearly much farther from reality in these economies than it is in western economies. Second, *externalities* are significant, and should not be ignored. Third, the *income distribution* aspect is an important constraint on the speed and scope of reforms in transition countries.

The most important is the competitive markets assumption. How does it affect the analysis of trade liberalization effects? Starting again with the simplest case of the removal of a tariff, the ensuing price effect reduces the profitability of import-substitution activities and drives out the least efficient producers. The resources released are assumed to shift to export sectors where profitability increases. This is simply a structural adjustment, evoked by a combination of demand and supply factors.

But for this adjustment to take place, several things have to happen. First, output has to decrease and unemployment to increase in the import-substitution sectors; this should be accompanied by a fall of real factor incomes in them. Second, the redeployment of resources to export sectors can in principle be done only through new investment, which may have to be quite substantial if the reallocation needs are large. For the new investment to materialize, capital resources have to be transferred because self-financing is likely to be insufficient. For the capital to be transferred, not only must the future profits prospects be attractive enough — i.e., the net present value of new projects must be substantially positive after allowing for the risk factor — but also reasonably efficient capital markets have to be in operation. In the meantime, manpower will also have to be reallocated, which requires a reasonably

---

[456] It should be noted however, that a vast body of empirical literature supports the view on the existence of a close causal relationship between trade liberalization and the rate of economic growth. But these conclusions are typically drawn on the basis of large samples of liberalization episodes, observed in various countries in various periods, and have a stochastic rather than a deterministic character.

[457] See e.g., D. Papageorgiou, A.M. Choksi, M. Michaely, "Liberalizing Foreign Trade in Developing Countries. The Lessons of Experience", *The World Bank*, Washington, D.C., 1990.

flexible labour market; moreover, unemployment effects have to be contained through an efficient social safety net. To sum up, for the adjustment mechanism to work efficiently three things are needed: time, money and market institutions. But none of them is in sufficient supply in transition economies.

(c) Specific conditions of transition economies

The scope of the transformation in eastern Europe obviously goes beyond and above any reform attempt in post-war world history. Latin American experience, so influential in shaping the new liberal orthodoxy, has been mostly limited to macroeconomic stabilization, market deregulation and reduction of the public sector. Similarly, the celebrated German reform of the late 1940s essentially consisted of removing widespread distortions from the economy, but its institutional dimension was relatively limited. Nevertheless, it still took years for these changes to yield the expected results.

The east European transformation is nothing less than a revolution. One should not be misled by the peaceful and largely non-violent character of this process. Deep economic reforms are being accompanied by fundamental political changes and also, in most cases, by the restoration of national sovereignty. The interaction of social, political and national aspects creates a highly unstable and novel environment for the economic reforms. In some cases they facilitate the change, whereas in others they impose additional constraints on the reform process. While the initial enthusiasm stemming from national emancipation and restored political freedoms may provide a conducive climate for truly radical reforms, it also creates unrealistic expectations for the rapid improvement of living standards. These aspects cannot be ignored if the economic reform is to avoid getting locked into political and social conflicts.

Emerging from the socialist era, the east European economies differed from other reforming non-socialist economies in many important respects. They differed even more from developed market economies. The most fundamental differences concerned their structural and institutional characteristics.

Unlike in other countries, the *private sector* in most transition countries was almost non-existent. The share of the private sector in GDP in the mid-1980s varied from some 20 per cent in Poland (where the bulk of it was in agriculture) to less than 3 per cent in the USSR, Bulgaria and Czechoslovakia, and its share in industrial output was generally even lower. The state sector itself was but an extension of the centralized management system, and its "enterprises" barely deserved that designation, having to work under a distorted incentive structure which produced perverse responses to market signals and macroeconomic policy measures. The conversion of such firms to more entrepreneurial entities (through corporatization or privatization) is proving a slow process (see section 3.6). In such conditions, adjustment is sluggish and not necessarily economically efficient.

Industrial structure in transition economies was geared for many years to the needs of the huge and uncompetitive Soviet market. It was also shaped by specific economic policy priorities which were concentrated on the military and heavy industry sectors, while the service sector was weak and underdeveloped. The key implication of this structural bias is that the scope of the required reallocation of physical and human resources is much larger than in other reforming countries. This means that for the necessary adjustment to take place in transition economies more funds and more time is needed.

The well-known feature of the Soviet-type economies is their high degree of *specialization and monopolization* − a direct consequence of central planning. Monopolistic firms' response to market signals is typically perverse, and possibilities of breaking up the monopolies and establishing competitive markets from "within" are limited in the short run.

Finally, the marketing and management *skills* necessary to operate in the market environment are in short supply in the centrally planned economies. Of course, they can be eventually acquired either through "imports" of managers and/or foreign firms, or by "learning-by-doing", but neither can be done overnight. On the other hand, a large part of the *human capital* accumulated in the industrial sectors becomes obsolete once the economy opens up to international markets and needs to be retrained.

Institutional differences are no less important. The *price structure* in centrally planned economies, being an outcome of widespread price controls and long-term isolation from international markets, displayed many distortions and differed considerably from the relative prices prevailing in market economies. The relative price shock resulting from general price liberalization and realignment drastically changes profitability patterns in the economy. The *legal framework* for market mechanisms is only in *statu nascendi* in such essential areas as property rights, the enforcement of private contracts, bankruptcy laws, the banking system, taxes, customs control, etc. *Capital and labour markets* are rudimentary and in most transition countries have to be built *ab ovo*. This is an important impediment on resource mobility, which in turn is required for structural adjustment. Also, the different *habits and mentality* of east European societies should not be forgotten: they were accustomed for decades to job security, or egalitarian income structure and generally low work discipline. But the transformations require exactly the opposite: jobs cannot be guaranteed, greater income and wealth differentiation is inevitable and work effort and productivity have to increase. To change this "socialist" mentality will take some time, and the process of acquiring new ethical and work standards may result in severe social unrest.

(d) Implications for trade liberalization

This sketchy review is not meant to give a full description of the "distorted world of Soviet-type economies"; it aims simply to recall that conditions in the transition economies are quite different from those in market economies. The necessary adjustment may be expected to assume considerably larger proportions because the change in relative prices is indeed dramatic, and the structural distortions inherited from the communist past are so very significant. However, the transition countries have not much time left for a successful adjustment and money and market institutions are in limited supply.

New investments on the required scale will be difficult to obtain because of the limited amounts of capital in those countries. The restrictive macroeconomic policies followed in 1990-1991 reduced the level of domestic demand, thus diminishing prospects of future profits and business confidence. They also made domestic credit relatively expensive and difficult to obtain. Furthermore, capital markets are underdeveloped and an efficient mechanism for transferring the limited savings available to the production sector is missing. The most important is, however, the risk factor: not only are the economic prospects of these countries are uncertain, but the social and political situation is also becoming less and less stable.[458]

Under such conditions, the necessary capital transfers will not simply materialize. Not only will the time span needed for adjustment have to be extended, but the adjustment costs will be much higher. Many enterprises which could potentially be viable under efficient markets, will probably disappear altogether. A large proportion of physical and human capital resources, which could otherwise be effectively redeployed, will be irrevocably lost. The costs will be particularly substantial in manufacturing (and especially in high technology) industries, where the need for restructuring is greatest, and where large and prolonged unemployment may lead to an irreparable loss of skills. These are important *externalities*, which so far have been disregarded or belittled in the design of economic transition policies, and the outcome may be considered a kind of *market failure*.

The limitations of the traditional model of gains from trade and the specific unfavourable conditions and high social aspirations in the transition countries call for a very careful application of liberal policies. While these observations do not question the need for transparency and neutrality of trade policies, they suggest that the opening of the economy should not be abrupt, because adjustment possibilities in transition economies are much more limited than in market economies. If this is the case, then active trade policies should be applied — on a selective basis and with limited scope — in order to strengthen the market mechanism wherever it is too weak, and correct it wherever it is significantly distorted.

Of course, the possibility of *government failure* cannot be excluded either. Under still distorted prices it is difficult to determine a priori the areas of comparative advantage and to pick out "winners" in terms of sectors, branches or individual firms. Besides, the danger of favouritism, corruption and profit-seeking is still considerable in the transition countries. On the other hand, however, democratically appointed governments seem generally less prone to these dangers than are totalitarian governments. A possible solution would be to avoid a highly differentiated, made-to-measure protective trade policy structure, and rather start with a uniform, moderate tariff on manufactures, lower (or zero) tariffs on production inputs and, possibly, no quotas. Initially, it could also be accompanied by a uniform export subsidy, although budget constraints may make this option unfeasible. In all cases, a competitive exchange rate and a large measure of current account convertibility are necessary conditions for successful trade reform.

It is important to note that the specific conditions call for a "specific" package of trade liberalization measures. While large initial price distortions, market shortages and monopolization provide strong arguments for an early opening to international trade, especially imports, other factors such as the scope of structural adjustment required, low competitiveness, lack of market institutions and great uncertainty provide arguments for temporary protection. It is, however, essential for the adjustment mechanism to be set in motion as early and as efficiently as possible. Hence, market signals should be strong and undistorted from the very beginning of transformation. But undistorted prices and competition cannot be established internally overnight, because of high monopolization, the weakness of domestic markets and the sluggish response of the state sector. Strong policies to foster competition are therefore important, but these can be aided significantly by the "import" of correct relative prices and competition from abroad. For this to happen, transparency and neutrality should be implemented at the outset of the programme. This means the replacement of quotas and administrative restrictions by tariff measures, because the latter do not blunt the spur of competition, and the removal of the notorious import-substitution bias in the structure of protection. This also means the elimination of the most binding restrictions on currency convertibility and the introduction of a transparent, i.e., market-based, system of foreign exchange allocation, coupled with a neutral — i.e., neither undervalued nor overvalued — exchange rate.

By contrast, the opening of the economy — in terms of lowering import tariffs and export subsidies — will probably have to take some time. Large initial price differentials as compared with international markets and inherently slow adjustment responses call for the gradual removal of protection. Too abrupt an opening will result in a deep output contraction, large-scale unemployment and the danger of wiping out potentially viable industrial capacities on a large scale. The costs of such radical liberalization are likely to be excessively high and these negative consequences could

---

[458] Social instability and fragility of political structures has been repeatedly demonstrated in countries as different as Poland, Romania, Bulgaria and Russia. In Czechoslovakia, where the transformation started in a climate of remarkable national unity, the recent split into two separate states was prompted at least in part by the unexpectedly high costs of economic reform and their uneven distribution between the two republics. Even in Hungary, the political situation deteriorated in 1992.

damage the social and political stability of transition countries.

Of course, the level of initial protection must not be too high either. As mentioned earlier, a preferable solution would be to start with a uniform and moderate tariff for manufactures and very low tariff on intermediates.[459] Then, tariffs could subsequently be scaled down gradually according to a predetermined schedule extending over 5-7 years.[460] Together with a competitive real exchange rate, the system would thus provide incentives for export-oriented restructuring and allow the avoidance of unnecessary adjustment costs.

(e) Trade liberalization in transition: from the "new liberal orthodoxy" to a revival of protectionism?

The radical liberalization policy course first adopted in Poland, and then also in other east European countries, is a direct outcome of an important evolution in mainstream economic thinking observed over the last 10-15 years. A new free-market orthodoxy emerged at the beginning of 1980s, resulting from dissatisfaction and criticism of traditional economic policies which had been based for several decades on fundamental Keynesian ideas. The change, initiated by debates over the causes for the marked slow-down of economic growth in industrialized countries in 1970s, received additional support from the experience of the Latin American countries. That region was plagued for a number of years by high inflation, external deficits and sluggish rates of growth. Observers linked this poor performance to the excessive and pervasive role of governments, and especially to the systematic inward-orientation of economic development strategies. These were contrasted with the outward-oriented policies of some east Asian countries, which developed much faster and maintained control over domestic and external balances. On the other hand, the gradual decay of socialist countries and their declining competitiveness provided yet another proof of the inefficiency of excessive "etatism" in the economy.

On the theoretical front, the new orthodoxy was buttressed by the revival of monetarism, the theory of rational expectations and profit-maximizing agents and, above all, by the theories of second-best and of domestic distortions. The latter provided strong arguments against clumsy interventionism by politically-motivated governments and in favour of liberal economic policies.

The new ideas were quickly absorbed by the international financial institutions, which also actively contributed to the new intellectual trend. The involvement of the IMF and the World Bank in supporting economic stabilization and guiding reform programmes in developing countries in the 1980s was based essentially on three main pillars: macroeconomic discipline, microeconomic deregulation and liberalization, and outward orientation. Whenever asked for assistance in fighting internal and external imbalances, the IMF invariably suggested drastic cuts in government spending in order to achieve fiscal and monetary balance, extensive price decontrol, devaluation of domestic currencies and foreign trade liberalization.[461] These "orthodox" programmes were in some cases supplemented by less orthodox measures, such as wage controls or credit ceilings, if these were considered likely to work in favour of macroeconomic stabilization.

It is of course not the purpose of this *Survey* to attempt an evaluation of the IMF-supported adjustment and stabilization programmes. Suffice it to say that they have been based on a coherent economic logic and their contribution to bringing the macroeconomic situation under control is undisputed. But it should also be noted that these programmes suffered from insufficient attention to structural impediments and to sociopolitical conditions. Moreover, some crucial assumptions underlying the standard model of financial programming are increasingly considered problematic (e.g., the assumption of constant velocity of money circulation or of the profit-seeking motives of economic agents). This is probably why these programmes, although conceptually consistent, have been aborted or have failed in many developing countries.

Nevertheless, it was quite natural to think that essentially similar stabilization-cum-reform programmes would have to be adopted in the transition countries. Not only did the mainstream economic policy thinking and the unquestioned reputation of the IMF play a role, but the approach also fitted perfectly well into the socio-political intellectual atmosphere of the east European countries. Disastrous experience with central planning had led to a wholesale rejection of government involvement in the economy, while the high welfare levels achieved in developed western economies offered a demonstration of the efficiency of market mechanisms. Finally, the transition countries had little choice: the availability of external financial assistance hinged critically on their adoption of IMF-supported reform programmes.

When economic reform started in 1989-1990, uncertainties surrounding the measures undertaken and the results to be expected were substantial. A number of important policy dilemmas had to be solved without sufficient theoretical or empirical underpinnings. In the area of trade liberalization, four problems seemed to be of particular significance: the timing of the introduction of currency convertibility, the level of protection during transition, the level of the exchange rate (the size of the initial devaluation) and, in a more general context, the speed of liberalization. With the benefit of hindsight it could be concluded that some decisions, such as those

---

[459] A specific system of agricultural protection could also probably be put in place, depending on local conditions.

[460] It is important to make this scheme credible and independent of the changing political environment. Some experts speak in this context about a "hard tariff path". W.M. Corden, "Trade Policy and Exchange Rate Issues in the Former Soviet Union", op.cit.

[461] J. Williamson, "The Eastern Transition to a Market Economy: A Global Perspective", *Centre for Economic Performance*, Occasional Paper No.2, London, March 1992.

referring to convertibility, were proven right, while others now appear more questionable.

Specifically, the abrupt removal of trade barriers exposed domestic producers to strong international competition. Initial devaluation and low wage levels allowed them to withstand the first period of transformation without mass bankruptcies and mass unemployment, but at the same time they were unable to restructure and modernize because of the acute lack of financial resources. Later, when domestic currencies appreciated and comparative labour costs rebounded, their situation deteriorated sharply. Industrial and agricultural lobbies started to call for more protection, and in many cases they were successful. Tariffs were increased in Poland in 1991, while in Hungary new quotas were imposed in 1992. A retreat from liberalization can be also observed in Russia in the second half of 1992.[462]

(f) The optimum pattern of liberalization during transition: transitional tariff protection

If the logic of the argument in favour of a transparent system is accepted, then the main trade policy instrument to be used during transition should be a tariff. Quotas are acceptable, and sometimes even necessary, only in so far they are designated to reflect external obligations connected with trade restrictions imposed by other countries. For domestic policy reasons, quotas are in all cases inferior to tariffs. The case of energy exports from the former Soviet Union is sometimes used as an argument in favour of quantitative restriction on exports. But in that case also an export tax is preferable for a number of reasons — as discussed earlier.

While an opening towards international markets is an essential element of the transformation, it does not mean that tariffs should necessarily be eliminated or drastically reduced, as was done in Poland in 1990. There are important reasons for tariffs to be maintained at "moderate" levels (i.e., *not* lower than the tariffs prevailing in industrialized countries) for at least several years. The arguments in favour of such a tariff policy fall into several categories.

The most controversial is the argument of the need for a *transitional protection*. The core of the argument is that because the production sector in postcommunist economies is inefficient and structurally distorted, it cannot compete internationally, and thus has to be protected for some time in order to allow for the necessary adjustment to take place. Once the restructuring is made and the industries become competitive, the protection can and should be removed.

This reasoning has been attacked on many grounds. First, it is argued that the exchange rate policy can be used instead: in principle, if the exchange rate is sufficiently low, it can make most loss-making industries efficient. But this may not be feasible. Not only would it lead to an excessive reduction of real wages throughout the economy but, because of large disparities in the profitability of individual enterprises, the exchange rate would have to depreciate to absurdly low levels. But the most important argument against using the exchange rate to deal with structural deficiencies is the presence of "value-added subtractors". Under new relative prices some activities may yield negative value added, especially in energy-intensive sectors.[463] In such a situation, devaluation cannot help, and these activities would have to be discontinued.

Another argument against temporary protection draws on the experience of developing countries. Protection offered to infant industries in those countries very rarely resulted in an increase of international competitiveness which eventually would bring the removal of protection. As a result, "infant industries" never matured. It should also be added that the ability of governments to determine potentially efficient sectors and firms a priori is widely disputed.

Actually, as pointed out by some observers, the case of the transition economies concerns the protection not of "infant" but of "senile" industries.[464] These industries exist already, but have to restructure if they are to survive. They cannot receive credits for restructuring because of the weakness of capital markets, generalized lack of capital and the high sociopolitical risks in transition countries. They may, however, offer important externalities which should be taken into consideration — especially in manufacturing. Direct subsidies may be infeasible for budgetary reasons. These industries are also prone to rent-seeking. A second best solution might therefore be a transitional tariff protection. This solution has well-known disadvantages: protection of one industry always occurs at the expense of others. Moreover, it introduces an important price distortion. These costs should be weighted against current and future benefits.

However, scepticism is in order with respect to the possibility of establishing sector-specific or even product-specific protection structures with tariff rates varying according to the value-added content, as advocated by some economists. Since under distorted prices and costs it is difficult to determine a priori areas of comparative advantage and to pick out sector or branch "winners", what remains is a less ambitious but probably more realistic option of imposing a uniform, moderate tariff on all manufacturing industries.

Trade taxes are also a very important source of *government revenue*. This aspect should not be forgotten, because fiscal systems in the transition countries are generally obsolete and inefficient, which sooner or later results in a fiscal crisis. Since the traditional tax

---

[462] See United Nations Economic Commission for Europe, *Economic Bulletin for Europe*, Vol.44(1992), New York, 1993, pp.56-58.

[463] See R. McKinnon, *The Order of Economic Liberalization: Financial Control in the Transition to a Market Economy*, Baltimore, Johns Hopkins University Press, 1991.

[464] J. Williamson, "The Eastern Transition to a Market Economy: A Global Perspective", *Centre for Economic Performance, Occasional Paper No.2*, London, March 1992.

base shrinks because of prolonged recession and widespread tax evasions, alternative sources of budget revenues have to be sought. Trade taxes are easier to collect and to enforce, and thus should not be forsaken. Theoretically, a tax on imports can be equivalent to a tax on exports, but in practice, export taxes should be used only in special situations (e.g., exports of underpriced energy and fuels) because they have strong disincentive effect.

By contrast, tariffs (and other trade policy measures) should not be used for balance of payments reasons. External equilibrium is best assured by a competitive exchange rate and prudent financial policies. Tariffs should remain instruments of structural policies, independent or current stabilization goals.

3.6 PROGRESS IN PRIVATIZATION, 1990-1992

(i) Introduction

Little by little, privatization is forging ahead in all transition economies. Marked progress was made in agriculture, in the privatization of small businesses and the urban housing stock. Progress has been less rapid in the core: the medium and large-sized state-owned enterprises (SOEs), and virtually nothing has happened in the banking sector. This unevenness is understandable — in previous editions of this *Survey* it was argued that privatization is an immensely complex task with a long time horizon under the best of circumstances. However, the the changes which occurred over the past three years taken together fell short even of "realistic" expectations.

First, although institution-building and the creation of the necessary legal framework have accelerated, big gaps remain.[465] In several Soviet successor states and also in Albania, "large" privatization was formally suspended in the course of 1992 or early 1993. True, these suspensions were relatively short-lived almost everywhere — Ukraine being an important exception[466] — but they nevertheless indicate the underlying tensions. Other countries — including the three "advanced" reformers of Czechoslovakia, Hungary and Poland — have moved ahead on most fronts, but the speed of change remained behind declared intentions in these economies as well.

The second general observation is that the overall sociopolitical conditions have been significantly less favourable than expected. In early discussions on the modalities of privatization, a certain consensus was reached that "foundations must be laid first,"[467] including elected governments with popular support, competent public administrations, functioning legal instruments, liberalization of economic activity, a viable banking system and — last but not least — macroeconomic stability.

By now, however, many governments find that these requirements are not satisfied yet and that there is no hope of achieving them during their expected tenure. Yet, privatization was for them also political promise — hence their desire to move ahead in spite of unfavourable circumstances.[468] Moreover, it was often argued — not without some justification — that swift advancement on the privatization front would help to achieve many other goals as well. Acting under such pressures, governments find themselves in a situation where the range of options is limited to a choice between bad and worse.

The third and perhaps unexpected development is that the actual changes occurring in the transition economies are more similar to each other than could have been assumed two years ago. It seems that priorities, policies, instruments and proportions are converging in spite of the fact that countries had started from rather different positions and the first privatization programmes had shown a great deal of diversity. Now, it seems that practically all conceivable privatization mechanisms have already been applied or seriously contemplated (table 3.6.1). There is no such thing as a mainstream method. Of course, the relative proportions vary from country to country, but not as much as the original privatization scenarios had indicated.

Fourth, it is interesting to observe that so far privatization has met relatively little open public opposition, given the magnitude and importance of the underlying changes.

The objective of this section is to provide an overview of the progress achieved in privatization; discussion of the merits and demerits of various methods of privatization will therefore be kept to a minimum.[469] The next section recalls some of the salient points on these issues made in last year's *Survey*. This is followed by a description of privatization institutions as they developed in 1990-1992 (sections (iii) and (iv)). The focus of section (v) is on restitution. The main elements of the privatization process — small privatization, land reform and large privatization — are discussed in sections (vi) to (viii). The discussion is rounded off by a review of data on the size of the pri-

---

[465] For a recent overview of the principal laws on privatization, see United Nations Economic Commission for Europe, Committee on the Development of Trade, *Guide on Selected Legal Issues Related to Privatization and Foreign Direct Investment in the Economies in Transition: A Comparative Analysis*, Geneva, 25 November 1992. TRADE/WP.5/R.9/Rev.1. Annex 3.

[466] The dates of suspension were: Albania (25 April 1992-22 July 1992), Belarus (12 November 1992-27 January 1993), Estonia (27 November 1992-4 December 1992), Kazakhstan (13 November 1992-??), Lithuania (12 December 1992-15 January 1993), Russia (in the oil industry, as from 7 February 1993), Ukraine (November 1990-December (?) 1992).

[467] S. Zecchini, "Critical issues in privatization" in OECD, *Methods of Privatising Large Enterprises*, Paris, 1993, p.81.

[468] As two observers noted on recent developments in Russia, "Privatization is now emphasized for the perfectly legitimate political reason that it is about the only piece of positive news that the government can offer". S. Lainela and P. Sutela, "Russian Privatization Policies", paper prepared for the 2nd EACES Conference in Groningen, The Netherlands, 24-26 September 1992.

[469] For a comprehensive survey of these issues, as well as for a review of the literature, see J. M. van Brabant, *Privatizing Eastern Europe: The Role of Markets and Ownership in Transition*, Kluwer Academic Publishers, Dordrecht, 1992.

TABLE 3.6.1

The main mechanisms of privatization

Free distribution	Usufruct divestment	Sales of assets
To workers' councils	Franchises	Stock flotations
To workers	Management contracts	Auctions
Mass distribution	Leases	Negotiated
Shares	Sub-contracting	Management buy-outs
Investment funds		Workers' buy-outs
Mutual funds		Unsubscribed capital expansion
Decentralized government agencies		Debt-equity swap
Financial institutions		
Holding companies		

Source: United Nations Economic Commission for Europe, *Economic Survey of Europe in 1991-1992*, New York, 1992, p.231.

vate sector and its relative share within the individual transition economies at the end of 1992.

(ii) The framework of analysis

The present survey is a continuation of previous work by the ECE secretariat; the definitions and concepts used here are the same.[470]

Privatization is understood as all those actions which take the state out of decision-making concerning *existing* capital assets. Hence the start-up of new private ventures is not covered here.[471]

Three of the main conclusions discussed in the last *Survey* appear to have been borne out by experience over the past 12 months.

— First, privatization is an issue of political economy *par excellence*. This was reflected from the onset both in the motivations and the goals of privatization (table 3.6.2). Present governments cannot forget that the inherited capital assets were mainly created by the work of a generation which is politically still active. It is little wonder that these people want their views to be heard and their interests represented when the future of "their" capital is decided.[472] When an entire country is restructured according to the needs of a market economy, jobs and power positions are redefined at all levels. It is hardly necessary to re-emphasize that this "creative destruction" leads to rapidly growing unemployment — that was to be expected and a great deal of attention has been devoted to this. But it is less frequently mentioned that restructuring existing capital assets also means a redistribution of economic power and political influence. New elites are created, consisting partly of self-made men, partly of political appointees. It is a common experience in the region that governments are inclined to help their own clients in scaling the new heights (top management positions, seats on supervisory boards, etc.), which inevitably gives rise to resistance form the opposition and among ordinary citizens. Finally, it needs to be understood that, under the conditions of systemic transformation, privatization means not merely a redistribution of ownership rights among existing economic agents. New agents, new institutions are created day by day. The stakes are high: once institutions have become rooted in society, they are likely to persist for a long period of time.[473] This is one reason for the resistance met by governments in parliaments when ownership issues are under debate.

— Secondly, estimating the asset value of the very large number of state-owned enterprises (SOEs) is a hazardous undertaking under the best of circumstances, let alone in the inflationary environment which prevails in many transition countries; the high costs of such assessments can in general be justified only on behalf of owners or potential owners.[474] It is erroneous to believe that the market value of an SOE is a given magnitude, somehow intrinsically frozen in land, tangible assets, etc., so that once the right accounting expertise is acquired (or hired from abroad), these figures can be worked out with unquestioned objectivity. Ultimately, the market value of the SOE must be determined by the interplay of supply and demand forces.

— Policy has been focused too much on the divestment of large state firms, while the possibilities for improving their performance were not sufficiently explored. Systemic change in itself has already reduced or eliminated many growth-retarding factors: the *nomenklatura* system, the priority treatment of the military sector, ideological commitments to full employ-

---

[470] See United Nations Economic Commission for Europe, *Economic Survey of Europe in 1990-1991*, New York, 1991, pp.124-183, and *Economic Survey of Europe in 1991-1992*, New York, 1992, pp.191-256.

[471] But see sections (ix) and (xi) below on this important subject.

[472] As a Hungarian observer noted, people are deeply suspicious about privatization when somebody else gets something: "*He runs away with my national wealth.*" J. Köllö, "This is mine, this is yours", *Beszélő*, No.24, 1990 (in Hungarian).

[473] On this point, see F. Targetti, "The Privatization of Industry with Particular Regard to Economies in Transition", in F. Targetti (ed.), *Privatization in Europe: West and East Experiences*, Aldershot, Dartmouth Publishing Company, 1992, pp.1-32.

[474] A useful review of accounting and valuation issues, based on recent experience in some east European economies, can be found in OECD, Centre for Co-operation with European Economies in Transition, *Valuation and Privatisation*, Paris, 1993.

TABLE 3.6.2

The main motivations for and goals of privatization

Economic	Political	Social	Trade union	Wealth distribution
Efficiency	De-etatization	Integrate	Less wage pressure	People's property
Profit maximization	Strengthen democracy	Social peace	Reform consensus	Portfolio diversification
Shareholding	Social stability			
Budget easing	Remove *nomenklatura*			
Finance growth				
Foster competition				
Enhance stabilization				

Source: United Nations Economic Commission for Europe, *Economic Survey of Europe in 1991-1992*, New York, 1992, p.231.

ment and CMEA trade, etc. After the removal of these impediments, state-owned companies may have a good chance to improve their performance relative to the past.

(iii) Agents of privatization

One of the most striking observations that a cross country comparison of privatization practices reveals is the similarity in institutions. It seems that more and more countries are finding it necessary to create a structure which rests on the same three main pillars (table 3.6.3).

At the government level, a *privatization ministry* is in charge of the development of strategies for industrial and service enterprises. Privatization in agriculture is typically run by a special unit of the ministry of agriculture. In some countries – notably in Russia and Hungary – privatization is carried out by a privatization or reform ministry. In Ukraine, privatization is largely controlled by the Ministry of the Economy, although until March 1992 the Ministry of Destatization and Demonopolization of Production was also involved.[475] As often happens, the name of the privatization ministry indicates the basic policy orientation or at least the orientation that the lawmakers had originally intended. In Poland, for example, the designation "Ministry of Ownership Changes" (rather than Ministry of Privatization) served to indicate that privatization would be only one of several forms of ownership change which ought to be taken into consideration.[476] There are only few countries where policy issues were left in the hands of the branch ministries. This seems to be the case in Romania and Latvia, and to some extent in Bulgaria.[477]

The operational work of privatization is usually given to a *privatization agency*. In addition to the practical work, these agencies are responsible for the elaboration of the governments' annual privatization plans in which a detailed account is given of the intended course of privatization, the number of firms involved, the utilization of privatization revenues, etc. Typically the agency is directly subordinated to the government (Bulgaria, Hungary) or to the ministry of the economy (Estonia). The Hungarian authorities for a while (six months) gave the privatization agency total independence, subject only to parliamentary oversight, but this organizational set-up was quickly abandoned and – as far as is known – was never followed by any other country.

There is more diversity with respect to the legal status of the privatization agency. One approach, first initiated by the then still existing east German government, is the *Treuhand* model. The Treuhand agency is an independent public body which reports to the Ministry of Finance but is not subordinated to it; it is the sole proprietor of all state-owned capital assets and sole caretaker of assets temporarily placed under state trusteeship. When it comes to divestment, reorganization or replacement of management, the decisions of the agency cannot (except on formal grounds) be challenged in the courts by the government, the enterprise concerned or individuals affected by the decisions. So far, the Treuhand model has been followed only by Hungary, but only for a while and with certain limitations.[478]

In all other countries, the state and the branch ministries – as founding organs – preserved their legal ownership rights over existing SOEs and the privatization agency is just one of many government offices. In several Soviet successor states, notably in Russia, but

---

[475] In 1991, the Ministry of Privatization played an important role in the development of the privatization concept, but its role dwindled later on (S. Johnson and S. Eder, "Prospects for Privatization in Ukraine", *Radio Free Europe/Radio Liberty Research Report*, Vol.1, No.37, 18 September 1992).

[476] For a detailed account of the 1990 controversy between the Mazowiecki government and the Sejm on these issues, see Z.M. Fallenbuchl, "Polish Privatization Policy", *Comparative Economic Studies*, Vol.33, No.2, Summer 1991.

[477] At one stage, Bulgaria's privatization programme was entrusted to the Minister of Industry I. Pushkarov. When he was forced to resign, a National Agency for Privatisation, the second institution under the same name, was formed (*East European Privatisation News*, Vol.1, No.9, September 1992, p.7). More recently, the new Minister of Industry R. Bikov has taken over many aspects of the privatization process, including the licensing of joint ventures (see Mr. Bikov's statements in the National Assembly on progress in privatization quoted by BTA news agency: BBC, *Summary of World Broadcasts*, EE/W0267, 4 February 1993).

[478] In Hungary, a *Treasury Property Management Office* was created to administer separately property left behind by the Soviet army, the communist party, the workers' militia and the trade unions, etc. In the second half of 1992, some functions of the privatization agency were taken over by a state-owned asset called *State Holding Company* (see below).

TABLE 3.6.3

State institutions involved in the privatization process

	Ministry or Minister responsible for privatization	Privatization agency	State asset-management fund(s)
Albania	-	Preparatory Commission for the Process of Privatization; National Privatization Agency	-
Bulgaria	-	Privatization Agency	State Fund for Reconstruction and Development
Czechoslovakia			Federal Fund of National Property
Czech Republic	Ministry of National Property Administration and its Privatization	-	Fund of National Property
Slovak Republic	Ministry of National Property Administration and its Privatization	-	Fund of National Property
Hungary	Minister without portfolio	State Property Agency	State Holding Company
Poland	Ministry for Ownership Changes	-	-
Romania	Branch ministers	National Agency for Privatization and Development of Small and Medium-Sized Enterprises	State Ownership Fund
Yugoslavia (SFR)			
Bosnia-Herzegovina			
Croatia	-	Agency for Restructuring and Development [a]	Fund for Development [a] Retirement Fund for Employees Retirement Fund for Agricultural Workers
Macedonia	-	Transformation Agency	-
Slovenia	-	Privatization Agency	Privatization (Development) Fund
Yugoslavia (FR)			
Serbia	-	Privatization Agency	Development Fund of the Republic
Montenegro	-	Agency of Montenegro for Economic Restructuring and Foreign investment	Development Fund of the Republic
CIS			
Armenia	Committee of Privatization and Management of State Property	Under consideration	-
Azerbaijan	-	-	-
Belarus	Committee on the Management of State Property	State Property Agency	-
Kazakhstan	Ministry for Privatization State Committee on Property	-	-
Kyrgyzstan	-	-	-
Moldova	-	-	-
Russia	First Deputy Prime Minister [b]	State Committee for the Management of State Property (Goskomimushchestvo)	Fund of the Assets of the Federation
Tajikistan	Committee on State Property	-	-
Turkmenistan	State Committee on Privatization	-	-
Ukraine	Minister of Economy Ministry of Destatization and Demonopolization of Production Minister of Privatization	-	State Property Fund
Uzbekistan	-	State Committee for the Management of State Property (Goskomimushchestvo)	-
Georgia	State Committee on Privatization	-	-
Estonia	Ministry of Privatization State Property Department of the Ministry of Finance	Government Agency for the Privatization of State Property	Under preparation
Latvia	Branch ministries	-	-
Lithuania	Minister of Economy Department of Privatization	Central Privatization Commission of the Government	-
Ex-GDR *Länder*	-	Trust Agency (Treuhandanstalt)	-

Sources: Official reports, national newspapers.

[a] Merged in early 1993, under the name "Croatian Fund for Privatization".
[b] Head of the privatization agency.

also Lithuania and Georgia, the agency is called a state committee or commission.

The privatization agencies' disposal rights are usually constrained by upper and lower limits. In Bulgaria, for example, small firms with a book value of less than 10 million leva are privatized by branch ministries, while the disposal of very large SOEs with assets exceeding 200 million leva requires case-by-case approval of the Council of Ministers.[479] In Hungary, state assets

---

[479] For details, see the full English-language text of the relevant legislation (Transformation and Privatization of State-Owned and Municipally-Owned Enterprises Act) in *Bulgarian Economic Review*, 6-19 May 1992.

below a certain value can be freely divested by the SOE's own management, which allows the agency to concentrate on large transactions.[480] In Croatia, the lower limit is based on asset values expressed in deutsche mark (maximum 5 million), which is understandable given the high level of "dollarization" of that economy.[481] Another consideration which limits the agencies' disposal rights pertains to the branch affiliation of SOEs. The *Russian Privatization Programme for the Year 1992*, for example, specifically mentions that firms in the defence sector cannot be privatized without the approval of the Russian government. This is hardly surprising, but given the size and the fluid boundaries of this sector this limitation could be used to head off privatization by many thousands of firms. The restriction applies also to firms in the fuel-energy complex, to mining enterprises in general, to commercial banks, foreign trade companies, publishing houses, health resorts, and others.[482]

Countries follow different principles concerning the size of the privatization agency as well. The *Treuhandanstalt* had more than 4,000 employees working in Berlin and 15 regional agencies. Similarly, the Russian privatization agency, *Goskomimushchestvo*, was conceived as a nationwide organization with 91 regional offices, staffed by over 10,000 workers; it has its own research institute and runs a newly-established university.[483] In Hungary, the State Property Agency (SPA) has a staff of 300 only and much of its routine work is contracted out to private consulting and auditing firms.[484] In Bulgaria, the agency has a staff of about 120,[485] while in Albania, after almost one year of existence, its counterpart had no more than seven people working in Tirana and another two or three in the provinces.[486]

Size in itself, of course, though not decisive, has important implications. The smaller the organization, the greater the need to delegate responsibilities to outsiders. Typically, these outsiders are the enterprises themselves: they prepare their own privatization programmes, they negotiate with foreign partners, etc. In other cases, consulting firms are asked to do the same. Both solutions have obvious drawbacks. Enterprise managers and consultants are not independent civil servants; they may pursue their own interests and do not have the same degree of accountability.

The third pillar of the privatization construct is, in most cases, still in the development phase. Governments or parliaments have concluded that there is need for special *funds* or *asset management agencies* to act as a temporary owner-managers of those capital assets which are not yet divested or which are not planned to be privatized in the foreseeable future. The justification for this has been similar everywhere. One institution cannot be made responsible for two types of business activity. Selling assets requires different organizational forms and a different type of staff than the management of assets in large firms, where corporate governance is needed. The fear of corruption has also been quoted as an argument in favour of such systems of dual control.

With the exception of the former GDR, where divestment of the large majority of SOEs is to be nearly completed by the end of 1993, fund-type institutions are likely to emerge soon in all countries. In some countries a single fund is to be created to own and manage all state assets. So far there is only one case (Croatia), where large numbers of firms were entrusted to decentralized funds. Such funds are also likely to emerge as an outcome of the voucher privatization technique (see below). In countries where there is no privatization agency (e.g., Ukraine), the state asset management fund has become responsible for the preparation and execution of privatization.

(iv) Establishing property rights

Whatever good arguments exist for a three-tier system (privatization ministry, privatization agency, asset management fund), it inevitably will pose problems of coordination. Since there is no clear-cut way of distributing responsibilities among the three tiers, a certain amount of overlapping, departmentalism, institutional and personal rivalries are probably inescapable.[487]

Such risks appear, for example, in the Romanian mass privatization programme, for which six investment funds were created. According to their statutes, these companies will become holders, not owners, of assets, which they are bound to administer on behalf

---

[480] The threshold varies between 20 and 50 million forint, depending on the circumstances of divestment and the type of assets concerned.

[481] N. Cuckovic, "Privatization Practices in Croatia, Poland, Hungary and Czechoslovakia: Evaluation of the Results", Institute for Development and International Relations, *Macroeconomic Policy Reform and Private Enterprise in Central and Eastern Europe*, International Centre for Economic Growth, Zagreb, December 1992, pp.45-65.

[482] Other types of assets (such as forests, monuments, roads, the central bank, universities, broadcasting stations, etc.) cannot be privatized at all. *Ekonomika i zhizn'*, No.29, 1992.

[483] A.D. Radygin and A.D. Krasnosel'ski, "The Process of Privatization in Russia", Institute for East-West Studies, *The Uncertain State of the Russian Economy*, published in cooperation with the Institute of Economic Policy (Moscow), New York, 1992, p.63.

[484] The SPA was established in 1990 with a staff of 56. By 1991, 126 people worked there. In 1992 the number of employees reached 300 and for 1993 a further 100 people are considered necessary. *PRIVINFO*, Vol.2, No.2, January 1993, p.35.

[485] *Bulgarian Economic Review*, 18 November-1 December 1992.

[486] *Albanian Economic Tribune*, August 1992, pp.32-33.

[487] Some countries have taken explicit note of this risk. When Kazakhstan's Minister for Privatization was asked in an interview why his country did not have a privatization agency, he said: "The ministry for privatization combines both functions. Privatization would come to a halt if we set up a privatization agency because there would be a clash between the two". *Central European*, December 1992/January 1993, p.12.

of the population under parliamentary supervision.[488] But such observations can be made about almost every country.

The situation is potentially more confusing if more than one organization operates within one level (Ukraine) and/or if the three levels are subordinated to bodies which are in conflict with each other. This would appear to be the case in Russia, where *Goskomimushchestvo* is a government-run institution without ownership rights, while the state asset management agency, the ultimate owner, is subordinated to the Supreme Soviet (parliament). As often happens in such situations, a way out of the conflict can be sought by amalgamating the tiers. So far this has happened only in one country (Croatia), where the privatization agency and the asset management fund were merged in early 1993. But it cannot be excluded that, as present governments are replaced, other countries will emulate this example.

However, leaving these problems aside, it is still true that over the past two year most countries have taken large steps toward improving the transparency of property rights relations.

First of all there was a need in many countries to designate the state itself as the owner of capital assets inherited from the previous regime and to entrust to it authority to guide the privatization process. This is particularly important in those transition economies which are new-born or re-born states. The imposition of *de jure* ownership rights over enterprises, land and other types of tangible asset took considerable time because it had to be preceded by complex inter-state negotiations and legislative work. With the adoption of basic privatization laws, this process was completed practically everywhere by the end of 1992.

From this generalization, however, the Russian Federation may have to be excluded, because it has inherited very blurred ownership rights from the socialist era. In view of the fact that Russia itself is a federation of 20 autonomous republics and 11 autonomous regions, each of which exerts or at least claims some ownership rights over the assets located on its territories,[489] the process of establishing clear property rights is bound to be protracted. In the successor states of the former Yugoslavia, similar problems are compounded by the heritage of the self-management system, which in a legal sense had been based on the concept of "social ownership". In practice, a solution required the introduction of state ownership prior to privatization.

Secondly, the abandonment of central planning necessitates the delegation of ownership rights over vast assets to municipal authorities, notably the housing stock. These assets, however, are often in a physically dilapidated shape, or heavily burdened with debt. Indeed, this is one reason why many countries were unable to move beyond the stage of a political declaration or the passing of a general law on this subject.

It is also worth noting that some governments had planned to endow autonomous public bodies (social security, health and pension funds, charitable organizations, etc.) with property but, except in Croatia, these intentions have so far not been implemented.[490]

Thirdly, it is generally agreed that "corporatization" is an important element of the privatization process, as a means of establishing clear property rights within the state sector.[491] Two extreme cases can be envisaged. One is to convert all existing state-owned firms (including agricultural enterprises) into joint stock companies at a very early stage of the privatization process without any restructuring. Alternatively, it is feasible to postpone changes of the legal form until the appearance of new owners and the adoption of a specific programme of restructuring – and then to consider corporatization as a simple administrative issue. So far, most countries appear in practice to have followed the second option, and the transformation of non-privatized SOEs has therefore been slow.

However, in the former GDR, virtually all Treuhand companies were transformed into joint stock companies by 1 July 1990.[492] In Romania the transformation also took place early. State enterprises were reorganized into some 7,600 commercial companies (joint stock and limited liability companies), 390 *régies autonomes* and 338 joint-ventures with foreign participation. State commercial companies can all be privatized, while the *régies autonomes* – mainly public utilities and strategic industries – are to remain state-owned and subordinated to branch ministries – although these companies may also improve their financial performance by leasing or selling assets. The entire process was virtually completed by the end of 1990. In Czechoslovakia, the corporatization of SOEs

---

[488] D. Ionescu, "Romania: Testing Large-Scale Privatization", *Radio Free Europe/Radio Liberty Research Report*, Vol.1, No.2, 10 January 1992.

[489] Cities like Moscow or Nizhnii Novgorod have their own privatization procedures which substantially differ from the rules set by legislators for Russia as a whole.

[490] In *Bulgaria*, 20 per cent of the shares of privatized enterprises must, according to the Privatization Law, be transferred to a specially created fund, the State Fund for Reconstruction and Development. The resources of this fund may then be used – inter alia – to capitalize the social security fund. In 1991, the *Hungarian* Parliament, in legislation creating a Social Security Fund, endowed it with assets worth Ft 200 billion – an amount equivalent to roughly 10 per cent of the total assets suitable for privatization – but this has so far not been implemented. After protracted negotiations on the size and composition of this portfolio, end-1993 was set as a new deadline (*Magyar Hirlap*, 28 January 1993). In *Poland*, the Constitutional Tribunal recently urged the government to propose a scheme by which 18 per cent of the shares in firms included in the mass privatization programme would be set aside in a fund to compensate 3.5 million pensioners and employees who had lost benefits owing to budgetary austerity (*Radio Free Europe/Radio Liberty Research Report*, No.35, 4 September 1992, p.37). In *Slovenia*, 10 per cent of shares of all companies must be transferred to the Old-Age Pension and Disablement Insurance Fund (*Central European*, February 1993).

[491] The term "corporatization" refers to the change of the firms' legal status from a state agent to a joint stock company, the ownership of which is initially entrusted to a state agency.

[492] The change of legal form was not required for firms that were already earmarked for liquidation, nor for "traditional" state companies such as postal services, the railway system and the administration of public roads and waterways. Firms under the supervision of local and regional authorities were also exempted.

TABLE 3.6.4

Restitution and compensation measures approved or planned

	Bulgaria	Czecho-slovakia	Hungary	Poland	Romania	Estonia	Latvia	Lithuania	Slovenia	Ex-GDR Länder
Eligibility										
Restitution to:										
Resident citizens	Yes	Yes	No	Yes	Yes	Yes	Yes	Yes	Yes	Yes
Individuals including foreigners	No	Yes	No	Yes	?	Yes	Yes	No	?	Yes
Resident institutions	Yes	+	Yes	Yes	?	?	No	?	+	Yes
Compensation to:										
Resident citizens	Yes	Yes	Yes	Yes	Yes	?	Yes	Yes	Yes	Yes
Individuals including foreigners	Yes	Yes	Yes	Yes	No	Yes	Yes	No	?	Yes
Resident institutions	Yes	+	No	Yes	No	?	No	?	+	Yes
Forms of compensation										
Allocation of a similar asset	Yes	No	No	Yes	No	?	No	?	Yes	No
Cash payments	Yes	Yes	No	No	Yes	No	Yes	No	Yes	Yes
Tradable vouchers, life annultites, pension supplements, securities, etc.	Yes	No	Yes	No	No	Yes	Yes	Yes	Yes	No
Non-tradable privatization vouchers	No	No	No	No	No	Yes	Yes	Yes	No	No

Sources: Official reports, national newspapers.

Note: (+) Under consideration.

was a pre-condition for their inclusion in the voucher privatization programme. In Croatia and Latvia, laws or government decrees set firm deadlines for the compulsory corporatization of all SOEs — June and November 1992, respectively — but many enterprises were unable to complete the required changes by the date given.[493]

Similar deadlines were promulgated in Hungary (December 1992, then prolonged until June 1993) and in Ukraine (June 1993). In Poland, almost 500 companies had been converted by September 1992, and another 1,000 SOEs are expected to undergo corporatization in 1993.[494] In Russia, the end of October 1992 was fixed as a target date for the corporatization of some 6,000-7,000 SOEs,[495] but later the privatization agency pushed this deadline back to end-1993. However, mandatory corporatization is to apply only to those 4,948 large and medium-sized enterprises which were not exempted from privatization by the government's 1992 privatization programme.[496]

(v) Restitution: a long detour

An important concomitant of privatization is the present governments' promise to restitute property confiscated by previous regimes. Indeed, it has often been argued that restitution also belonged to the "fundamentals", not only as a matter of justice, but also a technical pre-condition. Sales cannot start, it was argued, as long as the ownership of a given asset is under dispute.

Marked differences initially existed between countries in this respect, but later the dividing lines appear to have become somewhat blurred. Generally speaking, the east European countries and the Baltic states made a strong commitment to undo the injustices of the past. The case was somewhat different in Germany, where the Unification Treaty put explicit limitations on the restitution of assets confiscated under Soviet military rule during the post-war years. In Russia and in other successor states of the Soviet Union, the possibility of doing justice after 75 years was considered unfeasible and the issue of restitution or compensation was thus never formally placed on the agenda.

However, in some of these successor states de facto Soviet rule had a shorter history. Therefore, the desire to restore the *status quo ex ante* may be just as strong as in eastern Europe or in the Baltic countries. Thus, it is likely that the distribution of agricultural land, which is fairly advanced in the CIS countries, is partly based on hereditary rights. When local authorities are in charge of the operation, they have to harmonize many objectives and interests, and previous ownership is most probably one of them.[497] Secondly, there are many examples in eastern Europe where divestment to foreigners is also a form of restitution, when the "foreigners" are expatriates whose motivation is as much emotional as commercial and this fact is taken into account by the national authorities.

A comprehensive and just codification of restitution procedures has proved to be a daunting task even in the relatively more developed transition countries. In

---

[493] In Croatia, for example, 2,700 SOEs out of 3,900 were reported to have missed the deadline (The Economist Intelligence Unit, *EIU Country Report: Croatia*, No.3, 1992, p.18).

[494] Figures are taken from a report by J. Lewandowski, Minister of Privatization (BBC, *Summary of World Broadcasts*, EE W0263, 7 January 1993). At end-1992 there were some 7,300 state enterprises as against 8,200 a year earlier.

[495] The modalities are described in a presidential decree issued on 1 July 1992.

[496] BBC, *Summary of World Broadcasts*, SU/W0261, 18 December 1992.

[497] The same logic may apply to the reprivatization of the housing stock.

countries where agriculture had been socialized and where the new governments were elected on a pro-agrarian platform (e.g., Bulgaria, Hungary and Romania), the problem was tackled head-on with the adoption of laws on land reform and land restitution. In contrast, in Poland and in many Yugoslav successor states, where agricultural land was predominantly in private hands even prior to 1989, politicians and law makers enjoyed a "grace period" and real work on these thorny issues did not start until late 1992.[498]

At the start, many countries tried to follow a simple reasoning: only property still in existence could be restituted. No compensation can be given for objects and real estate which for one reason or another no longer exist. Nor are large companies and banks subject to restitution.

But this reasoning begs several questions, and these doubts soon led to modifications in the basic principles. It is very difficult to draw strict borderlines on which laws and regulations can be formulated. If houses are given back, why not castles? If shops can be returned to the original owner, perhaps factories and mines should be given back as well. Questions can arise from the legislative ceilings imposed on restitution and compensation.[499] This question was raised in the context of such discussions in the Baltic states — "why should the feckless 1939 owner of a bankrupt or heavily-mortgaged farm (...) receive his acreage back in 1992, while a fellow-citizen, who kept his wealth in government bonds, bank deposits, or a stamp collection, receives nothing?"[500] If natural persons are compensated, why not legal persons, e.g., churches, which were not only expropriated but banned as well, but are now restarting activities for which they evidently need premises and resources in general? These issues are finally being resolved in different ways in the various countries concerned (table 3.6.4). However, it is still too early to assess final outcomes, since in some of the countries listed — e.g., Albania, Poland and the former GDR — the relevant legislative work is incomplete.[501]

Once all the political, moral and technical issues began to be taken up seriously by lawmakers, it turned out that a Pandora's box had been opened. As the first set of laws were passed — in Germany concerning claims in the former GDR, in Czechoslovakia and later in Hungary — it became evident that restitution claims numbered in millions (table 3.6.5). In other words, relatively large segments of the population are affected by these laws. This raised the political stakes for governments and caused logjams in administration as well as in the law courts.

Lawmakers also had to realize that there was no way to regulate the entire issue simply by passing one or two legislative acts. Every new regulation opens the door for another one. In Czechoslovakia, for example, the law allowing for the restitution of urban villas took no account of the fact that, in Prague alone, 170 villas were occupied by foreign embassies and consulates. So Law No.403/1990 had to be amended by Law No.458/1990, which ruled that in such cases only financial compensation could be claimed. But the final settlement of this problem was only achieved with Law No.137/1991, which allowed rightful claimants to get

TABLE 3.6.5

Restitutions and compensation claims
(Numbers)

Bulgaria	
Claims filed in for agricultural land by mid-August 1992	1 705 731
of which from abroad	10 000
Completed restitution cases in other sectors by May 1992	10 000
Czechoslovakia [a]	
Total claims filed by end-1991	50 000
of which settled by mid-March 1992	10 000
Hungary	
Total claims filed by end-1992	1 500 000
of which settled	400 000
of which from abroad	106 000
Poland	
Total claims filed by mid-1992	Over 100 000
of which from abroad	30 000
Romania	
Claims filed in for agricultural land by mid-July 1992	5 200 000
of which settled	3 900 000
Latvia	
Claims filed in for agricultural land by Latvian citizens	50 879
by former Latvian citizens by end-July 1991	1 000
Ex-GDR *Länder*	
Number of claims on real estate	1 700 000
of which settled	170 000
Number of claims on productive property	15 456
of which settled	6 438

Sources: Official reports, national newspapers.

[a] Incomplete data. Restitution claims are chiefly handled at the municipal level and there is no systematic reporting on nationwide aggregates.

---

[498] Note, however, that such a law was adopted in *Serbia* in 1991 and a commitment was made to return 0.2 million ha of land to farmers by the end of April 1992 (BBC, *Summary of World Broadcasts*, EE W0228, 30 April 1992). In Vojvodina province alone 60,000 ha had been returned by the end of May (BBC, *Summary of World Broadcasts*, 4 June 1992). In *Slovenia*, the 1991 Law on Denationalization took the position that "denationalization" means primarily the restitution of property in kind. By mid-1992, most claims pertaining to forest and farmland were settled, while cases pertaining to industrial assets were delayed until the elaboration of privatization laws. In *Poland*, a draft reprivatization law was submitted to Parliament by late summer 1992, but there was no deadline for its consideration (*Central European*, September 1992, p.80). So far, the Polish authorities have had to make decisions on the basis of pre-war laws. For example, 118 out of the total number of 130 flour mills claimed were returned on this basis (Z.M. Fallenbuchl, "Polish Privatization Policy", *Comparative Economic Studies*, Vol.33, No.2, Summer 1991, p.65).

[499] In Czechoslovakia, for example, the upper limit of compensation has been fixed at Kcs 60,000. In Hungary, the limit was set at Ft 5 million per claimant.

[500] *The Baltic Independent*, 18-24 September 1992.

[501] For a comprehensive overview of the relevant legislation, see United Nations Economic Commission for Europe, Committee on the Development of Trade, op.cit., pp.46-56.

their property back after a 10-year waiting period.[502] In Estonia, the commitment to land restitution pushed the government to pay compensation for cattle and equipment forcibly taken over by collective farms in the 1940s. However, once such claims were recognized, a *numéraire* was needed to price these diverse assets and the issue was not resolved until early 1993.[503]

It was also necessary to decide how far back into the *past* a country should go when redressing injustice. Evidently, violations of rights in general and expropriations in particular did not start with the communist takeover in 1945 (or 1917) and did not necessarily end in the 1950s (or the 1920s), as is generally believed.

In Hungary, rivalry between the coalition parties has led to the adoption of no less than four major "compensation" acts and five other enabling instruments, each of which addresses different types of injustice (e.g., the Holocaust, imprisonment after the Second World War, etc.), setting different time limits and offering different remedies. In Germany, a law passed in September 1990 extended restitution principles to expropriations on east Germany territory under the Nazi regime between 1933 and 1945. In Bulgaria, where the political party of the Turkish minority is the leading opposition force, the government felt obliged to compensate more than 300,000 Bulgarian Turks pressured into leaving the country in 1989. So, in July 1992, a law was adopted which allowed Bulgarian Turks to resume ownership of their property — mostly housing — by April 1993, while offering compensation to present owners.[504]

In Czechoslovakia, moral and legal difficulties were the opposite in nature: most large companies were nationalized under a democratic government between 1945 and 1948. These properties were as a rule declared non-returnable. But even the Czechoslovak law makers had second thoughts. In April 1992 — i.e., two years after the main restitution laws — parliament approved an act on the return of some property seized from ethnic Germans during this period.[505]

Another aspect of the same problem is the delimitation of property claims in *territorial* terms. Borders have changed many times over the past decades and there is no international authority to solve conflicts arising from this fact. As a general practice, countries have denied the validity of such cross-border claims. They do not compensate foreigners for losses which occurred within currently existing borders and they do not compensate their own citizens for losses occurred on territories which now lie outside present borders.

But it is interesting to note that in Poland lawmakers are contemplating compensation in kind for property losses incurred in the present territories of Ukraine.[506] Lithuania is doing the same for those of its citizens who suffered property losses prior to 1945, when parts of Lithuania were under Polish occupation.[507] Hungary is exceptional in another sense. It provides compensation to foreign citizens if they were Hungarian citizens when their property was expropriated. But, to make matters more complicated, Hungarian lawmakers had to take into account the fact that foreign citizens whose property had been unlawfully confiscated during early post-war years had in many instances already been compensated through intergovernmental agreements concluded by previous communist governments with the major western countries.[508]

It has proved extremely difficult to agree on fair principles in defining citizenship for the purposes of restitution laws. The initial position of the governments concerned was that only those persons qualified who were citizens of the country and who had lived there permanently when the law came into effect. This means, however, that property cannot be given back to political refugees living abroad, although in many ways the very same governments are keen on re-establishing close political and economic ties with their diaspora. The way out of this contradiction was usually to liberalize naturalization and double-citizenship rules. Albania outstripped other countries in this respect: it offers an Albanian passport to anyone wishing to take it.

(vi) Small privatization

The term "small" privatization was first introduced in Czechoslovakia, but it has since gained broad acceptance everywhere as an appropriate designation of government actions to divest itself of such assets as service work shops, retail outlets, catering establishments, taxis, car rentals, etc. Often, though not always, privatization of state housing is included in the definition and this practice is also followed here.

By 1992, privatization of *shops, catering establishments*, etc., had gathered momentum everywhere from

---

[502] For more details, see M. Kupka, "Transformation of Ownership in Czechoslovakia", *Soviet Studies*, Vol.44, No.2, 1992 and more specifically on the above-mentioned issues J. Burger, "Politics of Restitution in Czechoslovakia", *East European Quarterly*, Vol.36, No.4, January 1993.

[503] According to the law, the size of compensation is based on the present price of a cow (1 cow = 500 Estonian kroons), and all other claims will be valued in terms of cow-equivalents.

[504] *Radio Free Europe/Radio Liberty Research Report*, Vol.1, No.32, 14 August 1992.

[505] This measure applies to Czechoslovak citizens of German origin who still live in Czechoslovakia. The new law does not, however, recognize the property rights of some three million ethnic Germans who were deported from the country and also excludes people who collaborated with the Nazis. *Radio Free Europe/Radio Liberty Research Report*, Vol.1, No.18, 1 May 1992, p.57.

[506] Although the Polish Parliament is still labouring on a new set of restitutions laws, in 1991 alone 59,824 such claims were submitted to county administrations. *Central European*, September 1992, p.80.

[507] A government decree to this effect was adopted in December 1992. BBC, *Summary of World Broadcasts*, SU/1567, 18 December 1992.

[508] Between 1949 and 1983, Hungary concluded 16 such settlements.

TABLE 3.6.6
Progress in small privatization

	Number of units designated for privatization (1)	Number of units privatized (2)	Sales revenues (3)	Cut-off date for information in in col. (2) and (3) (4)
Albania	..	15 000	Lek 360 million	December 1992
Bulgaria	30 000-40 000	..	..	..
Czech Republic	31 914	22 001	Kcs 28.5 billion	1.1.1991-31.10.1992
Slovak Republic	11 099	9 020	Kcs 13.7 billion	14.2.1991-31.10.1992
Hungary	10 000	7 637	Ft 11 billion	1.9.1990-31.12.1992
Romania	..	1 400	Leu 15 billion [a]	Jan.-Dec. 1992
Russia	84 000	28 000	..	Jan.-Dec. 1992
Estonia	1 212 [b]	676 [b]	Est.kroon 38.2 million	1991-1992
Latvia	8 000	302	Latv. R. 270 million	Jan.-Dec. 1992
Ex-GDR *Länder*	30 000	22 300	DM 340 million	mid-1991

Sources: National statistics, newspaper reports.

Note: Countries have different reporting habits with regard to the definition of "units" privatized. In some cases, the number refers to retail trade outlets, in other cases to retail trade enterprises with dozens or more outlets. This reduces the cross-countries comparability of data.

a The figure may refer to the book value of assets, as opposed to actual revenues.
b Including "larger" units.

Albania to Yugoslavia. In most countries, the privatization ministry and/or the privatization agency are responsible for process, but there are important counter examples (e.g., Russia or Poland) where the design and implementation of the process has been put into hands of the municipalities. This also explains the lacunae in the summary data, presented in table 3.6.6.

Besides missing information, one more factor reduces the cross-country comparability of data in table 3.6.6. In some countries, reported data refer to trade or service outlets *stricto sensu*, in other cases the numbers refer to the trading or servicing organization itself which may have dozens or more outlets. This can be a major source of confusion. In Russia, for example, the statistical authorities as well as the privatization agency *Goskomimushchevstvo* appear to cling to the latter definition (according to which 28,000 enterprises were privatized under the small privatization programme), while in other sources one can find reports which mention that 80,500 trade outlets were privatized in Moscow alone.[509] According to a well-documented study on Poland, a similar confusion exists with regard to the numbers published by various officials of the privatization ministry.[510]

The auction method appears to be used most extensively, though with important variations in modalities. One source of these variations stems from the fact that the dividing line between "small" and "large" privatization cannot be set in a hard and fast way. In Hungary, the scope of "small" privatization (or "pre-privatization" as it was officially called) was very narrowly defined. Its declared aim was to privatize retail units employing a maximum of 10 persons (15 in the case of restaurants) through public auctions. Originally, the government assumed that out of the 54,000 units statistically recorded, some 37,000 would fall under the jurisdiction of the pre-privatization law. However, by the time the auction mechanism was actually set up in early 1991, this number has been reduced to 10,000 only. Given the very low limit in the number of employees (and some other loopholes,[511] this was almost inescapable.

Small privatization in practice touches on restitution, decentralization and large privatization as well. In Czechoslovakia, the first factor explains why in many cases shops were not auctioned together with the land in and/or on which they were situated. In east Germany, as well as in Hungary and Poland, the situation was often the same, but for a slightly different reason. After state ownership of buildings and other real estate had been transferred to municipalities, they were often unwilling to give up potentially valuable assets. In all these cases, leasing arrangements were made to regulate the use of immovable property. In the eastern part of Germany, this solution posed no problems with regard to the auction process because leasing of immovable property was the pattern generally followed in the western part of the country. Polish shopkeepers were also satisfied with such leasing arrangements. In Hungary, by contrast, leasing met strong resistance on the part of would-be buyers who preferred everything or nothing.

---

[509] Mayak radio, reporting on 4 December 1992. BBC, *Summary of World Broadcasts*, SU/W0261, 18 December 1992.

[510] Figures, indicating the number of privatized retail stores vary from 30-80,000. R. Frydman, A. Rapaczynski, J. S. Earle et al., (eds.), *The Privatization Process in Central Europe*, Central European University Press, Budapest, London, New York, 1993, p.202.

[511] E.g., shops constituting a part of a chain were exempted. In a similar way, the law did not have an effect on sales units operating in factories, army barracks, prisons, etc. Some 2,000 pharmacies, a few hundred hard currency shops, travel agencies and a few dozen pawnshops were excluded on the grounds that their management requires special qualifications.

Although the auction method was frequently hailed as the chief method of small privatization, in practice Russia and many other countries have preferred to turn these auctions into manager or employee buy-outs. As a means of payment or supplementing cash and loans, vouchers issued for "large" privatization may be used.

Speed and the fear of resistance were perhaps the most important explanations of these decisions. In Moscow, for example, 90 per cent of "working collectives" were given permission to buy out the retail shop in which they worked.[512] A study in Poland, surveying 10,000 municipally owned shops rented to individuals in the first half of 1990, also found that only 9.3 per cent were allocated by auction.[513] In Hungary and Czechoslovakia, manager buy-outs were more common – chiefly because managers-on-the-spot had been engaged in long-term leasing agreements in 1990 or before, so a buy-out was the logical solution to avoid a gridlock. In Ukraine the same obstacle arose, although here it was typically not managers but the entire work force which was involved in leasing agreements.[514] In Lithuania, the process of small privatization was actually stopped in early 1992, when workers in the service sector went on strike and engaged in street protests in protest against auctions. As a result of their actions, Parliament changed the procedures and gave workers a priority right to acquire 30 per cent of the shares in their establishment.[515]

As noted above, small privatization accelerated in 1992 but is still incomplete as yet. The exception is Germany, where the special Treuhand subsidiary – *Gesellschaft zur Privatisierung des Handels mbH* – set up to organize privatization of domestic trade, hotels, etc. had largely completed its job by mid-1991. In Hungary, by contrast, the time limit set for the privatization agency expired in September 1992 with about half of the job done. In this country, as well as in the Czech Republic, end-1993 is often mentioned as the final deadline. Elsewhere, the end of the process is still further away.

Total revenues from small privatization – as tabulated in table 3.6.6, wherever possible – were generally modest. In east Germany the sum concerned (DM 340 million) turned out to be even lower than the amount the Treuhand agency was forced to pay to employees whose contracts were not renewed (DM 440 million).

Due to the low level of urbanization, the traditional pattern of *housing* in eastern Europe and the former Soviet Union has been based on individual ownership, except in major cities where the state owned a large percentage of the housing stock. In many countries, previous governments had opened the way for tenants to buy their apartments as early as the 1970s. Such transactions, however, remained the exception until recently.[516] Although divestment of the state-owned housing stock is a relatively easy exercise, its sheer size entails considerable administrative and logistic work. In the Russian Federation, for example, more than 400 privatization agencies were established by August 1992 strictly for this job.[517] In Germany, the Treuhand also created a special company to carry out these tasks.[518]

As a general policy, apartments are given free of charge to tenants; only in a few countries (e.g., Hungary, Slovenia), at least symbolic down-payments are required from them. Such give-away policies, of course, smack of restitution and quasi-compensation for losses suffered during the past decades. Therefore, the dilemmas discussed above under those headings appeared here as well.

In the former GDR, for example, more than 3 million dwellings transferred from state into municipal property included about 750,000 flats which were claimed back by previous owners.[519] In Ukraine, municipal flats are supposed to be automatically transferred to present tenants free of charge. There is a complicated scheme to determine eligibility thresholds for individuals and families, beyond which surplus square meters have to be purchased. Ukrainian citizens who already own their apartments – and therefore cannot expect to benefit from these reforms – are entitled to privatization vouchers. In Belarus, the same logic led the authorities to introduce from 1 July 1992 a so-called "apartment quota", which linked compensation and the divestment of the state-owned housing stock. In this scheme, not only tenants but every citizen is promised an equal amount of rouble-denominated vouchers in the form of a transferable cheque-book. These vouchers can buy not only apartments but also land and other state property and may be accepted as a cash deposit to supplement housing

---

[512] BBC, *Summary of World Broadcasts*, SU/W0261, 18 December 1992.

[513] R. Frydman et al., op.cit., p.202.

[514] In some cities, as many as nine out of ten small businesses were reported to have such a legal status. *The Economist*, 27 February 1993.

[515] *Baltic News*, No.11, 1992, p.16.

[516] These legal provisions aimed to reward the *nomenklatura* twice. First, these people were put into low rent-high quality apartments, then these apartments were sold to them at bargain prices. This extreme form of favouritism, however, met with so much resentment by the public that it was never widely used.

[517] A.D. Radygin and A.D. Krasnosel'ski, op.cit. p.67.

[518] More precisely, divesting about 200,000 apartments which previously belonged to east German industrial firms and state agricultural farms is only one of the tasks of the *Treuhand Liegenschaftsgesellschaft GmbH* (TLG). TLG has been also responsible for selling off 50,000 buildings and construction plots of other kinds as well. *Frankfurter Allgemeine Zeitung*, 2 December 1992.

[519] Another critical issue which has remained unsolved so far pertains to the question of debt arising from the legacy of the old state housing management companies. This debt is now valued at DM 51 billion and there is no agreement between the central authorities and the municipalities on how to share this burden. *Frankfurter Allgemeine Zeitung*, 2 December 1992.

TABLE 3.6.7

Sales of municipally-owned dwellings

	Number of apartments sold to tenants or other citizens	Cut-off or reference period for information
Hungary	163 500	1989-1991
Lithuania	389 058	18 August 1992
Romania	2 000 000	End-May 1992
Russia[a]	2 600 000	Jan.-Dec. 1992
Slovenia	80 000	End-1992
Latvia	10 000	End-1992
Ukraine	349 000	Jan.-Dec. 1992
Ex-GDR *Länder*	7 000	Oct. 1992-Dec. 1992

Sources: National statistics, newspaper reports.

a Given to private ownership free of charge.

loans.[520] A similar scheme is proposed in Albania. According to a draft law, introduced in October 1992, all citizens would receive a voucher, worth lek 2,600 towards buying a house or a flat. A family would become the owner of a home when half its value has been paid off. Former political prisoners and their families would be able to acquire their homes free of charge.[521] In Estonia, where sales started only in September 1992, the authorities main concern was to guarantee preferential treatment to indigenous Estonians, as opposed to "immigrants" (chiefly Russians). In this spirit, it was decreed that the right to buy flats could be given to tenants who had lived in Tallinn for at least five years. When buying their apartments, tenants can pay — *inter alia* — with vouchers, the availability of which depends on the number of years the bearer has worked in Estonia.[522]

Relative to the saleable stock, the privatization of flats appears to be fastest in Lithuania: 89 per cent of tenants were reported to have bought their flats or to have become owners as a result of restitution (table 3.6.7).[523] In other countries, for instance Romania, half of the housing stock or even less has been divested so far.

(vii) Fundamental changes in agriculture

The dismantling of *kolkhoz-* and *sovkhoz-*types of large state farms has been forging ahead with remarkable speed and peacefulness in most east European transition economies. In the successor states of the Soviet Union, progress is slower, less rounded and less pervasive, but it would be a mistake to underestimate the importance of the achievements so far. By the end of 1992, there were more than 1 million peasant farms in the CIS countries, which is a relatively small but certainly not negligible number.[524]

By any standards the task is large. In eastern Europe almost 60 per cent of total agricultural land is likely to be involved (table 3.6.8). With the exception of Poland, the envisaged changes pertain to a land area which is significantly larger than the amount of land involved in the post-1945 land reforms. For this reason alone, it is not difficult to predict that land-related problems will continue to shape the socio-economic life of the region for many years. It will take a couple of years to sort out title claims and prepare proper cadasters on the basis of the rapid changes which occurred in 1990-1992 alone. Then a process of land concentration will start, whereby fragmented land holdings will be amalgamated into larger, more viable units.

Orderliness and speed appear to be inversely related. In some Balkan countries, land distribution and the liquidation of old structures went ahead almost immediately. In the former GDR, by contrast, land privatization is not even planned to start until 1995.

The process is highly complicated. In the absence of proper cadasters and established land prices, land assessment is indispensable — but it is bound to be at best an administrative short cut which is always open to dispute. There is a great deal of uncertainty regarding heavily indebted cooperatives, state farms and food processing enterprises. Where restitution is the main element of the land reform, numerous problems have emerged from the fact that applications are based on affidavits instead of missing title deeds. Given the slow progress in decision making, millions of farmers are now cultivating their lands without proper title deeds, while tens of thousand of cooperatives and state farms continue to operate in a "no future" environment. Table 3.6.9 presents an incomplete summary — overview of the institutional changes taking place in the transition economies.

Only a year or two ago there was great interest in private farming, but recently that interest has abated. Market prospects are poor, government subsidies are minimal and the banking system is not strong enough to support long-term investments. Among many peasants a wait-and-see attitude prevails. There is more willingness to take out land on a contract basis, as opposed to owning it, because ownership is often linked to cultivation obligations. Moreover, the marketability of land is generally limited. Foreigners cannot buy it at all,[525] and even domestic investors are constrained if

---

520 *Ekonomika i zhizn'*, No.24, 1992.

521 Economist Intelligence Unit, *Albania — Country Report No.4, 1992*, p.42.

522 *The Baltic Independent*, 11-17 September 1992.

523 Minister of Economics A. Simenas quoted by ELTA news agency. BBC, *Summary of World Broadcasts*, SU/W0261, 18 December 1992.

524 This figure was reported by V. Nefedov, Deputy President of the Statistical Committee of the CIS. *Izvestiya*, 2 February 1993.

525 The former GDR is an exception in this respect. For a comprehensive overview of land ownership rights from the vantage point of foreign investors, see United Nations Economic Commission for Europe, Committee on the Development of Trade, op.cit., pp.110-114.

TABLE 3.6.8
Past and planned land ownership changes in eastern Europe

	Total agricultural land around 1990 (million ha) (1)	Expropriated in post-Second World War land reforms (million ha) (2)	of which: Owned/cultivated by cooperatives and state farms around 1988 (million ha) (3)	Share of land likely to be affected by post-1989 changes ((3):(1))100 (percentage) (4)
Albania	0.70	0.30	0.70	99
Bulgaria	6.16	0.17	5.54	90
Czechoslovakia	6.77	4.50	6.36	94
Hungary	6.50	3.20	5.58	86
Poland	18.74	13.80	4.15	22
Romania	15.09	1.40	13.66	90
Yugoslavia (SFR)	14.17	1.57	2.45	17
Bosnia-Herzegovina	2.53	..	0.15	6
Croatia	3.22	..	0.46	14
Macedonia	1.32	..	0.59	45
Slovenia	0.87	..	0.06	7
Yugoslavia (FR)	6.24	..	1.18	19
Ex-GDR Länder	6.18	2.00	5.58	90
Total	74.32	26.93	44.00	59

Source: Columns (1), (3): CMEA Statistical Yearbook 1989, p.194, supplemented with national statistics for Albania and Yugoslavia. Column (2): N. Spulber, *The Economics of Communist Europe*, Technology Press of MIT, John Wiley and Sons, Inc., New York, 1957, pp.224-242, supplemented with data for Albania and the former German Democratic Republic from United Nations Food and Agriculture Organization, *European Agriculture*, Geneva, 1954, p.50.

they intend to create large estates[526] or wish to discontinue cultivation. The cautious approach of lawmakers is explained by their concerns about speculation, which might be very frequent under present circumstances, but at the same time politicians have to bear responsibility for the unwarranted negative consequences of the newly-imposed restrictions. Land is a basic factor of the entire production process. Limitations aiming to regulate the farming sector can soon backfire through their indirect impact on industry or the service sector (e.g., tourism), if they deter foreign capital due to the artificially low supply of development sites and complications arising from different treatment of real estates as opposed to buildings or equipment.

In *Albania* most collective farms were abandoned spontaneously in 1991 and their assets — including land — were taken over by the members. Division of the land according to the law did not start before June 1992,[527] but by then the authorities were confronted with a *fait accompli*: 77 per cent of the cultivated land was already in private hands.[528] In August of the same year, the government entrusted the newly-created Agency for the Restructuring and Privatization of the Agricultural Enterprises to work out a privatization programme for the remainder of the state agricultural sector.[529] This process, as well as the granting of title deeds, is to be completed in 1993.[530] It has also been decided that the country's 33 motor-tractor stations — at least one in each district — will be sold by the end of 1992 and the remaining ones closed.[531]

Final data released by the *Bulgarian* authorities suggest that claims were filed on 95 per cent of collectivized land. This means that 5.6 million ha of agricultural land will have to be given back if all claims are validated. However, given the structure of these claims, this is not likely to happen. More than 90 per cent of applications were filed in by urban dwellers. It is estimated that in one third of the land-use areas, applications have been filed for more land than was available, while in other areas the opposite is the case.[532] By end-1992, only 10 per cent (0.5 million ha) of the country's arable land had been restored to its original owners. Specialists in the Ministry of Agricultural Development, Land Use and Land Ownership Restitution office expect that three quarters of arable

---

[526] In Hungarian policy discussions, 200-250 hectares are mentioned as a likely upper limit. *Magyar Hirlap*, 20 February 1993.

[527] Decision of the Council of Ministers No.255, dated 2 August 1991.

[528] According to a later report, released by the Ministry of Agriculture and Food, by end-1992 0.4 million ha of land, 90 per cent of the available stock, was disposed of. Eighty per cent of peasant families have become land owners. BBC, *Summary of World Broadcasts*, EE/W0267, 4 February 1993.

[529] *Albanian Economic Tribune*, No.8, 1992. p.25.

[530] L. Kora, Deputy Minister of Agriculture, quoted by ATA new agency. BBC, *Summary of World Broadcasts*, EE/W0239, 16 July 1992.

[531] BBC, *Summary of World Broadcasts*, EE/W0242, 6 August 1992.

[532] BBC, *Summary of World Broadcasts*, EE/1464, 20 August 1992.

land will be returned to its owners by the end of 1993 and the land reform will be complete by 1994 only.[533]

In *Czechoslovakia* land had formally remained in private hands even in communist times. As industrialization progressed over the decades, peasants left their land in the cooperatives when they moved into towns. This resulted in a situation where approximately 50 per cent of land used by the cooperatives were formally owned by city dwellers who were not members.[534] The law set end-1992 the deadline for peasants to decide whether to leave or stay and for the cooperatives to return land to its rightful owners. In the Czech Republic, all cooperatives implemented these requirements and they were all reregistered by that date,[535] while in Slovakia the progress appears to have been slower, which probably explains the lack of any comprehensive data. Privatization of state farms was scheduled to begin in 1993 in both countries, following the resolution of restitution claims.

Privatization of the farm economy in *eastern Germany* is facing many difficulties. The liquidation and/or transformation of the former cooperatives (LPGs) was relatively simple. In the former GDR, as in Czechoslovakia, the agricultural cooperatives had largely maintained the formal land ownership rights of their members. In other words, after unification these cooperatives were not considered as state assets, so they were not placed under the supervision of the Treuhand.[536] But this did not mean that they could remain untouched: LPGs (as well as state farms) were transformed into new legal entities. Of the 512 state farms, only 271 remained intact by the end of 1992 — the rest were liquidated. According to plans finalized in December 1992 between the federal and *Land* authorities on the one hand and the Treuhand agency on the other, 1.2 million ha of agricultural land were to be leased to farmers for a period of 12 years, with an option to purchase after 1995. However, when land is actually up for sale, persons excluded from the restitution (because their land was expropriated between 1945-1949) will be given preferential treatment. Farmers in the former GDR whose land had been forcefully taken away in later years can count on similar treatment as well. The forestry fund (2 million ha), by contrast, is envisaged for sale in 1993. So far, restitutions claims for 0.8 million ha of forest have been filed. Up to the end of 1992, the Treuhand agency had disposed of 0.9 million ha of farm land under leasing contracts, the majority of which (0.54 million ha) went to companies formed from cooperatives. Only 26 per cent was taken over by individual farmers. In other words, collective land cultivation continues to dominate in the five new *Bundesländer* of Germany.[537]

As happened with so many of the other building blocks of a genuine market economy, *Hungarian* lawmakers opened the way for transforming cooperatives into private farms as early as 1989. In that year, it was decreed that for 50 per cent of all assets title deeds must be given to members. However, not very much happened until 1990 when Parliament suddenly decreed a moratorium on all transformations in order to pre-empt changes before new laws came into effect. In January 1992, Parliament finally gave only 12 months to the country's 1,400 agricultural cooperatives for restructuring. This tight deadline was met, nevertheless. With the exception of 65 cooperatives which missed it, 1,207 cooperatives were remoulded according to the new requirements — i.e., title deeds for land and other assets were given out to members and other eligible individuals[538] — but these cooperatives remained together as a voluntary association of members. To the great surprise of lawmakers, only 119 cooperatives have been liquidated so far and only 10 per cent of the membership has opted for individual farming.

This seemingly smooth process, however, is likely to become rougher soon. First of all, about a quarter of the cooperatives are under bankruptcy procedures. As soon as their cases are taken up by the courts, they will all have to make further changes. The second important development to come pertains to land auctions among voucher title holders. In the past, cooperatives owned 5.6 million ha of land, of which 1.9 million was set aside for compensation to previous landowners.[539] By the end of 1992, however, only 3 per cent of the land stock earmarked for such auctions has actually been sold.[540]

The country's 120 state farms were ordered to present detailed plans for restructuring and to transform themselves into joint stock companies by end-1992. Only one farm was able to meet the deadline. Important developments occurred, nevertheless. First, one-hectare land plots were given free of charge to some 90,000 workers.[541] This figure is interesting, not because

---

[533] BBC, *Summary of World Broadcasts*, EE W0263, 7 January 1993.

[534] R. Frydman et al. (eds.), *The Privatization Process in Central Europe*, Central European University Press, Budapest, London, New York, 1993, p.54.

[535] *Lidové noviny*, 11 February 1993.

[536] However, 30 per cent of the land (1.2 million ha), previously owned by the cooperatives was taken over by a Treuhand subsidiary company, called Bodenverwertungs- und Verwaltungsgesellschaft mbH (BVVG). This is a recent development — BVVG was created on 1 July 1992. *Frankfurter Allgemeine Zeitung*, 16 February 1993.

[537] Figures quoted above were given by G. Rexrodt, an official of the Treuhand agency, as reported in *Neue Zürcher Zeitung* and *Frankfurter Allgemeine Zeitung*, 9 December 1992.

[538] The relevant regulations gave ownership rights to retired cooperative members, to heirs of expatriate land owners, etc.

[539] *Magyar Hirlap*, 9 February 1993.

[540] The numbers quoted above were taken from *Figyelő*, 21 January 1993. According to an other source, by early February 1993, 30,000 voucher holders were able to buy land through these auctions.

[541] Report from the press conference of Mr. T. Szabó, minister responsible for privatization. *Magyar Hirlap*, 23 December 1992.

TABLE 3.6.9

Restructuring in the former state agricultural sector
(Number of units)

	Around 1989/1990	Around end-1992
Czech Republic		
Cooperatives	1 197	1 223
Joint stock companies	-	39
Other	-	5
Hungary		
Cooperatives (all kinds)	1 333	1 498
Enterprises (including state farms)	139	159
Economic associations	163	-
Joint stock companies	-	38
Limited liability companies	-	1 507
Romania		
Cooperatives	4 260	-
State farms	411	411
Agricultural associations	-	11 499
Agricultural societies	-	4 050
Agricultural groups	-	1 515
Private farms	..	2 000 000
Estonia		
"Kolkhoz"-type cooperatives	200	300
State farms	126	-
Private farms	-	10 000
Latvia		
"Kolkhoz"-type cooperatives	372	-
State farms	240	-
Registered companies	-	668
of which:		
Joint stock company	-	12
Partnership	-	459
Limited liability company	-	191
Cooperatives	-	2
Other	-	4
Private farms	-	50 000
Lithuania		
"Kolkhoz"-type cooperatives	835	..
State farms	275	..
Private farms	..	15 000 *
Russia		
"Kolkhoz"-type cooperatives	12 900	6 000
State farms	13 000	-
Re-registered cooperatives and farms	-	7 000
Incorporated business		
of which:		
Joint stock companies	-	300
Cooperatives	-	1 700
Other	-	8 600
Private farms	-	183 700
Ex-GDR Länder		
"Kolkhoz"-type cooperatives	4 530	-
State farms	580	-
Cooperatives	-	1 475
Joint stock companies, etc.	-	1 560
Partnerships	-	926
Individual farms	3 558	13 707

Sources: National statistics, newspaper reports.

of its magnitude, but because free distribution runs counter to all earlier announcements. Second, out of a total Ft 100 billion of assets, buildings and equipment worth Ft 35 billion were earmarked for immediate sale and a third was actually disposed of. This revenue was used to service old debts. If there were none, the State Property Agency sequestered it. So in fact, what remains to be transformed are heavily indebted rump-farms. At this stage, the State Property Agency is trying to market them through leasing schemes, whereby the land stock would remain in state ownership for the foreseeable future.[542]

In *Poland*, collective ownership in agriculture was essentially limited to 1,500 state farms and the role of cooperatives was negligible. Between June and December 1992, the Agricultural Property Agency of the State Treasury took over about half of the farmland belonging to state farms — about 1.2 million ha of land and thousands of buildings. So far 50,000 ha from this stock only has been released and only 7,000 was sold to private owners. In order to speed up the divestment process, the agency plans to sign managerial contracts with applicants.

Privatization in *Romania* involves the redistribution of 9.3 million ha of land. As one of the first measures, the provisional government of 1989 distributed land to members of cooperatives and in early 1991 shares in state farms were allocated to locals. The 1991 land law had a dual-track approach. It regulated both the restitution process and the distribution of unclaimed areas. In those regions where demand for land was high, both restitution and distribution were limited to 10 ha per beneficiary. By October 1992, 7.9 million ha of land had been given back to or distributed among 5 million citizens, of whom 3.9 million received property titles as well.[543] Private claims have also been placed on about 20 per cent of the state farms. Based on the legislation completed in May 1992, these applicants will receive compensation in kind on a *pro rata* basis from the output of the farms.

In *Slovenia* the reorganization of cooperatives has stalled, although the relevant legislation has already been passed in mid-1992. The difficulties are caused by the generous 2-year deadline for filing restitution claims under the provisions of the 1991 denationalization law.

Among the successor states of the former Soviet Union, *Lithuania* appears to have taken the biggest steps towards private farming: four fifths of the state property in the agricultural sector was reported to have been privatized through restitution and free distribution.[544] In *Latvia*, peasant holding were reported to have tripled during 1992, chiefly through restitution. Peasants now hold 0.8 million ha, or 21 per cent of the total area. The largest proportion of land — 1.1 million ha — remains under the control of joint-stock companies amalgamating collective farms and state agricultural enterprises. The average size of a peasant holding

---

[542] *Figyelő*, 4 February 1993.

[543] P. Marculescu, Minister of Agriculture, as quoted in BBC, *Summary of World Broadcasts*, EE/W0255, 5 November 1992.

[544] S. Simenas, Minister of Economics, as quoted by ELTA news agency BBC, *Summary of World Broadcasts*, SU/W0261, 18 December 1992; K. Prunskienė, "Economic Consequences of the Revolution", *Lithuania Today*, January-February 1993.

TABLE 3.6.10
Private farms in the CIS republics
(As of end 1992)

	Number of farms (1 000) (1)	Average size (hectares) (2)	Total size a (million ha) (3)	As a percentage of total agricultural land (4)
Armenia	243	2	0.49	37.38
Belarus	2	19	0.04	0.41
Kazakhstan	8.5	400	3.40	1.72
Kyrgyzstan	8.6	43	0.37	3.66
Moldova	500	3	1.50	60.00
Russia	183.7	42.5	7.81	3.65
Tajikistan	..b	..	..	1 *
Turkmenistan	100	10	1.00	2.79
Ukraine	14.4	20	0.29	0.70
Uzbekistan	6	8	0.05	0.18
CIS	1066.2	14.01	14.94	2.75

Sources: 1992 data are complied from an article published by B. Nefedov (Deputy President of the Statistical Committee of the CIS) in *Izvestiya*, 2 February 1993. 1990 data on arable land: *Narodnoe khozjaistvo SSSR v 1990 g.*, p.468.

Note: Some of the figures in columns (1) and (2) are rounded. See also text for interpretation of the data.

a Columns (1)*(2).
b "Not many", as the source says.

is 16.4 ha.[545] The planning to privatize *Estonian* agriculture started early, but scarcely any practical measures have been taken. In January 1993 Parliament was still working on a law to determine the value of collectivized property for the purposes of restitution.[546] Thus, the majority of state farms and cooperatives continued to operate in 1992 and their liquidation is foreseen only for the first half of 1993.[547] In *Georgia*, the process of land distribution gathered momentum in the first quarter of 1992, when some 50 per cent of the country's cultivated land was distributed to eligible village residents free of charge. Land was provided with transferable title, but has to be retained for two years before being sold.

Information is more uneven and detail is lacking on ownership changes in the *CIS countries* (table 3.6.10). The main hiatus is the ambiguity concerning the "conditionalities" of people's right to own land. This is a hotly debated issue in most of the CIS republics.

In *Russia,* President Yeltsin has been trying to introduce constitutional amendments since 1990, but so far changes have done little to advance the principle of marketability of land. Some loosening was achieved in December 1992, but these modifications appear to guarantee full ownership rights only to the owners of private garden plots. So far, the real issue, the question of buying and selling larger tracts of agricultural land, was not addressed at the level of the constitution. Although the lower level legislation on land-related issues has become increasingly complicated and often contradictory, it seems that land acquired without payment cannot be sold for ten years and there is a five-year moratorium on the sale of land once bought.[548] In many CIS republics, land reform so far has been proceeding primarily through leasing arrangements. For a number of reasons (overpopulation, inter-ethnic strife, shortage of water) the central authorities continue to maintain state ownership of land. In terms of transferring property rights to farmers Turkmenistan has gone furthest, but even in that country ownership is conditional on production for state procurement.[549]

As already noted, the information provided in table 3.6.10 should be understood as tentative. The numbers contained therein refer to the number and size of peasant holdings, irrespective of the question of ownership. As far as the Russian data are concerned, it is known from a recent census among private farmers that only 39 per cent of the land was actually owned by the heads of the farms, 32 per cent was hereditary but non-alienable land and 29 per cent was leased (*arenda*).[550] In other words, the vast majority of agricultural land continues to remain in the ownership of cooperative and state farms. True, these large production units are also in the process of change. By the end of 1992, they were obliged to restructure themselves from pseudo-cooperatives into voluntary associations and — at least on paper — many of them successfully implemented

---

[545] BBC, *Summary of World Broadcasts*, SU/W0263, 8 January 1993.

[546] *The Baltic Independent*, 5-11 February 1993.

[547] BBC, *Summary of World Broadcasts*, SU/W0263, 8 January 1993.

[548] S. Marnie, "The Unresolved Question of Land Reform in Russia", *Radio Free Europe/Radio Liberty Research Report*, Vol.2, No.7, 12 February 1993.

[549] For a thorough analysis, see G. Gleason, "Central Asia: Land Reform and the Ethnic Factor", *Radio Free Europe/Radio Liberty Research Report*, Vol.2, No.3, 15 January 1993.

[550] Information based on a sample of 78,000 farms, as reported in an article written by V. Nefedov, Deputy President of the Statistical Committee of the CIS. *Izvestiya*, 2 February 1993.

the government's order.[551] Similar information, however, is lacking for the other CIS countries.

Even these imperfect data give certain insights. There is a clear differentiation among the CIS countries both with respect to the speed of privatization and the principles underlying the ongoing redistribution process. Moldova and Armenia appear to be pace-setters: in these countries peasant holdings already account for a very large share of the farm economy. In the remaining countries, Russia included, this share is still negligible (less than 3 per cent on average).[552] Table 3.6.10 suggest that there are also large variations in average farm size among the CIS countries. This is probably a reflection of real differences in the distribution principles,[553] but it can be suspected that the data presented in the table might be somewhat distorted if the differences in the treatment of arable and pasture land by those who prepared the census in the different countries.

(viii) Commercialization of large firms

Placing medium-sized and large firms in private hands is obviously the biggest challenge in the transition countries. Theory offers a plethora of procedures for achieving this goal (table 3.6.1) — especially if mixed solutions are counted separately — and indeed experience shows that virtually all possible privatization techniques have been tried by one country at least.[554] In this sense, therefore, the picture is fragmented.

In principle, it should be possible to collect enough information about the relative weights of each of the privatization procedures applied in individual countries and thus arrive at a precise assessment for the region. However, necessary information is lacking, and what is available is often imprecise — chiefly because of confusions in terminology. The same terms appear to be used in different ways even in the same country. Another important reason is that privatization transactions at the micro-level represent mixed cases in themselves — a fact which is no way objectionable but poses classification problems for the analyst.

Governments in the transition economies are under enormous pressure to show that privatization in their country is proceeding swiftly and fully in line with western recommendations. In September 1990, for example, Hungary made headlines with her commitment to divest large SOEs through stock flotations and auctions. This "truly market-conforming" method was highly praised abroad and helped the country to maintain its reform image, in spite of the fact that during the first 15 months not a single deal was completed according to these principles.[555] In a similar way, 1991 and 1992 were the years of the Czechoslovak voucher privatization, which similarly attracted vivid international attention, although in those two years no ownership change took place by that method. The year 1993 is likely to become the year of the Russian voucher privatization. What it will achieve it still too early to say, but the first reactions of western observers were favourable. This was particularly important after reform-minded Acting Prime Minister Gaidar was replaced: voucher privatization was viewed as *the* evidence for the continuation of economic and political reforms.[556]

Two to three years have been enough to see some common trends emerging. Interestingly, they appear to result from different mechanisms. As shown below, privatization techniques which are distinctly different on paper can, under the pressure of similar socio-economic circumstances, yield to very similar results.

Perhaps the most obvious procedure for privatizing medium-sized and large SOEs is to sell them to *foreign investors* without restructuring of any kind. Two or more birds are thus killed with one stone. The country earns foreign currency, improves its image in the eyes of western governments and the international agencies, while the company concerned is provided with management assistance and capital investment. It is not by chance that the transition economies are promoting this solution and many such deals have been concluded.

To the great disappointment of the transition countries, however, foreign investors appear to have been unwilling to bid for many of the hundred thousands of SOEs put on sale. As a sweeping generalization for the region as whole, it is probably accurate to state that the demand is sufficiently large in eastern Germany, but its intensity falls as progressively as it moves eastward. Leaving eastern Germany aside because of the special nature of its "foreign" investors,[557] Hungary has so far provided the most attractive acquisitions in the eyes of western businesses. But even there

---

[551] See the Russian government's decree of 29 December 1991. *Zakon*, March 1992, pp.83-84.

[552] This figure refers to independent producers only. If account is taken of production on the personal plots allocated to farmworkers by their state or collective farm alone accounted for 34 per cent of total agricultural output in 1991.

[553] For example the very small average farm size in Armenia reflects the fact that, in 1991, land was distributed in half-hectare lots through a village-organized lottery system. These lots cannot be sold for three years.

[554] For a detailed presentation of these techniques, see United Nations Economic Commission for Europe, *Legal Aspects of Privatization in Industry*, New York, 1992.

[555] On the failures of the so-called *First Privatization Programme*, see É. Voszka, "Not even the contrary is true: The transfiguration of centralization and decentralization", *Acta Oeconomica*, Vol.44., Nos.1-2, pp.77-94.

[556] When regular auctions started on 14 January 1993, R.V. Burke, deputy director of the US Agency for International Development's Office in Moscow, was quoted in a front page story as saying: "This is by far the highest priority the US government has. Everything else depends on getting these big enterprises on a market footing". *International Herald Tribune*, 15 January 1993.

[557] In the period 1 July 1990 to 31 January 1993, foreign investors acquired altogether 556 companies and factories. This is about 5 per cent of the grand total. In other words, 95 per cent of the privatization demand has come from investors in western and eastern Germany. There is no direct information about the

the share of foreign capital the economy remains fairly modest — around 3 per cent according to official estimates[558] — which shows that this alternative has not led to quick results. Without any detailed investigation, it can be safely said that this is *a fortiori* true for Russia and most successor states of the Soviet Union.

The second equally direct and simple solution is *liquidation*. The company in question is wound up, its assets are sold at scrap value and its name is deleted from all registers, including those of taxpayers. Governments have to solve a dilemma of great complexity. On the one hand, writing off assets on a mass scale is a very expensive solution for society as a whole. On the other hand, governments are aware that they do not have the means to inject financial or intellectual capital into thousands of ailing firms and — in the absence of a workable industrial policy — they can not even know where to start.

In Poland, privatization through liquidation has been a widely pursued practice so far.[559] Liquidation was originally meant for smaller companies in Poland, while larger ones were expected to go through a transformation procedure. In practice, the liquidation route has become more popular for SOEs of all sizes, as it offers two major advantages over the more formal transformation method.[560] In a liquidation, the assets are directly invested in the new enterprise, while under the transformation procedure, the proceeds of the sale go to the Treasury and not to the enterprise itself. The second advantage relates to the transformation route's requirement that 20 per cent of the equity of the new firm must be offered to the employees on preferential terms. This requirement, however, does not hold for liquidated firms.[561]

Liquidation also appears to be a widespread form of privatization in Moldova, although in this country this is against the law. According to a Russian report, such practices are known in all branches of the economy, but the Ministry of Industry has been worst hit: 800 collectives privatized equipment, vehicles and immovable funds worth tens of millions of roubles.[562]

In Croatia, where large privatization has hardly started yet, liquidations appear to be one of the favoured solutions, although it meets very strong opposition from the trade unions.[563] Contrary to widespread belief, complete dissolution of companies is a very rare phenomenon in the former GDR. So far, 2,340 Treuhand companies have been forced into liquidation or bankruptcy, but the process has been completed for only 31 of them.[564]

What is meant by liquidation in some countries is not very different from changes occurring in Hungary under the name of *transformation*.[565] Indeed, this avenue has technically been open in Hungary since 1984, but it was hardly used until large SOEs were forced by their rapidly deteriorating financial situation to streamline their organization structures and break themselves up into smaller, more transparent units. This technique became widespread after 1987, as more and more managers realized that the government was unable to control the process and that legal constraints on asset stripping were weak and ill-defined. The Transformation Act of 1989 and the creation of the privatization agency (SPA) in 1990 brought some tightening of state control, but at the same time political changes have given new interpretations of what should be considered as "fair" business practice.

In Hungary, the transformation of companies involves the co-opting of new owners: foreign investors, local authorities (in exchange for the use of real estate), banks (in debt-equity swap deals), the workers and managers of the company itself. As the Transformation Act was flexible enough to accommodate all these claims, it finally became the main legal avenue for the privatization of existing SOEs. By the end of 1992, 602 transformations had been approved by the State Property Agency, while the remaining 1,000-1,500 SOEs are required to change their legal forms by mid-1993.

The privatization plans in the successor states of the former Yugoslavia are very similar to current Hungarian practice. The only difference is that the blueprints generally define precise proportions among the co-opted new owners, something the Hungarian lawmakers preferred to leave open for bargaining. In Slovenia, for example, where political and economic conditions are probably most conducive to reform, the

---

total value of foreign acquisitions, but on the basis of published figures on investment and employment commitments available to German and other investors, a figure not higher than DM 3-4 billion can be accepted as a rough estimate. See Treuhandanstalt, *Monatsinformation der THA*, 31 January 1993.

[558] This figure is obtained as a weighted average of foreign ownership in the statutory capital of all transformed companies in 1992. This share should not be confused with the average share of foreign capital in joint ventures arising from the transformation procedure, which is above 30 per cent. *PRIVINFO*, Vol.2, No.3, 1993.

[559] In the period 1 January 1990 to November 1992, 1,474 out of 1,955 potentially privatizeable companies have chosen this method. Główny Urząd Statystyczny, *Informacja o sytuacji społeczno-gospodarczej kraju w roku 1992*, p.59.

[560] These two main avenues are often referred to as privatization according to article 18 (transformation) or article 37 (liquidation) of the Office of the Minister of Ownership Changes Act. For a full English-language text of the law, see *Legal Aspects ...*, op.cit., pp.115-130.

[561] J. Cieslik, "Privatization in Poland: Constraints and Impacts", Paper presented at a UNDP Expert Group Meeting held in Geneva on 17-21 August 1992.

[562] Russia's Radio, quoted by BBC, *Summary of World Broadcasts*, SU/1627, 3 March 1993.

[563] In the view of M. Mesic, President of the Confederation of Independent Trade Unions of Croatia, "This model of privatization brings robbery of the social, worker's and state property unknown in history. (...) First, managers bankrupt a firm in order to make its value low. Then, the managers obtain management loans through banking tricks. (...) This is fake privatization and criminal of the highest calibre". *Vjesnik*, 18 February 1993, pp.20-21.

[564] *Treuhandanstalt*, op.cit.

[565] For an in-depth discussion of the Hungarian privatization process, see United Nations Economic Commission for Europe, *Economic Survey of Europe in 1991-1992*, New York, 1991, p.233; É. Voszka, "Privatization in Hungary: debates, developments and risks" in Targetti, op.cit., pp.209-233; Y. Kiss, "Privatization in Hungary - Two Years Later", *Soviet Studies*, vol. 44. No. 6. 1992. pp.1015-1038.

TABLE 3.6.11
Voucher privatization programme in the transition economies

Country	Status	Designation	Law passed	Distribution started	Eligibility	Number of vouchers or "points" per person	Approximate nominal par value
Albania	Planned	-	-	-	Citizens 18+	-	-
Czechoslovakia	In progress	Investment coupon	February 1991	October 1991	Citizens 18+	1 000 points	1 point = 35Kcs
Hungary	In progress	Compensation voucher	June 1991 a	March 1992	Hungarians including foreign nationals	F(C) Max. Ft 5 million	-
	Planned	Credit voucher	..	..	All citizens	..	Ft 1 million
Poland	Planned	Certificate of the National Investment Fund (NIF)	-	-	Citizens 18+	1 share in each of the 20 NIFs	-
Romania	In process	Certificate of the Private ownership Fund (POF)	August 1991	June 1992	Citizens 18+	1 share in each of the 5 POFs	1 certificate = Leu 5 000
Slovenia	Planned	Ownership certificates	December 1992	-	All citizens	F(E) max. DM 4 000	-
Armenia	Planned	-	-	-	-	-	-
Belarus	Planned	Personal privatization voucher	?	-	-	F(A) 10-50	-
Kazakhstan	In process	Voucher	?	Early 1992	All citizens	F(E)	-
Moldova	Planned	National assets bonds	November 1992	April 1993	Citizens	F(E)	R5 000
Russia	In progress	Privatization cheque	June 1992	October 1992	All citizens	1 person	R10 000
Ukraine	In progress	Privatization	March 1992	-	All citizens	-	30 000 karbovanets
Estonia	In process	Public capital bonds	-	September 1991	All citizens	F(E)	1 sq m
	In process	Compensation voucher	-	-	All citizens	F(C)	housing space
Latvia	Planned	Privatization accounts	November 1992	-	Most residents b	F(T)	½ sq m housing space
Lithuania	In process	Special investment accounts	February 1991	February 1991	All citizens	F(C,A)	-

Country	Nominal issue price	Validity	Acceptable as cash for buying	Payment for buying equity in state-created holding companies	Number of medium and large SOEs targeted	Share obtainable through vouchers in any SOEs (per cent)	Secondary trading allowed	Auctions started
Albania	-	-	-	In special funds	-	-	Yes	-
Czechoslovakia	Kcs 1 035	10 months	Shares in firms or equity in investment funds	Not planned	4 200	15-97	No	May 1992
Hungary	Free	Unlimited	Land, housing, life annuity, equity	Planned	10-20	0-100	Yes	January 1992
	Ft 20 000	1-2 years	Shops, firms etc.	Not planned	-	49	No	..
Poland	10 per cent of monthly wage	4 years	-	In each of the 20 NIFs	600	60	Yes	-
Romania	Free	-	Shops, firms (details unspecified)	In each of the 5 POFs	5 931	30	Yes	-
Slovenia	-	-	-	Planned	-	20	Not decided	-
Armenia	-	-	-	-	-	-	-	-
Belarus	-	-	-	-	-	-	-	-
Kazakhstan	Free	-	Housing, firms, etc.	About 20 planned	-	10-15	No	-
Moldova	Free	2 years	Shares	..	-	-	-	-
Russia	R25	1 year	Shops, firms, or equity in investment funds, housing, land	Not planned	40 000	35-80	Yes	January 1993
Ukraine	Free	2 years	Shops, firms, or equity in investment funds, land (planned)	Planned	-	0-70	No	-
Estonia	Free	4 years	Houses, firms, etc.	Not planned	-	-	After 4 years	Sept. 1991
	Free		Houses, firms, etc.	Not planned	-	-	-	-
Latvia	Free	-	Firms, shops, houses	Not planned	-	-	-	-
Lithuania	Free	-	Houses, shops, land equity in investment funds	Not planned	-	-	-	September 1991

Source: Official reports, national newspapers.

Notes: F(C), F(E), F(A), F(T) = Eligibility is a function of the assessed losses under Communist rule (C); Years of employment (E); Age (A); Time lived in the country during the post-war period (T).

a Similar laws were also passed in April, May and June 1992, enlarging the scope of eligibility.
b Members of the Soviet armed forces, the KGB and high level party functionaries are excluded. For non-citizens five vouchers will be deducted for using the country's infrastructure.

privatization law guarantees 20 per cent share to workers, 20 per cent to investment funds, 10 per cent to pension funds and 10 per cent to an indemnification fund for the former co-owners.[566] This distribution scheme leaves 40 per cent available for sales to domestic or foreign bidders. At the same time, however, a SOE is also allowed to sell its assets in their entirety to a domestic or foreign buyer, provided he makes the required allocations for pension, indemnification and investment funds.[567]

In contrast to the case-by-case operations examined above, more and more countries are contemplating the use of *mass privatization* methods. In terms of the degree of progress so far, Czechoslovakia, Lithuania, Romania and Russia have taken significant and irreversible steps: in these countries 80-95 per cent of those eligible to take part in such schemes have actually bought vouchers or collected them from the distribution points. Similar programmes are in the making in Hungary, Poland, Slovenia and in virtually all the successor states of the Soviet Union (table 3.6.11).

In essence, they all involve giving away to the population at large the lion's share of state enterprises. Distinctions should be made, however, between the design of the programmes according to differences in eligibility criteria. In most countries, citizenship and age (18 plus) is the only criterion, but in Estonia, Moldova and Slovenia the number of years in employment is also taken into account. In some countries, notably Estonia and Hungary, two different types of privatization vouchers are envisaged – one for compensation purposes and another for which all adults are eligible.

Given the large number of countries and the continual changes in the plans themselves, it is not easy to form a comprehensive overview of all programmes. At the risk of over-generalization, the mass privatization plans seem to fall into three distinct categories: computerized initial public offerings, the holding concept and direct public auctions.

Computerized initial public offering is an original method launched by the Czechoslovak authorities.[568] In this scheme, individuals bid for shares with voucher points in country-wide computerized auctions. These voucher points are not transferable and bids have to be repeated until a satisfactory level of supply-demand equilibrium is achieved: i.e., until the vouchers are used up and all of the firms have found buyers.

At the outset, it was envisaged that this method would guarantee a fast solution for the future of 1,500 corporatized SOEs.[569] Progress was considerably slower than expected, in spite of the fact, that the so-called "first wave" of bidding was shortened from six to five rounds (table 3.6.12). From the conceptual decision in September 1990 it will have taken two and half years until the completion of the first wave. Investors will receive their shares during the spring of 1993, and the first meeting of shareholders will probably not be held until the summer, thus company restructuring will not start before the autumn of 1993. From the shareholders' point of view, the time factor is also important, because their choice was made 12-15 months earlier on the basis of 1991 balance sheets of the companies concerned. In the meantime, absolute and relative prices have changed substantially and the country has been split into two. This latter development means that Czech and Slovak investors will now possess shares in foreign companies, without having a clear idea of future rules concerning cross-border capital transactions, exchange rates, etc.

Another important development was the sudden appearance of privatization investment funds (*investiční privatizační fond*) or IPF. An overwhelming majority (72 per cent) of Czechoslovak citizens invested their voucher points through these institutions. At the end of the first wave, more than 400 such funds were registered. The nine largest controlled almost 50 per cent of all investment points. Of these, six are subsidiaries of well-known Czechoslovak state banks, one belongs to an Austrian bank, one is owned by an American expatriate and only one is a domestic private venture. Of all voucher points placed, 37 per cent are in the hand of the IPFs created by state-ownd commercial banks.

Poland and Romania opted for mass privatization through closed-end investment funds or *holdings* created by government *fiat*.[570] First, a large number of SOEs were selected (600 in Poland, almost 2,000 in Romania) for inclusion in the programme and converted into shareholding companies. In the second step, the shares of these companies are distributed

---

[566] In contrast to earlier drafts, the late 1992 version of the privatization law has obtained the blessing of the Slovenian parliament and once the administrative details are in place the process should start in May 1993. *East European Markets*, 19 February 1993.

[567] *East European Markets*, 11 December 1992.

[568] The theoretical foundation of this model was laid down by Polish authors. For an English-language work, see J. Szomburg and J. Lewandowski, "Property Reform as a Basis for Social and Economic Reform", *Communist Economies*, No.3, 1989. As far as implementation in the CSFR is concerned, see United Nations Economic Commission for Europe, *Economic Survey of Europe in 1991-1992*, New York, 1992, (pp.237-238), or the English-language translation of the information handbook prepared by the Czechoslovak Federal Ministry of Finance ("Coupon Privatization: An Information Handbook", *Eastern European Economics*, 1992, Vol.30, No.4). For a more up-to-date analysis, see *PlanEcon Report*, Vol.VIII, Nos.50-52, 25 January 1993 and Vol.IX, Nos.3-4, 16 February 1993.

[569] More precisely, 1,491 SOEs were involved in the first wave, representing all sectors of the economy, including banks. Among the firms, 943 were located in the Czech Republic, 437 in the Slovak Republic and 61 were of a federal nature. On the basis of 1991 data, these firms had a 17 per cent share of employment in the Czechoslovak economy. For the second wave (starting in autumn 1993) more than 2,100 companies are earmarked.

[570] The origins of the holding model go back the preparatory discussions of the Hungarian New Economic Mechanism of 1968, when several experts argued for the need to separate ownership rights from disposal rights in order to induce capital flows among SOEs. A proposal to corporatize SOEs, through compulsory transformation into shareholding companies had been prepared, but it was finally rejected on political grounds and the publication of such ideas was forbidden. For a review of the pre-1968 discussions, see R. Hoch, "Changing formation and privatization", *Acta Oeconomica* Nos.3-4. 1991. A few years later, the same suggestion re-emerged in a seminal article by M. Tardos, "The problems of economic competition in our country", *Közgazdasági Szemle*, No.2, 1972 (in Hungarian).

TABLE 3.6.12
The scenario of voucher privatization in Czechoslovakia

September 1990	Scenario of Economic Reform approved by Parliament
26 February 1991	Basic legislation on voucher privatization enacted
1 October 1991-15 February 1992	Registration of coupon books
First wave	*(Czech and Slovak Republics)*
17 February-27 April 1992	Pre-round of auctions
18 May-8 June 1992	First round of auctions
8 July-28 July 1992	Second round of auctions
November 1992	Fourth round of auctions
23 November-22 December 1992	Fifth round of auctions
22 December 1992	Completion of first wave
29 March 1993	Start of physical distribution of shares (planned)
Second wave (planned)	*(Czech Republic)*
Summer 1993	Registration of coupon books
December 1993	Completion of voucher privatization
No decision yet	*(Slovak Republic)*

Sources: Press reports.

among 5-20 investment funds. Parallel with this process the country's citizens are entitled to buy shareholding certificates representing in these holdings in the expectation of dividends.

As noted above, these certificates have been already distributed in Romania, where progress has also been made in setting up five holdings. One certificate contains five coupons, one each for the five holdings. The nominal value of the certificates will be announced at a later date. By April 1993, 400 commercial companies, including banks, will be allocated to each of the five Private Ownership Funds (POF) created on sectoral criteria (i.e., there is a "metallurgy" fund, a "chemical industry fund", etc.). At the same time, this allocation process aims at a certain territorial balance, too. POFs are supposed to become regional economic centres. Each POF will be directly involved in the activity of these firms, both in their restructuring and in their privatization. Recent reports indicate that POFs may be considering the retention of profits for as much as for four years to finance restructuring investments. Against this background, it is not surprising that so far most "investors" took a wait-and-see position, the demand for shares in these funds has been very restrained, while secondary transactions among voucher holders also appears to be marginal.[571]

In Poland, government and Parliament were still struggling over different modalities in March 1993. There is no final decision as yet concerning the number of holdings to be created, nor on the precise list of SOEs to be involved in the scheme. Laws and statutes relating to the operation of the holdings are also missing.

The latest wave of large privatization in Russia, which started in December 1992,[572] combines elements from schemes that have already been adopted elsewhere. However, the scheme as a whole is certainly unique and includes untested elements. First of all, divestment of enterprises is envisaged for implementation in the form of *direct public auctions*.[573] Firms on sale by virtue of the law and the decision of the management itself are auctioned directly in a session lasting several hours. There is no computerized bidding or detailed comparison of written documents.

Second, when buying shares in auctions, bidders may pay in cash or with vouchers. Originally, the minimum share of vouchers was set at 25 per cent, but this limit was later raised to 80 per cent in the case of state-owned firms and to 90 per cent in the case of municipally-owned enterprises.[574]

Third, in contrast to Czechoslovakia, foreigners can participate openly at the auctions. Given the size of the Russian economy, it is unlikely that foreign investors will carve out a considerable share of it, but cherry-picking, of course, is a real possibility. Probably, this is a reason why the list of non-privatizeable enterprises has turned out to be so long from the very outset and also explains why the entire oil industry was suddenly added to the list later on.[575]

Fourth, Russian vouchers issued with R10,000 nominal face value, were made freely tradeable immediately on issue. This raises the question of whether vouchers can be regarded as surrogate money and — in the case of a positive answer to the first question — whether the privatization agency had acted legally when issued them.[576]

Fifth, similarly to the privatization blueprints of the post-Yugoslav states, the Russian scheme is deliberately biased toward protecting the interests of insiders. For those 5,500 medium-sized companies (i.e., with less than 10,000 workers) which were allowed to participate

---

[571] There is a very thin free market of vouchers, where they sell within the leu 1,500-3,000 range, as opposed to their unofficially suggested face value of leu 30,000. *Heti Világgazdaság*, 13 March 1993.

[572] The pilot auction of shares under the voucher programme was held on 9 December 1992 at the Bolshevik bakery and confectionery plant in Moscow. K. Bush, "Industrial Privatization in Russia: A Progress Report", *Radio Free Europe Radio Liberty Research Report*, Vol.2, No.7, 12 February 1993. But there are also reports indicating that in some provinces – e.g., in the Yaroslav Oblast – privatization vouchers were exchanged for shares in SOEs as early as October 1992. (Russia's radio, quoted in BBC, *Summary of World Broadcasts*, SU/1509, 12 October 1992.)

[573] Note, however, that the relevant legislation, entitled "State Programme for Privatization of State and Municipal Enterprises in the Russian Federation for 1992" also envisaged the use of such standard methods as closed tenders, liquidation, leasing, etc. For the full text of the law, see *Ekonomika i zhizn'*, No.29, July 1992, or in English, BBC, *Summary of World Broadcasts*, SU W0240, 24 July 1992.

[574] This corrective measure was introduced when voucher prices on the secondary market fell to very low levels and the authorities wanted to arrest the fall. Prices on this market were also helped by another announcement of the privatization agency, hinting that in 1994 large companies with more than 10,000 employees will be sold off as well.

[575] According to an Interfax news agency report dated 7 February, Goskomimushchestvo has sent a telegram to all regional branches instructing them to cease accepting and approving privatization applications in the oil and oil refining industries. BBC, *Summary of World Broadcasts*, SU/1608, 9 February 1993.

[576] A. Gluseckiy, "Voucher – genezniy surrogat" (Voucher as surrogate money), *Ekonomika i zhizn'*, No.6, February 1993.

TABLE 3.6.13

Methods of privatization in Russia for medium-sized enterprises

Variant	Preferred (non-voting)		Common shares	
	Shares	Constraint	Shares	Constraint
1	Free to all employees up to 25 per cent of charter capital	One employee cannot receive more than 20 times the minimum wage	Employees can buy shares up to 10 per cent of the charter capital with a payment period of 3 years and a 30 per cent discount and Chief officials have an option to buy shares up to 5 per cent of charter capital at nominal price	Downpayment minimum 15 per cent One official cannot buy more than 2,000 times the minimum wage
2	-	-	All workers can buy shares up to 51 per cent of charter capital at nominal price	-
3	-	-	A group of employee can buy shares up to 20 per cent of charter capital with 30 per cent discount and Employees can buy shares up to 20 per cent of charter capital with 30 per cent discount	The firm has to be privatized in a year
4	-	-	Employees can buy shares up to 90 per cent of charter capital with 30 per cent discount in a payment period of 3-5 years	-

Source: State Programme for Privatization of State and Municipal Enterprises in the Russian Federation for 1992; newspaper reports.

Note: Variants 1-3 from the State Programme, Variant 4 is from newspaper reports. See also text for explanation.

in the 1992 round of auctions, *Goskomimushchestvo* has defined three alternatives, offering a 51 per cent stake for management and employees as the maximum (table 3.6.13). Prior to, but also after, the enactment of the relevant Supreme Soviet decree, the Russian Fund for the Assets of the Federation, the *de jure* owner of all SOEs, waged an active campaign for a fourth alternative, pushing up the limit of insider ownership up to 90 per cent. This option was, however, discarded by lawmakers.[577]

Last but not least, an important feature of the Russian scheme is that it allows SOEs to buy vouchers from the public and to use them to acquire partly or fully the outstanding equity remaining on sale beyond the 51 per cent purchased by the staff.

This latest wave of privatization has started only very recently, so it is too early for any assessment. Some startling signs, however, are visible. The ordinary citizen often had no idea what to do with the vouchers; so an immediate reaction was to sell them for whatever price they would fetch.[578] So far, the downward trend on the secondary market has been steep. After five months, the price of a R10,000 voucher had fallen to below R4,000. Other people with more understanding of the rules of the game have opted to deposit their vouchers into one of the 300 investment funds, as in Czechoslovakia. In the Russian case, however, the inherent risk of this avenue has manifested itself at a very early stage. The first reported case of fraud by an investment fund occurred in St. Petersburg, inflicting losses of more than R3 billion to thousands of citizens.[579]

Up to mid-February 1993 only a few dozen voucher auctions have been held. It seems that two thirds of the selected SOEs have chosen Version 2, which guarantees a 51 per cent stake to the working collective, albeit without any discount. This is understandable. Since the law was adopted in mid-1992, the purchasing power of the vouchers has increased astronomically, because asset valuation is based on 1991 book keeping values. Thus, price has become virtually irrelevant for the first 51 per cent of shares allotted to workers without auction. According to analysts, there is a tendency for the

---

[577] However, a motion to include Variant 4 among the options remained on the agenda of the Supreme Soviet, which is supposed to approve the 1993 privatization programme. *Izvestiya*, 19 February 1993.

[578] According to a public opinion poll carried out in October 1992, 58 per cent of respondents were not interested in investing their vouchers and most of them planned to sell them for cash. Only 15 per cent wanted to buy additional vouchers. The potential buyer tended to be under 25, male, and in a good private- or state-sector job. Interest in buying vouchers was also markedly higher among high earners. *Radio Free Europe/Radio Liberty Research Report*, Vol.1, No.48, 4 December 1992.

[579] The private investment fund in question had offered the following deal: in exchange for every voucher deposited, it promised to pay R12,000 per month and pledged to give back the vouchers without notice at the owner's request. For the first month, the fund honoured its commitments, after the second it declared bankruptcy and its managers disappeared with 300,000 vouchers. Three thousand aggrieved clients demonstrated on the streets of St. Petersburg demanding stricter regulations and better control. For the full details, see *Izvestiya* 13 February 1993.

management and staff of firms put up for auction to aim at acquiring 90 or 100 per cent ownership, as was envisaged by the proponents of the so-called "fourth alternative". They hire brokerage firms to buy up the required number of vouchers and, when it comes to bidding, these insiders can easily squeeze out other investors.[580]

(ix) The new owners

Following this review of privatization practices in eastern Europe and the countries of the former Soviet Union, one important question remains to be answered: who are the new owners of privatized state assets?

One answer can be easily discarded. It is mistaken to believe that foreigners are the main players. Existing provisions generally restrict the sale of land to foreign individuals or enterprises. By and large, the same is true for the housing stock. Foreigners have wider access to assets divested in the service sector, but so far their interest has proved to be relatively limited in most east European countries and, *a fortiori*, in the CIS states. Large privatization is still in its initial phase in most countries, so no firm conclusions can yet be drawn. On the whole, there are signs of a certain foreign interest, but after corporatization the lion's share of assets has fallen firmly into the hands of domestic firms, including the various state asset management funds.[581]

It is more difficult to discard a second claim frequently met, namely that the new owners are essentially representatives of the old *nomenklatura*, supplemented by criminals and adventurists.[582] Once again, this allegation is obviously groundless in the case of land, housing and service workshops. The sheer size of the asset stock proves that there is a much larger circle of people involved in the privatization process. None the less, there is probably some element of truth in these assertions when it comes to large privatization. It is certainly true that, prior to 1989, there was virtually no possibility of accumulating significant private wealth — in other words, starting capital — in any of the countries concerned. For this reason, the concentration of private wealth observed by early 1993 in many countries does give rise to questions. It is also undeniable that crime — including organized crime — is on the increase everywhere and that it is possible, indeed likely, that there is no strong barrier between the world of organized crime and the world of business.[583] On the other hand, it should not be forgotten that the turbulent changes of the past three to four years have opened numerous legal avenues to enrichment, starting from street trading to foreign exchange and real estate transactions. As markets were gradually liberalized, risk-taking entrepreneurs have had ample opportunities to find activities where extra profits could be earned without transgressing any legal barrier. Restitution may bring riches overnight and — at least in eastern Europe and the Baltic countries — there is a significant inflow of private capital arising from the return of the *diaspora*.

But beyond this general comment, it should be noted that the privatization process is driven by far more complex motives than ownership alone. In those cases where a large SOE is split into 10-15 smaller units, new executive positions open up. When foreign partners are involved, this holds out prospects of increased travel possibilities, access to state-of-the-art technologies, etc. After the political changes in late 1989, successful SOE managers could aim at positions in the new state administration or at better paid jobs with western firms. In addition to these possibilities, the establishment of hundreds of joint stock companies has created new types of management positions.

The second problem with the view which sees the inherited old regime behind every change is that it fails to put the issue in the right historical and cultural context. In the transition economies, there are few people available with sufficient technical and commercial competence to do a better job than the previous management cadres. After a fundamental change in a country's political system, it proved easier to find a certain number of qualified intellectuals who were ready to assume political and government functions in the new framework. But there is no back-up for the thousands or tens of thousands of managers and officials running the existing SOEs and the state machinery. This is a much broader circle than the *nomenklatura* — i.e., officials whose appointments in the past were directly controlled by the central party apparatus.[584] Independent experts can, of course, be invited from abroad, but this is not a cheap solution and very often these "foreigners" are in fact expatriates themselves, so

---

[580] *Finansovye Izvestiya*, 4-10 February 1993.

[581] It is worth noting, however, that both the Czechoslovak and the Russian voucher privatization programmes have opened the door for foreign investors to buy stakes in the privatization funds and thus obtain property rights in large SOEs without lengthy negotiations with company directors and the state authorities. But there is no way of knowing to what extent foreign investors actually exploited this possibility.

[582] For a passionate exposition of this view, see the article by G. Popov, the prominent Russian economist and former mayor of Moscow, in *Moscow News*, 18 February 1993.

[583] This issue appears to be particularly grave in Russia and perhaps in some other CIS republics. Addressing an all-Russia conference on the problems of combating organized crime and corruption, President Yeltsin estimated that "over half the criminal groups have corrupt links. Bribes for obtaining licences and the allocation of quotas, for the granting of credits at privileged rates, for creating special conditions for commercial structures, for acquiring at reduced prices state property which is being privatized and many other things — all of this is the reality of our life." BBC, *Summary of World Broadcasts*, SU 1613, 15 February 1993.

[584] At the onset of communist rule, nationalization campaigns went along with a ruthless removal of previous managements. Qualified engineers and businessmen were sacked (if not arrested) and reliable party workers were put in their place. In terms of lost efficiency, the price of this policy was enormous. Today, this lesson is still very much in the memory of decision-makers in many countries and they are trying hard not to repeat the same mistake.

their independence can also be questioned.[585] Experience has shown that the increase in objectivity they bring is not sufficient to outweigh their lack of detailed knowledge. Western accounting firms appear to have a particularly poor record in this respect. In Hungary, for example, their involvement in the privatization process has been limited to rubber-stamping after Hungarian accountants have done the real job. This problem has arisen in a very direct manner in Poland too, where the planned state investment funds were supposed to be headed by foreigners. Given the obvious language requirements, external recruitment is basically limited to expatriates.

As shown in the previous sections, none of the transition countries are following a coherent, linear path to private ownership. First of all, corporatization in many instances is not followed by commercialization. This is often the case when the legal transformation is done in a uniform, centralized manner — as in Romania.

Without an in-depth analysis on a case-by-case basis, it is very difficult to arrive an impartial assessment. In one country, the state authorities are deeply dissatisfied with the conduct of the management running the corporatized firms and try to slow down privatization in order to restore discipline and control.[586] But in another country, a similar situation is often interpreted as an evidence against the government, which wants nothing else but to spread itself out upon the enterprises — directly or indirectly through the newly created state asset management funds — to seize their revenues and to fill the bottomless pits of the Treasury.[587]

It might be noted that in critical areas, private entrepreneurship is developing on the basis of leaseholds (agriculture, retail trade), where the right of use and *de jure* ownership are not in the same hands. In the sphere of medium-sized and large SOEs, corporatization and privatization have given rise to various forms of *cross-ownership*. Due to the overall economics depression prevailing in the transition economies, the newly-transformed companies are forced to look for unconventional sources of financing. As a means of avoiding bankruptcy, they seek ownership links with their creditors (state-owned banks in most cases) or their suppliers, which may or may not be in private hands. In other cases, subsidiaries and production units are sold or leased to outsiders. The rise of financial intermediaries (holdings) has had a similar impact. Many Czechoslovak investment funds, for example, were originally created by commercial banks, which in turn are partly owned by enterprises.

Finally, note should be taken of the fact that privatization schemes are often designed to give preferential treatment to the work force and the management. This is a prime characteristic of the Russian privatization programme, but it frequently occurs in other countries as well.

(x) Privatization revenues

In one important respect, the three main avenues appear to have been leading to a common end-result. From a budgetary point of view, large privatization has yielded relatively small revenues. This was, of course, intentional in the case of the mass privatization programmes, but it was certainly not envisaged in Germany or Hungary.

At the time of unification, it was believed that east German assets were worth over DM 600 billion. In autumn 1992, when publishing the audit of the 8,500 firms on its books reflecting the situation as of July 1990, the Treuhand agency estimated their net worth as a *negative* DM 200 billion. Revenues from privatization by 31 January 1993 amounted to a mere DM 41 billion.[588] The cost of restructuring — in the form of debt relief, environmental and employee compensation, etc. — which all appear on the expenditure side of the Treuhand's budget — are much larger than anybody expected. According to current estimates, when the Treuhand winds up its activities in 1994, instead of contributing to the German state budget it will leave a DM 250 billion debt.[589] In Hungary, the balance sheet of the privatization agency (SPA) is less transparent. On the one hand, there are fairly detailed reports on the revenue side according to which more than Ft 100 billion ($1.3 billion) was cashed in during 1991 and 1992. But there is virtually no information about the true cost implications of the SPA-approved transformations.[590] It is also known that companies are often sold to foreign investors with multi-year tax priv-

---

[585] For example, in Estonia, the head of the privatization agency was removed because he was accused of breaching impartiality in favour of the country in which he grew up. Similar doubts were raised in Hungary when the government nominated an American citizen as the head of State Property Management Ltd.

[586] This appears to be the case in Romania, as the following quotation of President Iliescu indicates: "We have very many companies whose leadership survives because of their popularity. Certainly, I do not want to say what kind of business deals are made, but they are anything but clean, since any kind of contract relations have disappeared; thus all these control instruments have disappeared; there are independent people who have complete freedom — it is they who are the owners, not the state. (...) These people manage the assets of the given enterprises but they are not the owners of the social capital which belongs to the state." BBC, *Summary of World Broadcasts*, EE 1544, 21 November 1992.

[587] This is how, e.g., the eminent Croatian economist Branko Horvat assesses the situation in his country: "Privatization ought to be stopped momentarily and the government must stay out if it. Instead the Parliament ought to form a commission of independent, internationally renowned experts that should take care after the implementation of the reform." ("Stopirati pretvorbu" (Stop privatization), *Vjesnik*, 3 March 1993.) For a similar view, see N.4Cuckovic, "Privatization in Croatia: What went wrong?", paper presented at the Third Conference of the International Society for the Study of European Ideas (ISSEI), Aalborg, Denmark, 24-29 August 1992.

[588] *Treuhandanstalt*, op.cit., p.12.

[589] *The Wall Street Journal*, 16-17 October 1992.

[590] One of the rare examples when both pluses and minuses were reported was the case of the country's largest television manufacturer VIDEOTON. To the state, liquidating the firm cost Ft 20-30 billion in the form of unpaid debt. To the new owners it cost a mere Ft 4 billion and "many nice promises". É. Voszka, "Kerülőutak, ösvények, zsákutcák" (Roundabouts, trails, cul-de-sacs), *Közgazdasági Szemle*, 1992. No.6.

TABLE 3.6.14

Official estimates of the contribution of the private sector to GDP
(Percentages)

	1988	1989	1990	1991	1992
Bulgaria	..	..	..	5*	10
Czechoslovakia	..	4.1	5.2	9.3	..
Czech Republic	..	..	..	..	20
Slovak Republic	..	..	..	..	20-21[a]
Hungary (1)	..	..	10	27	35
Hungary (2)	..	..	..	41	45
Hungary (3)	14.3	14.9	15.8	18.2	25
Poland	..	28.6	30.9	42.1	45-50
Romania	..	..	..	..	26
Russia	..	..	..	..	6.7[b]
Yugoslavia FR	..	..	..	40[c]	..

Sources: Official statistics, unless indicated otherwise. Bulgaria: Statement of President Zhelev, as quoted in BBC, *Summary of World Broadcasts*, EE 1577, 4 January 1993; CSFR: *FSU Bulletin* 1992, No.5, p.A34; Czech Republic: *Rudé Právo*, 20 February 1993; Slovak Republic: Slovenského Statistického Úradu, *Bulletin* No. 12. 1992. p.2; Hungary (1): Ministry of Finance, *Economic Policy 1993-1994*, Budapest, 1992, p.3, (1989-1991), B. Kádár (Minister of International Economic Relations) in *Magyar Hirlap*, 23 January 1993; Hungary (2): Estimates of the State Property Agency. *Hungarian Review*, Vol.2. No.12, 29 March 1993; Hungary (3): CSO data. Share of small enterprises, including economic activites of households. *National Accounts — Hungary 1985-1991* Budapest, February 1993, p.17, 1992 data from "Gazdasági-társadalmi áttekintés az 1992. évről" (An economic and social assessment of the year 1992). *Statisztikai Hírek* (undated); Poland: *Rocznik Statystyczny 1992*, p.xiii, 1992 data from the press conference of Deputy Prime Minister H. Goryszewski, held on 20 January 1993, as reported in BBC, *Summary of World Broadcasts*, EE 1594, 23 January 1993; Romania: *Buletin de Informare Publica* No. 12. 1992. Russia: *O razvitii ekonomicheskikh reform v Rossiiskoi Federatsii v Janvare-Octobre 1992 goda*, p.138.

a Contribution to national income.
b Contribution to industrial output in the January-October period.
c Includes cooperatives and enterprises in mixed ownership.

ileges, but these items are not reflected in the SPA's balance sheets. Thus, it is impossible to state whether any net revenue from privatization has materialized in Hungary. But there is a strong possibility that there will be hardly any when the whole process is completed sometime in the late 1990s.

(xi) The size of the private sector

Structural changes are difficult to measure in any case, but the task is even more demanding when it comes to measuring something as fundamental as ownership rights.

In the east, statistics on the size of the private sector are subject to a number of biases and deficiencies in basic data recording. Thus statisticians are at times under pressure to produce numbers showing a high share of the private sector in GDP, employment or in the capital stock. At the same time, however, the changes in statistical reporting systems and data collection rules are still under way, and many small enterprises are probably still escaping the official statistical net. Inadequate tax collection systems and strong incentives to avoid or reduce tax liabilities are another reason for incomplete coverage of the new private sector.

It might also be noted that the propagation of business-like behaviour is not necessarily measured by the extension of majority private ownership. In the transition countries there are important examples — such as the Czech car producer Skoda or the previously quoted Hungarian example, the VIDEOTON factory — where private (including foreign) owners are in a minority. None the less, these companies act under a hard budget constraint and behave like truly private business entities. Obviously, contrary examples could also be cited.

The generic growth of the private sector should also be taken into consideration. Privatization is unfolding against a backdrop of mushrooming new ventures, green field foreign direct investment and the appearance of hundreds of thousands of small, non-incorporated businesses. Although a large majority of these new initiatives have some roots in the old production structures, statistically the connections are hardly identifiable. For this reason the process is sometimes labelled as "hidden privatization", which is an apt characterization of its essentials. This practice appears to be rather widespread in the sphere of business services, where former state employees are creating their own private ventures with a minimum of fixed assets and are trying to make their living by capitalizing on their personal business connections and market knowledge. The other side of the coin, of course, is that a state-owned enterprise deprived of its intellectual capital is likely to slide into bankruptcy soon and its remaining assets are taken over at bargain prices by outsiders — perhaps by the very same people who ran the enterprise before.

Given these factors, any attempt to quantify the distribution of ownership rights on a nationwide level is very approximate at best (table 3.6.14). In the case of mixed ownership, which occurs rather frequently, as shown above, there is no unambiguous method of deciding which company should be termed private and which state-owned. Table 3.6.14 is essentially based on information provided by the national statistical authorities of a few east European countries and Russia. From a methodological point of view, it is interesting to note that the Hungarian authorities are working with three different sets of numbers. Two sets of estimates, produced independently by the Ministry of Finance and the privatization agency, attempt to sort out private and non-private firms on the basis of tax returns. The third series, reported by the country's central statistical office, is based on the size of the work force. It is assumed that size thresholds can be established — not necessarily uniform throughout the economy – below which statisticians that below a certain size firms are unlikely to be state owned. Obviously, both methods are imperfect and only experience will show which will be more suitable for the purposes of measurement.

(xii) Outlook

Slow progress in privatization reflects the ponderous heritage from the past – four decades in the case of eastern Europe and seven decades in the case of the successor states of the Soviet Union – cannot be regarded as a closed chapter. Vested interests, deeply-established forms of thinking, legal traditions and informal connections do not disappear overnight. This tension between the heritage from the past and the move towards a democratic consensus will be mitigated and probably overcome as the the democratic process moves forward – but all this needs a lot of time.

The list of open questions is fairly long. Here only the three most salient ones can be mentioned. First, restitution and compensation cause delays and create obstacles to privatization as well as to new private business developments. Second, there is no mechanism in place to ensure adequate corporate governance of those enterprises which, for one reason or another, remain in state ownership. There is no shortage of determined statements,[591] but there is much less action. The same question can be also raised with respect to the emerging financial intermediaries, the holdings or investment funds. There is a potential danger that they could become vehicles of state intervention, especially in those countries where they are led by government-appointed officials. The third problem is that privatization in many countries is viewed with increasing discomfort and anger by the electorate and the population at large. The fear of crime and the identification of privatization with theft is a palpable (and sometimes not an entirely groundless) sentiment.

All in all, the balance is positive. In the context of overall systemic changes, privatization has already become irreversible in the entire region. True, the path of privatization followed so far has been different from the one envisaged by parlamentarians, governments and academic experts. But this does not count for too much. The scene has changed and will doubtlessly continue to evolve in a promising direction: away from central planning and closer to a market-based, democratic system.

---

[591] For example, speaking on state property which remains in state ownership, Deputy Minister and Chairman of Goskomimushchestvo A. Chubays said the following on Russian television: "(it should be subject) to strict, extremely strict, dictatorial state control over the effective use of that state property, no matter what ideas there are about full economic management or anything like that. State property with state control, with the appointment of the director, with the state determining his release from work and establishing his wages. This is the normal scheme of things that we shall come round to sooner or later." BBC, *Summary of World Broadcasts*, SU 1612, 13 February 1993.

3.7 COUNTRY NOTES

(i) Eastern Europe: the central eastern countries

Czechoslovakia

In 1992 Czechoslovakia maintained macroeconomic stability after the radical economic reform introduced early in 1991. The rate of inflation was the lowest in all transition countries, the fiscal balance did not deteriorate, the exchange rate of the koruna remained stable and the current account ended with a surplus. Czechoslovakia was the only transition country where the unemployment rate fell in 1992. On the other hand, output contracted more than expected. GDP fell in 1992 by an estimated 7-8 per cent. None the less, an upward trend was registered in construction and private sector activities, and in the last quarter of the year industrial output rose both on a month-to-month basis and, for the first time since 1990, above prior-year levels. Domestic absorption increased in 1992, to judge from the estimated 5 per cent rise in real private consumption and stagnating investment. However, that favourable macroeconomic situation of the country is now affected by the split of Czechoslovakia into two independent states. The negative impact of separation on trade between the Czech and Slovak republics, and hence on output and employment, could be important, particularly for Slovakia, which is more dependent on mutual trade.

But behind these generally positive results there are problems of a structural character. Their resolution has progressed slowly and will influence the future economic development of the country. Privatization of the big state enterprises is at its initial stage, and the effects of their restructuring will be felt only in 1993 or later. Inter-enterprise indebtedness is very high, and the application of the bankruptcy law, which was due to come into full effect in October 1992, was postponed by 6 months to avoid a chain of enterprise failures during the run-up to the split of the country.

The basic objectives of economic policy for 1992 were formulated as a minimization of the fall in economic activity, reduction of inflationary pressures and stabilization of the exchange rate. The macroeconomic policies pursued in the course of 1992 combined a restrictive fiscal stance with neutral monetary policy, and an easing of the incomes policy which contributed significantly to increased consumer demand. However, due to the impending parliamentary elections in mid-1992 and prospects of a breakup of the country after the June elections, there was no comprehensive government programme for 1992. In the second half of the year the role of the federal government weakened substantially as the policy focus shifted to the republics. A system of intergovernmental agreements provided for policy coordination, and the central bank pursued an independent monetary policy. Social peace was maintained through tripartite agreements and there have been no important social disturbances.

In 1992 progress was made in institutional and structural reforms for which the legal foundations had been laid earlier. Small-scale privatization gained momentum, and is to be completed in 1993. The privatization of large companies progressed also, particularly by means of the voucher privatization scheme, the first round of which was held with five auctions (see section 3.6 above). Restitution of nationalized property to former owners was important in the field of small businesses. The banking system expanded rapidly, particularly through the foundation and expansion of private and foreign banks. Basic institutions for the emerging labour and capital market were created. Legislation was adopted for a new system of taxation (including VAT) and of social security. This system came into effect from 1 January 1993.

The basic challenge facing the newly independent Czech and Slovak Republics in 1993 can be seen in minimizing the losses stemming from the breakup of the country and in maintaining the macroeconomic stabilization reached in the previous two years.[592]

The outlook for economic growth in 1993 in the area of the former Czechoslovak federation is quite clouded and depends to a large degree upon level of cooperation the two new states can maintain.

The economic situation is better in the Czech Republic, due to its higher level of economic development, its more diversified economic structure and the economic policies which have been implemented since 1991. The new government led by V. Klaus intends to continue policies maintaining macroeconomic stabilization. Structural policies, particularly privatization, are to be promoted, but the role of direct state intervention is to be limited. However, the impact of a reduced internal market and the process of industrial restructuring in the wake of accelerated privatization of big companies, as well as the coming into force of the bankruptcy law, are likely to result in a further decline of industrial production, albeit at a substantially slower rate than in 1992. It is not certain that this will be compensated by improved performance of other sectors

---

[592] The two governments concluded several agreements to make the transition smooth and to mitigate the adverse consequences of the separation. In order to maintain trade between the two republics, a customs union was agreed upon, as well as a very hedged arrangement for the temporary maintenance of common currency which collapsed within slightly more than a month. At the beginning of February 1993 the two countries introduced separate currencies. A clearing system was then established, with end-year balances to be settled in convertible currencies.

of the economy. The Czech government is currently predicting GDP growth of 1-3 per cent, with inflation at 13-17 per cent and an unemployment rate of about 5 per cent (as against less than 3 per cent in 1992 on average). The state budget is to be balanced. The actual developments might well be worse than government expectations, however.

The Slovak economy will be hit harder both by the costs of systemic transformation and by the consequences of the split-up of Czechoslovakia. Current projections for 1993 indicate a further fall in GDP (by some 3 per cent) and industrial output, estimates of accelerating inflation ranging from 17 to 27 per cent, and a further rise in unemployment.[593] The Slovak government headed by V. Meciar envisages an active structural policy and a greater involvement of the government in the economy. However, in contrast to the Czech Republic, where the coalition government has a slight majority in the parliament, the current political situation in Slovakia is more complex. This can complicate the process of formulation and implementation of coherent economic policies under difficult internal and external conditions.

Eastern Germany

The steep fall in overall output levels has bottomed out and real GDP rose by about 6 per cent in 1992 compared to the previous year. It is doubtful, however, whether this improved performance is the precursor of an imminent self-sustainable recovery. Endogenous growth forces are still fragile and large parts of industry endure a deep adjustment crisis. Major support for domestic activity levels continues to be provided by fiscal policy support payments. Recessionary forces prevailing in western Germany and in other foreign markets will tend to dampen activity levels in 1993 and create a less conducive environment for restructuring in the enterprise sector.

East Germany has always been treated as a special case among the transition countries. The legal and institutional framework of a market economy was taken over *uno actu* with unification, and financial and technical assistance has been available on a much larger scale than east European countries can expect. Monetary stabilization has been ensured with the introduction of the deutsche mark, and – given the overall strength of the German economy – there is also no external (balance-of-payments) constraint. But unification has created its own severe adjustment problems for east Germany. These reside *inter alia* in the fact that the introduction of the deutsche mark was tantamount to a huge appreciation of the former domestic currency (the GDR mark), which dealt a devastating blow to the already feeble competitiveness of industry.

The basic approach of German economic policy to the required radical restructuring of the east German economy has been to create a conducive environment for private-sector investment. Massive public spending programmes have been launched to improve the poor infrastructure and a plethora of generous investment subsidy schemes has been made available to stimulate business investment. A key ingredient of this approach has been rapid privatization of state-owned assets. In addition, a generous social safety net was created to cushion the adverse labour market consequences of the emerging adjustment crisis.

More than two years after unification, the first positive results of this approach are clearly visible, but at the same time there has been a more sober assessment of the enormous costs involved and the time required to catch up with west German real income levels. The infrastructure, notably as regards telecommunications, has been improving, though not surprisingly it still leaves much to be desired. Enterprise investment has shown remarkable strength in spite of prevailing important obstacles to investment. Privatization of enterprises has advanced much more rapidly than expected. More than 75 per cent of all enterprises have already been transferred to private ownership, and the *Treuhand* intends to wind down its selling activities by the end of 1993. Noteworthy progress has been made in creating a more efficient public administration. However, the protracted west German planning and decision-making procedures have often proved to be too complex and costly for the particular circumstances prevailing in eastern Germany. Significant administrative bottlenecks prevail, notably pertaining to the settlement of the more than 1 million claims for the restitution of the property of expropriated former owners.

Positive developments largely contrast with a number of serious problems which have emerged over the last two years and which constitute a major challenge for economic policy. The situation in the labour market is very gloomy. By the end of 1992 the number of persons employed in east Germany had fallen by nearly 4 million compared to the final quarter of 1989. Restructuring and privatization in industry has gone along with a fall in output and employment levels by about two thirds compared to the end of 1989. Although a shrinking of the industrial sector had been expected, the curbing of capacity has taken on dimensions which have given rise to fears that a process of de-industrialization is under way which may be difficult to reverse. The severe adjustment problems in large parts of industry have been exacerbated by wage agreements concluded between trade unions and employers in 1991 which aim at establishing wage parity between east and west Germany by 1994. The implied annual wage increases of more than 20 per cent by far outpace productivity gains and render it even more difficult for many industrial firms to attain or maintain competitiveness. A considerable number of the more than

---

[593] "Memorandum of the Government of the Slovak Republic on the aims and structure of economic policy for 1993", *Trend*, 10 February 1993, pp.9-10; *Lidove noviny*, 27 February 1993. The "Memorandum" of the Slovak government submitted to the IMF states that a balanced state budget is planned for 1993. Monetary policy is to be anti-inflationary and will aim at maintaining internal convertibility of the Slovak currency. Credits are to grow by 6 per cent, with an application of credit and interest ceilings possible. Privatization is to be broadened by freeing various forms of privatization and by the de-etatization of state companies. Integration of the Slovak economy into the global economy and increased inflows of foreign capital are seen as important factors of economic restructuring and future economic growth.

2,500 companies still owned by the *Treuhand* at the end of 1992 will be difficult to privatize, at least in the short or medium term. This pertains notably to those, often very large, manufacturing firms which were heavily oriented towards eastern markets and which find it exceedingly difficult to capture markets in other regions.[594]

This provides the broad background for the economic policy problems which are high on the agenda in 1993. First, there are the worries about the de-industrialization trend. The existence of a strong industrial base is generally recognized to be an important condition for a successful catching-up process in eastern Germany and, related to this, for substantially higher employment levels. There is, secondly, the need to develop a policy approach to companies which have not been privatized by the end of 1993 when the *Treuhand* intends to cease its selling activities. These two problems are interrelated. Thus the government announced in late 1992 its intention to preserve existing "industrial cores" to halt the decline of industrial capacity. It was also decided to increase investment allowances for small and medium-sized manufacturing firms. The concrete details of how firms which cannot be privatized rapidly will be handled in the future have still to be decided. In this connection there have been suggestions that the government should pursue a more active industrial policy. But other observers — notably the five leading economic research institutes — have already warned against the emergence of various state industrial holdings with an open-ended commitment to cover losses from the public sector budget. The key issue appears to be the design of a policy which will give these companies a fair chance to become viable in a competitive environment but which at the same time establishes a credible mechanism for closing them down if restructuring efforts fail to establish profitability after a reasonable adjustment period has elapsed.

As regards the striving for wage parity by 1994, these agreements are now more clearly seen as a major labour market failure. Against the background of the dismal state of the labour market and the crisis in industry, there are now strong pressures on trade unions to accept an extension of the adjustment period. In February 1993, employers in the metals industry revoked the wage parity arrangement concluded in 1991 which would have led to an automatic upward adjustment of wage rates by 26 per cent on 1 April. At the time of writing, negotiations to arrive at a compromise solution on this issue were deadlocked. In the chemicals industry, where no such parity agreement exists, the union have recently accepted a much smaller wage increase (9 per cent) which would broadly stabilize real incomes. It remains to be seen whether this outcome will influence the wage negotiations in other sectors of industry.

Hungary

After two years of austerity aimed at increasing the country's international reserves, Hungarian policymakers have turned to other priorities in 1991 and 1992. First and foremost, domestic demand was kept under continuous restraint in order to moderate the underlying inflationary forces. For the same reason, price liberalization and exchange rate adjustments were dosed cautiously and timed with extreme care. Little action was taken to help firms and farmers in their struggle for survival as export markets disappeared, and even non-exporting firms were hit by the secondary impact. Indeed, large and small producers in all segments of the economy went bankrupt by the thousand and the surviving ones were forced to cut production and shed labour.

This new policy stance, however, does not mean that fiscal and monetary policies were excessively restrictive. Both in 1991 and 1992, the budgetary deficit was much larger than planned. In real terms, the demand generated by state expenditures contracted much less than total demand in 1991 (by 6 per cent versus 10.5 per cent).[595] The growth of money supply (M2) was in both years significantly faster than the nominal increase of GDP. In other words, government policies on both fronts were cautiously calibrated not to deepen the depression caused chiefly by the external shocks of 1991 and to give markets time to adjust to the new circumstances.

Two factors helped the Hungarian government to maintain such a relaxed policy line. First, it was able to capitalize on its good international image as a stable partner.[596] Hence, through stepped-up borrowing and foreign direct investments, a positive capital inflow could be secured without recourse to *de jure* debt rescheduling. This helped the government to maintain political stability at home and thus to improve the country's external image as well. It was also important that foreign direct investment went to a large extent into the privatization process, so the government was able to show relatively good results on that front, too. Second, uncertainty as well as other factors drove households to increase their savings at an unprecedented rate. Market balance was thus not threatened by excess demand originating in consumer spending. Instead, households were willing to keep their savings in bank deposits,[597] which in turn were used by the banks to buy state securities and satisfy the government's borrowing requirements.

---

[594] There are, in particular, some 60 companies, each employing more than 1,000 workers, with a combined labour force of about 220,000 persons, which have not yet attracted investors. These companies, even though currently loss-making, will be difficult to close down for overriding regional economy and labour market reasons.

[595] As reported by Finance Minister M. Kupa in *Figyelö*, 14 January 1993, p.15. Data for 1992 are not yet available, but are probably similar.

[596] As the Prime Minister noted in a recent interview, he was the only head of government of a transition economy who has been serving since 1990 (*Pesti Hirlap*, 22 January 1993)

[597] Though not necessarily in forint-denominated deposits. In fact, 20 per cent of savings are held in foreign-currency necouints and another 25 per cent is tied up in various bank debentures (share of savings net of cash holdings as of 31 December 1992).

However, this overall strategy — usually associated with the name of Finance Minister M. Kupa (the "Kupa programme")[598] — has not always been supported by everyone in the Hungarian government, let alone by the opposition. Objection followed three main lines. Some fear that the decline of industrial production cannot be halted without direct government intervention and/or additional demand stimulation. Others argue that the country's exports cannot be maintained without a good dose of devaluation and/or direct government assistance to major producers. A third type of objection concerns the government's handling of foreign investments. Critics would like to see support of "fast-track" methods of privatization to help domestic would-be entrepreneurs and to prevent what they call "selling the family silver". Indeed, the government has been under such pressures since 1990, and it cannot be precluded that it will give in on one or more fronts as the 1994 elections are approaching.

But even without these doubts, 1993 will be another year of uncertainties. First of all, both the government and the country badly need an upturn in industrial production. This may or may not happen. There is a general concern about the burden on the banking sector stemming from bankruptcies and curtailment of activity in industry and agriculture. At the end of 1992, a portfolio cleaning operation was launched, but its results will not be seen until the middle of 1993. Another major concern is the savings behaviour of households. Savings worth about Ft 1,200 billion (about 45 per cent of nominal GDP in 1992)[599] are kept in highly liquid form, which can endanger the stability of consumer markets as well as the fragile health of the banking sector. Moreover, it is indispensable that savings continue to increase by the same amount as in 1992, lest the budgeted 1993 deficit will have to be covered either by internal money creation or by foreign borrowing.[600] The recourse to such solutions may succeed in the short term, but could do a lot of harm in future years.

Poland

After more than two years of deep recession which followed the implementation of the ambitious stabilization programme in 1990, the Polish economy showed the first symptoms of recovery in 1992. The growth of industrial output which started in the second quarter has been accompanied by an upswing in the construction, trade and telecommunication sectors and in some non-material services. With GDP in 1992 now estimated to be somewhere between 0.5 and 2 per cent above the 1991 level, Poland has emerged as the first transition country poised for a modest upturn in 1993.

The key reason behind these favourable developments appears to be the early start of market reforms, a year or two ahead of many other countries in the region. On the other hand, the macroeconomic policy pursued throughout 1992 cannot possibly be considered as an important recovery factor: while the large fiscal deficit enhanced consumer demand, real money supply declined, interest rates remained high, and the domestic currency appreciated in real terms. As a result, the financial position of enterprises further deteriorated, and by end-1992 more than 42 per cent of enterprises operated at a loss.

Stabilization and reform policies in Poland did not follow a clear course in 1992 as they reflected changing priorities of successive governments. While restraint and relative stability generally dominated the monetary policy, the budgetary and fiscal policy was more accommodative, leading to a larger-than-planned deficit of some 6 per cent of GDP. Compared with previous years, growing emphasis on structural and industrial policies (as manifested by the proliferation of budget-financed restructuring programmes for coal-mining, steel industry, aircraft industry and some high-unemployment regions) seems to suggest that a gradual shift away from free-market orthodoxy is taking place.

Institutional reforms in 1992 proceeded at a hesitant pace, draft legislation being frequently altered or rejected by the politically divided Parliament. Specifically, this was the case for the draft laws on mass privatization, on restitution of nationalized property and on "bad debts" of banks and enterprises.

The interpretation commonly given to the 1992 upturn in the Polish economy is that it demonstrates that market reforms surely pay off at last if only they are pursued in a consistent and sufficiently determined manner. This is probably true, but optimism and hopes which hinge on lessons of the Polish case may none the less turn out to be premature. A more detailed scrutiny reveals that the Polish recovery has a still rather weak and fragile basis. The growth of industrial output has probably been driven primarily by expansion of exports to the west, while two other key components of final demand — investment and consumption — remained stagnant or even declined further in 1992. The fall of investment is of particular concern because it has continued for four successive years already: no solid foundations for economic growth and restructuring can be established without substantial new capital formation.

The main challenge facing the Polish economy in 1993 is therefore the need to strengthen the growth basis and make the recovery sustainable. Further expansion may be expected to come from the rapidly developing private sector, which accounts for almost half the country's GDP and employs more than 60 per cent of total workforce. But export growth may slow down in 1993 when the initial impact of lower EC tariffs will fade away, and import competition may increase. Also, domestic investments cannot be expected

---

[598] See Government of the Republic of Hungary, "A Programme of Transition and the Development of the Hungarian Economy – Stabilization and Convertibility," in *Magyar Hirlap*, 17 April 1991, Supplement.

[599] Data refer to 31 December 1992 and include cash holdings.

[600] The 1993 budget, as presented to Parliament, worked with a Ft 260-270 billion increase in household savings. This is virtually unchanged relative to the 1992 forecast (Ft 255-265 billion). See Hungarian Ministry of Finance, *Economic Policy 1993-1994*, Budapest, 1992, Table 4.

to stimulate growth significantly because the large budget deficit will crowd out enterprise borrowers and keep interest rates on high levels. On the other hand, a consumption-led demand expansion can easily translate into higher inflation with no sizeable effect on output.

One possible way of addressing this dilemma would probably be to apply a combination of policies aimed at lowering lending rates on longer-term credits, establishing strong fiscal incentives for productive investments, and maintaining the real exchange rate on a competitive level. This has to be accompanied by a strengthening of the revenue base for the budget, mainly through the introduction of a value-added tax (planned for July 1993). At the same time, the banking reform, privatization and restructuring of enterprises should proceed at much faster pace than in 1991-1992, which would call for a much greater degree of policy coordination at the central levels of state authority.

However, the possibilities for more decisive policy manoeuvres seem to be rather limited under the current political situation in Poland. The government, which commands only a very small and fragile majority in the Sejm, recently suffered several setbacks in votes on the 1993 budget; moreover, the coalition itself is deeply divided over a number of important and politically sensitive issues, ranging from privatization and restitution schemes to "lustration" and abortion. The "small constitution", which was pushed through the Sejm at the end of 1992, has given the government the possibility to issue decrees. This may accelerate the reform process and allow the government to finalize talks with the IMF on resuming financial assistance within the extended facility arrangement which had been suspended in the spring of 1991.

(ii) **Eastern Europe: the south-eastern countries**

Albania

Albania was the poorest country in Europe even before the dissolution of its communist regime. From this low starting point, its collapse was none the less the steepest registered among the transition countries. The cumulative fall of NMP/GDP in 1990 and 1991 can be estimated at some 40 per cent and that of industrial production at some 45 per cent. Agricultural output appears to have fallen by one quarter in 1991 alone. As production collapsed, exports almost halved in 1990-1991 whereas imports increased by a quarter, resulting in a substantial external debt, about $500 million, most of it in the form of arrears – which effectively freezes the country out of all credit markets. There are severe shortages of energy, material inputs and consumer goods, including foodstuffs. The fiscal and monetary system fell into disarray, with a fiscal deficit of almost one half of GDP in 1991 and inflation at some 100 per cent in that year (December 1991 over December 1990). Albania's economy is thus in a deep crisis.

Sketchy but largely non-statistical information on developments in 1992 indicates that the downward trend of the economy continued, albeit at a slower pace in the second half of the year after a reform programme was instituted. GDP probably fell by another 8-10 per cent for the year as a whole. Industrial output may have stopped contracting in the second half. However, macroeconomic imbalances remained severe. The fiscal deficit, although reduced, is still over one fifth of GDP, and – as in other transition economies during early reform phases – inter-enterprise arrears started to mount rapidly. With imports more than eight times larger than exports, the balance of payments difficulties have further intensified and no reduction of external arrears was possible. The rate of inflation accelerated; prices were some 250 per higher at the end of the year than at the beginning (but this included a large price liberalization shock in August-October, with much lower inflation thereafter). Unemployment is estimated to have risen to one quarter of the labour force.

Albania is probably the country least advanced in the transition to the market system as specific features make the transition process more difficult than elsewhere. Among them are inherited large structural distortions and the long-term isolation of the country due to the former development strategy which stressed national self-reliance and economic autarky. The collapse of Albania's highly centralized system of planning and control has been accompanied by a general breakdown of government institutions. The measures taken in 1990 and 1991 lacked a consistent framework, their legal and institutional preparation was insufficient and their implementation met many difficulties.

The first comprehensive reform programme was adopted only in July-August 1992 by the new government headed by A. Meksi. It stressed the establishment of essential institutions, price reform, the liberalization of the exchange and trade system, and improved fiscal balance. It is too early to assess its implementation, but it should be noted that progress has been made in price and trade liberalization. In August 1992 almost all prices were freed, and the prices of few basic food products remaining under administrative control were raised by 300-400 per cent. A limited currency convertibility and floating exchange rate of the lek were introduced in July 1992. Financial policies have been tightened.

In the construction of market institutions, initial preparations have been made in developing a financial sector. The legal framework for a two-tier banking system is in place. Structural reform progressed in agriculture, where the distribution of land to private farmers has been almost completed. However, the spontaneous and rapid disintegration of agricultural cooperatives caused many difficulties. The restructuring of industrial enterprises will be more complicated.

No official forecasts for 1993 are available and given the poor state of statistics and information on economic developments in 1992, none can be ventured from the outside. An economic programme for 1992-1993 elaborated with the assistance of the IMF aims at the stabilization of output in 1993, with emphasis on an early recovery of agriculture. A reduction of inflation to 2-3 per cent a month by mid-1993 and the gradual building up of foreign exchange reserves is also envisaged. The fiscal deficit is to be further reduced, and an austere incomes policy is to be implemented. The macroeconomic stabilization necessitates relatively large external assistance which has started to be provided in different forms, including emergency aid, by the IMF, the World Bank and the G-24 countries.

Bulgaria

Developments in 1992 indicate that the transition process in Bulgaria has appeared to be more difficult than in some other east European countries. Overall output and industrial production continued to fall at a rate only moderately below that in 1991. The supply side constraints caused mainly by compression of imports have eased, and depressed domestic demand appeared to be mainly responsible for the large loss of output. However, external shocks also contributed to output contraction. Due to the spottiness of statistical data, it is difficult to assess the developments on the demand side in detail, but it seems that domestic absorption fell substantially less than in 1991 when private consumption dropped by more than a quarter and investment by almost a half. Given the strong downturn in economic activity, the rate of unemployment further increased, reaching almost 16 per cent by the end of the year. The social costs of the transformation process are thus large and have resulted in growing social tensions.

The reform programme which was initiated in February 1991 and continued throughout 1992 relied on tight monetary, fiscal and income policies. However, the macroeconomic stabilization has progressed only slowly. The rate of inflation remains high. Since the post-liberalization price jump in February-March 1991, inflation continued to run on average at about 5 per cent per month. The government reduced budget spending, mainly on subsidies and investment outlays. In spite of that, due to weakened revenues, the budget deficit remains large, particularly if the unpaid accumulated interest on Bulgarian external debt is taken into account. A restrictive stance in monetary policy brought about a substantial decline in real money supply. A negative real interest rate for depositors, particularly in the last quarter of 1992 when inflation accelerated, has reduced the propensity to save. On the other hand, interest rates remain high for borrowers, and together with restricted credits this contributed to the difficult financial situation of many enterprises and mounting inter-enterprise arrears. Sustainable macroconomic stability has thus not been achieved, probably owing to the lags in the implementation of structural change, particularly in the financial sector and industry. In the external sector, the process of adjustment continues to be painful, given a large external debt and insufficient capacity of Bulgaria for its servicing. Nevertheless, Bulgaria has introduced a unified exchange rate and created an interbank foreign exchange market. The foreign exchange reserves of the Bulgarian National Bank increased substantially in 1992. The exchange rate was broadly stable — at about 23 lev/dollar — until December 1992 when it depreciated to 24-25 lev/dollar.

The legal framework for structural change was put in place in 1992 with the adoption of the laws on restitution, small and big privatization, and the amendments to the Land Act which allow the restitution of land to former owners. In addition, the law on protection of foreign investment was designed to attract foreign investment, which remain very low. The development of the banking system was stimulated by the law on banks and credit activity adopted in March 1992. The process of merging of smaller commercial banks with bad loans portfolio into stronger banks is under way. However, legal acts on bankruptcy and a new taxation system are still under preparation. None the less, the basic preparatory work for economic restructuring has been done, and it is to be hoped that the structural change will progress faster in 1993.

It is likely that 1993 will be another difficult year for the Bulgarian economy. There are many uncertainties and contradictory forces which cloud forecasts. Output will probably continue to decline, but at a substantially lower rate. Privatization and the dynamic growth of the private sector will contribute to economic growth, but its role in 1993 will remain limited in the initial stage reached by the privatization process. Since Bulgaria is to pursue macroeconomic stabilization policies, it cannot be expected that strong impulses for economic recovery will come from the side of domestic demand. Growing exports can stimulate production, but this needs improved access to western markets and western support for Bulgarian transformation process. The association agreement recently concluded with the EC may be of crucial importance in that respect as it can open room for increased Bulgarian exports and new patterns and regional orientation of its foreign trade and payments.

The principal macroeconomic problems facing the government are twofold: to improve the internal and external balance of the country by the continuation of macroeconomic stabilization policies, and to prepare conditions for economic recovery through structural change. The reconciliation of these objectives is difficult with the persisting large imbalances and given that reforms are only at the initial stage. External assistance seems thus very important. The new government led by L. Berov, which has been in office since December 1992, has confirmed its commitment to the adjustment programme designed to stabilize the economy and to transform it into a functioning market economy. Arrangements with the international community in respect of the external debt which could ensure access to external financing are an important aim of the government.

The implementation of reforms in Bulgaria can be complicated by the current political situation. No political party has a majority in parliament, and the successive coalition governments have had only a fragile parliamentary basis. The difficulties of transition sparked the last government crisis. Moreover, important issues of transformation, such as the tightness of stabilization policies and the scale and methods of restitution and privatization, continue to be at dispute. The lack of social and political consensus can thus hamper more decisive actions of the new government.

Romania

Although only a limited number of economic objectives for 1993 have been set by the new Romanian government which took office after the election in autumn 1992, there is no reason to expect any significant departures from previous policies. The previous administration carried out a programme of far-reaching institutional change while exercising a high level of government intervention until fully-fledged market mechanisms were in place.

The government's economic programme for 1992 was approved by the IMF and submitted to Parliament by the Prime Minister in late April of that year. Its institutional aspects were to be implemented from 1 May. The programme laid down six main quantified objectives: to reduce the rate of output decline, to reduce inflation to an annual rate of 40 per cent by end-1992, to reduce the budgetary deficit to 2 per cent of GDP for the year as a whole, to limit the current-account deficit to $1.3 billion, to build up foreign currency reserves to about $500 million, and to limit the external debt to $3.5 billion by the end of 1992.[601]

These objectives were to be attained by tight fiscal, monetary and credit policies, limitation of the growth of the money supply to a maximum of 89 per cent and domestic credit to 68 per cent over the year while forcing enterprises to pay off debt arrears to other enterprises, and by the passage and enforcement of new bankruptcy laws. Price liberalization was to be extended to natural gas. Consumer goods subsidies were to be removed and the corresponding wholesale and retail prices allowed to adjust to demand gradually up to the end of 1993. Public investment priorities were to be reviewed and social welfare criteria examined to provide more effective support to needy social groups without allocating extra funds. On the institutional side, some 6,244 small-to-large state enterprises were scheduled for privatization. In addition, exporters were to be allowed to retain all their foreign-currency earnings within the framework of a flexible exchange rate policy.

The programme was jeopardized over most of the year by failure to arrest the fall in production – one of its key objectives. GDP, which had fallen by 14 per cent in 1991, contracted by another 15½ per cent in 1992, reflecting an acceleration in the fall of industrial output (from 20 per cent in 1991 to 22 per cent in 1992) as well as a 9 per cent drop in agricultural production. Production falls reflected shortages of domestically-produced but especially imported fuel and other inputs.

Inflation decelerated between the first and third quarters but picked up strongly again at the end of the year. The underlying inflationary impetus was reinforced by the removal of most price controls in the first half year and by subsidy reductions in the third quarter. Monetary policy remained tighter than anticipated, credits to the enterprise sector rising by some 30 per cent only during January-November. For public finance, the central budget deficit (lei 131 billion) can be estimated to have been some 4 per cent of GDP, but half of this was offset by surpluses on local authority and social security budgets.

The domestic objectives of the economic programme have thus been largely missed. The main concerns are the accelerated decline in production, the continuing decline of investment, and the recent sharp upturn in inflation (which, in the final quarter of 1992, averaged 12 per cent a month, more than four times the rate envisaged for that period). Only the fiscal balance and credit expansion targets appear to have been met, and progress has been made towards the rationalization of the price system. The external targets, by contrast, appear to have been largely achieved: the current account deficit is estimated at just under $1 billion, external indebtedness (including short-term debt) at about $3.5 billion, but foreign currency reserves, at some $240 million, were apparently some way below the target.

Substantial steps to breach the system of central planning and management in 1991 were followed by an increased pace of legislation to create market institutions in 1992. Rapid progress was registered with privatization in agriculture, where it covered over 80 per cent of the area farmed by the end of the year, but advance was slower in other sectors. The relatively simple process of selling off small state-owned retail shops, restaurants and other establishments was slow to begin, but accelerated through the year after a timetable was published. This and the newly-created small business sector account for over a million workers, including some cooperative enterprises still in the process of privatizing their ownership structure. If private farmers are included, employment in the private sector now totals some 4 million, or over 35 per cent of a total work force of nearly 11 million workers. Their output probably accounts for some 26 per cent of GDP.

Parliamentary approval was also given to a bankruptcy law and to the privatization of large enterprises in July 1991. However, no actual bankruptcies have yet been reported. Privatization vouchers have been distributed to virtually all of the eligible population and the various funds through which the process will be implemented are in place. The sale of the first state-owned enterprise took place in the autumn, and a total of 15 had been privatized by the end of 1992.

The new government has published a number of economic objectives for 1993, emphasizing continued restructuring, stabilization of output in the first half year and the resumption of growth thereafter; a budget deficit somewhat bigger in absolute terms than in 1992 has been forecast, assuming a further fall in GDP no larger than 5 per cent. Inflation is targeted at no more than 40 per cent. Attainment of these objectives is expected to leave room for more equitable tax systems, but remaining consumer subsidies are to be phased out in the spring, and the situation of the most vulnerable population groups is to be ameliorated. A more detailed government programme for 1993 is expected to be published later in the spring.

---

[601] *Romania Economic Newsletter*, Vol.2, No.1 (April-June 1992), p.3.

(iii) The countries of the former Yugoslavia

Croatia

Falls of industrial output in Croatia of 11 and 28 per cent in 1990 and 1991, respectively, were followed by a further decrease of 15 per cent in 1992. The largest falls were registered in metallurgy, whereas the production of electricity, fuels and the chemical industry had positive rates of growth. Domestic retail sales fell 39 per cent in volume in 1992. There was also a contraction in foreign trade: exports fell 5 per cent in value and imports 10 per cent. The foreign trade deficit narrowed, but was still some $300 million.

Although military hostilities in Croatia ceased and the cease-fire brokered by the United Nations largely held throughout 1992, the economy of Croatia was under significant strain. The government was faced with a long list of pressing concerns that were asking for increased expenditure. The problems included institution and capacity building to consolidate independence after the disintegration of the former Yugoslav Federal State, the need to reconstruct the country's war-damaged economy, a decline in industrial production and investment, support for the unemployed (18 per cent of the labour force) and refugees (around 600,000), soaring inflation, the contraction of home and loss of large parts of the former Yugoslav markets, low capacity utilization, need to overhaul the banking system, a fall in receipts from tourism and road transport, and the transition from the self-managed to a market-type economy. The decline in the economic activity shrank the tax base. The government ran a budget deficit that was financed by money emission, which generated rapid inflation. While this reduced the real value of the public debt, it discouraged saving.

The most immediate task for the government is the reconstruction of the war-damaged part of the country and the transition to a market-type economy. To do this, Croatia will have to curb its budget deficit and inflation, and to take serious steps towards restructuring and privatizing enterprises. These are difficult tasks, since a restrictive monetary policy may deter investment, job creation, consumption and growth in the short term.

A stabilization programme for 1993 was announced by Prime Minister H. Sarinic at the end of 1992.[602] The main objectives are the control of inflation, the revival of economic activity and of exports, transition towards a market-type economy, and improving social welfare. Policy instruments include a restrictive monetary policy, a reduction of fiscal pressure, and an active incomes policy. The budget deficit is to be financed from savings and foreign funds. The incomes policy aims to hold wage growth at least 15 per cent below the pace of inflation. While this will further diminish the already very low living standard, without control of inflation, domestic investment and job creation, no significant inflow of foreign direct investment can be expected.[603]

The Government expects that the programme will increase the social product in 1993 by 15 per cent, an expectation considered to be highly unrealistic. For deteriorating circumstances, the government has a contingency plan involving the introduction of central material balancing, production assignments and rationing.[604]

The Croatian privatization programme opted for a rapid sale of shares. While firms initiated privatization, the final approval of buyer, price and the conditions of sale was under the control of the government's Fund for Privatization. The deadline for the privatization of all socially-owned enterprises was June 1992. In accordance with the privatization law, all shares not sold by that date (around half of the value of all socially-owned firms) were transferred to three state funds.[605] Some observers characterized that development as an "etatization" of the economy, as the government had obtained those assets without compensation.[606] The procedure has aroused considerable hostility on the part of trade unions,[607] opposition parties and academic circles.[608]

Home resources for investment that must be the engine of growth in the future are in very short supply, as are transfers from abroad;[609] hence attracting foreign investment is one of the priority policy areas for the government, especially as private investors are looking for less volatile locations for their investment.[610] Foreign investors, among other things, seek either a secure and protected market or free access to a large market. Since Croatia is a small economy, an inward-looking policy will be counter-productive. However, a boost to the

---

602 Vlada Republike Hrvatske, "Osnove stabilizacijskog programa", Zagreb, 22 December 1992.

603 F. Kiseljak, "Place i cijene", *Vjesnik*, 31 January 1993, p. 17.

604 Radio Zagreb, 30 December 1992, as reported in BBC, *Summary of World Broadcasts*, 4 January 1993.

605 Retirement Fund for Employees, Retirement Fund for Agricultural Workers, and Croatian Fund for Development.

606 N. Cuckovic, "Privatization practices in Croatia, Poland, Hungary and Czechoslovakia" in IRMO, *Macroeconomic Policy Reform and Private Enterprise in Central and Eastern Europe*, Zagreb, 1992, p.62.

607 L. Markovic,"Tko ce kupiti svoj otkaz", *Vjesnik*, 18 February 1993, pp.20-21.

608 "Stopirati pretvorbu", *Vjesnik*, 6 March 1993, p.4.

609 German aid to Croatia of more than DM 4 billion in 1991 was probably a one-time transfer, at least on that scale, although it may have created a false expectation that it will become a permanent feature. See J. Dempsey, "German business slow to follow state on Croatia", *Financial Times*, 5 February 1992.

610 In order to be attractive as a location, Croatia has to offer prospects for higher profit than alternative locations. Unfortunately, the attitude towards foreign investment is not always friendly and the potential investors face not only much paperwork, but also the resistance of the current managers of the socially-owned firms that to be privatized (D. Herceg, "Ulagaci na potjernici", *Vjesnik*, 29 January 1993, pp.36-7).

Croatian economy may come from foreign demand. In view of the lack of signs of recovery in Croatia's main foreign trading partners, a case can be made for efforts to re-establish the reduced or broken economic ties with the former Yugoslav republics.

Slovenia

Slovenia is the only one of the new internationally recognized states emerging from the former Yugoslavia which obtained its independence without extended military conflict. Owing to this, a sounder budgetary position than in the rest of the former Yugoslavia (there was a surplus in public expenditures of some 2 per cent of the GDP in 1991, and a much smaller surplus was projected for 1992),[611] its higher level of development and outward-oriented economy, Slovenia was also the only successor state to register some advance towards macroeconomic stabilization in 1992.

Output none the less contracted in 1992. GDP fell by 6½ per cent in 1992, less than the 9 per cent decline registered in the year before. The volume of industrial production was down by 13 per cent, after falls of 10-12 per cent in each of the previous two years. Contraction of domestic demand and of trade with the other states of former Yugoslavia were the main factors behind the fall. Exports increased by 8 per cent in dollar value, whereas imports were stagnant in 1992, and a small surplus was registered in the trade balance. The current account showed a surplus of some $830 million (mainly due to a surplus in non-factor services like transport, travel and processing), which facilitated the implementation of internal convertibility of the national currency, the tolar.

The main problems of the Slovenian economy are the steep fall in manufacturing production and rising unemployment. Some positive signs are emerging on industrial production and inflation, but the employment picture remains bleak.

Industrial production in 1992 was just two thirds of what it had been in 1985. None the less, on a seasonally-adjusted basis output has been almost steady since the middle of the year, raising the hope that the fall may have reached bottom.

Linked with the contraction of production, there was sharp growth of unemployment, which rose from 10 per cent at the end of 1991, a twofold increase in relation to 1990, to 13 per cent of the labour force at the end of 1992, increasing throughout the year. The rate would probably have been significantly higher but for a freeze on bankruptcies and the lags in the privatization process.

On the positive side, the main achievements of the macroeconomic policy were a good export performance and a check on inflation. Although the average year-over-year price level rise in 1992 was higher than that registered in 1991, the rate of inflation came down from double-digit monthly changes in the first quarter to just above 2 per cent per month on average in the second half of the year. The Bank of Slovenia implemented a relatively restrictive monetary policy and allowed only a small monthly depreciation of the tolar in 1992. This harmed exporters because of the resulting real appreciation value of the tolar. The Slovenian labour force is at current exchange rates the most expensive among all transition economies (DM 600 per month on average); hence arguments have been raised in favour of a devaluation of the tolar.[612]

Transition to a market economy asks for a privatization of the formerly "socially-managed" enterprises. A privatization law was adopted only in November 1992, providing for a combination of free distribution and sale of vouchers. Implementation rules, however, will not be settled before the middle of 1993. The outlook for the privatization process is clouded by rather sharp divisions in parliament about the details of the implementation provisions, and by concerns about the impact on unemployment of far-reaching privatization including rationalization of production. An additional factor is the existence of a moratorium on bankruptcies.[613]

In a small open economy like Slovenia, employment depends to a very large extent on foreign demand. Slovenian goods and services were highly competitive in the (protected) market of the former Yugoslavia. Before the disintegration of the country, one quarter of Slovenian sales went to the other Yugoslav republics, with which Slovenia always registered a trade surplus.[614] At present, these markets are no longer outlets on the previous scale. Trade with Serbia and Montenegro falls under the embargo, and a trade conflict has developed with Croatia. In 1992, Croatia introduced a 23 per cent *ad valorem* tariff on imports from Slovenia, and Slovenia retaliated with an "equalization tax" of 8.5 per cent. As a result, trade between the two states fell by a third.[615] Relations between the two countries are also encumbered by a dispute over private Croatian savings deposited in Slovenian banks.[616] None the less,

---

[611] CICD, "Economic developments in Slovenia in 1992", a report prepared for the ECE secretariat, Ljubljana, January 1993, p.10.

[612] S. Sicarov, "Pesimisticke ocene dr Aleksandra Bajta: Procerdane prednosti Slovenije u Jugoslaviji", *Borba*, 12 February 1992, p. 13.

[613] CICD, "Economic developments in Slovenia: A country report", a paper prepared for the ECE secretariat, Ljubljana, November 1992, p.38. High on the list of bankruptcy candidates are three steel plants that were once important suppliers of the Yugoslav Federal army. If these were to be shut down, unemployment could double to the level of over 200,000 people ("Slovenia's economic prospects", *East European Markets*, 11 December 1992, p.13).

[614] CICD, "Economic developments in Slovenia: A country report", op.cit., p.40.

[615] D. Valenticic, "S Hrvatskom se mora trgovati", *Novi Vjesnik*, 26 November 1992, p.7a; "Privreda u nebranom grozdu", *Ekonomska politika*, 16 November 1992, p.18.

[616] It is estimated that Ljubljanska Banka, the main commercial bank in Slovenia, owes around $600 million to savers in Croatia. That is not a negligible amount bearing in mind the size of the economy and transition difficulties (*Ekonomska politika*, 18 January 1993, p.20).

official intentions are to negotiate a free trade agreement between the two states.[617]

In this situation, in spite of its relatively successful export performance in the western markets, Slovenia has to double its efforts to maintain competitiveness on international markets, in particular that of the European Community. But as there are no signs of recovery in the main trading partners in the west and a deep recession prevails in Russia, efforts to re-establish the trade flows with the rest of the former Yugoslavia assume added significance.

In the near future the Slovenian government will have to deal with a number of problems. There are substantial worries that privatization will bring an increase in unemployment, which would impose additional social expenditures that might have an inflationary impact. A freeze of wages and pensions at the end of February 1993 and reports of increased demand for hard currency because of the fear of devaluation[618] are signs that the transition process may be costlier than expected.

Federal Republic of Yugoslavia (Serbia and Montenegro)

Gross material product (GMP)[619] of the Federal Republic of Yugoslavia (Serbia and Montenegro) fell by 27 per cent in 1992. On the origins side, value added in manufacturing industry, transport and tourism contracted by almost a quarter relative to 1991, and that of distribution sector enterprises was nearly halved. The decline in the output of both agriculture and construction was around 15 per cent. Among final uses, current consumption fell some 20 per cent, while investment was halved. Exports volume also declined by one half, while the fall in import volume was 37 per cent, and the trade deficit rose to $1.3 billion.

The output of Serbia and Montenegro had begun to contract even before the violent disintegration of the former Yugoslavia in 1991. In that year, GMP fell by 11 per cent. The contraction accelerated sharply in the second half of 1992, after economic sanctions were imposed on the country by the United Nations Security Council at the end of May, including a trade and financial embargo. In industry, the production of the metal and electrical goods were the sectors most strongly affected, whereas output increased in iron ore mining and the extraction of coal, oil and natural gas.

The need to finance a war-type economy coupled with the costs imposed by the sanctions and expanding social outlays, including the care of around 600,000 refugees, in the face of the collapse of economic activity that shrank the tax base, induced the Government to cover its budget deficit through money emission. The resulting inflation averaged almost 60 per cent a month throughout 1992 (and accelerated steeply in the first months of 1993, rising to 212 per cent in February).

Officially recognized unemployment changed very little during the year. The number of unemployed rose from 720,000 in January to 750,000 at the end of the year (23 per cent of the labour force). However, this apparent stability is misleading, as would be expected in the face of the magnitude of the output fall: it reflects rather the government's decision that no enterprise closures or massive labour shedding would be permitted while the sanctions are in place. It is estimated that around one million workers are kept artificially in employment by that ban.[620] Although there was an increase in the establishment and activities of private firms, predominantly in the services (trade), it was not large enough to absorb a significant part of the surplus labour force.

Nominal wages increased around 30 times in Yugoslavia in 1992, but this rise was outpaced by inflation; real wages declined some 50 per cent, following a fall of 5 per cent in 1991.

The new Federal Government of R. Kontic inherits a difficult economic situation. It is compounded of hyperinflation, fall in output and investment, huge unemployment, budget deficit, company losses, devastated financial system, transition problems, proliferating grey economy and economic sanctions. A shortage of capital and of imported inputs presents the most serious constraint to the operation and development of the economy.

Economic sanctions pose a significant impediment to the introduction of stabilization and anti-inflation policies in the FR Yugoslavia let alone to the badly-needed recovery of its economy. Therefore the new federal government intends to endeavour, for its part, to obtain their lifting. In the conduct of economic policy the government intends first to reach a consensus around the basic options presented in its recently announced programme,[621] and then to harmonize economic policies of the Federation and the constituent republics.

The basic policy objective is to arrest the negative economic trends and to invigorate the economy. It intends to revive the economy "by every necessary measure", with the goal of bringing GMP back to the 1992 level. It also plans to reduce inflation below the current hyperinflation pace and to stabilize prices. The application of monetary policy will be such as to further the recovery of production. After the lifting of the sanctions, price and import liberalization will be the main

---

[617] D. Valenticic, op.cit.

[618] S. Sicarov, "Slovenci masovno kupuju devize", *Borba*, 1 March 1993.

[619] Value added, including depreciation, of the material sphere.

[620] Institut Ekonomskih Nauka, "Ekonomska situacija u SR Jugoslaviji", Belgrade, February 1993, mimeo, pp.9-10.

[621] "Programme of the Federal Prime Minister Designate, Dr. Radoje Kontic, presented at the Joint Session of the Chambers of the Federal Assembly", *Yugoslav Daily Survey*, 3 March 1993.

options to be followed, but the prices of basic goods and services that influence the living standard will remain controlled. Public spending is to be reduced and deficit financing to be brought within the legal limits. Economic relations with Europe and the rest of the world, primarily with the neighbouring countries and Russia and Ukraine are also to be restored quickly. Preparations will be taken in hand for the re-establishment of relations with the IMF and other international and regional organizations. The aim of an active social policy, including incomes policy, will be to ease social tensions. Selective support will be given to promising sectors of the economy which rely on domestic demand and have a low import dependence. The highest priority will be put on agriculture, not only because of the concern with food safety and the creation of jobs, but also to further the country's export potential.

The deceleration in the unfavourable macroeconomic trends and a subsequent improvement in the situation, as well as economic transition, depend to a large extent on the lifting of sanctions. As long as they are in place, the government can only attempt to mitigate their impact. None the less, a further decline of 15 per cent in industrial output will be difficult to avoid in 1993, in spite of some resilience in sectors which are not import-dependent and are primarily oriented towards the domestic market. Among the policies that have some chance for success in the short term is an improvement in the collection of taxes. The government is aware that the major sources of inflation are the deficit financing of the budget and the monopolistic pricing of the domestic enterprises. It is not clear, therefore, why it intends to put an extra burden on already very low incomes (real wages declined around 50 per cent in 1992) through an incomes policy, rather than tackling first and foremost the major sources of inflation.

Long-term development structural change and systemic transformation at present are not in the forefront of economic policy. Nevertheless, the government intends to prepare for the ownership transformation and the overhaul of the banking sector because of its necessity for the smooth functioning of the economy. In any case, in order to conduct its policy and to avoid a shift to a command economy, the government intends to reach a common stand on its policies with all the political forces in the country.

(iv) The CIS countries

Belarus

The Belarus government did not put forth a comprehensive programme for economic stabilization and structural change for 1992. Discussions with the IMF in mid-1992 resulted in a memorandum specifying a limit on the fiscal deficit of 12-15 per cent of expenditures or 6 per cent of GDP − up from an original target of 2 per cent of GDP.[622] No other objectives were published. Direct government intervention in the economy has continued to be relatively pervasive. Even so, a privatization programme has begun and institutional reforms have been initiated.

The liberalization of prices at the beginning of 1992 is the most decisive step yet taken in the direction of a market economy. The price liberation in Russia had to be followed, but the share of goods for which liberalization was postponed was bigger than in Russia and the prime minister announced his intention of re-introducing price controls on basic food and necessities as late as November 1992.[623]

Official data show much smaller NMP falls in Belarus than in most other CIS countries in both 1991 and 1992. Given the high import intensity of the economy,[624] the output loss was surprisingly modest. None the less, the contraction became much more severe in the latter year.

Following a small budgetary surplus in 1991, fiscal policy fell into disarray in 1992 as state revenues declined under the impact of the steepening output fall in the enterprise sector. The budget deficit apparently widened to some R2 billion in the first quarter alone, and to R40 billion − estimated at 6 per cent of GDP − for the first half year (no information on the full-year deficit for 1992 is available yet).

Little information is available on the conduct of monetary policy in 1992. Monetary emission was however some 18 times higher than in 1991, the biggest component of which was probably short-term credits to enterprises. As in the other CIS countries, this device for heading off large-scale enterprise closures augmented the inflationary effects of the price liberalization. Despite the retention of some price controls, post-liberalization inflation rates in Belarus have been about the same as in other CIS countries.

The spread of market relations was constrained by attempts to maintain output by variants on traditional methods of state intervention − notably the use of "state orders", considered necessary to offset the breakdown in traditional flows to customers and from suppliers within Belarus and elsewhere in the former Soviet Union. The government also resorted to other direct controls, including a coupon system for use with identity documents in November 1992 to "protect the internal consumer market" by preventing non-residents from buying up goods which were in short supply. Roubles continue to circulate in parallel with the coupon. It was also announced that a Belarus *rouble*

---

[622] See J. Odling-Smee et al., "Belarus", *IMF Economic Review*, Washington, D.C., April 1992, pp.9-10. See also *Sovetskaya Belorussia*, 14 August and 23 October 1992.

[623] BBC, *Summary of World Broadcasts*, SU/1535, 11 November 1992, p.C3/1.

[624] In 1990, the ratio of total imports (sum of imports from abroad and in inter-republican trade) to NMP was two thirds, one of the highest among the CIS countries. See United Nations Economic Commission for Europe, *Economic Bulletin for Europe*, Vol.44 (1992), New York, 1993, p.85.

would be introduced as a completely separate currency[625] and that currency is now regularly quoted on the Belarus currency exchange.[626]

The development of a privatization framework has been much slower than in Russia and, though the sale of state enterprises has begun, activity in this area has been limited mainly to leasing. It was originally planned to sell state productive assets amounting to less than 1 per cent of the total capital stock of state-owned enterprises in 1992. Towards the end of the year, 200 state enterprises, employing 1.4 per cent of the labour force, had been disposed of.[627]

No economic forecasts or programmes have yet been published for 1993. A revised draft bill on privatization and an anti-corruption law, submitted to the Supreme Soviet for consideration before the summer of 1992, were both rejected. A new legal framework is to be submitted in 1993. It is to include legislation on private property rights which is still lacking at present. However, it appears that this legislation is controversial: a first draft bill to create a voucher system as a basis for the distribution of ownership rights in state firms was rejected by parliament in January 1993.

Moldova

An unstable political situation in Moldova was caused by attempts of some regions with Slavic majorities, notably the Trans-Dniestr district, to seek autonomous status in response to the central government's desire for closer political and economic links with Romania. This brought the country to the brink of civil war in 1992. Economic links with the region, where more than 60 per cent of Moldova's industrial potential is concentrated, were broken.

A "Draft Economic Reform Programme of the Government of Moldova" was prepared in 1991 which laid down the broad objective of establishing a market economy based primarily on private ownership. In the view of IMF experts, a disproportionate role was, however, reserved for government intervention. Few quantified objectives were given; the IMF experts recommended that such targets should be given and set within a short- to medium-term time frame.[628]

The economy of Moldova was affected by two policy-induced shocks at the beginning of 1992. First, following the Russian lead, prices for most commodities (except bread, milk, energy, utilities and transportation) were liberalized in January 1992. The second stemmed from decisions to carry out a rapid withdrawal from the rouble area and, ultimately, to reorient trade towards the west.[629] Payments problems with Russia and other CIS partners contributed to a drastic contraction of trade ties with the Soviet successor states in 1992. In the first nine months of 1992, Romania alone accounted for as much as 70 per cent of Moldovan exports to countries outside the former Soviet orbit and 48 per cent of the corresponding import flow. The short-run costs of this apparent shift in the geographical distribution of trade must have been large, given the high ratio of inter-republican imports to NMP in the past (53 per cent in 1990). Falling supplies from Russia and other CIS countries could not be replaced by increased hard currency imports. Severe shortages of production inputs and other supplies developed.

By the middle of 1992, most of Moldova's industrial enterprises were close to ceasing production as Russian fuel supplies dried up. The first post-independence government of Prime Minister V. Muravski had to resign. In November 1992, the new government, headed by A. Sangheli, obtained from parliament the right to declare a state of economic emergency, which seemed still to be in force at the beginning of 1993. Preliminary data indicate that Moldovan NMP dropped in 1992 by over one fifth though the gap compared with the same period of 1991 narrowed over the year.

The 1992 budget outlook was adversely affected by several factors, notably uncertainties concerning revenues stemming from the output decline and the civil disturbances in the Trans-Dniestr district. The draft state budget made provision for a R164 million deficit in the first half of 1992. No data on the outturn for that period are available, but by October the deficit already amounted to some R12 billion and was expected to reach R15 billion for the full year.[630] This is probably substantially higher than the budget forecast, estimated at about 8 per cent of GDP.[631]

Little information is available on monetary developments in 1992. Monetary emission however totalled some R5.3 billion over January-November and was 53 times higher than in 1991 as a whole — the highest increase among the CIS countries. However, in November itself, the money supply fell slightly, probably resulting from the withdrawal of roubles from cash circulation in that month. A temporary coupon currency used for cash payments only was introduced in June 1992 to alleviate the rouble cash shortage, but roubles continue in use for banking and intra-CIS transactions. A decision has been made in principle to introduce a Moldovan currency (the *leu* or the *dukat* — no final decision has been taken on its name) and withdrawal from the rouble area is expected to occur in late 1993 or in 1994.

---

[625] BBC, *Summary of World Broadcasts*, SU/1509, 12 October 1992, p.C4/2.

[626] See, for example, *Izvestiya*, 5 January 1993 — when it was quoted at 1.45 to the Russian rouble.

[627] *Financial Times*, 16 November 1992.

[628] See J. Odling-Smee et al., "Moldova", *IMF Economic Review*, Washington, D.C., May 1992, p.5.

[629] *Rossiiskaya gazeta*, 17 July 1992.

[630] *Izvestiya*, 30 October 1992.

[631] "Moldova", *IMF Economic Review*, Washington, D.C., May 1992, p.6.

Only modest progress has been made in the task of building market institutions. The Moldovan parliament had adopted a number of laws on economic reform in 1991, including provisions to regulate the leasing of land and other state assets. A privatization programme initiated in January 1992 laid down the conditions for the sale of such assets. Later in the year a programme for 1992-1993 was adopted to implement this legislation, beginning with the housing stock and small enterprises. The total value of the assets concerned was given as R21 billion, or about 53 per cent of all state property holdings.[632] A voucher system for the purchase of assets by the public is to be operated from April 1993.[633] Other legislation includes guidelines for foreign economic activities (including the right to set up wholly foreign-owned businesses), a bankruptcy law, and laws to permit the leasing of all types of assets in all sectors of the economy and to allow the free grant of agricultural land and its resale. Legislation to limit monopoly profits has also been passed.

No quantified economic targets nor an economic policy statement for 1993 were available at the time of writing.

Russian Federation

At the end of 1991 the Russian government committed itself to the rapid introduction of a market system. A number of miscalculations were made and the disruptions which inevitably accompanied the process were greater than initially foreseen. Corrective policies were not consistently applied in 1992, partly because parliament, elected under the previous regime in 1990, was reluctant to accept them.

In November 1991, when the Soviet Union was still in existence, parliament granted temporary powers to the President of Russia. The government announced its decision to liberalize prices from January 1992 and a stabilization package to support economic reform. These were incorporated into a programme, drawn up in collaboration with the IMF, which was approved by the government of independent Russia on 27 February 1992.[634] It included a reduction of the budget deficit to zero in the first quarter of 1992, notably by large cuts in social programmes, military procurement and investment, and the introduction of a value-added tax to replace enterprise profits as a main revenue source. Sharp cuts in the money supply and the volume of credit were also targeted. Restrictive monetary and fiscal targets were re-emphasized in a revised programme worked out in accordance with IMF stabilization guidelines and approved by the Russian cabinet on 30 June 1992.[635] The budget deficit was to be restricted to 5 per cent of GDP in the second half of the year, inflation brought down to 9 per cent monthly and credit creation limited to R700 billion up to the end of the year.

The liberalization of most prices in January 1992 gave a severe shock to the existing system. The disappearance of the organs of central economic administration accelerated during the year. Thus economic agents faced the task of rapid adjustment to new circumstances in the near absence of the necessary market institutions – modern banking and other financial organs, sales and purchasing networks, insurance and legal services. Against this background, an output fall similar to or steeper than that in the two east European countries where equally concentrated reform policies had been introduced (Czechoslovakia and Poland) was recorded. This contraction gathered pace during the year.

Given that the ultimate objective of reform was to restructure output in line with real demand patterns and away from those imposed by central planning, large production falls were inevitable due to the normal lag in adjusting the production structure to changes in demand – particularly in an economy unaccustomed to such abrupt change. The process was further delayed by declining imports from abroad and from other CIS countries, which together were nearly nearly a third as large as Russian NMP in 1990.

Prices failed to stabilize following their initial post-liberalization surge. Hence the biggest departures from the specific objectives of the two programmes as the year advanced occurred in the monetary and fiscal fields. It was claimed that, as planned, the budget deficit was reduced to zero in the first four months.[636] The rise in credit and the money supply was not excessive in relation to the price explosion which immediately followed liberalization. But implementation of tight monetary and fiscal restraint which should have followed price liberalization generated considerable parliamentary opposition, since deputies feared that tight money to choke off inflation would lead to large-scale enterprise closures, mass unemployment and social problems and an intensification of the output slump. Parliament therefore decided to increases industrial subsidies, social payments and other public expenditures before its summer recess. In August, the newly-appointed chairman of the Russian central bank publicly stated his disagreement with an IMF-agreed credit limit. The monetary components of the programme were thus in effect abandoned even as they were put into operation. Inter-enterprise debt sky-

---

632 BBC, *Summary of World Broadcasts*, SU/1271, 7 January 1992.

633 In contrast to the Russian system, the Moldovan vouchers will have differing face values to take into account the period of employment on Moldovan territory of the holder and they will not be transferable. BBC, *Summary of World Broadcasts*, SU/1473, 31 August 1992.

634 "Memorandum on the Economic Policy of the Russian Federation", *Ekonomika i zhizn'*, No.10, March 1992.

635 "Programme for deepening economic reforms (up to 1995-1996)". For the summary presented to the Russian parliament, see *Rossiiskie vesti*, No.29, 11 July 1992. A full draft was published under the same title by the Working Centre for Economic Reform, Centre for Conjunctural Economics and Forecasting of the Ministry of Economics, Moscow, June 1992.

636 Such a turnround from the large deficit of 1991 is surprising and may at least partially represent different accounting concepts. Analysis of the claimed shift will have to await publication of detailed information on the outturns for 1991 and 1992.

rocketed in the first half year as enterprises, expecting a bail-out, entered commitments beyond their resources and sales revenues — and credits rose by almost sixfold in the first half to accommodate them. After a temporary slow-down in April-June 1992, credit again rose — by fivefold between the middle and the end of the year and, despite the bail-out, inter-enterprise debt doubled again. The new credits enabled enterprises to settle tax arrears and the general government (federal plus local authority) budgetary deficit was thus held to 5 per cent of GDP — in line with the IMF target but largely cosmetic — even if it was not, at that time, considerably understated. However, in early 1993 it was reported that the federal budget deficit for the whole year reached 7.8 per cent of GDP, and if extra-budgetary resources in the form of credits to the Ministry of Finance and an advance drawing on January funds are added, the total deficit reached R2.6 thousand billion or 17 per cent of GDP.[637]

The continuation of high rates of inflation after price liberalization suggests failure to deal in advance with, or even to anticipate, the price-setting behaviour of the semi-monopolistic enterprises which still accounted for the bulk of output.[638] Nor was any clear attempt made to extend credit selectively to potentially profitable firms, though data on which such decisions could be based appear to be available.[639] Monetary discipline was also weakened by the lack of the Russian central bank's control over credit creation in the other rouble-area countries. The de facto distinction made by the central bank of Russia (and some others) between roubles originating in different states presented in settlement of trade debts did not prevent the build-up of a large net Russian creditor position with the other successor-states, which fed a rapidly-accelerating inflation reaching 25-26 per cent monthly in November and December 1992.

Loss of monetary control, a real wage fall during the year of over 50 per cent, steeply falling levels of trade with CIS and other countries and the virtual collapse in the rouble/dollar exchange rate on the Moscow Interbank Currency Exchange in the last quarter of 1992 and the first quarter of 1993, show that economic stabilization is no nearer now than it was a year ago, before the reforms began. Even so, institutional reform progressed rapidly. Some 7 per cent of industrial output originated in the private sector in the first three quarters of the year and well over one third in agriculture. The issue of privatization vouchers has put in place a mechanism for the distribution of shares in 6,000 large and medium-sized state companies due to be sold off in 1993. For almost all of them the preliminary stage of transformation into joint stock companies was achieved on time before the end of 1992.[640] Over 185,000 private farms have been created though no final decisions on genuine land ownership rights have yet been taken.

When the Congress of People's Deputies was reconvened towards the end of the year, it refused to extend the president's temporary powers. It also forced President Yeltsin to replace acting prime minister Ye. Gaidar, whose economic responsibilities had lasted for just over 12 months, half that time as deputy prime minister. The new government of V. Chernomyrdin, who took over an otherwise almost unchanged ministerial team from Gaidar in December 1992, published a detailed economic policy programme in January 1993.[641] Specific targets include reducing inflation to 5 per cent monthly and the budget deficit to 5 per cent of GDP by the end of the year and stabilization of the rouble exchange rate for the second half. The production decline is targeted to decelerate and output to stabilize.[642] Unemployment growth is to be curbed, though social security spending is to rise as a share of total state spending on consumption, which is, however, expected to fall in relative terms. The programme also contains detailed proposals presented under three main headings: financial stabilization and strengthening of the rouble; the overcoming of distortions and restructuring; and increasing the efficiency of foreign relations and improving the foreign currency position.

The experience of 1992 underlines the need for careful and detailed advance analysis of the likely impact of major policy innovations. The success of the new programme could also depend on improved output performance, which cannot be excluded. The impact of the two-thirds fall in defence spending in 1992 is unlikely to continue on the same scale in 1993 and supply disruptions could begin to ease as arrangements developed in 1992 in response to the breakdown of central planning settle into place. The main task for 1993 however remains as in 1992 — resolution of the dilemma between restructuring output while avoiding an unmanageable rise in unemployment — though it is still extremely low by western standards given an output fall of the magnitude recorded (see section 2.2). The specific targets noted above already contain some inconsistency in this respect but appear to recognize that lax monetary policies cannot be sustained now that high inflation is threatening to turn into hyperinflation. After the repeated, and much sharper, confrontations between President Yelsin and the Congress of People's Deputies in the second week of March 1993, however, it does not seem that technical details of various aspects

---

[637] A. Khandruyev, Central Bank of Russia, *Rossiiskie vesti*, 2 March 1993.

[638] Enterprise profitability nearly doubled between 1991 and 1992 to a rate of 40 per cent despite the output slump. See Russian Federation Goskomstat, *O razvitii ...*, 1993, op.cit., p.19.

[639] Nearly one firm in five wrote a loss in 1992. CIS Statistical Committee, *Ekonomika Sodruzhestva Nezavisimykh Gosudarstv v 1992 godu*, Moscow, 29 January 1993, p.5.

[640] *The Economist*, 28 November 1992, p.77.

[641] "On Financial-Economic Policy in Russia in 1993", *Izvestiya*, 26 January 1993.

[642] The latest quantified forecast for 1993 by the Russian Central Bank foresaw further declines of 5-7 per cent for GDP and 12-15 per cent for industrial production and an inflation rate of 12-15 per cent monthly. See *Radio Free Europe/Radio Liberty Research Report*, Vol.1, No.48, 4 December 1992.

of economic policies are at the top of the agenda in Russia. Both for an improvement of the country's economic situation and for its systemic transformation it would appear essential that fundamental uncertainties surrounding future political structure of the country be removed and that political consensus and public support for the basic strategic options chosen be established.

Ukraine

After Ukraine declared its sovereignty in August 1991, some hesitant steps were taken to implement the transition to a market economy. Price liberalization in January 1992 was not thoroughly prepared but Ukraine had to follow the Russian decision. The economic reform programme, adopted by the Ukrainian parliament on 23 March 1992, contained few, if any, short-term policy aims and thus had little impact on the day-to-day conduct of economic policy in 1992. Its most important objectives were longer-term: widespread privatization, the foundation of a national capital market, the introduction of a national currency, rapid withdrawal from the rouble area and, ultimately, a shift to settlements with other CIS states on the basis of world prices.

The Ukrainian economy was subjected to major disruptions and a steep fall in output in 1992 following the collapse of old structures. The output fall compared with the previous year slackened slightly in the spring and summer, but gathered pace in the autumn and up to the end of the year. The inflationary effects of the price liberalization were worsened by deteriorating terms terms of trade in transactions with other CIS countries, which are important for the Ukrainian economy (in 1990, inter-republican trade came to the equivalent of about one third of the country's NMP).

The stabilization effort in 1992 relied more on administrative methods of economic management rather than on the market. At the end of January, steps to control the post-liberalization price rises were taken through a presidential decree to limit profit margins for all goods for which prices had been freed and another one to limit wage increases. Ceilings on some food prices were imposed. Other presidential decrees concerned compensation for losses incurred by the public from the depreciation of savings and financial losses incurred by agricultural enterprises and banks. The inevitable concomitant of these measures was an early upward pressure on already large state budgetary deficits. Continued reliance on them, on state orders and readiness to provide unlimited credit to finance increased production costs may have mitigated output losses in the short term but clearly proved unsustainable by the year's end.

With regard to policies toward institutional change, since the beginning of 1992 the Ukrainian parliament has adopted laws concerning the privatization of state firms and other small enterprises and the setting-up of a voucher system for the privatization of large state firms. Other acts include a land code, which does not, however, establish full rights to the private ownership of agricultural or other types of land; laws to limit monopoly profits, to allow the privatization of housing by tenants and a liberalized law on foreign investments. Though the foundations have been laid, implementation has been slow. Large-scale privatization has scarcely begun and high taxation rates on new enterprises have discouraged private initiatives. Legislation to date does not amount to a comprehensive framework for transition to a market economy.

The deteriorating economic situation in the early months of 1992 brought to a head deeply-rooted differences in policy-making circles over the pace of economic reform. President Kravchuk, parliament and the prime minister took a more cautious line than the responsible minister, V. Lanovoy, who was dismissed at the beginning of the summer. Thereafter, central intervention in production and supply decisions and higher state spending as tax revenue fell short raised the full-year budgetary deficit to over a third of GDP by the end of 1992. Credit granted since the beginning of the year rose 20-fold between January and November and the monthly inflation rate rose to 30 per cent in December. The effects on foreign exchange markets have been severe. The Ukrainian temporary currency, the *karbovanets*, which replaced the rouble hitherto used for banking transactions and the cash-transaction coupons following the withdrawal of Ukraine from the rouble area in November, became one of the few currencies to have deteriorated against the rouble in the final weeks of 1992 and in early 1993.

The economic debate in parliament intensified during the year and in November Prime Minister V. Fokin was replaced by a manager of large state enterprise, L. Kuchma. His announced policy was to implement "evolutionary change" along a specifically Ukrainian path while avoiding Polish- or Russian-style shock therapy. His government has been granted special powers by parliament to rule by decree until May 1993 and the prime minister has already used them to speed up the pace of change. One of the first economic acts of the Kuchma government was the abolition of subsidies on almost all agricultural products in December 1992 — a measure which resulted in overnight increases in food prices in state stores by 300-500 per cent.

The new government's economic programme, adopted by parliament in February 1993,[643] acknowledges the need for urgent measures to stabilize the economy while at the same time moving towards a free market. Stabilization measures include a reduction in the budgetary deficit to 6 per cent of GDP in 1993, involving cuts in and a more selective approach to industrial subsidies. It also aims to bring inflation down to 3-4 per cent monthly by the end of the year. These measures are to be supplemented by a pay freeze beginning in March and the imposition of penal taxation on enterprises which pay unacceptably high wages. A government decree has also been issued restricting credit creation to 80 per cent of the government's in-

---

[643] See *Financial Times*, 27 January 1993 and *The Wall Street Journal*, 27 January and 9 February 1993.

flation expectations. Finally, the programme recommends cuts in social welfare payments and other public expenditures. No indications have been given by the new government on likely output developments in 1993,[644] but unemployment is expected to rise to some 6 per cent of the labour force by the end of the year (from a rate of 0.3 per cent at the end of December 1992).

Failure to implement radical market innovations in 1992 has clearly prolonged both the period needed to stabilize the economy and the time-scale required for the economy to adjust to new market institutions and policy instruments. The breakdown of old certainties in 1992 at least gave an impetus to intellectual debate on the Ukrainian reform process during the year and this could bear fruit in 1993. But while the new programme indicates an appreciation of the problems, it is not yet clear how much freedom of manoeuvre the government has; one of the first acts of the parliament in 1993 was to reject a decree issued in January to allow the sale from July of medium- and large-scale enterprises.

(v) The Baltic republics

Estonia

After secession from the USSR, it took two years for the Republic of Estonia to re-establish the main pillars of independent statehood. It was on 21 August 1991, when the country formally declared its independence, but multiparty elections were not held until 20 September 1992. In the interim period, the country had three legislative bodies competing with each other,[645] while "daily" matters were handled by two successive governments. The long-awaited elections produced a fragmented Parliament with almost 30 parties and formations, from which a coalition government emerged with a slim and fragile majority.

In the sphere of economics, 1992 brought much faster progress than the confusing and slowly evolving political consolidation process would suggest. Still, in January 1992 the overall situation in Estonia was not much different from any other successor state of the USSR. Production and trade collapsed due to pervasive shortages, bread was rationed, and the place of the rouble was taken over by western currencies, gold, tobacco, locally-issued currencies, etc. In face of this situation and against the advice of the IMF, the caretaker government of Mr. Vähi opted for a monetary reform at the earliest possible moment.[646] On 20 June 1992, Estonia became the first country to achieve monetary independence from the rouble zone.

The timing of the currency reform was carefully chosen. First, Estonia was able to secure the restitution of its pre-war gold stock, partly in kind and partly in convertible currency.[647] Thus the new currency had full backing (worth about $60 million) without outside assistance. Second, the timing was fortunate because rouble inflation temporarily moderated in the run-up period.[648]

The Estonian stabilization programme, later approved and supported by the IMF as well,[649] rests on two anchors. The exchange rate of the kroon was set at a rate of 10:1 against the Russian rouble and then pegged to the deutsche mark at 8 kroon to the mark. The acceptance of these exchange rates amounted to a very deep "devaluation", squeezing the average wage below $40 per month.[650] Relying on the country's own reserves, a currency board mechanism was put in place to keep the growth of the domestic money supply in line with the inflow of hard currency. The second anchor was the statutory fixed minimum wage, to which salaries in the state sector were rigidly linked.

In spite of this careful design, the introduction of a convertible currency has not put an end to inflation as fast as it was envisaged,[651] with monthly rates averaging 7 per cent in last third of the year. The increase was 3 per cent in December 1992 — i.e., almost 50 per cent on an annualized rate. However, the overall supply situation has improved considerably on most consumer markets. The currency board system has proven its efficiency, too. The kroon's external value remained un-

---

644 The previous cabinet had reviewed "optimistic" and "pessimistic" scenarios of zero growth and a further 8 per cent fall in NMP in October 1992.

645 The 105-member *Supreme Council* had come into being after the March 1990 elections — i.e., still under Soviet rule. It functioned as a parliament and its Chairman as Head of State. The *Congress of Estonia* was a kind of "alternative" parliament built up from grassroots in 1989. Its 490 deputies were elected by citizens of pre-1940 Estonia and their descendants. The third legislative body, the *Constitutional Assembly*, was formed in September 1991 with 30 delegates each from the Supreme Council and the Congress of Estonia.

646 The role of the IMF is still the subject of controversy among those who actually participated in the preparation of the monetary reform. According to the account of the Estonian government's two advisors, the Fund had first tried to delay the introduction of the new currency, arguing that the country was not ready and it should be introduced only in late 1992 or 1993 (see A. Hansson and J. Sachs, "Crowning the Estonian Kroon", *Transition* (A World Bank Publication), Vol.3, No.9, October 1992). This interpretation of events was rejected by the IMF (ibid., Vol.3, No.10, p.9), although it was acknowledged that Fund representatives had believed that the reform could have benefited from a standby agreement which was scheduled for a later date.

647 Although precise details are not known, it is believed that the United Kingdom and Sweden had held 4.8 and 2.9 tons of Estonian gold, respectively. The Bank of International Settlements in Basle had also stored Estonian pre-war gold for decades, but the size of this holding was never disclosed (see *Magyar Nemzet*, 31 August 1991; *The Baltic Independent*, 8-14 January 1993).

648 The rate of inflation of rouble-denominated prices in Estonia had come down from 74 per cent in February to 5 per cent in May.

649 On 14 September — i.e., one week *before* elections — a 12-month standby agreement went into force, backed by a loan of SDR 27.9 million (*IMF Survey*, 14 December 1992).

650 For comparison it is worth recalling that in 1990 the Polish stabilization programme started with a 64 dollar/month average wage level.

651 The limits on the domestic convertibility of the kroon are similar to those imposed by the Polish or Hungarian monetary authorities.

changed throughout 1992 and the central bank was able to double its international reserves.[652]

The agenda for institutional reforms and industrial restructuring in 1993 is necessarily heavily loaded. In this respect, Estonia has no particular advantage over other Baltic countries or other successor states of the Soviet Union. Given its weak parliamentary position, the hands of the government are quite tied. This is well characterized by the fact that it is forced to operate with a six-month budget (still not approved in mid-February 1993) and that it could not sign a $20 million loan agreement with Japan without prolonged negotiations with Parliament. According to observers, these developments and the discontent of the population have already raised fears of early new elections.[653]

Latvia

Latvia declared its independence from the Soviet Union on 4 May 1990, but it was internationally recognized only after the failed coup in the Soviet Union in August 1991. The first free parliamentary elections, held in March 1990, were won by the Popular Front (a loose coalition of various political groups sharing common objectives of national sovereignty, political democracy and mixed economy) which received 66 per cent of the seats. Together with the two other Baltic states, Latvia was admitted to the UN (and ECE) on 17 September 1991 and joined the IMF and the World Bank on 19 May 1992.

Since the first coalition government was established after the 1990 elections, the overall political situation in Latvia has displayed a remarkable degree of stability. The government's policy immediately concentrated on three broad issues considered of top priority: (a) building up new state institutions; (b) macroeconomic stabilization and creating a market economy; and (c) restoring balance in political relations with Russia. But as the economic situation of the country sharply deteriorated in 1992, economic problems became of primary importance.

After a relatively mild downturn in 1991, the Latvian economy contracted sharply in 1992, with a 44 per cent fall in GDP and a fall of industrial output of some 35 per cent. Like the other Baltic republics, the country participated in the inflation spurt of the rouble zone in the early part of the year, but price advances remained high also after the introduction of the national currency, averaging 14 per cent per month in the second half of the year. Like the other Baltic states, Latvia suffered mostly from the abrupt decoupling of its economy from the former Soviet Union. The task of building up an independent economy proved very difficult for a small country which until 1991 had conducted 90 per cent of its foreign trade with the rouble zone and had practically no foreign reserves. Under conditions of output specialization pushed to the extreme, so characteristic of the ex-Soviet state, the break-up of trade and production links existing between the Latvian enterprises and their counterparts in other republics produced tremendous shocks on both the demand and supply sides, which inevitably translated into a steep fall of living standards.

The severance of trade ties with Russia and other CIS countries in 1991 and 1992 was particularly felt in two areas: imports of fuels from Russia and exports of Latvian industrial and agricultural products to ex-Soviet markets. Cuts in oil and gas supplies led to prolonged stoppages in industrial production, while inability of Russian importers to pay for Latvian exports put the Latvian producers into deep financial crisis and forced them to reduce deliveries. The shock was exacerbated by the lack of immediate possibilities of switching to alternative, i.e., non-CIS markets, due to the structural and technological backwardness of Latvian industry, scarcity of financial resources, lack of convertibility, and weaknesses of the banking and monetary systems.

Under these difficult external and internal conditions, Latvia nevertheless persevered with a broad spectrum of market reforms. While most of domestic prices were decontrolled by the end of 1991, Latvia in fact went much farther with price liberalization than the CIS states, lifting controls also from prices of fuels and basic foodstuffs in the course of 1992. The reform of the fiscal system from January 1992 included the substitution of VAT for a mosaic of various turnover taxes and subsidies, as well as numerous fiscal incentives for investors. Latvia was also one of the first post-Soviet countries to leave the rouble zone and to introduce its own currency. The monetary reform, which was spurred by the acute shortage of cash roubles in the spring of 1992 and by growing monetary chaos in the former Soviet Union, started in May 1992 with the introduction of a parallel currency, the Latvian rouble ("roublis"). On 20 July, the Russian rouble was entirely withdrawn from circulation and Latvia acquired monetary independence. The move resulted initially in a blockage of trade settlements with Russia, but after the payment rules had been agreed upon, trade was resumed in the autumn, although most of the trade and payments preferences characteristic for rouble-zone transactions had been revoked. The "roublis" is conceived as a transition money, although it fulfils all the functions of national currency. The Latvian authorities introduced the permanent national currency — the "lats" — on 5 March 1993 but it will coexist with the Latvian roublis for some time.

Along with the reconstruction of its monetary and price system, Latvia started with a large-scale privatization programme in 1992. While the so-called "small" privatization had begun in Latvia already in 1990, the rules of privatization of state-owned enterprises and of restitution of confiscated property were defined only in a series of laws between June and November 1992. According to the regulations, all citizens and all permanent dwellers of the Latvian republic will have the right to a share of state property and will be

---

[652] Between July and December, reserves grew from 1.2 billion kroons ($101 million) to 2.5 billion kroons ($192 million), or 4-5 months of import coverage.

[653] See *The Baltic Independent*, 22-28 January 1993, p.7.

given special privatization certificates. The number of certificates for each citizen will vary, however, according to their labour record in the republic, amount of property nationalized under the Soviet rule, and some political considerations. The privatization will start with the conversion of state enterprises into joint-stock companies (for large units) and limited liability companies (for smaller enterprises). A certain amount of shares will be earmarked for employees.

Latvia was the first country of the former Soviet Union to sign a standby arrangement with the IMF in June 1992. A loan of $86 million was approved by the IMF board in September, and the first tranche of $22 million has already been disbursed. The arrangement provides crucial financial support for the Latvian economy in the transition period. The IMF's support, which opens way to international capital markets, is even more important in view of the unresolved political, military and trade issues straining Latvia's relations with Russia.

The essence of the economic problems in Latvia is, however, of a structural character. The adjustment of the economy to new relative prices and reorientation away form the ex-Soviet to international markets will necessarily take some time. The Latvian economy will probably remain in recession in 1993 under the impact of the stabilization programme. If the government succeeds in keeping inflation at low levels through restrictive budget and monetary policies, the medium-term prospects will be enhanced and the inflow of foreign capital may increase. This policy could be put in danger, however, by growing unemployment, falls in living standards, and resulting social tensions. As reform "fatigue" increases, voices calling for less austere and more protective policies are more and more common.

Lithuania

Lithuania, the largest and most populous of the Baltic states, was the first Soviet republic to have declared its independence from the Soviet Union (in March 1990). International recognition, as for the others, did not come until September 1991. The first free parliamentary elections in early 1990 were won by non-communist forces led by the Sajudis, a coalition movement for national independence. However, political tensions between nationalist and post-communist forces, unfavourable (and largely unexpected) consequences of breaking the trade and production links with the former Soviet Union, and a lack of consensus over the required course of economic policy and the speed and sequence of reforms, led to a rapid deterioration of political and social climate in the republic since late 1991. Within less than one-and-a-half years the country has seen four governments, as many ministers of finance and two presidents of the central bank. This political instability has probably been one of the reasons for the much worse economic performance of Lithuania in 1992 as compared with other Baltic states and even most other post-Soviet republics.

Like the other two Baltic states, Lithuania had been fully integrated into the Soviet economy, and the establishment of economic sovereignty will require a considerable effort of structural change. Soviet supplies had accounted for 89 per cent of Lithuanian imports and the Soviet market absorbed 94 per cent of exports in 1990. The precipitate collapse of the Soviet state and the crisis in the post-Soviet economy in 1991-1992 thus led to a sharp decline of traditional export markets and disruption of import supplies.

Preliminary data indicate that Lithuania's industrial production dropped by a staggering 52 per cent in 1992, the largest fall among all post-Soviet states except for war-torn Armenia. With other sectors also contracting sharply, the republic's GDP is estimated to have fallen by 30 per cent relative to 1991. At the same time, year over year the price level increased by 1130 per cent, inflation accelerating in the last quarter to 27 per cent per month. The main cause behind this unprecedented fall of output is the collapse of trade with the rouble zone, and in particular of energy imports from Russia. Deliveries of Russian oil and gas were suspended several times for rather long periods during 1992, and Lithuania had neither technical nor financial possibilities of switching to alternative sources of fuels supplies. However, the impact of the Soviet trade shock, though dominant, has been compounded by other factors, such as inconsistent macroeconomic policies, delays in monetary and banking reform, the slow pace of the privatization of state enterprises, and a volatile political situation.

Lithuania started its market reforms in late 1991 with the adoption of legislation on privatization and a new fiscal system, coupled with a sweeping liberalization of foreign trade activities a broad decontrol of prices. The impact of these measures on output and efficiency was muted, however, as the republic remained in the rouble zone and could not effectively control its own money supply. A parallel currency was introduced in May 1992 to alleviate a shortage of cash roubles, but the complete separation from the rouble zone came only in October. A temporary currency (the "talonas") was substituted for the Russian rouble. The permanent national currency (the "litas") is to be introduced only after the economic situation stabilizes. Although the "talonas" quickly strengthened against the rouble, it depreciated against the dollar at much faster pace than the Latvian roublis. This was not only because of the delayed departure from the rouble zone, but also because of weaknesses of the Lithuanian banking sector[654] and rather expansionary budget policy. Lack of full monetary controls and a deterioration of government finances towards the end of 1992 led to higher and accelerating inflation.

Among many urgent policy issues faced by Lithuania at the beginning of 1993 three seem to be of particular importance: restoring trade and political re-

---

[654] The Lithuanian central bank transferred its commercial and merchant functions to a newly-established commercial bank only in December 1992, thus establishing a two-level banking structure.

lations with Russia, on which Lithuania is vitally dependent for its fuel and energy supplies, curbing government expenditures, and maintaining restrictive credit policy. If the new government succeeds in these three areas, the fall of GDP may be halted in 1993. These are also necessary conditions for receiving further assistance from the IMF in the form of subsequent tranches of standby credit (totalling $81 million) on the basis of the agreement signed in September 1992.

Chapter 4

AID AND TRADE: WESTERN RESPONSE TO THE TRANSITION

This chapter focuses on external finance and western assistance to the transition countries. In section 4.1 information is provided on the flow of private and official capital into these countries during 1992. Bilateral and multilateral flows are presented separately, the latter in terms of the transition countries' relations with the international financial institutions. A discussion of the special financing obtained by four transition countries follows. Then estimates of resource transfer into the transition countries are presented. The section ends with some observations on the adequacy of the finance available to the transition countries. Section 4.2 presents an overview of developments in east-west trade agreements and policies, focusing on eastern access to western markets.

4.1 FINANCIAL FLOWS AND COMMITMENTS OF ASSISTANCE

(i) Private capital flows

Although official financing has become the major source of new credits to the eastern countries in transition, *private finance* has been important for several countries and, in relative terms at least, has been gaining in importance. Capital markets and foreign direct investment provided these countries with over $4 billion in 1992, somewhat more than in the previous year. Hungary and the former Czechoslovakia were again the chief recipients of these funds, which give them room for manoeuvre far beyond that available to the other eastern countries. In the longer term, private sources are being counted on to supply the bulk of the foreign capital required by all eastern countries.

Total funds raised by the east in the international *capital markets* continued to decline in 1992 (table 4.1.1). This is due to the fact that most eastern countries lack significant access to the international financial markets, and reduced borrowing by Czechoslovakia.[1]

Current account surpluses, healthy inflows of foreign investment and borrowing from various official sources tended to reduce Czechoslovakia's and Hungary's needs for private credits.

Hungary was again the leading borrower, raising $1.4 billion, chiefly through the issue of bonds.[2] This total includes a DM 600 million ($365 million) bond flotation, initially scheduled for 1993, but moved up to November to benefit from a fall in DM interest rates on long maturities.[3] However, the funds were to be drawn only in January 1993. Czechoslovakia arranged a one-year stand-by credit of DM 100 million with Commerzbank of Luxembourg (not shown in table 4.1.1).[4]

In the first quarter of 1993 the flotation of bonds by Hungary and the Czech Republic picked up sharply. Hungary issued a DM 1 billion bond, its largest ever.[5] This was followed in early March by a $300 (later increased to $375) million bond issue, the first launched

---

655 Currently, Moody's Investor Services rates the debt of Hungary as Ba1, one notch below investment grade. The same rating had been applied to the former Czechoslovakia. In March Moody's assigned the new Czech Republic an investment grade rating of Baa3. *Financial Times*, 12 March 1993. The higher rating should enable the country to obtain a better price, to issue longer maturities of debt and further broaden its access to securities markets.

656 Over the years, Hungary has issued a steady stream of bonds, the outstanding stock of which totalled $6.6 billion at the end of December 1992 (table 4.1.1)

657 The issue has a maturity of seven years and a yield of 250 basis points (2.5 percentage points) over German bunds. *Financial Times*, 11 November 1992.

658 According to J. Tosovsky, Governor of the State Bank of Czechoslovakia, the facility has been negotiated to provide additional credit lines in view of the forthcoming division of the country and the second stage of its privatization programme. *East European Markets*, 27 August 1992.

659 The bond carries a yield of 265 basis points over German bunds and a seven-year maturity. *Financial Times*, 25 February 1993. The National Bank of Hungary intends to borrow a total of $1.5-$2 billion in 1993.

TABLE 4.1.1
European transition countries: Medium- and long-term funds raised on the international financial markets, 1988-1993
(Million US dollars)

	1988	1989	1990	1991	1992	January-March 1993	International bond issues amounts outstanding [a]
Bulgaria	194	580	-	-	-	-	200
Czechoslovakia	330	334	438	278	40	375	700
Hungary	1 016	1 708	987	1 378	1 446	621	6 600
Poland	-	163	-	5	9	-	-
Romania	-	-	-	-	-	-	-
Eastern Europe	1 540	2 785	1 425	1 661	1 494	996	7 500
Former Soviet Union	2 679	1 858	3 250	-	-	-	1 800
CMEA Banks	75	75	-	-	-	-	-
Total above	4 294	4 718	4 675	1 661	1 494	996	9 300
of which:							
Bank loans [b]	1 050	2 047	2 993	86	235	-	-
Foreign bank loans [c]	1 652	358	-	60	9	-	-
Other [d]	232	75	-	-	-	-	-
Bonds	1 360	2 239	1 682	1 516	1 250	996	-

Sources: OECD, *Financial Statistics Monthly, Part I*, February 1993 and previous issues; BIS, *International Banking and Financial Market Developments*, Basle, February 1993. Press reports.

a End-December 1992.
b International bank loans in Eurocurrencies, excluding officially guaranteed loans and rescheduled debt.
c In domestic currency of lending countries, excluding guaranteed loans.
d Other bank facilities, including bankers' acceptances.

by the Czech National Bank after the creation of the Czech state.[660]

Foreign direct investment is widely expected to play a major role in the transformation of the economies in transition. In 1992 inflows of FDI into eastern Europe increased to nearly $3 billion (table 4.1.2).[661] However, as in 1991, the investments were concentrated in Hungary and the former Czechoslovakia. In both cases, they accounted for the bulk of private capital inflows, which was also the case in Slovenia.

FDI into the former *Czechoslovakia* increased substantially in 1992. The $1.1 billion may have helped to finance the boom in capital goods imports discussed in section 3.3. The result is also noteworthy since the inflow in 1991 was already considered high, because of the exceptional investment by Volkswagen AG in Skoda. The increased activity in 1992 reflects the execution of contracts concluded early in the year in conjunction with the country's privatization programme. However, the authorities of the former republic indicated that external investment slowed in the wake of the decision to divide the country. Foreign surveys also suggest greater caution on the part of external investors.

Data for the first nine months of 1992 show that the Czech Republic was the destination of the bulk of these flows (some 92 per cent).[662] The leading investor countries were the United States and France, followed by Germany, which ranked first in 1991. The Czech authorities hope that foreign investment will pick up somewhat in 1993, to an estimated $1.3 billion. However, the prospects for the Slovak Republic appear less sanguine.

Hungary obtained $1.5 billion of FDI in 1992, about the same as in 1991. There had been concern that these inflows would decline because of, *inter alia*, the belief that the most attractive enterprises had been privatized already in 1991, uncertainty concerning the future government, and exchange rate policy. Experience suggests that FDI flows cannot be counted on to rise steadily simply because the necessary legal and business environment of a country is well advanced.

Of all of the eastern countries, Hungary has progressed the furthest in broadening its access to various sources of foreign funds. To that end, the National Bank of Hungary has authorized the sale of forint (domestic) bonds to foreigners. So far the strategy has met with only limited success since the government's Ft 7

---

660 The yield is to be set at 272 basis points over comparable US Treasury notes and carries a maturity of three years for the first $300 million and 260 points for the $75 million tranche. *Financial Times*, 2 March 1993 and 12 March 1993.

661 It should be borne in mind that foreign investment statistics shown in table 4.1.2 are reported on a balance of payments (*cash*) basis. Thus they cover only payments remitted through the national banking system. This cash is available for balance of payments support. The figures are narrower in coverage than balance of payments data reported on an *accrual* basis, which also reflect reinvestment of profits, equity investments in kind and intercompany loans. However, such data are not available on a current basis, and for some eastern countries not at all. Annual aggregations of the *value of newly announced projects* yield the highest foreign investment figures of all. However, they normally represents plans to invest, to be implemented over several years. Moreover, as a rule any such data should be treated with considerable caution since projects may be held up, the value of investment modified, or the venture may be cancelled outright.

662 *Hospodarske noviny*, 1 December 1992 and *Financial Times*, 9 December 1992.

TABLE 4.1.2
European transition countries: Joint ventures and foreign direct investment, 1990-1992
(Number of projects, million dollars)

	Joint ventures a			Net flows of FDI		
	1990	1991	1992	1990	1991	1992
Bulgaria	140	900	1 200	4	56	42
Czechoslovakia	1 600	4 000	5 995	188	592	1 054
Czech Republic	..	..	3 120	..	..	983
Slovak Republic	..	..	2 875	..	..	71
Hungary	5 693	9 117	13 000	311	1 459	1 471
Poland	2 799	4 796	10 100	88	117	185 b
Romania	1 501	8 022	20 684	-18	37	73 c
Total above	11 733	26 835	50 979	573	2 261	2 825
Former Yugoslavia	3 918	..	..	189 d	137 d	..
Slovenia	..	1 000	1 650	-2	41	113
Former Soviet Union	2 905	3 920	15 292	..	200	..
Belarus	..	283	710	..	..	..
Estonia	..	1 100	2 662	..	..	..
Latvia	..	295	1 300	..	..	..
Lithuania	..	220	2 300	..	..	..
Russia	..	2 022	3 500	..	..	100
Ukraine	..	..	900	..	..	..

Sources: ECE joint venture data base; national balance of payment statistics. World Bank, *Russian Economic Reform*, Washington, D.C., September 1992.

a Number of foreign direct investment projects registered, end of year.
b January-September.
c January-November.
d January-October.

billion domestic bond issue in December 1992 attracted only Ft 20 million ($250,000) in foreign funds.[663]

Aside from *Slovenia,* direct investment into the other eastern countries remains disappointing. In *Romania,* the rate of foreign investment registrations has picked up but the size of individual investments and the total inflow of capital has remained small. The same is true of Poland, given the size of the country.

For various reasons, FDI into the *republics of the former USSR* have been very low. However, a decision by the World Bank to waive its "negative financing"[664] rule for the financing of projects in the transition economies should be beneficial. The Bank is reported to be in the process of defining a category of project which would automatically qualify for a waiver. It would comprise only those investments which would produce a new or increased stream of income from which repayments could be made. Projects in the energy sector may be the most favourably affected, including those involving joint financing with the EBRD.

Overall, there are indications that some western companies are re-examining their investment commitments in the eastern region. They point to the delayed economic recovery[665] and the risk of selling their products to enterprises in chronic financial difficulty. The western recession has also caused many firms to postpone new investments.

The attitude of *commercial banks* toward lending to the countries of eastern Europe remains cautious, although the deciding factors vary from borrower to borrower. In consequence, there have been few syndicated loans (table 4.1.1) and little new commercial lending by banks. Data for the first nine months of 1992 indicate that BIS banks' commercial claims on the eastern countries continue to decline.[666] Balance of payments statistics indicate that some countries continued to make net repayments of short-term credits (table 3.3.6).

(ii) Official financing

Commitments of international assistance from bilateral and multilateral sources to the transition countries amount to ECU 119 billion (some $145 billion; table

---

[663] *International Herald Tribune*, 16 January 1993. In this case, analysts felt that the coupon of 16 per cent was not high enough given the risk of larger forint devaluations.

[664] *Financial Times*, 22 January 1993. The rule prohibits member states (or their state enterprises) participating in World Bank programmes from giving equal or more favourable terms to other investors than to the Bank. That is to say, a state cannot pledge the future income from a project to other banks where a pledge has already been given to the IBRD. The World Bank gets paid before all other creditors.

[665] *Wall Street Journal*, 30 November 1992.

[666] BIS OECD, *Statistics on External Indebtedness*, New Series, No.10, January 1993 and BIS, *International Banking and Financial Market Developments*, February 1993.

TABLE 4.1.3

Commitments of international assistance the countries of eastern Europe and the independent states of the ex-USSR [a]

(Million ECU)

Donors, creditors/recipients	Albania	Bulgaria	Czech and Slovak Republics	Hungary	Poland	Romania	Ex-Yugoslavia	Estonia	Latvia	Lithuania	Regional NSP [b]	Eastern Europe	Ex-USSR [c]	Grand total
Community total	570	846	2 455	3 681	6 595	1 604	2 127	33	56	64	3 012	21 043	52 416	73 459
EEC Member States	363	197	1 753	2 112	5 576	818	797	8	22	20	1 310	12 977	48 903	61 880
Belgium	-	-	-	-	-	-	-	-	-	-	174	175	337	512
Denmark	-	-	6	3	41	2	-	4	8	6	357	425	975	1 400
France	2	73	173	294	650	342	69	3	1	5	196	1 809	1 961	3 771
Germany	34	84	1 287	1 408	3 430	248	572	-	-	-	432	7 496	39 218	46 713
Greece	47	3	-	-	-	1	-	-	-	-	1	51	72	124
Ireland	-	-	-	-	1	-	-	-	-	-	-	2	1	3
Italy	271	21	98	191	570	129	46	-	12	8	65	1 410	4 315	5 725
Luxembourg	-	-	-	-	1	2	-	-	-	-	20	23	1	24
Netherlands	7	10	93	124	24	15	37	2	1	1	58	372	372	744
Portugal	-	-	2	-	2	-	-	-	-	-	4	4	67	71
Spain	-	1	78	84	165	76	73	-	-	-	3	478	1 085	1 564
United Kingdom	2	5	17	9	693	3	1	-	-	-	2	731	499	1 231
EC Institutions	206	649	702	1 569	1 019	787	1 329	25	34	45	1 702	8 066	3 513	11 579
EC	206	534	617	1 284	754	762	49	25	34	45	187	4 496	3 513	8 009
EIB	-	115	85	285	240	25	1 280	-	-	-	1 340	3 370	-	3 370
ECSC	-	-	-	-	25	-	-	-	-	-	175	200	-	200
EFTA	10	112	191	189	1 096	109	85	75	48	31	1 771	3 717	1 186	4 904
Austria	2	20	68	47	257	23	15	-	-	-	1 234	1 667	600	2 267
Finland	-	17	47	81	75	9	1	56	33	18	22	359	73	432
Iceland	-	-	-	-	1	3	-	-	-	-	-	3	-	3
Norway	-	11	14	10	33	15	11	-	-	1	34	129	272	400
Sweden	2	36	19	18	534	24	34	19	14	12	81	792	35	828
Switzerland	6	29	43	34	196	35	24	-	-	-	400	767	206	973
Other G-24 countries	114	240	647	1 154	6 187	335	61	29	28	29	367	9 191	11 359	20 550
Australia	-	-	1	14	137	26	1	-	-	-	-	178	273	451
Canada	-	12	312	93	1 230	87	2	7	7	7	8	1 766	1 336	3 102
Japan	1	75	225	732	1 212	75	-	-	-	-	153	2 473	2 036	4 509
New Zealand	-	-	-	25	25	-	-	-	-	-	-	51	-	51
Turkey	55	78	1	91	110	40	-	-	-	-	1	375	877	1 252
United States	57	76	108	199	3 473	107	59	22	21	22	204	4 348	6 837	11 185
G-24: total above	694	1 198	3 293	5 025	13 878	2 048	2 273	137	132	125	5 150	33 951	64 962	98 913
Other countries	-	-	-	-	-	-	-	-	-	-	-	-	5 544	5 544
Total bilateral	694	1 198	3 293	5 025	13 878	2 048	2 273	137	132	125	5 150	33 951	70 506	104 457
Total multilateral	34	930	1 790	2 863	4 297	2 127	939	-	-	-	33	13 013	1 314	14 327
of which:														
EBRD	1	40	80	226	124	222	67	-	-	-	33	727	83	810
IMF	-	655	1 181	1 687	2 215	1 102	-	-	-	-	-	6 906	720	7 626
World Bank	32	235	528	950	1 958	803	873	-	-	-	-	5 380	511	5 380
Grand total	727	2 128	5 082	7 888	18 175	4 175	3 212	137	132	125	5 183	46 965	71 821	118 785

Source: EC, *G-24 Scoreboard*, Brussels, 13 January 1993 and EC, *Breakdown of World Assistance to the Independent States of the ex-USSR*, Brussels, 5 November 1992.

[a] For eastern Europe commitments are cumulative from beginning of QI 1990 to end of QII 1992; for the ex-USSR, from QI 1990 to November 1992.
[b] NSP = Non specified.
[c] Excludes Baltic states.

4.1.3).[667] Of this, some ECU 47 billion ($60 billion) was committed to eastern Europe (including the Baltic states) and ECU 72 billion ($84 billion) to the CIS countries.

Bilateral finance

Bilateral assistance accounts for the bulk of these commitments, some ECU 34 billion in favour of eastern Europe (including the Baltic states) and ECU 71 billion for the CIS (table 4.1.3). In this category of financing, pledges to Russia are believed to dominate, although this cannot be confirmed for lack of more detailed data. These figures cover emergency assistance, grants, credit and investment guarantees, and various official loans. They exclude assistance from private sources (e.g., emergency aid from private humanitarian agencies) and relief on commercial bank obligations.[668]

Overall, the EC and its member states are the major donors/creditors, Germany being the largest. The latter's commitments of ECU 47 billion include, among many other things, financing for housing in the former Soviet Union for the military leaving the ex-German Democratic Republic. The United States (ECU 11 billion), Italy (ECU 6 billion) and Japan (ECU 5 billion) are the next largest individual contributors.

Within the G-24 programme, Poland and Hungary, the first countries to embark on economic reform and to request assistance, have received the greatest share of specified commitments (41 per cent and 15 per cent respectively). In the case of Poland, this figure also includes the considerable commitment made by Paris Club members to reduce the country's official debt by 50 per cent (see below). Pledges to the Baltic states and Albania, the latest countries to benefit from G-24 assistance, amount to some ECU 400 million and ECU 700 million, respectively. Most of the funds made available to Albania have been in the form of grants.

Reasonably comprehensive and current information for the *disbursement* of bilateral assistance is not available.[669] Statistics provided by BIS/OECD reporting institutions show that officially backed claims on *eastern Europe* changed little in the first half of 1992, after some increase in the previous year.[670] Claims on the states of the former Soviet Union rose by some $7 billion in the first three quarters of 1992.[671] In addition to any new credits, this increase reflects the capitalization of significant amounts of unpaid interest.[672] It should also be noted that these figures are net and thus may give little information on the drawing of new credits. Moreover, they do not cover all guaranteed credits, government-to-government loans and grants.

Miscellaneous statements and estimates suggest that Russia drew some $11-12 in new guaranteed and official loans during 1992.[673] The country's borrowing was held back for various financial and legal reasons. For example, tranches of the EC ECU 1,250 million loan (originally offered to the former Soviet Union) could not be released until the Russian government provided guarantees required under EC procedures. Similarly, United States' deliveries of grain were suspended in 1992 because Russia (and other republics), constrained by shortages of convertible currencies, failed to make overdue payments on loans for previous purchases of US grain.[674]

Multilateral financing

In 1992, the drawings of the *east European* countries on IMF facilities declined to $1.3 billion, about one third of their borrowing in the previous year (table 4.1.4). The arrangements concluded in 1992[675] (table 4.1.5) have been supported by smaller stand-by credits than those agreed in 1991.[676] However, in the case of Czechoslovakia, the authorities announced that they would forego the last two drawings, because of the favourable development of the country's financial position.[677]

In other cases, the failure of certain countries to meet performance targets agreed with the IMF (see chapter 3) essentially precluded access to IMF re-

---

[667] The figures for eastern Europe include commitments made between January 1990 and end-June 1992 (through November 1992 for the republics of the former Soviet Union), see table 4.1.3.

[668] As noted below, the four eastern countries have benefited greatly from relief on commercial debt obligations.

[669] G-24 donor countries have agreed to provide annual data on disbursements to the OECD, by recipient and type of assistance. Among other factors, significant delays in reporting by several major donors have precluded publication of estimates of flows in 1991. Preliminary and incomplete data for 1990, made available by the OECD, were reproduced in United Nations Economic Commission for Europe, *Economic Survey of Europe in 1991-1992*, New York, 1992, table 5.2.3.

[670] BIS/OECD, *Statistics on External Indebtedness*, New Series, No.10, January 1993.

[671] Ibid., and BIS, *International Banking and Financial Market Developments*, February 1993.

[672] BIS, *The Maturity and Sectoral Distribution of International Bank Lending*, first half of 1992, Basle 1993.

[673] Estimates made by the World Bank indicate that Russia drew some $7 billion in the first half of 1992.

[674] Under the US guaranteed loan programme, a country must repay all arrears before fresh grain loans can be extended. Wheat shipments by Australia to Russia had been stopped for the same reason, and the decision was reversed only in March 1993. *Financial Times*, 4 March 1993.

[675] A full list of agreements concluded by the eastern countries with the IMF during 1990-1992 is presented in United Nations Economic Commission for Europe, *Economic Bulletin for Europe*, vol.44 (1992), New York, 1993, table 3.3.2.

[676] For Albania, the agreement reached with the IMF in 1992 was its first since becoming a member in 1991. The Bulgarian government has been negotiating a three-year extended agreement with the IMF, which would supersede the current stand-by agreement. The programme aims at further stabilization of the economy and structural reforms, and will give Bulgaria the option to buy some of its external debt.

[677] According to Josef Tosovsky, Governor of the State Bank of Czechoslovakia. *Hospodarske noviny*, 24 September 1992.

TABLE 4.1.4

Selected multilateral financing to the European transition countries, 1990-1992
(Commitments, million US dollars)

	Bulgaria				Czechoslovakia			
	1990	1991	1992	Cumulative	1990	1991	1992	Cumulative
IBRD	-	267	-	267	-	450	246	696
IBRD [a]	-	58	92	150	-	200	131	331
EBRD	-	-	90	90	-	38	44	83
EIB	-	-	140	140	-	-	105	105
G-24 [a,b]	-	203	170	373	-	288	150	438
Total	-	261	491	752	-	526	492	955
IMF [a,c]	-	396	279	675	-	1 256	333	1 588
Total	-	657	770	1 427	-	1 782	761	2 543

	Hungary				Poland			
	1990	1991	1992	Cumulative	1990	1991	1992	Cumulative
IBRD	516	400	390	1 306	1 081	1 140	390	2 611
IBRD [a]	227	297	317	841	54	349	343	746
EBRD	-	128	126	254	-	100	226	326
EIB	150	144	85	379	119	181	61	361
G-24 [a,b]	438	450	130	1 018	1 000 [d]	-	-	-
Total	1 104	1 122	732	2 957	2 200	1 421	677	3 298
IMF [a,c]	173	963	162	1 298	485	327	-	812
Total	1 276	2 084	893	4 254	2 685	1 748	677	4 110

	Romania				Former Yugoslavia			
	1990	1991	1992	Cumulative	1990	1991	1992	Cumulative
IBRD	-	230	500	730	692	300	-	992
IBRD [a]	-	3	211	214	274	367	52	693
EBRD	-	178	98	276	-	-	-	-
EIB	-	31	-	31	228	-	-	228
G-24 [a,b]	-	225	342	567	-	-	-	-
Total	-	664	940	1 604	920	300	52	1 220
IMF [a,c]	-	774	479	1 253	89	-	-	89
Total	-	1 438	1 419	2 857	1 009	300	52	1 309

	Eastern Europe [e]				Russia		Belarus	
	1990	1991	1992	Cumulative	1991	1992	1991	1992
IBRD	2 289	2 787	1 567	6 643	-	760	-	-
IBRD [a]	555	1 274	1 148	2 284	-	1	-	-
EBRD [f]	-	500	648	1 097	-	3	-	39
EIB	496	356	390	1 243	-	-	-	-
G-24 [a,b]	438	1 166	838	2 441	-	-	-	-
Total	3 223	4 758	3 443	11 424	-	764	-	39
IMF [a,c]	747	3 716	1 266	5 640	-	989	-	-
Total	3 970	8 473	4 709	17 152	-	1 753	-	39

Other transition countries, 1992:	Albania	Estonia	Latvia	Lithuania	Moldova
IBRD	41 [g]	30	45	60	-
IBRD [a]	2	1	-	1	-
EBRD	10	51	39	19	1
EIB	-	-	-	-	-
G-24 [a,b]	46	-	-	-	-
Total	96	82	83	80	1
IMF [a,c]	14	11	36	25	-
Total	110	93	119	105	1

Sources: EBRD, EC, World Bank; press releases and direct communications to ECE secretariat.

Note: Totals include commitments except for G-24 and IMF disbursements.

a Disbursements.
b Exceptional balance of payments assistance only.
c Purchases.
d Zloty stablization fund, which was not accessed. It is not included in the total for eastern Europe.
e Excludes Yugoslavia.
f Excludes regional projects.
g IDA.

TABLE 4.1.5
European transtion countries: IMF arrangements in force in 1992
(Million SDRs)

	Duration		Type of agreement	Amount
	Commencement	Expiration		
Albania	26 Aug. 1992	25 Aug. 1993	Standby	20.0
Bulgaria	17 Apr. 1992	16 Apr. 1993	Standby	155.0
CSFR	3 Apr. 1992	2 Apr. 1993	Standby	236.0
Hungary	20 Feb. 1991	19 Feb. 1994	Extended	1114.0
Poland	18 Apr. 1991	17 Apr. 1994	Extended [a]	1224.0
	8 Mar. 1993	Apr. 1994	Standby	476.0
Romania	29 May 1992	28 Mar. 1993	Standby	314.0
Former USSR:				
Estonia	16 Sept. 1992	15 Sept. 1993	Standby	27.9
Latvia	15 Sept. 1992	14 Sept. 1993	Standby	54.9
Lithuania	21 Oct. 1992	20 Oct. 1993	Standby	55.6
Russia	5 Aug. 1992	4 Jan. 1993	Standby [b]	719.0
Total 4 Republics				857.1

Source: IMF, *International Financial Statistics* and press releases.

a Superceded by stand-by arrangement of March 1993.
b First credit tranche.

sources. *Hungary* voluntarily refrained from drawing the second instalment[678] of the funds available for 1992 under a three-year extended arrangement because its budget deficit significantly exceeded the agreed target.[679] The size of the budget deficit for 1993 contributed to the failure of recent negotiations which were intended to restore Hungary's access to the IMF resources.[680] *Poland* missed the targets for the budget deficit, the inflation rate and foreign reserves in 1991. As a result, it could draw only $327 million of a three-year extended arrangement before further disbursements were suspended in mid-1991. Negotiations on a new 14-month stand-by arrangement (intended to replace the three-year accord) were concluded in November 1992, but examination of the Polish letter of intent by the IMF was postponed until prospects for the passage of the new budget by the Parliament became clearer.[681] The IMF Board approved the new stand-by agreement, accompanied by a credit of $655 million, in March 1993.

In 1992, Czechoslovakia, Romania and Hungary were authorized by the IMF to make further drawings totalling some SDR 200 million under the oil import element of the compensatory and contingency financing facility (CCFF). This facility was intended to cover excess costs of imports of crude petroleum, petroleum products, and natural gas and was attached to the stand-by credits arranged by the countries in transition in 1991.[682] Sixty-five per cent of the oil import element allocations were disbursed immediately on approval by the Executive Board in 1991.

In December 1992, the Executive Board of the IMF determined that the former Yugoslavia had ceased to exist and that Czechoslovakia would cease to exist at the end of the year. Therefore both would cease to be members of the IMF.[683] The Czech Republic, the Slovak Republic, Croatia and Slovenia became members of the IMF in December 1992. In its first operation with the new Czech Republic, the IMF extended a $240 million preventative credit line, intended to support the new Czech currency and strengthen the republic's access to the international credit markets.[684]

Of the republics of the former Soviet Union, *Russia* was the first to conclude a stand-by arrangement with the IMF authorizing a drawing in the *first tranche* of up to SDR 719 million ($1 billion).[685] The stand-by had a duration of five months and was intended to support the Russian government's economic programme, which built upon the framework set out in the *Memorandum of Economic Policies* submitted to the IMF in March 1992. The Russian authorities earmarked these funds

---

[678] *Heti Vilaggazdasag*, 29 May 1992.

[679] The government was able to draw SDR 79.6 million at the end of the first quarter of 1992.

[680] *Wall Street Journal*, 16 February 1993.

[681] Mr. J. Lipszyc, an advisor to the Minister of Finance, BBC, *Summary of World Broadcasts*, EE/1586 A1/2, 14 January 1993.

[682] In November 1990 the oil import element was introduced temporarily into the CCFF in response to the fact that the IMF had not had the ability specifically to compensate for economic shocks caused by sharp increases in the price of oil imports. Later tranches were to be released, provided that the recipient countries were satisfactorily implementing agreed energy policy actions. In addition, credits were to become available through the external contingency mechanism if oil and gas prices rose above the levels envisaged in their programmes. However, fuel prices did not rise further and this facility remained unused.

[683] IMF, *Press Releases*, Nos. 92/92 and 92/96. Succession to membership has been subject to special conditions. In assessing the ability of each successor state to meet its membership obligations, the IMF would, *inter alia*, take into account the effect of sanctions imposed by the Security Council of the United Nations.

[684] BBC, *Summary of World Broadcasts*, EE/1612 A1/2, 13 February 1992. The credit was to be recommended by the IMF without undergoing the usual approval procedures.

[685] IMF, *Press Release*, No.92/60, 5 August 1992.

for strengthening the country's foreign currency reserves and drew down the facility by the end of 1992.

Progress on a full stand-by arrangement for Russia has come to a standstill. Its finalization, which is conditional upon a credible economic programme, would trigger the release of an additional $3 billion in IMF funding.[686] When conditions are deemed appropriate, a third stage, involving the pegging of the rouble exchange rate and activation of the promised $6 billion stabilization fund, is to be implemented. Assembly of this stabilization fund has also been entrusted to the IMF.

During September and October, *Estonia, Latvia* and *Lithuania* became the first republics of the former Soviet Union to conclude full stand-by agreements with the Fund. The other republics are much further behind both in their reforms and in their talks with the IMF, and agreements are not expected until well into 1993.

World Bank lending commitments to *eastern Europe* reached almost $2.8 billion in 1991 (table 4.1.4). The $1.6 billion approved by the board in 1992 indicates that the pace of lending has eased. Disbursement of World Bank funds – over $1 billion in 1991-1992 – has progressed more slowly (table 4.1.4). Projects take some time to start up after they are approved.[687] Structural adjustment loans (SALs) can be disbursed quickly, generally in two tranches, provided that recipients meet the conditions attached. The Bank has now approved loans to all east European countries except Albania, which has obtained funding through the Bank's affiliate, the IDA.

The World Bank's first financial commitment to *Russia* involves a $600 million "rehabilitation" loan which will support the first phase of a broad-based programme of reforms to be implemented by the government.[688] Of this, $250 million is for imports by the growing private sector and $350 million for priority imports by the health care, agriculture, transport and energy sectors. Also in the pipeline for Russia are loans of up to $1 billion covering the oil and agricultural sectors. Altogether the World Bank envisages lending $2.5 billion to the republics of the former Soviet Union in the current fiscal year (ending June 1993), provided that economic reforms progress. Annual lending to the republics (excluding the Baltic republics)[689] is expected to rise within two years to $4.5-5 billion.[690] However, through the end of 1992 only $2 million had been disbursed to the CIS.

Between June 1991 when it commenced operations and end-1992, the *European Bank for Reconstruction and Development* committed around ECU 1 billion to the countries in transition (table 4.1.4). The slower pace of disbursements, which amounted to some ECU 200 million in 1992,[691] reflected the length of time taken to implement projects. Also, the number of good projects which the Bank has found to invest in has fallen short of its financial resources.[692] EBRD lending has been concentrated in Hungary and Poland, which account for 44 of the 71 projects approved to date. The bulk of the financing is in loans and some ECU 71 million in equity investment. Most projects are in the telecommunications sectors of the eastern countries.[693]

Five eastern countries are currently eligible for *European Investment Bank* credits. Together Hungary and Poland are entitled to borrow a maximum of ECU 1 billion during 1990-1993 while the combined borrowing of Bulgaria, the former Czechoslovakia and Romania is not to exceed ECU 700 million. The European Commission has proposed the extension of the EIB's activities to the Baltic States and to Albania, with a combined credit ceiling of ECU 200 million.[694] Through end-1992, all those east European countries currently eligible to borrow from the EIB arranged loans amounting to ECU 820 million, chiefly for energy and infrastructure projects.[695] Czechoslovakia and Hungary obtained global loans to finance small and medium-sized ventures. The data for 1992 indicate a somewhat higher level of new commitments than in 1991 (table 4.1.4)

Complementary exceptional financing from the Group of 24[696] was conceived to fill projected financing gaps in the eastern transition countries' balance of payments. Such gaps are estimated by the IMF for the

---

[686] Full initial funding of the full stand-by agreement would be some $4 billion.

[687] This comment also applies to the lending of the other development banks, for which only data on disbursements are available.

[688] *World Bank News*, August 7, 1992.

[689] The World Bank has approved its first loans (totalling $135 billion) to the Baltic republics. Among other things they will be used to finance supplies of energy this winter and other essential imports. *World Bank News*, 23 October 1992.

[690] According to Tim Cullen, chief spokesman of the World Bank, *Wall Street Journal*, 16 September 1992. A number of former Soviet republics will qualify for loans from the International Development Agency, the concessional lending arm of the IBRD.

[691] *Financial Times*, 2 March 1993.

[692] According to Mr. A Ljungh, vice-president for finance of the EBRD. Ibid.

[693] EBRD, *Annual Report*, London, April, 1992. Direct communication to ECE secretariat, and press releases.

[694] *Agence Europe*, 28 December 1992. In order to accelerate infrastructure projects in Albania, the granting of interest rate subsidies for EIB loans, using funds from the PHARE programme, is under consideration.

[695] EIB, *Annual Report 1991*, Brussels, 1992 and press releases.

[696] Only those eastern countries benefiting from the G-24 coordinated assistance have been eligible for G-24 complementary financing. This group includes all east European countries (although Poland has not received such funds), Slovenia (since September 1992), the Czech Republic, the Slovak Republic and the Baltic states. The exceptional financing needs of the other ex-Soviet republics have been considered within the framework of the G-7 or the World Bank consultative groups.

year ahead. The estimates take into account the availability of multilateral financing, any commitments of bilateral official grants and credits, including export credits, debt restructuring agreements with both private and official creditors and expected flows from the private sector.[697] G-24 creditors have made almost all of their contributions to these loans in the form of untied credits, although each contribution must be negotiated bilaterally. Disbursements (generally in one or two tranches) are conditional upon compliance with the conditions attached to the recipient's IMF stand-by arrangement and/or World Bank structural adjustment loans. The EC has been the largest source of this type of assistance, providing approximately one half of the financing target agreed for each eligible transition economy.

In 1991 the G-24 mobilized exceptional balance of payments support in favour of Bulgaria, Czechoslovakia, Hungary and Romania. The operations carried out in favour of the former Czechoslovakia and Hungary were successful (table 4.1.6).[698] Although some delays were encountered, the pledged funds became available and were drawn to the extent required. Overall, it appears that these loans provided backing for the reforms at a crucial time (even if they were not fully drawn upon) and helped the recipients to strengthen their financial positions. On this basis and their favourable balance of payments prospects, it was decided in early 1992 that neither country would need further G-24 complementary financing.

Mobilization of exceptional balance of payments financing for Bulgaria and Romania has proved more difficult. In 1991, pledges by the G-24 fell short of agreed targets,[699] and only a fraction of those funds was released (table 4.1.4). IMF projections for the balance of payments of Bulgaria and Romania in 1992 indicated a persistence of financing gaps. It was also calculated that Albania and the three Baltic republics would require similar macroeconomic support. The pledging exercise for 1992 in their favour has also been only partially successful (table 4.1.6), and early 1993 the agreed targets still had not been reached.[700] Also, Bulgaria and Romania had not obtained all of the funds committed for the 1991 exercise, the carry over of which had been counted on in their financial programmes for 1992.

For 1991-1992 G-24 disbursements to Bulgaria amounted to some $350 million, compare to identified financing gaps of $1,040. For Romania the comparable figures are $567 million and $1,180 million, respectively. Albania obtained its first G-24 funds ($46 million)[701] in December 1992 while the Baltic states were scheduled to receive their first disbursements in early 1993 (some $210 million). In both cases, the EC accounted for the bulk of the released funds.

Overall, the *combined disbursements* of the G-24, EBRD, World Bank, and EIB to eastern Europe amounted to some $2-2½ billion in both 1991 and 1992. At the same time, the east European countries' purchases from the IMF fell from $3.7 billion in 1991 to $1.3 billion in 1992.

Prior to the scheduled expiry of the *Polish Stabilization Fund* at the end of 1991, seven donor countries committed some $500 million of their original contributions to the support of the restructuring and privatization of Poland's banking system. The $1 billion fund was established in January 1990 to back the zloty with the introduction of internal convertibility. Its lifetime had been extended through 1991 and again through 1992, but the development of the exchange rate of the zloty made its use unnecessary.

TABLE 4.1.6

Complementary financial assistance provided by the G-24 and EC to the European transition countries
(Million US dollars)

Programme for:	Requested target	Pledged [a]
Albania		
1992-1993 [b]	165	89
Bulgaria		
1991	800	606
1992	240	152
Czechoslovakia		
1991	1 000	937
Hungary		
1990	1 000	1 000
1991	500	500
Romania		
1991	1 000	737
1992	180	121
Baltic republics: 1992-1993 [b]		
Estonia	105	..
Latvia	210	..
Lithuania	285	..
Total	600	523

Sources: Commission of the European Communities, documents submitted to G-24 Meetings of April and July 1992 and direct communications from the EC.

[a] As of 3 February 1993.
[b] July 1992-June 1993.

---

[697] Commission of the European Communities, *Common Guidelines for Coordinated G-24 Exceptional Balance-of-Payments Assistance to Central and East European Countries*, Brussels, 14 June 1991.

[698] In 1990 Hungary had already benefited from a $1 billion restructuring loan from the EC, which was intended to help it to meet its high repayments obligations and to rebuild reserves.

[699] According to EC authorities, this was partly due to the more general questions concerning the sharing of the financial burden between individual donors. Commission of the European Communities, *Complementary G-24 Balance-of-Payments Assistance to Bulgaria*, prepared for the G-24 Consultative Group on Bulgaria, Brussels, 23-24 April 1992.

[700] In addition to the 50 per cent normally contributed by the EC, the EFTA countries participated heavily in the mobilization of the funds pledged so far for the Baltic republics.

[701] This includes one half of the EC's ECU 70 million grant to the G-24 complementary financing effort for Albania.

(iii) Special financing

With three eastern countries and the successor states to the former Soviet Union now benefiting, special financing[702] has become the major form of external financing and international support for the eastern area. In 1991, Albania, Bulgaria and Poland received some $10 billion in special financing,[703] more than the total commitments disbursements made by the multilateral financial institutions to all of eastern Europe (table 4.1.6). In 1992 special financing obtained by these three countries fell to an estimated $5 billion, chiefly because unpaid repayments of principal by Poland fell sharply. By contrast special financing benefiting the successor states of the former Soviet Union rose to around $14 billion in 1992.

The rapid increase in *Albania's* debt (to some $600 million) in recent years has chiefly been due to the accumulation of arrears. Albania managed to secure external financing in late 1990 and early 1991 by *buying* foreign exchange from a number of European banks, without delivering the countervalue.[704] Since Albania did not service any of its bank debt, it continued to benefit from special financing in 1992.[705] By most criteria Albania is, in relative terms, the most heavily indebted eastern country. It is doubtful that it will be in a position to service this debt in the years to come and for that reason is likely to become a candidate for debt reduction.

Since *Bulgaria's* Foreign Trade Bank declared a moratorium on the servicing of its obligations in March 1990, the country has paid little interest and obtained deferrals and rescheduling of repayments of its $12 billion debt (of which some 80 per cent is owed to commercial banks). As a result, Bulgaria has obtained special financing amounting to $2-2.5 billion annually during 1990-1991. Preliminary estimates for 1992 indicate a smaller amount.

The Paris Club agreed to a restructuring of official debt on 17 April 1991 and a second accord was reached on 14 December 1992.[706] Negotiations between Bulgaria and international banks on the country's commercial debt ($9-10 billion) are reported to have made progress.[707] The prospective agreement is to include a substantial reduction of debt and debt servicing and a menu of options for implementing the cuts under the Brady plan.[708] Such reductions are essential as medium-term projections indicate that the country's debt burden is unsustainable. Even with substantial forgiveness, the country's debt will continue to rise.

Poland has benefited from debt restructuring and forgiveness agreed with the Paris Club and from arrears on commercial debt. Formal debt arrangements and arrears in 1991 amounted to nearly $7.6 billion, comprising $2.5 billion in unpaid interest (including $864 million cancelled by Paris Club members) and $5.1 billion in unsettled repayments of principal. On the basis of data for the first three quarters of 1992, it appears that special financing obtained by Poland fell sharply. Unpaid interest amounted to $3 billion (of which $2.2 billion was cancelled by official creditors) and no repayments of principal were reported as falling due.[709]

During the first stage of the reduction in Poland's Paris Club debt package, to be undertaken between 1 April 1991 and 1 April 1994, Poland's $33 billion official debt is to be reduced by 30 per cent of its net present value.[710] All Paris Club creditors are to reduce interest payments by 80 per cent and Poland is to make no repayment of principal. Under this agreement, Poland committed itself to seek from its commercial bank creditors a reduction and reorganization of its debt on comparable terms.

The Polish government is seeking an agreement on debt restructuring and reduction with the *London Club of Commercial Banks* this year. Negotiations, which have been suspended since June 1991, are expected to intensify with recent IMF approval of a new stand-by arrangement. Poland hopes to get terms comparable to those obtained from its official creditors in March 1991.[711] The country ceased servicing its commercial obligations in the autumn of 1989, except for a $1.1

---

[702] Special financing comprises the sum of debt write-offs, rescheduling, deferrals and arrears.

[703] See the United Nations Economic Commission for Europe, *Economic Bulletin for Europe*, vol.44 (1992), New York, 1993, table 3.4.1, p.103.

[704] EBRD, *Quarterly Economic Review*, London, June,1992

[705] EC, IMF and IBRD, *Albania: Orientations for a G-24 Support Programme for the Restructuring of Albania's Economy*, presented to the G-24 Meeting in Tirana, 17 July 1992.

[706] The new agreement reschedules $256 million which was to fall due in the five months December 1992-end April 1993 (the terminal date of the current IMF stand-by arrangement). It carries a 10-year maturity, with a six-year grace period (instead of the normal five). IBRD, *Financial Flows to Developing Countries*, Quarterly Review, January 1993, p.16. Total Paris Club debt amounts to some $2 billion.

[707] Bulgaria started to repay some overdue interest from 15 September 1992.

[708] The Bulgarian authorities have offered to buy back 50 per cent of its debt at $0.10 in the dollar. The remainder would be exchanged for bonds – either 30-year collateralized par bonds with a 12-month rolling interest guarantee or 30-year discount ($0.20 in the dollar) bonds. Interest would be LIBOR plus 13 16 per cent. IBRD, *Financial Flows to Developing Countries*, Quarterly Review, January 1993, p16.

[709] National Bank of Poland, *Information Bulletin*, 9-10 1992.

[710] This agreement was reached in April 1991. For details, see UNCTAD, *Trade and Development Report, 1992*, United Nations, New York, 1992, p 60. When fully implemented, creditors will agree to a 50 per cent reduction in the net present value (NPV) of the entire stock of Poland's eligible debt. This reduction applies to all debt contracted before the cut-off date (including arrears). This feature, as well as the reduction in size, is unprecedented in debt rescheduling operations in recent years. The amount of interest payments reduction, discounted back to the base date using the "appropriate market rate", will be credited against the NPV reduction target.

[711] According to Mr. K. Krowacki, Poland's new chief debt negotiator with the London Club, *Financial Times*, 20 January 1992.

TABLE 4.1.7

**European transition countries: Two measures of net transfer of
resources and changes in reserves, 1990-1992**

(Million US dollars)

	Bulgaria			Czechoslovakia			Hungary		
	1990	1991	1992	1990	1991	1992 [a]	1990	1991	1992
Net transfer of resources									
Capital inflows less income payments	-555	87	-108	10	-18	-18	-2 103	1 122	-780
Trade balance plus non-capital services	464	49	-528	788	-431	-431	-1 541	-1 598	-1 541
Net change: reserves [b]	-888	45	400	-1 102	898	413	-562	2 720	760
of which:									
IMF (net) [c]	-	386	217	-	1 275	283	-145	905	-7

	Poland			Romania			Slovenia		
	1990	1991	1992 [a]	1990	1991	1992	1990	1991	1992
Net transfer of resources									
Capital inflows less income payments	-2 419	-902	-1 240 [d]	137	639	800	..	-273	-172
Trade balance plus non-capital services	-4 045	-637	-1 286	1 787	1 364	848	..	-499	-947
Net change: reserves [b]	1 938	-1 188	801	-1 644	-693	-122	..	83	603
of which:									
IMF (net) [c]	479	322	-33	-	772	261	-	-	-

Sources: National statistics; IMF, *Russian Federation, Economic Review*, April 1992; World Bank, *Russian Economic Reform*, September 1992; ECE estimates.

a January-September.
b A positive sign indicates an increase in reserves.
c A positive sign indicates net borrowing from the IMF.
d Short-term capital flows include errors and omissions.

billion loan originally extended as a revolving trade facility in 1983.[712] As a result, arrears have raised the level of bank debt to $12.1 billion.

The republics of the *former Soviet Union* benefited from about $14 billion in special financing in 1992. According to IMF estimates, $15.6 billion in interest and principal were due, and $7.2 billion in debt relief was originally foreseen by western creditors.[713] However, only $2 billion was remitted by Russia. Agreed debt relief and arrears on interest and principal payments accounted for the difference.[714] In December 1992, the Group of Seven (G-7) offered to reschedule about $15 billion of the approximately $17 billion in principal and interest due in 1993 on the external debt of the former Soviet Union.[715] Thus Russia would be obligated to pay about $2.75 billion. The maturity proposed is 10 years with a 5-year grace period.

(iv) Resource transfer

According to the two indicators[716] presented in table 4.1.7, for most countries of eastern Europe in 1992 there was a net outflow financial resources and a build-up of foreign reserves, tendencies already observed in 1991.[717] Measured by *capital flows less income payments* (the latter consisting chiefly of interest payments on debt), these negative transfers ranged from nearly zero in Czechoslovakia to over $1 billion in Poland. In most cases, they reflect small deficits on capital account and net payments of interest (see table 3.3.6).

---

[712] Recently the Polish authorities announced unilaterally that as of 4 February payments on this loan will be limited to only 20 per cent of the amounts falling due. Ibid., and *International Herald Tribune*, 19 January 1993.

[713] IMF, *Economic Review, Russian Federation*, Washington, D.C., April 1992, table 30.

[714] It has been reported that the former Soviet Union failed to settle $14 billion of debt service due in 1992. The total due amounted to $20 billion (less the $4 billion in arrears outstanding at the beginning of 1992 which were to be eliminated during the year), of which only $2 billion was paid. *Rossiiskie Vesti*, 29 January 1993.

[715] According to Mr. O. Wethington, US Assistant Secretary of the Treasury for International Affairs. *Bulletin of the United States Mission (Geneva)*, 14 January 1993, and *Wall Street Journal*, 12 December 1992. According to *Rossiiskie Vesti*, 29 January 1993, scheduled debt service obligations amount to $18 billion in 1993.

[716] Mathematical expressions for the two measures of net financial transfer in terms of standard SNA variables were derived in *World Economic Survey 1986*, (United Nations publication, Sales No. E.86.II.C.1), pp.163-164.

[717] All flows are on a net basis.

The outflow of resources, measured by the *trade balance plus non-capital services* (i.e., the current account excluding investment income payments and official transfers) are generally larger. This reflects the current account surpluses posted by most of these countries, which were used to further strengthen *reserves*. It should be noted that changes in reserves include net borrowing from the IMF. In both 1991 and 1992 IMF credits were used by Bulgaria and the former Czechoslovakia to increase reserves, which was also the case of Hungary in 1991.

Thus, overall, the east European countries have not obtained foreign resources to support domestic absorption, as can be seen by the fall in consumption and investment throughout the area. The boom in consumer goods imports from the west (section 3.3) was financed by export earnings rather than borrowing from abroad.

In the former Czechoslovakia, Hungary and perhaps Slovenia, the build-up of reserves is consistent with the stated objectives of the authorities. They have sought to strengthen their financial positions in order to improve their credit ratings, broaden their access to international capital markets, and attract FDI. The Polish authorities foresee the need for additional reserves to implement a debt reduction scheme which is being discussed with commercial creditors (see above).

The calculations for Bulgaria show a small outflow of financial resources. Bulgaria paid only a small part of its rather considerable interest obligations and obtained little new capital. The trade balance measure shows a considerably greater negative resource transfer (reflecting the country's trade surplus) a share of which went into reserve accumulation. Borrowing from the IMF and the EC also boosted reserves.[718] As argued below, this build-up was accomplished at a high cost in terms of foregone imports and output.

The situation of Romania differs sharply from this pattern, in that there was both a positive transfer of resources and a decline in reserves. During the past three years the capital inflow (on the first measure) has been positive, the result of increased official borrowing (net income payments were negligible). On the second, trade balance measure, the inflow of resources was even larger (but declining), due to the rundown of reserves. In 1991 and 1992, this chiefly reflected the use of IMF credits for balance of payments support, as opposed to the reserve build-up foreseen in the country's economic programme.

(v) Adequacy of short-term international assistance and prospects

Judgments about the adequacy of the overall levels of external financial assistance to a country in transition is virtually impossible without a clearly specified programme for stabilization, structural change and long-term growth. Such an assessment should take into account the special requirements of a reforming economy (i.e., creation of market institutions, business supporting infrastructure, social safety nets, upgrading human capital, etc.) as well as more traditional ones such as balance of payments support.

In practice, a large proportion of short-term assistance is calculated on the basis of forecasts of short-run economic growth, resources for certain structural adjustment projects, essential imports and export projections, which, together with projected inflows of grants, private capital, commitments of bilateral credits, and targets for reserve accumulation, yield an estimate of short-run balance of payments support (*financing gap*). A *financial package* is then drawn up, broken down by the type of assistance, and donors and creditors are then invited to make *commitments*. The contribution of the international financial organizations is largely determined by their own set procedures. Bilateral creditors (and perhaps commercial banks) are then called upon to make up any remaining difference. Obtaining such pledges is an important and often difficult part of the process. Finally, funds are *disbursed* after the necessary legal procedures are complete and the recipient meets certain conditions. The amounts and types of funds disbursed can differ considerably from the original financial package.

In general, the information available to the public about this process is incomplete and precludes serious independent assessment of a country's programmes and the supporting financial assistance. Even analysis *ex post* is frequently ruled out for lack of data (balance of payments, reserves, or disbursements from creditors and donors). Hence, the question addressed here is simply whether or not the transition countries obtained the necessary balance of payments support to meet their projected financing gaps. This is followed by a discussion of some of the problems facing economies which are severely constrained by shortages of convertible currency.

For *Czechoslovakia*, *Hungary* and *Poland* official balance of payments support has proved adequate.[719] Together with growing export receipts and private capital flows (marginal in the case of Poland), official financing has enabled these countries to boost imports substantially, strengthen foreign reserves and provide support for their currencies. In fact they have foregone access to certain facilities. In consequence, exceptional macroeconomic assistance of the type provided by the G-24 is no longer considered necessary.

It should be borne in mind, however, that the ultimate objectives of systemic transformation and the stabilization and structural adjustment programmes under way are a sustained growth of economic output and of improvement in standards of living. It would be premature to conclude that these three countries, in the vanguard of economic reform, will attain these goals

---

[718] Bulgaria's balance of payments accounts show borrowing from the EC under reserves (and not under long-term capital movements), presumably because the funds were used for reserve accumulation (see table 3.3.6).

[719] In the case of Poland this refers to the zloty stabilization fund and special financing.

soon. Even if there is an economic upturn in these countries, their formidable structural problems may continue to seriously constrain economic growth. Thus, beyond immediate balance of payments support, they will probably continue to require considerable amounts of technical assistance and financial help with investments in infrastructure. This is all the more true of those countries which started the transition process later and require continuing balance of payments support.

Albania,[720] Bulgaria and Romania have not been as fortunate as the three above-mentioned countries. Prior to the launching of their reform programmes, their economies were in a very poor state, foreign exchange reserves were low and export earnings were declining. However, as noted above, commitments of bilateral support to fully meet their estimated financing needs were not forthcoming and even firm commitments were disbursed only with a delay.[721] This resulted in severe shortages of convertible currency, the consequences of which are discussed below. In 1992, about 2½ years after the demise of the former governments, the external accounts of Bulgaria and Romania showed some small improvement, due to an upturn in their trade with the west.[722] Also, Bulgaria has rebuilt its reserves to a modest level, but those of Romania remain very inadequate. Neither country has much room for manoeuvre.

It had been estimated that the *successor republics of the USSR other than Russia* would require some $20 billion in external support in 1992.[723] However, for various reasons, little finance was actually committed and even less was disbursed. The Baltic states drew their first IMF credits in the fall of 1992 following approval of their economic programmes. Commitments of macroeconomic assistance by the G-24 have fallen short of target and the first disbursal of funds was scheduled only for early 1993.

As regards financial support for the *Russian Federation*, considerable public debate has centered on whether or not the $24 billion package[724] unveiled by the G-7 in March 1992 was implemented. Estimates of foreign assistance received by Russia range from over $20 billion[725],[726] to nothing at all, as claimed by at least one Russian official. Since a complete accounting of the relevant information has not been published, the public debate has been conducted on the basis of partial (sometimes contradictory) data and different or unspecified definitions.[727] However, even if Russia received the resources represented by the higher estimates (which seems likely) there were still signs of acute shortages of convertible currency — imports were cut back, arrears piled up, etc. (see features below). Also, many western observers indicate that the provision of assistance to the ex-Soviet Union has been woefully insufficient.

Characteristics of countries facing financing shortfalls

Transition countries which have experienced shortfalls of international assistance tend to share certain common features, although these features may also stem from other negative factors: systemic and external shocks, inappropriate macroeconomic policies, lack of a reform policy or hesitant implementation of reforms, and so on. Financial shortfalls appear to have a number of consequences:

— They have aggravated the decline in *domestic output*. These countries have experienced relatively steep drops in output, and they expect further declines in 1993. During the past few years, their national authorities have made numerous references to shortages of foreign inputs and their adverse affects on production, exports and employment. Currency constraints are also reflected in the fact that *ex post* current account deficits are smaller than those envisaged in the economic programmes approved by the IMF.

---

720 Albania received emergency assistance but little cash until the end of 1992.

721 Already in 1991 G-24 Ministers voiced their concern about delays in the release of balance of payments assistance for some countries in transition. Commission of the European Communities, *Press Release*, IP (91) 994, November 1991. Such problems apparently persisted in 1992, leading the Consultative Group for Bulgaria to declare: "The Bulgarian government, the international financial institutions, the EC and delegates, jointly declare their intention to facilitate and accelerate the utilization of the balance of payment support". Commission of the European Community, *Press Release*, IP (92) 316, 24 April 1992.

722 According to Bulgarian statistics; western mirror statistics show growth in Bulgarian exports since 1990 (see section 3.3).

723 According to Mr. M. Camdessus, Managing Director of the IMF. IMF, *Survey*, April 1992.

724 The package consisted of some $4.5 billion from the international financial institutions (chiefly a stand-by credit from the IMF); a $6 billion currency stabilization fund financed entirely through the activation of the Fund's General Arrangements to Borrow (GAB); $11 billion in bilateral credits and some $2.5 billion through the deferral of debt payments. *Bulletin of the United States Mission (Geneva)*, 1 April 1992.

725 ECE estimates show that the value of new loans, grants, and special financing (including arrears) obtained by the former USSR amounted to over $20 billion in 1992. See United Nations Economic Commission for Europe, *Economic Bulletin for Europe*, vol.44 (1992), New York, 1993. Russia obtained a $1 billion stand-by credit from the IMF and $760 million from the World Bank (although none of the latter was disbursed in 1992). Special financing obtained by Russia has been considerably larger than foreseen in the package (some $14 billion for all of the former USSR, including arrears) and $12 billion in bilateral credits has been reported.

726 On the occasion of a visit to the United States, Mr. Y. Gaidar, an advisor to President B. Yeltsin and former acting Russian Prime Minister, noted that much of the $24 billion package promised in 1992 has either never surfaced or is inappropriate for Russia at the current time. *Bulletin of the United States Mission (Geneva)*, 5 March 1993.

727 This issue is also clouded by the frequent failure of commentators to clearly state their terms of reference, or to distinguish clearly between: the proposed package, commitments, and actual disbursements; different definitions of assistance (some commentators focus only on new credits, overlooking debt deferrals and, especially, the substantial arrears built up during 1992); Paris Club and total debt; Russia and the whole former Soviet Union. As regards the latter distinction, Russia has accepted responsibility for the entire debt of the former USSR and some figures cited reflect this obligation. However, in other cases, it is assumed that Russia is responsible only for the 61 per cent allocation agreed in November 1991 (see section 3.3). The $24 billion package for Russia appears to have been based on this assumption.

Such smaller deficits have sometimes been interpreted positively, but they imply a cutback in the supply of essential resources during a key stage of the transition. *Foreign currency reserves* have tended to remain very low. This had also been true of Bulgaria, although it managed to strengthen reserves in 1992.[728] However, the cost of the build-up in terms of foregone imports and output has been very high.

— They have contributed to sharp *depreciations of domestic currencies*[729] When the Bulgarian and Romanian authorities introduced more liberal trade and payments systems in 1991, the exchange rates of the lev and leu fell to much lower levels than would have prevailed had reserves been adequate and had there been confidence in the governments' capacity to defend them. Devaluations, of course, raise the domestic price of foreign goods and reduce the quantity of goods which enterprises are able to import, with potentially serious consequences for the economy if they are too steep. Persistent currency depreciation tends to exacerbate *capital flight*, which can further weaken the domestic currency (e.g., Russia, Romania). Under such conditions and a system of obligatory surrender of foreign exchange, exporters avoid exchange rate risk by depositing hard currency receipts abroad. Through excessive devaluations, the potential impact of competition introduced by imports — virtually the only source of *competition* in these highly monopolized economies — is circumscribed. This enables domestic industries to raise prices in the wake of a general price liberalization, and, in the longer term, to avoid cost cutting pressures. In other words, the benefits of a more liberal and competitive trade regime are not fully realized.

— They may prompt the authorities to apply direct controls to limit imports. Romania was forced to suspend *internal convertibility* in May 1992[730] and to revert to currency allocation by an import licensing scheme.[731] This represents a setback for the reform since a system of administered foreign exchange allocation often denies access to enterprises which need it the most, curtails import competition, and deters foreign investment. Most reforming economies have not yet been in a position to introduce internal convertibility. Most CIS and, up until 1992, Bulgaria and Romania, have had to adjust to reductions in imports.[732]

Another characteristic of the transition countries is the propensity of their populations to *emigrate for economic reasons*. Recent figures indicate that the number of asylum seekers in western Europe originating in the transition countries has increased rapidly, as has the annual cost of the asylum systems ($6.4 billion in 1991 to $8.3 billion in 1992).[733] The special situation in the former Yugoslavia chiefly explains the large influx from that area, but those people coming from the other countries have made the move for economic reasons. The problem is not likely to subside quickly since the number of Romanians and Bulgarians seeking asylum increased in the second half of 1992,[734] and there are indications that trend continued in early 1993.[735] It is likely that Albania, Bulgaria, Romania are so heavily represented (table 4.1.8) because they have experienced the greatest deterioration of their economies.[736] These countries account for some 13 per cent, or $1 billion annually, of the total cost maintaining the asylum system. This money spent on asylum systems in the west could be far better used as assistance to the countries in transition.

In conclusion, inadequate external financial support aggravates the economic problems of a number of transition countries who are most in need of assistance. The cost of adjustment is raised and the implementation of the very reform measures essential to the functioning of a market economy is hindered. These additional pressures are likely to weaken support for the reform and the chances of survival of democratic systems of government. The adverse consequences for the international community include not only foregone exports, but also increased incentives for substantial numbers of the population to emigrate for western Europe.

---

[728] Although Bulgaria's foreign currency reserves increased sharply in the first eight months of 1992, as recently as end-April 1992 the Consultative Group for Bulgaria warned that the country was facing major constraints of external liquidity. Commission of the European Communities, *Press Release*, IP(92)316, 24 April 1992.

[729] In Russia, the depreciation of the rouble has been due chiefly to lax credit policies. Under these conditions additional foreign resources would have had only a temporary positive impact on the exchange rate.

[730] Internal convertibility, a key reform on the way to a market economy, had been introduced in November 1991.

[731] See United Nations Economic Commission for Europe, *Economic Bulletin for Europe*, vol.44 (1992), New York, 1993, p.69.

[732] As noted in section 3.3, Albania's imports of humanitarian aid from the west have increased, but industry still lacks crucial imported inputs.

[733] Inter-governmental Consultations on Asylum, Refugee and Migration Policies in Europe, North America and Australia, *Asylum-seekers in Western Europe (EC and EFTA) in 1992*, Geneva, 18 December 1992 and direct communications.

[734] Ibid.

[735] Germany, the largest western country of destination, has reported a sharp increase in the number of foreigners seeking asylum in the first two months of 1993. People from Romania, the former Yugoslavia, and Bulgaria comprise the largest groups. *International Herald Tribune*, 3 March 1993.

[736] The figures in table 4.1.8 may substantially understate the number of emigrants from transition countries. For example, they do not include Bulgarian economic asylum seekers in Turkey (ethnic Turks), which are reported to have migrated on a "mass scale" in 1992. BBC, *Summary of World Broadcasts*, EE/W0251 A/5, 8 October 1992. Similarly the figures exclude Greece, which may have been the destination of many of the 200,000 Albanians reported to have left the country.

TABLE 4.1.8

Total number of asylum-seekers originating from central and eastern Europe and arrived in western Europe, 1985-1992 [a]

(Thousands)

Country	Asylum-seekers
Albania	35.7
Bulgaria	53.4
Czechoslovakia	26.9
Hungary	24.9
Poland	148.5
Romania	265.5
USSR	17.6
Yugoslavia	414.6
Total	987.0
Memorandum item:	
Total asylum seekers	2 576.0

Source: Inter-governmental Consultations on Asylum, Refugee and Migration Policies in Europe, North America and Australia, *Asylum-seekers in Western Europe (EC and EFTA) in 1991*, Geneva, 18 December 1992 and direct communication to ECE secretariat.

[a] Reported by 13 west European countries.

Factors leading to shortfalls in assistance flows

The reasons for the shortfalls in commitments of assistance and the delays in their release are complex and vary from case to case. The following issues have arisen:

– The release of most committed funds to the eastern countries is conditional upon IMF approval of their economic programmes. The development of such programmes and others funded by other institutions takes time (especially in countries with no experience of formulating market-oriented programmes or obtaining their passage through the legislature). In some cases the process has been hindered by a lack of political commitment to reform or by of the simultaneous pursuit of counterproductive policies. In other cases armed conflicts have precluded any progress at all.

– Various commercial, legal and technical procedures or preconditions may delay disbursements: negotiation of terms, ratification procedures by one or both parties, guarantee of loans; settlement of arrears on outstanding loans, etc. The practice of tying loans to the purchase of narrowly-specified goods can also constrain disbursements.

– Some eastern countries simply have not received commitments of the bilateral assistance required to complement the resources provided by the international financial institutions. Western governments have faced budgetary pressures, and the pledging process has been hindered by the perception that the assistance burden has not been equitably shared among donors. Russia presents a special problem it that its assistance needs are particularly large.

– Capital flight has deprived certain transition countries of badly needed foreign exchange (e.g., Azerbaijan, Russia, Romania and others),[737] but it also discourages potential creditors from committing additional resources. The eastern countries affected have taken some steps to curb outflows, and in the case of Russia, to recover funds illegally transferred abroad, but the problem is still considerable. Appropriate macroeconomic policies and western technical assistance could yield high returnsin this respect.

Financing prospects for 1993 are bleak

Although comprehensive projections are lacking, it appears that the financial needs of the eastern countries in 1993 remain considerable, although there are large differences between them as regards prospective current account imbalances, reserves, and access to finance.

In most *east European* countries, the deterioration of current account balances in the second half of 1992 is expected to continue in 1993, and virtually all project deficits for the whole year (see section 3.3, including table 3.3.5). In Hungary, recent heavy borrowing in the international capital markets should cover a small current account deficit and refinance principal repayment falling due on its foreign debt. Financing is also not expected to pose a problem for the Czech Republic, where the current account is expected to be in balance. Poland, however, faces greater constraints. The country may have a much larger current account deficit (in 1992) than the first official estimates indicated. Even before this, the expectation of a further sharp deterioration in the current account (due in part to the drought) and a lack of access to financial markets had prompted the authorities to introduce a comprehensive package of measures to restrain the deficit, including a temporary tariff surcharge on most imports. Substantial payments to the country's official and private creditors are foreseen over the next few years.

Bulgaria and Romania will continue to require large amounts of international financial assistance in 1993. Both countries hope to finance current account deficits of over $1 billion and to strengthen their international reserves. As in the past two years, their financial plans also depend heavily on bilateral support which, as noted, fell short of the previous target. Furthermore, when agreement is reached on a badly-needed reduction in Bulgaria's bank debt, the country's assistance needs will have to be revised upward. Up-front costs will be large, and it is likely that bilateral funds beyond those currently under consideration will be required. At this stage, however, there remain considerable uncertainties about the availability of bilateral assistance for both countries. Hence, they will continue to lack virtually any room for manoeuvre.

---

[737] Although estimates of capital flight vary considerably, all of them indicate that significant amounts are involved. The IMF has estimated that $3-5 billion left the former Soviet Union illegally in 1991 and some $6 billion in the first half of 1992. In the case of Azerbaijan the flow is estimated at about $500 million in the first three quarters of 1992. IMF statement to consultative group on Azerbaijan. Paris, 14 December, 1992

Although comprehensive information on the financing needs of the *former Soviet Union* has not been released, the few figures available indicate that they will remain very large in 1993. Virtually all of the republics are entirely dependent on official financing and they lack adequate foreign currency reserves. A $30 billion financial package for *Russia* is reported to be under consideration. Presumably it includes bilateral and multilateral credits and debt relief. Some $2.5 billion in debt service payments is to be made by Russia. The mobilization of bilateral funds will require a determined effort on the part of the industrial market economies. For the time being, Russia has not made sufficient progress in reforming the economy and reducing macroeconomic imbalances to finalize a full stand-by agreement. In this regard, much will hinge on the course of political developments.

Another $10 billion is required in 1993 for the *other republics* of the former USSR. Although the situation varies from country to country, most continue to experience large shocks — deterioration in the terms of trade, loss of transfers and the disintegration of interrepublican trade. Balance of payments projections have identified financing gaps for most individual republics,[738] which can only partially be filled by the resources of the international economic institutions.

In these cases too, bilateral donors and creditors have been asked to cover the difference. It is important to bear in mind that the presentation of a request for a stand-by credit to the IMF Board is contingent upon the gap being closed through firm commitments by creditors and donors.

In the case of Kyrgyzstan, one republic for which some information is available, the IMF has drawn attention to the fact that the adjustment effort of the country has already been very substantial.[739] Failure to fill the financing gap would require a further adjustment, with severe adverse consequences for production and employment, and, ultimately, the success of the reform efforts. Moreover, there is an urgent need for generous assistance for several years. In view of the country's low income and limited near-term prospects for exporting outside the former USSR, it will be essential that much of this assistance be provided on highly concessional terms, including grants. The situation of other countries is likely to be similar, although they may not have progressed as far in preparing their reforms. In some countries, as was already the case in 1992, continuing military conflict make it difficult to imagine an effective assistance effort being put into place.

---

[738] For example, on the occasion of the meetings of consultative groups for four republics of the former Soviet Union (see section 4.1), the IMF cited its estimates of financing gaps for Kyrgyzstan ($400-500 million) and Kazakhstan (around $500 million). They assume that the republics have chosen the zero debt option, under which the Russian Federation has accepted responsibility for the entire debt of the former USSR. Not all of the republics are expected to be in deficit on current account. Azerbaijan, for example, is expected to remain in surplus.

[739] IMF Statement at the consultative group meeting on Kyrgyzstan, Paris, 15 December, 1992.

4.2 EAST-WEST COOPERATION AGREEMENTS AND MARKET ACCESS

Improving the transition countries' access to western markets has been one of the objectives of western policy makers since the international initiative to support their transition got under way. Initially, the reduction in western obstacles was looked upon to help promote these countries' systemic change, structural adjustment and integration into world markets, and to provide opportunities to quickly earn additional convertible currency. More recently, against the background of unexpectedly large declines in internal demand and CMEA trade, exports to the industrialized market economies have helped to contain the decline in output.

The exports of many transition countries to the west have grown in recent years at rates well above those of total western imports (see section 3.3). While there is little doubt that supply factors in eastern Europe have played a major role in this regard, the performance implies that western markets have been accessible and, perhaps, that they have recently become more open as well. None the less certain restrictive provisions still affect the transition countries' exports in the so called sensitive sectors.

Recent improvements in the transition countries' access to western markets have resulted from unilateral and bilateral policy actions, the latter including a host of economic agreements. The new policy actions are intended to promote fundamental change in the trade relations between the two areas. On the side of the eastern partners, this has involved the dismantling of an administered system of trade and replacing it with one based on market relations and the use of market instruments. On the western side, the actions involve the further elimination of measures which were designed to address the problems previously associated with state trading. The agreements differentiate between the transition countries, taking into account their degree of economic reform. In consequence, the trade policy framework governing east-west economic relations has been transformed since the process commenced in 1989.

The development of this east-west trade policy network is summarized in the matrix in table 4.2.1. Since it is the market-opening aspect of the various trade agreements that is of interest here, the dates provided are those on which the relevant trade measures entered, or were scheduled to enter, into force (as opposed to the dates on which the agreements were initialled or ratified).[740] Attention may be drawn to several developments:[741]

First, a further normalization of trade relations has occurred through the extension of MFN treatment, which now applies to the bulk of trade between the established market economies and the transition countries.

Second, most developed market economies have extended GSP benefits to the transition countries, although often only to the lesser developed ones.

Third, new types of agreement designed to upgrade and intensify economic cooperation between the two parts of the region have been completed or are under negotiation. This activity has also included the successor states of the former USSR and Yugoslavia. Western nations have developed new relations most quickly with the Baltic states and Slovenia. However, armed conflicts in a number of countries have ruled out agreements for the time being.

– *In the case of the EC,* this process has occurred at three levels. The trade and economic cooperation agreements with most east European countries have been replaced with *association agreements*, in the first instance with the former Czechoslovakia, Hungary and Poland. Interim agreements covering the trade provisions of those accords came into force on 1 March 1992.[742] The EC signed association agreements with Bulgaria and Romania at the end of 1992 and the respective interim accords were to come into force in July and May 1993, respectively. The first agreements between the EC and Albania, Estonia, Latvia, Lithuania, and Slovenia were concluded in 1992. They are *trade and economic cooperation agreements* of the type referred to above. As regards the twelve other republics of the former Soviet Union, the EC envisages relationships based on *partnership agreements*. These agreements are to include provisions which go beyond the traditional trade

---

[740] When these dates were not available to the secretariat a "yes" or "no" entry is shown, as appropriate.

[741] The trade provisions of the EC association, EFTA free trade, and other agreements are examined in more detail in United Nations Economic Commission for Europe, *Economic Survey of Europe in 1991-1992*, New York, 1992, p.187-189 and United Nations Economic Commission for Europe, *Economic Bulletin for Europe*, vol.44 (1992), New York, 1993, pp.72-74.

[742] Since the association agreements with Hungary and Poland have not yet been ratified, the terms of the respective interim agreements have been extended by the EC Council. For similar reasons the interim agreement of the former CSFR has been extended to the Czech Republic and Slovak Republic.

TABLE 4.2.1

Overview of east-west trade regimes: dates of entry into force

	Albania	Bulgaria	Czech and Slovak Republics a	Hungary	Poland	Romania	Ex-Yugoslavia b	Bosnia and Herzegovina	Croatia	Macedonia	Slovenia	Fed.Rep. Yugoslavia c	Ex-Soviet Union	Armenia
EC														
Most-favoured nation d	1.12.92	1.11.90	1.11.90	1.12.88	1.12.89	1.5.91	b	2.12.91	2.12.91	2.12.91	2.12.91	no	1.1.75	1.1.75
General system of preferences	1.1.92	1.1.91	1.1.91	1.1.89	1.1.89	1.1.74 e	b	1.1.92	1.1.92	1.1.92	1.1.92	no	no	1.1.93 f
Specific quantitative restrictions g	..	1.10.90	1.10.90	1.10.90	1.10.90	1.5.91	1.10.90	..	..	..	..	no	1.7.91	..
Quantitative restrictions g	..	1.10.90	1.10.90	1.10.90	1.10.90	1.5.91	1.10.90	..	..	..	..	no	..	j
TECA	1.12.92	1.11.90	1.11.90	1.12.88	1.12.89	1.5.91	b	2.12.91 h	2.12.91 h	2.12.91 h	5.11.92 h,i	no	1.4.90	..
Association agreement k	..	1.7.93	1.3.92	1.3.92	1.3.92	1.5.93	no	no	no	no	no	no	no	no
Partnership agreement	..													
EFTA														
Declaration of cooperation	10.12.92	10.12.91	13.6.90	13.6.90	13.6.90	10.12.91	b	no	no	no	20.5.92	no	17.10.55	no
Free trade agreement	no	1.7.93	1.7.92 m	1.7.93	1.4.93 n	1.5.93	no	no	no	no	no	no	1.4.93	1.4.93 q
Austria														
Most-favoured nation	20.8.92 o	18.8.74	yes	yes	yes	20.5.76	..	yes	yes	yes	yes	no	30.5.75 r	p.l
General system of preferences	1.1.92	1.4.72	1.7.91	1.7.88	1.1.90	1.4.72	..	1.5.92	1.3.92	1.9.92	1.3.92	no		
Free trade agreement	20.8.92 r,o	1.7.93 s	1.12.92 s	1.7.93 s	1.4.93 s,n	1.5.93 s			1.3.93 r	no	1.2.93 r	no		
Finland														
Most-favoured nation	yes	yes t	yes t	yes t	yes t	yes	yes	no	yes	no	yes	no	yes t	15.2.93
General system of preferences	1.12.90	4.7.73	no	no	no	10.12.73	1.1.72	no	1.5.93	no	1.5.93	no	no	no
Free trade agreement	6.9.86 r	1.7.93 s,u	1.12.92 s,u	1.7.93 s,u	1.4.93 s,u,n	1.5.93 s	8.5.85 r	no	no	no	10.6.92 r	no	1.12.47	no
Norway														
Most-favoured nation	yes	yes	yes	yes	yes	yes	yes	yes	yes	no	yes	no	yes	yes
General system of preferences	yes	yes	no	no	no	yes	..	8.5.92	24.1.92	no	24.1.92	no	no	no
Free trade agreement	no	1.7.93 s	1.7.92 s	1.7.93 s	1.4.93 s,n	1.5.93 s	..	no	no	no	no	no	no	no
Sweden														
Most-favoured nation	yes	yes	yes	yes	yes	yes	b	yes	yes	yes	yes	no	yes	no
General system of preferences	no	1.1.75	no	no	no	1.1.75	b	1.1.92	1.1.92	1.1.92	1.1.92	no	no	no
Free trade agreement	6.12.84 v,r	1.7.93 s	1.7.92 s	1.7.93 s	1.4.93 s,n	1.5.93 s	b	no	no	no	no	no	1.1.77 r	1.4.93 q
Switzerland														
Most-favoured nation	9.6.75	15.4.73	1.7.71	1.1.74	15.12.73	15.4.73	1.10.48	1.10.48	1.10.48	no	1.10.48	no	31.8.48	31.8.48
General system of preferences	1.1.92	9.10.81	no	no	no	9.10.81	9.10.81	1.9.92	1.9.92	1.9.92	1.9.92	no	no	no
Free trade agreement	28.10.74 r	1.7.93 s	1.7.92 s	1.7.93 s	1.4.93 s,n	1.5.93 s	1.10.48 p	1.10.48 p	1.10.48 p	no	1.10.48 p	no	31.8.48 p	31.8.48
Canada														
Most-favoured nation	no	7.1.74	20.4.48	9.9.73	18.10.67	14.11.71	25.12.66	25.12.66	25.12.66	25.12.66	25.12.66	25.12.66	29.2.56	29.2.56
General system of preferences	no	1974	13.12.91	1.12.89	1.12.89	1974	no w	no	no	no	10.4.92	no	10.4.92	10.4.92
Israel														
Most-favoured nation	no	10.5.92	yes	26.5.91	20.5.92	22.9.71	no	no	no	no	no	no	no	no
General system of preferences	no	no	no	no	no	no	no	no	no	no	no	no	no	no
Economic agreement	no	10.5.92	20.5.92	26.5.91	20.5.92	22.9.71	b	no	no	no	no	no	..	..
Japan														
Most-favoured nation	no	5.8.70	yes	9.9.76	26.10.80	19.7.70	20.7.59	no	yes x	no	yes x	..	9.5.58	yes x
General system of preferences	no	1.4.72	no y	1.4.86	1.1.90	1.4.72	1.8.71	no	no	no	no	..	no	no
Trade agreement v	30.8.88	5.8.70	17.10.92	9.9.76	26.10.80	19.7.70	20.7.59	no	1	no	1	..	9.5.58	1
Turkey														
Most-favoured nation	5.12.88	24.9.74	yes	yes	yes	yes	yes	no	no	no	no	no	8.10.37	no
General system of preferences	no	no	no	1.4.86	no	no	no	no	no	no	no	no	no	no
Trade agreement	12.2.86	11.6.74	29.8.75	12.11.74	23.4.74	27.10.70	14.4.71	no	no	no	no	no	no	..
United States														
Most-favoured nation	2.11.92	22.11.91	17.11.90	7.7.78	20.2.87 z	aa	yes	yes	yes	yes	yes	bb	no	4.2.92
General system of preferences	no	4.12.91	5.91	11.89	1.1.90	aa	yes	11.9.92	11.9.92	11.9.92	11.9.92	no	no	no
Trade agreement	26.8.92	yes		yes	yes	3.4.92	14.4.71							2.4.92 v

TABLE 4.2.1 (continued)

Overview of east-west trade regimes: dates of entry into force

	Azerbaijan	Belarus	Kazakhstan	Kyrgyzstan	Moldova	Russian Federation cc	Tajikistan	Turkmenistan	Uzbekistan	Ukraine	Georgia	Estonia	Latvia	Lithuania
EC														
Most-favoured nation	1.1.75	1.1.75	1.1.75	1.1.75	1.1.75	1.1.75	1.1.75	1.1.75	1.1.75	1.1.75	1.1.75	QII.93	1.2.93	1.2.93
General system of preferences	1.1.93 f	1.1.93 f	1.1.93 f	1.1.93 f	1.1.93 f	1.1.93 f	1.1.93 f	1.1.93 f	1.1.93 f	1.1.93 f	1.1.93 f	1.1.92	1.1.92	1.1.92
TECA	—	—	—	j	—	1.92 j	—	—	—	j	—	11.5.92 v	1.2.93	1.2.93
Partnership agreement	—	2.93 dd	3.93 dd	6.93 dd	—	11.92 dd	—	—	—	3.93 dd	1.1.93 ff	no	no	no
EFTA														
Declaration of cooperation	no	no	no	no	no	no	no	no	no	no	no	10.12.91	10.12.91	10.12.91
Free trade agreement	no	no	no	no	no	no	no	no	no	no	no	no	no	no
Austria														
Most-favoured nation	—	2.10.92 o	8.5.92 o	—	9.9.92 o	17.10.55	18.12.92 o	—	20.11.92 o	3.7.92 o	—	20.8.92 o	25.6.92 v	29.4.92 o
General system of preferences q	1.4.93	1.4.93	1.4.93	1.4.93	1.4.93	1.4.93	1.4.93	1.4.93	1.4.93	1.4.93	1.4.93	1.7.92	1.7.92	1.7.92
Economic agreement	—	2.10.92 o	8.5.92 o	—	9.9.92 o	dd	18.12.92 o	—	20.11.92 o	3.7.92 o	—	20.8.92 o,r	25.6.92 v,r	29.4.92 o,r
Finland														
Most-favoured nation	15.2.93	15.2.93	15.2.93	15.2.93	15.2.93	15.2.93	15.2.93	15.2.93	15.2.93	15.2.93	15.2.93	1.12.92 t	1.5.93 t	1.1.93 t
General system of preferences	no	no	no	no	no	no	no	no	no	no	no	no	no	no
Trade agreement	no	21.3.93	29.9.92 v	no	no	10.8.92	no	no	1.10.92 v	1.3.93	no	1.12.92 ee	1.5.93 ee	1.1.93 ee
Norway														
Most-favoured nation	yes	yes	yes	yes	yes	yes	yes	yes	yes	yes	yes	yes	yes	yes
General system of preferences	no	no	no	no	no	no	no	no	no	no	no	no	no	no
Free trade agreement	no	no	no	no	no	4.2.93 v,r	no	no	no	no	no	1.7.92 ff	1.7.92 ff	1.7.92 ff
Sweden														
Most-favoured nation	no	no	no	no	no	yes	no	no	no	no	no	28.10.91	28.10.91	29.10.91
General system of preferences	no	no	no	no	no	no	no	no	no	no	no	no	no	no
Free trade agreement	no	no	no	no	no	no	no	no	no	no	no	1.7.92	1.7.92	1.8.92
Switzerland														
Most-favoured nation	31.8.48	31.8.48	31.8.48	31.8.48	31.8.48	31.8.48	31.8.48	31.8.48	31.8.48	31.8.48	31.8.48	31.5.26	2.5.25	no
General system of preferences	no	no	no	no	no	no	no	no	no	no	no	no	no	no
Free trade agreement	no	22.1.93 o,gg	26.11.92 o,gg	no	no	31.8.48 p,l	no	no	30.11.92 o,gg	no	no	1.4.93	1.4.93	1.4.93
Canada														
Most-favoured nation	29.2.56	29.2.56	29.2.56	29.2.56	29.2.56	29.12.92	29.2.56	29.2.56	29.2.56	29.2.56	29.2.56	1.9.28	14.7.28	15.9.28
General system of preferences	27.11.92	27.11.92	27.11.92	27.11.92	27.11.92	10.4.92 hh	27.11.92	27.11.92	27.11.92	10.4.92	27.11.92	10.4.92	10.4.92	10.4.92
Israel														
Most-favoured nation	no	no	no	no	no	no	no	no	no	no	no	no	no	no
General system of preferences	no	no	no	no	no	no	no	no	no	no	no	no	no	no
Economic agreement	no	no	no	no	no	dd	no	no	no	no	no	no	no	no
Japan														
Most-favoured nation x	yes	yes	yes	yes	yes	9.5.58	yes	yes	yes	yes	yes	no	no	no
General system of preferences	no	no	no	no	no	no	no	no	no	no	no	no	no	no
Trade agreement	no	no	no	no	no	9.5.58	no	no	no	no	no	no	no	no
Turkey														
Most-favoured nation	yes	yes	no	no	yes	8.10.37	..	no	no	9.7.92	yes	no	no	yes
General system of preferences	no	no	no	no	no	no	..	no	no	no	no	no	no	no
Trade agreement v	1.11.92	22.1.93 o	26.9.91	23.12.91	19.6.92	12.3.91	..	3.12.91	19.12.91	4.5.92	30.7.92	no	no	27.8.92
United States														
Most-favoured nation	no	no	no	yes	19.6.92	yes	no	no	no	19.6.92	no	12.91	12.91	12.91
General system of preferences	no	no	no	26.9.91	no	no	no	no	no	no	no	22.2.92	22.2.92	22.2.92
Trade agreement	..	..	..	..	..	..	..	..	..	..	..	1.8.92 v	11.92 v	6.7.92 v

Source: Direct communications from governments to the ECE secretariat.

Note: Dates after 1 March 1993 are dates of scheduled or probable entry into force except as noted. GSP conditions may have been modified after the initial entry into force. For EC only, TECA = Trade and economic cooperation agreement and Partnership agreements = Partnership and cooperation agreements. Commonwealth of Independent States (CIS) = Armenia, Azerbaijan, Belarus, Kazakhstan, Kyrgyzstan, Moldova, the Russian Federation, Tajikistan, Turkmenistan, Uzbekistan and the Ukraine.

Footnotes: See following page.

FOOTNOTES TO TABLE 4.2.1

a Upon the separation of the CSFR on 1 January 1993, its western trade partners applied the provisions of existing agreements and preferences to the Czech and Slovak Republics as equal successors. The EC association and EFTA free trade agreements are being applied to both states on an interim basis. A protocol was scheduled to be signed 19 April 1993 extending the EFTA-CSFR agreement to the two successor states.
b Several western countries have taken formal actions concerning their agreements with the former Yugoslavia: the EC denounced the TECA of 2 April 1980 on 25 November 1991; Sweden terminated its trade agreement on 15 April 1992; Israel cancelled its extant economic agreement. EFTA suspended all cooperation under the Bergen Declaration (June 1983) in November 1991.
c Serbia and Montenegro are currently subject to a UN trade embargo.
d Reaffirmation of MFN status under the Trade and Economic Cooperation Agreements. The EC had extended MFN status to Hungary, Poland and Romania at the time of their accession to the GATT. Prior to that EC member states granted MFN status to these countries on a bilateral basis.
e Initially subject to restrictions.
f GSP granted on an "exceptional and temporary" basis.
g Actions involved removal of restrictions applying chiefly to "non-sensitive industrial products" undertaken as part of the G-24 initiative to support structural reform in the transition countries. See United Nations Economic Commission for Europe, Economic Bulletin for Europe, Vol.43(1991), New York, 1991, table 1.4.4 for relevant EC Council Decisions. Initially, removal of these restrictions was foreseen in the Trade and Economic Cooperation Agreements according to set timetables.
h Accepted obligations under former Yugoslavia-EC agreement; prolonged at 3 February 1992 and 21 December 1992.
i A new TECA was initialled on this date. Its trade provisions are those of former Yugoslavia-EC agreement, extended by an exchange of letters.
j Accepted obligations under the former USSR-EC agreement (also see text).
k Date of entry into force of Interim Agreement on Trade Provisions; following the separation of the CSFR into two independent countries, the EC Commission is seeking the approval of the EC Council of Ministers for draft negotiating directives for two separate agreements with the Czech and Slovak Republics. For Hungary, the Association Agreement was given assent by the European Parliament at 16 September 1992; Hungary ratified the agreement in November 1992, while ratification by all EC member countries is still required. For Poland, the Association Agreement was given assent by the European Parliament at 17 September 1992; Poland ratified the agreement at 6 July 1992, while ratification by all EC member countries is still required.
l Under consideration or negotiations on a new accord foreseen.
m First in effect in Norway, Sweden and Switzerland at 1 July 1992. All EFTA countries have agreed to apply the agreement on an interim basis as of 1 January 1993 while the final agreements with regard to the Czech and Slovak Republics are being discussed.
n Entry into force delayed.
o Date initialled.
p Economic agreement.
q Entry into force of preferences has been confirmed.
r Trade agreement.
s EFTA free trade agreement.
t Under free trade agreement, which provides for zero tariffs on industrial products and MFN rates on agricultural goods. See footnote u below.
u Agreements on the Reciprocal Removal of Obstacles to trade (free trade agreements) which entered into force 1 January 1975 (except for Poland, 1 April 1978) and will remain in effect.
v Date signed.
w GSP granted in 1974 and withdrawn 1 December 1991.
x De facto temporarily applied, except for Russian Federation.
y GSP applied 1 May 1992-1 January 1993.
z Reinstated, after suspension 1 November 1982.
aa Suspended November 1987. Originally granted through the wavier procedure 3 August 1975.
bb Suspended 20 October 1992. Originally received MFN status in 1967 upon accession to the GATT.
cc In may cases provisions of agreements with the former Soviet Union were extended to the Russian Federation.
dd Under negotiation.
ee Free trade agreement with the former Soviet Union was extended to the Baltic states following their declaration of independence in 1991. The arrangements were formalized through the Protocols Regarding Temporary Arrangements on Trade and Economic Cooperation which entered into force on these dates.
ff Provisional application.
gg Commercial and cooperation agreement.
hh Granted under agreement between Canada and the Russian Federation on Trade and Commerce, signed 29 December 1992.

---

and cooperation agreements currently in force,[743] but not as far as the association agreements. Negotiations with Russia, Belarus, Kazakhstan, and Ukraine have commenced and are scheduled with Kyrgyzstan (see table 4.2.1).

– EFTA has concluded *free trade agreements* with Bulgaria, the former Czechoslovakia, Hungary, Poland and Romania. Relations with Albania, Slovenia and the Baltic republics remain at the level of Declarations of cooperation, which do not include specific provisions for increased market access.

Fourth, some tendencies toward sub-regional cooperation are evident, e.g., in the free trade agreements concluded between the individual Nordic countries and the Baltic states.

In 1992, the transition countries continued to gain access to western markets. The major new improvements in market access involved Czechoslovakia, Hungary and Poland in the context of the association agreements with the EC. Also the free trade agreement between the former CSFR and EFTA became effective. Although the removal of obstacles agreed by EC is to progress over several years, the largest share of the concessions was implemented when the interim accord came into force on 1 March 1992 (see table 4.2.2). In the second year of the agreement (starting 1 March 1993), the contracted tariff reductions are smaller and concentrated on "non-sensitive" industrial products. No further tariffs cuts are scheduled in the second year in the textiles[744] and steel sectors. On the contrary, a

---

[743] These twelve republics of the former Soviet Union have accepted the obligations from the former EC-USSR trade and economic cooperation agreement through an Exchange of Letters. Only Azerbaijan and Uzbekistan have yet to initial the Exchange of Letters, while Russia does not wish to initial because it believes that there is automatic succession without confirmation or need for adjustment. Direct communication to the ECE secretariat from the EC Commission.

[744] Quantitative restrictions imposed by the EC on imports of textiles and clothing from the transition countries have been eased considerably over the past three years, most recently as part of the prolongation of the Fourth Multi-Fibre Agreement to 1992 and again during the negotiation of the Association Agreements. The number of restrained categories has been reduced and, more importantly, very large increases in quotas were granted for 1992, averaging about 200 per cent compared to the original MFA-IV quotas for 1991. Additional outward-processing quotas add a similar amount. See G. Pohl and P. Sorsa, *European*

TABLE 4.2.2
Improvements in access to the EC market under the EC association agreements with the former Czechoslovakia, Hungary and Poland
(Tariff reduction relative to the basic duty) [a]

	1 March 1992	1 March 1993
"Non-sensitive" industrial goods:	Quantitative restrictions are removed	-
Group 1:	Tariffs are removed completely	-
Group 2:	Tariffs are reduced by one half	Remaining half of tariffs is removed
Group 3:	Tariffs are reduced by 20 per cent of the basic rate	Tariffs are reduced by another 20 per cent
Group 4:	Miscellaneous tariff reductions [b]	-
Textiles-clothing:	Tariffs are reduced to 5/7ths of the basic rate	- [c]
Iron and steel:	Quantitative restrictions are removed	-
	Tariffs are reduced to 80 per cent of the basic rate	- [d]
Agriculture:	Duties are reduced on certain products, subject to quantitative limits.[f] Preferential access is given for some CAP products	- [e]

Source: Association agreements, which should be consulted for details.

[a] The basic duty considered for the reduction is the rate prevailing before the agreement entered into force.
[b] Tariffs on shares of certain product imports (determined by tariff quotas or increasing ceilings) were suspended.
[c] Further annual tariff reductions are scheduled to start in March 1995, by one seventh of the basic rate.
[d] Further, annual tariff reductions are to start in March 1994, initially at the rate of 20 per cent.
[e] Association agreements should be consulted.
[f] Above these limits Most-Favoured-Nation rates apply.

---

number of transition countries face new restrictions on their exports of certain steel products (see below).

In 1993, the market-opening process is to continue with the trade provisions of the EC association agreements with Bulgaria and Romania going into force. In the first year of these agreements, the benefits are similar to those enjoyed in 1992 by Czechoslovakia, Hungary and Poland. Following the separation of the CSFR into two independent countries, the EC Commission is seeking the approval of the EC Council of Ministers for draft negotiating directives for two separate Europe agreements with the Czech and Slovak Republics. Such agreements will be virtually identical to that concluded between the EC and the CSFR.[745]

Free trade agreements between EFTA countries and Bulgaria, Hungary, Poland and Romania are to come into force in the first seven months of 1993. The schedules for the reduction of tariff are similar to those incorporated in the EC association agreements (apart from agricultural products, the EFTA countries have set few quotas against the east European countries). With respect to the dissolution of the CSFR, the respective governments of the Czech and the Slovak Republics have declared that they assume the rights and obligations under the EFTA-CSFR free trade agreement. The EFTA states will apply the agreement on an interim basis until final arrangements for the application of the agreement can be agreed on.[746]

Is the tide turning?

When western countries took the initial steps in 1989 to improve market access for eastern goods, their domestic output and trade growth were buoyant. The favourable environment facilitated the acceptance in the west of such market opening measures, although in the sensitive sectors access was highly qualified. More recently, however, the developed market economies have slid into recession and trade has slowed. Historically, under such conditions, pressures for protection build-up, the present being no exception. Such pressures can be expected to strengthen in those western countries where economic activity weakens further.

Of all of the recent calls for relief from import competition, perhaps those from the steel sector have met with the most official response. Plagued by chronic and substantial excess capacity, increasing competition from mini-mills, and, recently, a downturn in demand, the sector has seen prices plunge and several major enterprises incur losses. In consequence the EC and the United States have taken actions against imports of selected products (see below). In western Europe, there has been a tendency to attribute much of the problem to the transition countries, even though it is officially recognized that these producers account only for a small (although growing) share of the western market.[747]

Currently, the steel sector holds a particular significance for the transition countries. Fostered by the

---

Integration and the Developing World, World Bank Policy and Research Series, No.21, Washington, D.C., 1992, pp.57-58. This study concludes that the east European countries will probably have substantial difficulties filling the huge increases in quotas granted already for 1992 (500 per cent, including outward processing, compared to the original MFA-IV quotas for 1991).

[745] EC Commission, *Overview, EC-East Europe,* Brussels, 29 January 1993.

[746] EFTA, *EFTA's New Partners,* December 1992.

[747] Sir Leon Brittan, *International Herald Tribune,* 25 March 1993.

specific development policies of the former regimes, iron and steel represents comparatively large shares of these countries' industrial output and exports, including those destined for the developed market economies. Given the capacities in place, steel products have been looked to for export growth, at least until domestic industrial restructuring — so far very slow — enables the transition countries to improve their product range. The collapse of the CMEA market, including that for arms, has made access to alternative outlets all the more important.

Policy measures taken in the west during the past year have led to some liberalization of steel trade. "Voluntary" export restraints (VERs) on steel expired in March 1992. Moreover, the association agreements between the EC and Czechoslovakia, Hungary and Poland led to the elimination of quantitative restrictions on steel from 1 March 1992. On that date EC tariffs on steel were reduced by 20 per cent of their basic level and further reductions are scheduled. However, the agreements contain safeguard clauses, in the event that a signatory finds that a product is being imported in such increased quantities and under such conditions as to cause, or threaten to cause, serious injury to domestic producers or serious disturbance in any sector of the economy.[748] A party which finds that dumping is taking place (within the meaning of GATT Article VI) may also take appropriate measures.[749]

During the past year western countries took a number of actions against steel imports from the transition countries. A surge in Czechoslovak steel exports in the early months of 1992 prompted the EC Commission to apply the safeguard clause under an emergency procedure against certain Czechoslovak steel products entering the French, German, and Italian markets.[750] The Commission determined that a 20 per cent limit on the growth of imports over 1991 levels was appropriate for 1992.[751]

In November 1992 the EC determined that several transition countries had engaged in *dumping* certain steel products and that injury to Community producers had occurred. On that basis the EC introduced provisional anti-dumping duties on seamless steel tubes of some 30 per cent on imports from Czechoslovakia, 22 per cent on those from Hungary, and 11 per cent on those from Poland.[752] The provisional duties were to be valid for four months. It may be added that the authorities of these three countries did not agree with this assessment and decision.[753]

As an element of the recent strategy adopted by the European Community to support community steel makers, the Czech Republic, Hungary, Poland and Slovakia will be asked to agree to tariff quotas on individual products which, according to the EC, were undermining EC markets. These are to be subject to annual review.[754] Quotas imposed in early 1993 on steel imported from the former Soviet Union are to be extended for 1994 and 1995.[755]

In January 1993 the United States Department of Commerce made a preliminary determination that 19 countries, among them Poland and Romania, had been dumping exports of flat-rolled carbon steel on the United States market. A final determination on whether dumping had occurred and whether there was injury to US industry remains to be made.[756]

The threat of new actions against imports from the transition countries in other sectors adversely affected by the recession must be taken seriously. As in the case of steel, other sectors face structural problems, falls in demand and lower prices. At the end of February, the European Commission set minimum import prices on white fish, to be in effect until end-June 1993. The measure was intended to stem the increase in imports from Iceland, Norway, Poland and Russia. However, according to Commission officials, if the situation were not stabilized quickly, the EC would resort to highly restrictive quotas on fish imports.[757] In response to increased Russian exports of aluminum, domestic producers have urged the EC to take restrictive measures. An imposition of quotas has been under consideration, which would essentially rule out new purchases of aluminum from Russia this year.[758]

In resisting the pressure from sectoral interests, western authorities need to stress that western nations and their exporters have already benefited significantly

---

[748] Article 30 of the association agreements.

[749] Article 29 of the association agreements.

[750] The EC decisions and the changes in the level of Czechoslovak exports to which the EC reacted are given in the United Nations Economic Commission for Europe, *Economic Bulletin for Europe*, vol.44 (1992), New York, 1993, p.72.

[751] *Council Decision* 92/434 ECSC. Subsequently the Dutch authorities called for the application of the safeguard clause throughout the Community to prevent Czechoslovak exports from entering the EC through member states not currently covered by the action. *Agence Europe*, 21 September 1992.

[752] The anti-dumping inquiry opened in December 1991. The action taken by the Commission against Croatia was suspended. *Agence Europe*, 19 November 1992.

[753] For example, Mr. Vladimir Petr, Director of the steel division of the Czech Ministry of Industry, contested the EC figures on the basis of which the determination was made and argued that the case for dumping was "not proven". *Financial Times*, 19 November 1992.

[754] This solution was adopted in preference to the negotiation of three-year tariff or volume restrictions on the imports from eastern Europe, as had been suggested by the EC Commission. *Financial Times*, 26 February 1993.

[755] Ibid.

[756] *Bulletin of the United States Mission (Geneva)*, 28 January 1993. Until the final determinations, the US Customs Service will collect cash deposit or bond on such imports. In the event of a negative determination, the money will be returned.

[757] *Financial Times*, 25 and 26 February 1993.

[758] *Financial Times*, 25 February 1993.

from the transformed relations with the transition countries. It was pointed out in the *Economic Bulletin for Europe* that Czechoslovakia, Hungary and Poland has raised their western imports by $11.5 billion between 1990 to 1992. This exceeds the $10.5 billion increase in the exports of these three countries to western markets.[759] Liberalization of markets in both parts of the region and international financial assistance for the three countries have thus yielded mutual benefits. There are signs that other transition countries will allow further increases in their western imports once their convertible currency constraint is loosened.

---

[759] United Nations Economic Commission for Europe, *Economic Bulletin for Europe*, vol.44 (1992), New York, 1993.

STATISTICAL APPENDICES

INTRODUCTORY NOTE

For the user's convenience, as well as to lighten the text, the *Economic Survey of Europe* includes a set of appendix tables showing annual changes in main economic indicators over a longer period. The data are presented in three sections, following the structure of the text: *Appendix A* provides macroeconomic indicators for the market economies of western Europe and North America for 1970-1992, *Appendix B* does the same for the east European countries and the successor states of the former Soviet Union for 1990 and 1991 and *Appendix C* collates time series on world trade and the development of foreign trade of the ECE economies between 1970 and 1991.

Except where otherwise stated, time series reflect levels or changes in *real* terms, i.e., at constant prices in case of series measured in value terms.

Data were compiled from international (United Nations, OECD, CMEA) or national statistical sources, as indicated in the notes to individual tables. *Regional aggregations* are ECE secretariat calculations, based on 1985 US dollar weights. All figures for 1992 are preliminary estimates, based on data available in the first weeks of March 1993.

Appendix A. Western Europe and North America

Data for this section were compiled from national and international [760] statistical sources, as indicated in the notes to individual tables. Volume figures underlying the data in tables A.1-A.6 reflect data at constant prices of the following years: Greece (1970), Yugoslavia (1972), Denmark, France, Netherlands, Switzerland (1980); Turkey (1982); Austria (1983), Belgium, Finland, Germany, Ireland, Italy, Portugal, Sweden, United Kingdom (1985); Spain, Canada (1986); United States (1987). Regional data in tables A.1-A.6, A.8 were aggregated from index series at 1985 US dollar exchange rates. Country data in tables A.9-A.10 reflect percentage changes expressed in national currencies.

---

[760] UN, OECD, EUROSTAT, IMF.

APPENDIX TABLE A.1
Gross domestic product
(Annual percentage change)

	1970	1975	1976	1977	1978	1979	1980	1981	1982	1983	1984	1985	1986	1987	1988	1989	1990	1991	1992
France	5.7	-0.3	4.2	3.2	3.4	3.2	1.6	1.2	2.5	0.7	1.3	1.9	2.5	2.3	4.5	4.1	2.2	1.2	1.8
Germany [a]	5.1	-1.4	5.3	2.8	3.0	4.1	1.1	0.2	-0.9	1.6	2.8	1.9	2.2	1.4	3.7	3.4	5.1	3.7	1.5
Italy	5.3	-2.7	6.6	3.4	3.7	6.0	5.0	0.6	0.2	1.0	2.7	2.6	2.9	3.1	4.1	2.9	2.2	1.4	1.2
United Kingdom	3.1	-0.8	2.8	2.3	3.6	2.8	-2.1	-1.3	1.7	3.7	2.3	3.8	4.1	4.8	4.4	2.1	0.5	-2.2	-0.6
Total 4 countries	4.8	-1.2	4.7	2.9	3.4	4.0	1.3	0.2	0.8	1.7	2.3	2.5	2.9	2.8	4.2	3.2	2.7	1.2	1.0
Austria	7.1	-0.4	4.6	4.5	0.1	4.7	2.9	-0.3	1.1	2.0	1.4	2.5	1.2	1.7	4.1	3.8	4.6	3.0	1.8
Belgium	6.4	-1.5	5.6	0.5	2.7	2.1	1.8	-1.0	1.5	0.4	2.1	0.8	1.5	2.2	4.9	3.6	3.8	2.1	1.2
Denmark	2.0	-0.7	6.5	1.6	1.5	3.5	-0.4	-0.9	3.0	2.5	4.4	4.3	3.6	0.3	1.2	0.6	2.0	1.3	1.0
Finland	7.5	1.2	0.3	0.1	2.2	7.3	5.3	1.6	3.6	3.0	3.1	3.3	2.1	4.0	5.4	5.4	0.3	-6.4	-3.5
Ireland	2.7	5.7	1.4	8.2	7.2	3.1	3.1	3.3	2.3	-0.2	4.4	3.1	-0.5	4.6	4.5	6.4	7.1	1.7	2.7
Netherlands	5.7	-0.1	5.1	2.3	2.5	2.4	0.9	-0.7	-1.4	1.4	3.1	2.7	2.0	0.9	2.6	4.7	3.9	2.2	1.7
Norway	2.0	4.2	6.8	3.6	4.5	5.1	4.2	0.9	0.3	4.6	5.7	5.3	4.2	2.0	-0.5	0.6	1.8	1.9	2.9
Sweden	6.5	2.6	1.1	-1.6	1.8	3.8	1.7	-	1.1	1.8	4.0	2.2	2.2	2.8	2.3	2.4	0.7	-1.8	-1.2
Switzerland	6.4	-7.3	-1.4	2.4	0.4	2.5	4.6	1.5	-0.9	1.0	1.8	3.7	2.9	2.0	2.9	3.9	2.3	-0.1	-0.6
Total 9 countries	5.6	-0.6	3.3	1.7	2.0	3.5	2.4	0.1	0.7	1.8	3.1	2.9	2.3	2.0	2.9	3.4	2.7	0.5	0.6
Total western Europe	5.0	-1.0	4.3	2.6	3.0	3.9	1.5	0.2	0.8	1.7	2.5	2.6	2.7	2.6	3.9	3.2	2.7	1.0	0.9
Greece	8.0	6.1	6.4	3.4	6.7	3.7	1.8	0.1	0.4	0.4	2.8	3.1	1.6	-0.7	4.1	3.5	-0.2	1.8	1.2
Portugal	9.1	-4.3	6.9	5.6	2.8	5.6	4.6	1.6	2.1	-0.2	-1.9	2.8	4.1	5.3	3.9	5.2	4.1	2.0	1.9
Spain	4.1	0.5	3.3	3.0	1.4	-0.1	1.2	-0.2	1.2	1.8	1.8	2.3	3.2	5.6	5.2	4.8	3.7	2.3	1.2
Turkey	4.9	7.5	8.7	4.3	2.8	-0.9	-0.7	4.4	5.0	3.7	5.7	5.1	8.3	7.4	3.6	1.9	9.2	0.3	5.4
Total southern Europe	5.0	1.9	4.8	3.5	2.4	0.6	1.2	0.7	1.8	1.8	2.3	3.0	4.1	5.2	4.6	4.1	4.4	1.8	2.1
Total Europe	5.0	-0.8	4.4	2.7	3.0	3.6	1.5	0.2	0.9	1.7	2.5	2.6	2.9	2.8	3.9	3.3	2.9	1.1	1.0
United States	-	-0.8	4.9	4.5	4.8	2.5	-0.5	1.8	-2.2	3.9	6.2	3.2	2.9	3.1	3.9	2.5	0.8	-1.2	2.1
Canada	2.6	2.6	6.2	3.6	4.6	3.9	1.5	3.7	-3.2	3.2	6.3	4.8	3.3	4.2	5.0	2.3	-0.5	-1.7	0.9
North America	0.1	-0.6	5.0	4.4	4.8	2.6	-0.4	1.9	-2.2	3.8	6.2	3.3	2.9	3.2	4.0	2.5	0.7	-1.2	2.0
Total above	2.1	-0.7	4.8	3.7	4.1	3.0	0.4	1.2	-1.0	2.9	4.7	3.0	2.9	3.0	4.0	2.8	1.6	-0.3	1.6
Cyprus	..	..	..	14.4	8.2	11.0	5.9	3.1	6.3	5.3	8.8	4.7	3.8	7.0	8.5	8.3	6.5	1.0	..
Israel	8.1	4.2	1.6	1.9	4.0	4.5	3.1	4.5	1.1	2.8	2.3	3.8	3.7	5.9	2.7	1.6	5.1	5.9	6.4
Malta	..	..	..	12.2	11.2	10.5	7.0	3.3	2.3	-0.6	0.9	2.6	4.0	4.2	8.7	..	..	..	..
Yugoslavia, SFR [b]	5.6	3.6	3.9	8.0	6.9	7.0	2.3	1.4	0.5	-1.0	2.0	0.5	3.6	-1.1	-2.0	1.0	-8.5	..	..

Sources: National statistics.

[a] Data refer to west Germany only.
[b] Gross material product.

APPENDIX TABLE A.2
Private consumption
(Annual percentage change)

	1970	1975	1976	1977	1978	1979	1980	1981	1982	1983	1984	1985	1986	1987	1988	1989	1990	1991	1992
France	4.3	2.8	4.9	2.7	3.7	3.0	1.2	2.1	3.5	0.9	1.1	2.4	3.9	2.9	3.3	3.3	2.9	1.5	1.9
Germany [a]	7.6	3.1	3.9	4.5	3.6	3.3	1.2	-0.8	-1.5	1.3	1.6	1.5	3.4	3.3	2.7	2.7	5.4	3.6	0.9
Italy	8.4	0.3	5.2	4.1	3.4	7.3	5.6	1.5	1.2	0.7	2.0	3.0	3.7	4.2	4.2	3.5	2.8	2.8	1.7
United Kingdom	2.9	-0.4	0.4	-0.4	5.6	4.3	0.1	0.1	1.0	4.5	1.9	3.8	6.4	5.5	7.4	3.3	0.7	-2.0	-0.3
Total 4 countries	5.7	1.6	3.6	2.8	4.0	4.3	1.9	0.6	0.9	1.8	1.6	2.6	4.3	3.9	4.3	3.2	3.0	1.5	1.0
Austria	4.2	3.2	4.5	5.4	-1.5	4.4	1.5	0.3	1.2	5.0	-0.1	2.4	1.8	3.1	3.6	3.5	3.8	2.4	2.5
Belgium	4.4	0.6	4.8	2.4	2.3	4.8	-0.7	-1.1	1.3	-1.7	1.2	2.0	2.4	3.2	3.1	3.3	2.6	2.7	2.2
Denmark	3.5	3.7	7.9	1.1	0.7	1.4	-3.7	-2.3	1.4	2.6	3.4	5.0	5.7	-1.5	-1.0	-0.4	0.6	1.2	1.5
Finland	7.6	3.1	0.9	-1.2	2.5	5.5	2.0	1.7	4.7	2.6	2.7	3.2	4.1	5.7	5.0	4.2	0.2	-3.7	-5.3
Ireland	-1.0	0.8	2.8	6.8	9.1	4.4	0.4	1.7	-7.1	0.9	2.0	4.6	2.1	2.3	3.6	3.7	1.1	0.5	3.5
Netherlands	7.4	3.3	5.3	4.6	4.3	3.0	-	-2.5	-1.2	0.9	0.9	2.5	2.7	3.2	0.8	3.5	4.1	3.3	1.7
Norway	-	5.1	6.1	6.9	-1.6	3.2	2.3	1.1	1.8	1.5	2.7	9.9	5.6	-1.0	-2.8	-2.8	2.9	-0.3	1.5
Sweden	3.5	2.8	4.2	-1.0	-0.7	2.4	-0.8	-0.5	0.7	-2.2	1.7	2.8	5.2	4.6	2.5	1.4	-0.3	1.0	-1.5
Switzerland	5.4	-2.9	1.1	3.0	2.2	1.3	2.6	0.4	-	1.7	1.6	1.4	2.8	2.1	2.1	2.2	1.5	1.5	-0.3
Total 9 countries	4.6	1.9	4.2	2.7	1.7	3.1	0.3	-0.6	0.5	0.9	1.6	3.2	3.5	2.7	1.8	2.3	2.1	1.4	0.6
Total western Europe	5.4	1.7	3.7	2.8	3.5	4.0	1.5	0.3	0.8	1.6	1.6	2.7	4.1	3.6	3.7	3.0	2.8	1.5	0.9
Greece	8.8	5.5	5.3	4.6	5.7	2.6	0.2	2.0	3.9	0.3	1.7	3.9	0.7	1.0	3.5	4.3	2.0	1.2	1.3
Portugal	2.6	-0.9	3.5	0.6	-2.0	-	3.7	2.9	2.4	-1.4	-2.9	0.7	5.6	5.4	6.6	3.3	4.9	4.9	4.0
Spain	4.2	1.8	5.6	1.5	0.9	1.3	0.6	-0.6	0.2	0.3	-0.4	2.4	4.1	5.8	4.8	5.6	3.7	3.1	2.6
Turkey	2.2	7.7	10.1	6.7	-3.9	-3.1	-5.2	0.6	4.2	5.0	6.8	1.3	11.5	6.5	2.6	3.9	10.4	2.1	10.6
Total southern Europe	4.2	3.1	6.3	2.8	0.2	0.4	-0.4	0.2	1.6	1.1	1.1	2.2	5.4	5.4	4.3	4.9	5.1	2.8	4.4
Total Europe	5.3	1.8	4.0	2.8	3.1	3.6	1.3	0.3	0.9	1.5	1.5	2.7	4.2	3.8	3.8	3.2	3.1	1.6	1.3
United States	2.4	2.1	5.2	4.0	4.1	2.4	-0.1	1.2	1.1	4.6	4.8	4.4	3.6	2.8	3.6	1.9	1.2	-0.6	2.2
Canada	2.0	4.7	6.5	3.2	3.4	2.9	2.2	2.3	-2.6	3.4	4.6	5.2	4.4	4.4	4.5	3.2	0.9	-1.7	1.0
North America	2.4	2.3	5.3	4.0	4.1	2.4	0.1	1.3	0.8	4.5	4.8	4.4	3.7	2.9	3.7	2.0	1.1	-0.7	2.1
Total above	3.5	2.1	4.8	3.5	3.7	2.9	0.6	0.9	0.8	3.3	3.5	3.7	3.9	3.2	3.7	2.5	1.9	0.2	1.8
Cyprus	..	..	..	19.8	11.5	6.8	5.7	1.7	11.5	7.3	4.5	5.9	1.7	5.5	10.5	6.9	8.2	6.2	..
Israel	..	0.3	4.8	4.8	8.1	8.0	-2.9	13.0	8.0	8.7	-7.0	0.7	14.8	9.0	4.2	0.3	5.0	7.6	7.9
Malta	..	..	..	14.9	4.4	3.0	6.2	3.3	2.9	1.4	4.0	5.0	2.8	1.2	10.6	..	..	..	..
Yugoslavia, SFR	14.6	3.4	4.4	7.0	7.0	5.6	0.7	-1.0	-0.1	-1.7	-0.9	-0.1	4.5	0.3	-1.3	1.0	6.0	..	..

Sources: National statistics.

[a] Data refer to west Germany only.

APPENDIX TABLE A.3
Public consumption
(Annual percentage change)

	1970	1975	1976	1977	1978	1979	1980	1981	1982	1983	1984	1985	1986	1987	1988	1989	1990	1991	1992
France	4.2	4.4	4.2	2.4	5.2	3.0	2.5	3.1	3.8	2.1	1.2	2.3	1.7	2.8	3.4	0.2	1.8	2.9	2.0
Germany [a]	4.3	3.9	1.6	1.4	3.3	3.4	2.6	1.8	-0.9	0.2	2.5	2.1	2.5	1.5	2.2	-1.7	2.4	0.5	2.5
Italy	4.4	2.5	2.3	2.8	3.5	3.0	2.6	2.3	2.6	3.4	2.3	3.4	2.6	3.4	2.8	0.8	1.3	1.7	0.7
United Kingdom	1.5	5.6	1.2	-1.7	2.3	2.0	1.6	0.3	0.8	2.0	1.0	–	1.8	1.2	0.6	0.9	3.2	2.7	0.5
Total 4 countries	3.5	4.2	2.2	1.0	3.7	2.9	2.3	1.8	1.3	1.7	1.7	1.8	2.1	2.1	2.2	-0.1	2.2	1.9	1.5
Austria	3.3	4.0	4.3	3.3	3.3	3.0	2.7	2.2	2.3	2.2	0.2	1.9	1.7	0.4	0.3	0.8	1.2	2.6	1.5
Belgium	3.1	4.5	3.7	2.3	6.0	2.5	-3.7	0.3	-1.4	0.1	0.2	2.4	1.7	0.3	-1.0	-0.8	0.9	1.1	-0.1
Denmark	6.9	2.0	4.5	2.4	6.2	5.9	4.3	2.6	3.1	–	-0.4	2.5	0.5	2.5	0.9	-0.6	-0.4	-0.2	0.7
Finland	5.4	6.9	5.7	4.2	4.1	3.8	4.2	4.3	3.5	3.7	2.8	5.2	3.1	4.5	2.4	2.6	4.4	2.3	-0.1
Ireland	11.3	8.7	2.6	2.1	7.9	4.6	7.1	0.3	3.2	-0.4	-0.7	1.8	2.5	-4.4	-5.3	-2.0	3.5	1.7	1.7
Netherlands	6.0	4.1	4.1	3.4	3.9	2.8	0.6	2.0	0.7	1.2	-0.7	1.4	2.0	2.1	1.4	1.5	2.1	1.6	0.7
Norway	6.3	6.4	7.4	4.9	5.3	3.5	5.4	6.1	3.9	4.6	2.4	3.3	2.2	4.0	0.5	2.6	2.0	2.3	2.7
Sweden	8.1	4.7	3.5	3.0	3.3	4.7	2.2	2.3	1.0	0.8	2.3	2.4	1.4	1.0	0.6	1.9	2.9	0.2	0.9
Switzerland	4.8	0.7	2.7	0.5	2.0	1.1	0.9	2.5	1.1	3.9	1.2	3.3	3.7	1.8	4.3	4.1	4.7	2.8	2.5
Total 9 countries	5.9	4.1	4.1	2.9	4.3	3.6	1.9	2.5	1.6	1.6	0.9	2.6	1.9	1.7	1.0	1.4	2.3	1.4	1.1
Total western Europe	4.1	4.2	2.7	1.5	3.9	3.0	2.2	2.0	1.4	1.7	1.5	2.0	2.1	2.0	1.9	0.3	2.3	1.8	1.4
Greece	5.9	11.9	5.1	6.5	3.5	5.8	0.2	6.8	2.3	2.7	3.0	3.2	-0.8	0.9	4.7	4.2	0.6	-0.7	0.7
Portugal	7.0	6.6	7.0	11.8	3.3	6.5	7.9	5.2	3.6	3.7	0.1	0.1	7.2	4.9	5.3	2.8	1.3	3.0	3.1
Spain	5.2	5.2	6.9	3.9	5.4	4.2	4.2	1.9	4.9	3.9	2.9	4.6	5.8	8.9	4.0	8.3	5.7	4.2	3.7
Turkey	3.6	13.4	10.8	3.2	9.9	1.7	8.8	0.9	2.0	1.7	2.1	3.1	6.5	5.0	3.0	4.4	17.0	0.9	1.9
Total southern Europe	5.3	7.3	7.0	4.9	5.4	4.4	4.3	2.9	3.9	3.4	2.6	3.8	4.8	6.8	4.1	6.7	5.8	3.0	3.0
Total Europe	4.1	4.4	2.9	1.7	3.9	3.1	2.4	2.0	1.5	1.8	1.6	2.1	2.3	2.4	2.1	0.8	2.5	1.9	1.6
United States	-2.7	1.2	-0.6	0.7	1.9	1.8	2.2	1.3	1.5	2.8	3.1	6.1	5.2	3.1	0.6	2.0	2.8	1.2	-0.3
Canada	9.4	6.5	2.0	4.6	1.7	0.6	2.8	2.5	2.4	1.4	1.2	3.2	1.6	1.7	4.1	3.7	2.9	1.9	1.6
North America	-2.0	1.6	-0.4	1.1	1.9	1.7	2.2	1.4	1.5	2.7	2.9	5.8	4.9	2.9	0.9	2.1	2.8	1.3	-0.1
Total above	0.1	2.7	0.9	1.3	2.7	2.3	2.3	1.7	1.5	2.3	2.4	4.3	3.8	2.7	1.4	1.6	2.7	1.5	0.5
Cyprus	⋮	⋮	⋮	0.3	-2.9	4.1	8.6	7.4	3.6	6.1	3.8	3.7	3.6	5.9	7.0	3.4	7.3	6.8	⋮
Israel	⋮	10.2	-9.7	-13.4	8.4	-8.7	9.1	6.5	-6.2	-4.0	5.9	3.8	-9.5	17.1	-1.7	-8.8	3.7	4.6	0.6
Malta	⋮	⋮	⋮	1.3	10.2	10.4	1.8	7.5	6.0	-1.1	-2.5	5.6	4.4	9.1	6.0	⋮	⋮	⋮	⋮
Yugoslavia, SFR	9.3	9.3	9.5	7.4	6.5	4.5	-1.0	-4.8	-0.7	-5.1	-0.2	-0.2	6.8	-1.5	2.5	-2.5	-2.0	⋮	⋮

Sources: National statistics.

[a] Data refer to west Germany only.

APPENDIX TABLE A.4
Gross domestic fixed capital formation
(Annual percentage change)

	1970	1975	1976	1977	1978	1979	1980	1981	1982	1983	1984	1985	1986	1987	1988	1989	1990	1991	1992
France	4.6	-6.4	3.3	-1.8	2.1	3.1	2.6	-1.9	-1.4	-3.6	-2.6	3.2	4.5	4.8	9.6	7.0	2.9	-1.2	-1.9
Germany [a]	9.2	-5.2	3.7	3.8	4.3	6.9	2.3	-4.9	-5.3	3.3	0.3	-	3.6	2.1	4.6	6.5	8.7	6.5	1.7
Italy	3.0	-7.3	-	1.8	0.6	5.7	9.1	-3.1	-4.7	-0.6	3.6	0.6	2.2	5.0	6.9	4.3	3.3	0.9	-0.1
United Kingdom	2.5	-2.0	1.7	-1.8	3.0	2.8	-5.4	-9.6	5.4	5.0	8.5	4.0	2.4	9.6	14.2	7.2	-3.1	-10.1	-2.0
Total 4 countries	5.3	-5.4	2.4	0.7	2.6	4.8	2.4	-4.5	-2.3	0.8	1.8	1.8	3.3	5.0	8.5	6.3	3.3	-0.2	-0.3
Austria	9.8	-5.0	3.8	5.1	-4.1	3.5	3.0	-1.4	-8.2	-0.6	2.1	5.0	3.7	3.1	6.0	6.1	5.8	4.9	2.3
Belgium	8.4	-1.9	4.0	-	2.8	-2.7	4.6	-16.1	-1.7	-4.4	1.7	0.7	4.4	5.6	15.2	14.5	8.3	0.3	1.4
Denmark	2.2	-12.4	17.1	-2.4	1.1	-0.4	-12.6	-19.2	7.1	1.9	12.9	12.6	17.1	-3.8	-6.6	1.0	-0.9	-4.2	-10.0
Finland	12.5	5.9	-8.8	-3.5	-6.9	3.0	10.4	2.2	4.4	4.1	-2.1	2.9	-3.1	5.4	10.5	14.1	-4.9	-19.8	-14.9
Ireland	-3.3	-3.6	13.6	4.1	18.9	13.6	-4.7	9.5	-3.4	-9.3	-2.5	-7.7	-3.1	-2.3	3.3	15.8	9.5	-5.5	0.5
Netherlands	7.5	-4.4	-2.2	9.7	2.5	-1.7	-0.9	-10.4	-4.1	2.1	5.4	6.8	7.4	1.1	4.5	4.9	3.6	0.1	0.9
Norway	14.9	11.9	10.1	3.6	-11.2	-5.0	-1.5	17.9	-11.0	5.8	10.9	-13.9	23.9	-2.1	1.6	-3.9	-26.6	1.0	5.6
Sweden	3.3	3.1	1.9	-2.9	-6.8	4.5	3.5	-5.8	-0.3	1.9	6.0	7.3	0.7	7.6	5.8	11.8	-0.5	-8.3	-8.7
Switzerland	8.9	-13.6	-10.5	1.6	6.1	5.1	9.9	2.4	-2.5	4.1	4.1	5.3	7.9	7.4	6.9	5.8	2.6	-2.5	-6.7
Total 9 countries	7.5	-2.9	1.1	2.1	-1.1	1.1	2.0	-3.7	-2.8	1.5	4.6	3.1	7.2	3.2	5.6	7.1	-0.3	-3.4	-3.3
Total western Europe	5.9	-4.8	2.0	1.1	1.7	3.9	2.3	-4.3	-2.4	1.0	2.5	2.1	4.3	4.5	7.7	6.5	2.4	-1.0	-1.0
Greece	-1.4	0.2	6.8	7.8	6.0	8.8	-6.5	-7.5	-1.9	-1.3	-5.7	5.2	-6.2	-5.1	8.9	10.0	4.8	-2.0	0.6
Portugal	11.5	-11.3	0.8	12.0	6.2	-1.3	8.5	5.5	2.3	-7.1	-17.4	-3.5	10.9	15.1	15.0	5.6	6.0	2.6	1.8
Spain	3.0	-4.5	-0.8	-0.9	-2.7	-4.4	0.7	-3.3	0.5	-2.5	-5.8	4.1	10.1	14.0	14.0	13.8	6.9	1.6	-2.7
Turkey	13.5	24.7	17.7	3.9	-10.0	-3.6	-10.0	1.7	3.5	3.0	0.5	16.7	11.0	5.6	-1.2	-1.0	14.0	-0.4	1.3
Total southern Europe	4.3	-0.8	3.2	2.0	-2.3	-2.3	-1.4	-2.1	0.9	-2.0	-5.9	5.9	8.3	10.3	10.5	10.0	7.8	1.0	-1.3
Total Europe	5.7	-4.4	2.1	1.1	1.2	3.3	2.0	-4.1	-2.1	0.7	1.7	2.4	4.6	5.0	8.0	6.9	3.0	-0.8	-1.1
United States	-2.9	-11.8	9.7	14.3	10.8	4.6	-8.1	0.6	-8.0	6.6	15.9	5.0	0.4	-0.5	4.2	0.1	-2.8	-8.5	5.6
Canada	0.3	5.8	4.6	2.1	3.1	8.5	10.1	11.8	-11.0	-0.7	2.1	9.5	6.2	10.8	10.3	6.0	-3.9	-3.7	-0.4
North America	-2.7	-10.3	9.2	13.2	10.2	4.9	-6.6	1.7	-8.3	5.9	14.5	5.4	0.9	0.6	4.8	0.8	-2.9	-8.0	5.0
Total above	1.6	-7.3	5.5	7.1	5.9	4.1	-2.7	-1.1	-5.5	3.4	8.6	4.1	2.5	2.5	6.3	3.5	-0.2	-4.5	1.9
Cyprus	13.6	5.3	..	56.2	25.5	20.9	5.2	-6.5	3.8	-2.2	21.5	-7.7	-7.1	4.5	7.9	23.0	-6.1	1.0	..
Israel	..	..	-12.1	-7.5	2.3	11.9	-13.3	-5.2	14.8	10.0	-7.1	-10.6	9.5	2.5	-0.4	-5.3	25.2	43.4	8.1
Malta	..	..	..	6.7	-3.3	11.4	0.8	17.2	14.7	15.8	-7.0	-4.0	-8.7	30.7	6.1	-	..	..	..
Yugoslavia, SFR	12.8	9.7	8.1	9.5	10.5	6.4	-5.9	-9.8	-5.5	-9.7	-9.6	-3.7	3.5	-3.9	-4.1	-	-18.0	..	..

Sources: National statistics.

[a] Data refer to west Germany only.

APPENDIX TABLE A.5
Volume of exports of goods and services
(Annual percentage change)

	1970	1975	1976	1977	1978	1979	1980	1981	1982	1983	1984	1985	1986	1987	1988	1989	1990	1991	1992
France	16.1	-1.7	8.2	7.4	5.9	7.5	2.7	3.7	-1.7	3.7	7.0	1.9	-1.4	3.1	8.1	10.2	5.5	3.6	4.7
Germany [a]	5.5	-6.4	9.6	3.9	2.9	4.3	5.2	7.2	3.7	-0.7	8.2	7.6	-0.6	0.4	5.5	10.1	10.4	12.8	3.4
Italy	-2.5	1.6	10.5	9.9	9.0	8.5	-8.1	7.9	-2.4	2.3	8.5	3.2	2.5	4.7	5.4	8.8	7.8	-0.8	4.5
United Kingdom	5.5	-2.8	9.1	6.9	1.9	3.8	-0.2	-0.7	0.8	2.0	6.6	5.9	4.7	5.6	-0.1	3.8	4.9	0.2	3.2
Total 4 countries	6.1	-3.2	9.3	6.4	4.3	5.6	0.9	4.5	0.7	1.5	7.6	5.1	1.0	3.0	4.6	8.4	7.5	5.5	3.8
Austria	16.4	-2.4	11.1	4.2	7.3	11.7	5.2	4.9	2.7	3.2	6.1	6.9	-2.7	2.4	9.0	10.3	8.1	8.2	4.0
Belgium	10.2	-8.2	12.9	2.2	2.3	7.0	10.0	2.6	1.3	2.7	5.3	1.1	5.6	6.6	8.6	7.3	4.6	3.1	1.1
Denmark	5.6	-1.8	4.1	4.1	1.2	8.4	5.2	8.2	2.5	4.9	3.5	5.0	-	5.1	7.8	4.2	8.5	7.9	3.7
Finland	8.7	-14.0	12.8	15.7	8.9	8.8	8.4	4.9	-1.1	2.5	5.4	1.2	1.3	2.6	3.7	1.6	1.6	-6.7	9.1
Ireland	18.8	7.6	8.1	14.0	12.3	6.5	6.4	2.0	5.5	10.5	16.6	6.6	2.9	13.4	8.7	10.1	8.8	5.7	9.5
Netherlands	11.9	-3.1	9.9	-1.8	3.3	7.4	1.5	1.5	-	3.5	7.3	5.5	2.9	3.2	9.0	6.6	5.5	4.6	2.6
Norway	0.1	3.1	11.3	3.6	8.4	2.6	2.1	1.4	-0.1	7.6	8.2	6.9	1.6	1.2	9.0	10.7	8.1	6.3	6.1
Sweden	8.6	-9.3	4.3	1.5	7.8	6.1	-0.6	2.0	5.7	9.9	6.9	1.4	3.2	3.9	3.0	3.0	1.5	-2.2	1.2
Switzerland	6.8	-6.6	9.3	9.7	3.7	2.5	5.1	4.6	-2.9	1.1	6.3	8.3	0.4	1.7	5.8	5.0	3.0	-0.7	3.3
Total 9 countries	9.3	-5.0	9.6	3.1	4.8	6.6	4.4	3.1	1.0	4.3	6.7	4.5	2.3	4.0	7.2	6.6	5.2	3.3	3.3
Total western Europe	7.3	-3.8	9.4	5.2	4.5	6.0	2.1	3.9	0.8	2.5	7.2	4.9	1.5	3.4	5.6	7.7	6.7	4.7	3.6
Greece	12.4	10.6	16.4	1.8	16.4	6.7	6.9	-5.9	-7.2	8.0	16.9	1.3	14.0	16.0	9.0	1.3	0.9	16.4	6.4
Portugal	-1.6	-15.6	-	5.9	9.1	33.0	2.2	-4.4	4.7	13.6	11.6	6.7	6.8	8.6	10.2	13.3	10.4	1.9	3.5
Spain	17.5	-0.4	5.0	12.1	10.7	5.6	2.3	8.4	4.8	10.1	11.7	2.7	1.6	6.3	5.1	3.0	3.2	6.6	5.8
Turkey	14.3	-1.1	37.5	-21.8	12.9	-12.3	7.4	62.2	36.9	14.6	20.4	11.3	-0.6	26.0	16.5	11.3	8.4	11.4	8.7
Total southern Europe	12.8	-1.4	9.2	5.6	11.5	7.2	3.3	8.8	7.0	11.0	13.7	4.5	3.2	11.2	8.4	5.8	5.0	8.2	6.3
Total Europe	7.5	-3.7	9.4	5.3	4.8	6.0	2.2	4.2	1.2	3.0	7.6	4.8	1.6	3.9	5.8	7.5	6.6	4.9	3.8
United States	9.1	-0.6	4.5	1.4	9.4	8.6	9.2	1.7	-9.0	-3.6	6.9	1.1	6.6	10.4	15.8	11.9	8.1	5.8	6.0
Canada	8.7	-6.8	10.6	8.9	13.6	5.0	2.7	4.4	-2.2	6.4	17.7	6.0	4.5	3.5	9.3	1.1	4.1	0.5	8.2
North America	9.0	-1.8	5.6	2.8	10.3	7.9	7.9	2.2	-7.7	-1.6	9.3	2.3	6.1	8.8	14.3	9.5	7.3	4.7	6.4
Total above	8.0	-3.1	8.2	4.5	6.5	6.6	4.1	3.5	-1.8	1.6	8.1	4.1	2.9	5.4	8.5	8.2	6.8	4.9	4.7
Cyprus	..	..	..	21.6	6.0	20.5	10.8	14.1	7.9	8.2	15.1	-1.1	-1.7	13.7	13.4	17.1	6.2	-8.0	..
Israel	9.6	1.8	14.6	11.7	4.6	3.0	6.0	5.4	-3.1	2.3	13.9	8.7	5.7	11.1	-1.6	3.9	2.2	-2.3	10.9
Malta	..	..	..	14.4	4.7	16.8	12.0	-11.4	-13.8	-1.8	4.0	7.4	5.7	12.1	4.9	..	..	..	..
Yugoslavia, SFR	3.3	-1.3	9.3	-3.1	-2.4	14.9	10.8	4.7	-9.3	-	6.4	7.3	6.3	6.8	-6.0	4.8	0.5	..	..

Sources: National statistics.

[a] Data refer to west Germany only.

APPENDIX TABLE A.6
Volume of imports of goods and services
(Annual percentage change)

	1970	1975	1976	1977	1978	1979	1980	1981	1982	1983	1984	1985	1986	1987	1988	1989	1990	1991	1992
France	6.3	-9.7	17.4	0.1	3.0	10.1	2.5	-2.1	2.6	-2.7	2.7	4.5	7.1	7.7	8.6	8.2	6.5	2.9	1.4
Germany [a]	15.9	1.2	10.8	3.4	5.6	9.3	3.7	-3.3	-0.9	1.6	5.4	4.7	2.8	4.2	5.1	8.5	10.2	11.7	2.8
Italy	12.4	-12.6	14.1	1.7	4.8	11.7	-0.8	-1.2	-0.3	-1.4	12.3	3.9	2.9	9.1	6.8	7.6	7.9	2.9	3.0
United Kingdom	4.0	-6.6	4.8	1.6	4.2	9.6	-3.6	-2.8	4.9	6.5	9.9	2.6	6.9	7.8	12.2	7.4	1.0	-3.1	6.5
Total 4 countries	9.9	-6.1	11.5	1.8	4.5	10.0	0.9	-2.5	1.3	1.1	7.0	4.0	4.8	6.8	8.0	8.0	6.5	4.3	3.3
Austria	16.9	-4.6	17.4	6.2	0.1	11.7	6.2	-0.8	-3.3	5.5	9.9	6.2	-1.2	4.7	9.4	8.5	7.8	8.9	4.1
Belgium	7.6	-9.0	12.2	4.8	2.6	9.0	5.8	-2.9	0.2	-1.3	5.6	0.7	7.6	8.8	8.0	8.8	4.2	2.7	1.7
Denmark	9.3	-4.8	15.6	–	0.1	5.0	-6.8	-1.7	3.8	1.8	5.5	8.1	6.8	-2.0	1.5	4.5	2.4	4.9	0.2
Finland	20.3	0.6	-2.0	-1.5	-3.7	18.4	8.3	-4.7	2.5	3.0	1.0	6.8	3.1	9.0	11.1	8.8	-0.9	-12.1	0.4
Ireland	8.6	-10.2	14.7	13.3	15.7	13.9	-4.5	1.7	-3.1	4.7	9.9	3.2	5.6	5.0	3.9	10.9	6.0	1.7	4.2
Netherlands	14.7	-4.1	10.1	2.9	6.3	6.0	-0.4	-5.9	1.1	3.9	5.0	6.5	3.6	5.1	7.6	6.7	4.9	3.7	1.8
Norway	13.6	7.0	12.3	3.4	-13.5	-0.7	3.3	1.5	3.7	–	9.5	5.9	9.9	-7.3	-1.7	0.9	2.3	1.3	1.0
Sweden	10.4	-3.5	9.0	-3.8	-5.5	11.6	0.4	-5.8	3.4	0.8	5.4	7.8	4.7	7.2	4.7	7.1	1.1	-5.9	0.5
Switzerland	13.9	-15.4	13.1	9.3	10.9	6.9	7.2	-1.3	-2.6	4.4	7.1	5.1	7.1	5.5	5.3	5.4	2.9	-1.7	-3.8
Total 9 countries	12.3	-5.4	11.3	3.2	1.7	8.1	2.4	-3.1	0.6	2.2	6.2	5.2	5.3	4.8	6.2	6.9	3.7	1.3	1.1
Total western Europe	10.8	-5.8	11.4	2.3	3.5	9.3	1.4	-2.7	1.1	1.5	6.7	4.4	5.0	6.1	7.4	7.6	5.5	3.3	2.6
Greece	6.2	6.3	6.1	8.0	7.2	7.2	-8.0	3.6	7.0	6.6	0.2	12.8	3.8	16.6	8.0	10.8	12.0	13.2	3.8
Portugal	0.9	-25.2	3.4	12.0	0.2	12.6	6.9	2.3	3.9	-6.1	-4.4	1.4	16.9	20.0	16.1	9.1	11.0	6.9	5.6
Spain	7.0	-0.9	9.8	-5.5	-1.0	11.4	3.3	-4.2	3.9	-0.6	-1.0	6.2	14.8	20.1	14.4	17.2	7.8	8.9	6.7
Turkey	22.0	26.1	32.1	-3.4	-31.8	-14.1	-0.4	12.5	13.0	18.6	14.9	6.9	14.5	15.6	3.5	13.7	32.0	-2.2	9.2
Total southern Europe	7.7	-0.2	12.5	-1.3	-6.2	7.0	1.6	-0.1	5.6	2.4	1.3	6.7	13.2	18.7	11.7	14.6	13.0	7.0	6.6
Total Europe	10.6	-5.4	11.5	2.0	2.7	9.1	1.5	-2.5	1.4	1.6	6.3	4.6	5.6	7.1	7.7	8.2	6.3	3.7	3.0
United States	3.8	-12.0	19.0	10.0	9.2	1.3	-4.7	4.9	–	12.5	25.0	6.3	6.6	4.6	3.7	3.7	3.0	-0.1	9.5
Canada	-1.7	-3.3	8.6	1.7	7.4	11.4	4.9	8.5	-15.2	9.0	17.1	8.7	7.6	7.0	13.8	6.4	1.9	2.2	5.5
North America	2.9	-10.3	16.8	8.4	8.9	3.1	-2.8	5.7	-3.3	11.8	23.6	6.7	6.8	5.0	5.5	4.3	2.8	0.4	8.7
Total above	8.2	-6.8	13.0	3.9	4.6	7.2	0.2	-0.1	-0.1	4.7	11.9	5.4	6.0	6.3	6.9	6.8	5.1	2.5	4.9
Cyprus	17.7	..	..	26.3	5.1	17.3	5.8	2.5	12.8	5.5	16.4	-4.5	-11.0	5.3	16.1	21.2	1.3	1.0	..
Israel	..	4.2	-3.0	-2.9	10.6	2.8	-6.3	10.2	4.2	6.7	-1.0	-0.5	9.5	15.5	0.8	-6.6	8.6	16.6	8.5
Malta	..	..	..	12.7	-3.9	9.7	11.8	-7.7	-2.0	-7.2	3.9	9.2	0.1	12.3	11.1	14.2	19.0	..	..
Yugoslavia, SFR	27.8	-5.6	-3.2	12.5	4.5	18.7	-10.1	-12.7	-14.1	-7.3	2.5	2.3	21.0	0.9	-13.7	14.2	19.0	..	..

Sources: National statistics.

[a] Data refer to west Germany only.

APPENDIX TABLE A.7
Current account balances
(Billion US dollars)

	1970	1975	1976	1977	1978	1979	1980	1981	1982	1983	1984	1985	1986	1987	1988	1989	1990	1991
France	-0.2	2.7	-3.4	-0.4	7.1	5.1	-4.2	-4.8	-12.1	-5.2	-0.9	-	2.4	-4.4	-4.8	-5.6	-13.8	-6.1
Germany [a]	0.8	4.4	3.7	4.0	9.1	-5.6	-13.8	-3.4	5.0	5.4	9.6	17.0	40.1	46.3	50.6	57.6	46.3	-19.5
Italy	0.8	-0.5	-2.8	2.5	6.3	5.8	-9.8	-9.6	-5.9	1.5	-2.4	-3.4	2.8	-1.6	-6.0	-10.9	-14.4	-21.5
United Kingdom	2.0	-3.5	-1.4	0.1	2.2	-0.8	6.9	14.1	7.9	5.5	2.1	3.9	-	-7.6	-28.8	-35.6	-29.4	-11.4
Total 4 countries	3.4	3.1	-3.9	6.2	24.6	4.6	-21.0	-3.7	-5.0	7.2	8.4	17.5	45.3	32.7	11.1	5.5	-11.3	-58.5
Austria	-0.1	-0.2	-1.1	-2.2	-0.7	-1.1	-1.7	-1.5	0.6	0.2	-0.3	-0.3	0.1	-0.4	-0.5	0.1	-	-0.3
Belgium	0.7	0.2	0.4	-0.6	-0.8	-3.1	-4.9	-4.2	-2.6	-0.5	-0.1	0.7	3.1	2.8	3.6	3.2	4.9	4.7
Denmark	-0.5	-0.5	-1.9	-1.7	-1.5	-3.0	-2.5	-1.9	-2.3	-1.2	-1.6	-2.8	-4.5	-3.0	-1.3	-1.1	1.4	2.2
Finland	-0.2	-2.1	-1.1	-0.1	0.7	-0.2	-1.4	-0.5	-0.9	-1.1	-	-0.8	-0.7	-1.7	-2.7	-5.8	-6.7	-6.7
Ireland	-0.2	-0.1	-0.4	-0.5	-0.8	-2.1	-2.1	-2.6	-1.9	-1.2	-1.0	-0.7	-0.7	0.4	0.6	0.4	0.9	1.6
Netherlands	-0.5	2.4	3.5	1.2	-1.2	0.1	-1.2	3.6	4.7	4.9	6.3	4.2	4.0	3.9	6.9	9.8	9.1	8.8
Norway	-0.2	-2.5	-3.7	-5.0	-2.1	-1.0	1.1	2.2	0.7	2.0	2.9	3.1	-4.5	-4.1	-3.9	0.2	3.9	4.9
Sweden	-0.3	-0.3	-1.6	-2.2	-0.3	-2.4	-4.4	-2.8	-3.4	-0.8	0.6	-1.2	0.6	-0.1	-0.7	-3.3	-6.9	-3.2
Switzerland	0.2	0.8	2.3	1.9	2.1	-0.2	-0.2	3.4	2.5	1.2	6.1	6.0	4.7	6.3	8.8	8.0	6.9	9.8
Total 9 countries	-1.2	-2.5	-3.7	-9.2	-4.7	-13.0	-17.3	-4.3	-2.5	3.5	12.9	8.2	2.0	4.0	10.8	11.5	14.5	21.9
Total western Europe	2.2	0.7	-7.6	-3.0	19.9	-8.4	-38.3	-7.9	-7.6	10.8	21.3	25.7	47.3	36.7	21.9	17.0	3.2	-36.6
Greece	-0.4	-0.9	-0.9	-1.1	-1.0	-1.9	-2.2	-2.4	-1.9	-1.9	-2.1	-3.3	-1.7	-1.2	-1.0	-2.6	-3.5	-1.5
Portugal	0.1	-0.8	-1.3	-1.0	-0.5	-0.1	-1.1	-2.6	-3.3	-1.6	-0.6	0.4	1.2	0.4	-1.1	0.2	-0.2	-0.7
Spain	0.1	-3.5	-4.3	-2.1	1.6	1.1	-5.2	-5.0	-4.2	-2.7	2.0	2.9	4.0	-0.2	-3.8	-10.9	-16.8	-16.0
Turkey	-	-1.6	-2.0	-3.1	-1.3	-1.4	-3.4	-1.9	-1.0	-1.9	-1.4	-1.0	-1.5	-0.8	1.6	1.0	-2.6	0.3
Total southern Europe	-0.3	-6.8	-8.5	-7.3	-1.0	-2.2	-11.9	-11.9	-10.3	-8.2	-2.1	-1.1	2.0	-1.8	-4.2	-12.4	-23.2	-17.9
Total Europe	1.9	-6.1	-16.2	-10.3	18.9	-10.6	-50.1	-19.9	-17.9	2.6	19.2	24.6	49.3	34.9	17.7	4.6	-19.9	-54.6
United States	2.3	18.1	4.2	-14.5	-15.4	0.2	2.3	5.0	-11.4	-43.6	-98.8	-121.8	-147.5	-163.4	-126.7	-101.2	-90.5	-3.7
Canada	1.0	-4.6	-4.2	-4.1	-4.3	-4.1	-1.5	-5.7	1.6	1.7	1.2	-2.3	-8.2	-8.8	-12.6	-19.3	-22.0	-25.5
North America	3.3	13.5	-	-18.6	-19.7	-3.9	0.7	-0.7	-9.8	-41.9	-97.6	-124.1	-155.7	-172.2	-139.2	-120.5	-112.4	-29.2
Total above	5.2	7.4	-16.1	-28.9	-0.8	-14.5	-49.4	-20.6	-27.7	-39.4	-78.4	-99.4	-106.5	-137.3	-121.5	-115.9	-132.4	-83.8
Japan	2.0	-0.7	3.7	10.9	16.5	-8.7	-10.8	4.8	6.8	20.8	35.0	49.2	85.8	87.0	79.6	57.0	35.9	72.9
Total above	7.2	6.7	-12.4	-18.0	15.7	-23.3	-60.2	-15.8	-20.9	-18.6	-43.4	-50.3	-20.6	-50.3	-41.9	-58.9	-96.5	-10.9

Source: IMF, *International Financial Statistics*, March 1993.

[a] Data as from the second half of 1991 include the five new *Länder*.

APPENDIX TABLE A.8
Industrial production
(Annual percentage change)

	1970	1975	1976	1977	1978	1979	1980	1981	1982	1983	1984	1985	1986	1987	1988	1989	1990	1991	1992 [a]
France	4.4	-7.2	8.8	1.4	2.4	4.5	2.2	-1.0	-0.8	-0.6	0.3	0.2	0.9	1.9	4.7	4.1	1.9	0.2	-1.5
Germany [b]	5.9	-6.2	6.9	2.6	1.9	5.1	-	-1.9	-3.2	0.7	3.0	4.5	2.1	0.4	3.6	5.0	5.1	2.9	-1.5
Italy	6.6	-8.8	11.6	-	2.0	6.6	5.2	-2.2	-3.1	-2.4	3.4	1.3	4.1	2.6	6.9	3.9	-0.7	-2.1	-0.5
United Kingdom	0.4	-5.4	3.4	5.1	2.9	3.9	-6.7	-3.2	1.7	4.1	0.1	5.5	2.4	3.2	3.6	0.4	-0.5	-2.9	-0.5
Total 4 countries	4.3	-6.8	7.5	2.3	2.2	5.0	0.1	-2.1	-1.6	0.5	1.8	3.1	2.4	1.9	4.6	3.5	1.8	-0.1	-1.0
Austria	8.6	-6.3	6.6	3.8	2.1	7.7	2.7	-1.6	-0.8	1.0	5.3	4.5	1.1	1.0	4.4	5.9	7.4	1.8	-1.0
Belgium	2.9	-9.9	7.7	0.6	2.4	4.5	-1.3	-2.7	-	1.9	2.5	2.5	0.8	2.2	6.5	3.4	3.8	-2.0	-0.5
Denmark	..	..	..	..	..	..	..	..	..	..	..	..	..	..	..	..	..	..	..
Finland	12.4	-3.8	0.9	0.6	5.1	10.7	7.8	2.6	0.9	3.3	4.6	4.1	0.8	4.2	5.8	2.4	0.4	-8.9	3.5
Ireland	1.6	-4.0	9.0	7.9	7.8	7.8	-1.3	5.4	-0.6	7.8	9.9	3.3	2.1	9.1	10.6	11.6	4.6	3.2	10.0
Netherlands	9.4	-4.8	8.0	-	0.9	3.6	-0.8	-2.0	-3.8	1.9	5.0	4.8	0.2	1.0	0.1	5.1	2.3	3.9	-
Norway	4.1	5.0	6.1	-0.6	10.2	7.3	3.1	2.6	-0.2	9.0	8.3	3.1	3.2	6.6	3.0	9.5	1.8	2.1	6.5
Sweden [c]	5.6	-1.7	-2.4	-5.5	-1.3	6.7	-0.4	-1.8	-1.7	4.0	7.2	3.0	0.4	2.6	1.3	3.6	-2.7	-8.0	-3.0
Switzerland	8.7	-12.4	-	5.9	..	2.2	5.4	-0.7	-3.7	-0.6	2.6	5.8	3.7	1.2	8.1	1.6	2.7	0.5	-
Total 9 countries [d]	7.1	-5.8	4.6	0.7	2.0	5.3	1.3	-1.0	-1.8	2.6	5.0	4.0	1.3	2.5	3.9	4.8	2.5	-0.4	0.8
Total western Europe [d]	4.8	-6.6	6.9	2.0	2.2	5.1	0.3	-1.9	-1.6	0.9	2.4	3.3	2.2	2.0	4.5	3.7	1.9	-0.2	-0.7
Greece	10.3	4.4	10.5	2.0	7.6	6.0	1.0	0.9	0.9	-0.3	2.3	4.2	-1.0	-1.5	5.1	1.9	-2.4	-1.4	-1.0
Portugal	6.4	-4.9	3.3	13.2	6.8	7.2	5.5	2.4	7.7	3.5	2.5	0.7	7.3	4.4	3.7	6.8	8.9	0.1	-3.5
Spain	10.5	-3.9	5.1	5.3	2.3	0.7	1.3	-1.0	-1.1	2.7	0.8	2.0	3.1	4.7	3.0	4.5	0.1	-0.9	-1.5
Turkey [e]	4.1	4.7	3.0	7.3	-0.7	-2.0	-4.3	12.3	8.6	9.8	11.4	4.2	11.0	10.7	1.0	2.2	9.5	1.8	5.5
Total southern Europe	8.9	-1.9	4.9	6.1	2.5	1.2	0.6	1.8	1.8	4.1	3.4	2.5	5.1	5.7	2.7	4.0	3.3	-0.1	0.2
Total Europe [d]	5.2	-6.1	6.7	2.5	2.2	4.6	0.3	-1.4	-1.2	1.3	2.6	3.2	2.6	2.5	4.2	3.8	2.1	-0.2	-0.6
United States	-3.4	-8.7	9.1	8.0	5.7	3.8	-1.9	1.9	-4.5	3.7	9.3	1.7	0.9	5.0	5.5	2.5	1.0	-2.0	1.5
Canada	-1.6	-7.3	6.7	3.4	3.4	4.8	-3.4	2.1	-9.9	6.6	12.1	5.6	-0.7	4.8	5.3	-0.5	-4.1	-4.1	1.0
North America	-3.3	-8.6	8.9	7.5	5.5	3.9	-2.0	1.9	-5.0	3.9	9.6	2.1	0.8	4.9	5.5	2.3	0.6	-2.2	1.5
Total above [d]	1.0	-7.3	7.7	4.9	3.8	4.2	-0.8	0.2	-3.0	2.5	6.0	2.6	1.7	3.7	4.8	3.0	1.4	-1.2	0.4
Cyprus	9.9	-14.7	20.7	14.3	8.7	8.3	7.3	6.0	2.6	3.1	5.3	-1.2	3.2	9.7	7.4	3.8	5.7	0.1	6.0
Israel	9.4	2.5	4.3	6.5	5.7	5.1	-3.0	6.3	0.9	3.5	4.9	2.9	3.6	4.8	-3.1	-1.6	6.3	6.8	9.5
Malta	..	..	15.7	18.7	4.8	6.5	18.7	1.8	5.0	2.7	8.0	1.3	6.0	3.8	1.5	6.6	..	..	..
Yugoslavia, SFR	7.5	5.0	3.2	10.8	8.3	9.0	3.5	3.4	-	1.1	5.4	3.1	4.0	1.0	-1.0	1.0	-10.5	..	..

Sources: OECD, *Main Economic Indicators*, Paris (monthly), and national statistics. Data for Norway, Switzerland and Yugoslavia are calculated from rounded index numbers (1985 = 100). National data are aggregated by means of weights derived from GDP originating in industry, expressed at 1985 US dollar purchasing parities.

[a] Country data rounded to the nearest 0.5 percentage point.
[b] Data refer to west Germany only.
[c] Refers to mining and manufacturing only.
[d] Excluding Denmark.
[e] Refers to manufacturing.

APPENDIX TABLE A.9
Consumer prices [a]
(Annual percentage change)

	1970	1975	1976	1977	1978	1979	1980	1981	1982	1983	1984	1985	1986	1987	1988	1989	1990	1991	1992
France	5.2	11.8	9.6	9.4	9.1	10.8	13.6	13.4	11.8	9.6	7.4	5.8	2.5	3.3	2.8	3.6	3.3	3.1	2.8
Germany [b]	3.3	5.9	4.3	3.7	2.7	4.1	5.5	6.3	5.2	3.3	2.4	2.0	-0.1	0.2	1.3	2.8	2.7	3.5	4.1
Italy	4.9	17.0	16.8	17.0	12.1	14.8	21.2	17.8	16.5	14.6	10.8	9.2	5.9	4.7	5.0	6.3	6.3	6.5	5.1
United Kingdom	6.4	24.2	16.5	15.8	8.3	13.4	18.0	11.9	8.6	4.6	5.0	6.1	3.4	4.1	4.9	7.8	9.5	5.9	3.8
Total 4 countries	4.8	14.0	11.2	10.8	7.7	10.3	13.9	11.9	10.2	7.7	6.1	5.5	2.7	2.9	3.3	4.9	5.2	4.6	3.9
Austria	4.4	8.4	7.3	5.5	3.6	3.7	6.3	6.8	5.4	3.3	5.7	3.2	1.7	1.4	2.0	2.5	3.2	3.4	4.0
Belgium	3.9	12.8	9.2	7.1	4.5	4.5	6.6	8.2	8.2	7.7	6.3	4.9	1.3	1.6	1.1	3.2	3.4	3.2	2.5
Denmark	6.5	9.6	9.0	11.1	10.1	9.6	12.3	6.8	5.4	3.3	5.6	3.2	1.7	4.0	4.6	4.8	2.7	2.4	2.1
Finland	2.7	17.4	14.4	12.5	7.7	7.1	11.7	12.2	9.6	8.3	7.1	5.9	2.9	4.1	5.1	6.6	6.2	4.1	2.6
Ireland	8.2	20.9	18.0	13.6	7.6	13.2	18.2	20.4	17.1	10.5	8.6	5.4	3.8	3.2	2.1	4.0	3.4	3.1	3.2
Netherlands	3.6	10.2	8.8	6.7	4.1	4.2	6.5	6.7	5.7	2.8	3.2	2.3	0.3	-0.2	0.7	1.1	2.4	4.0	3.7
Norway	10.7	11.6	9.2	9.0	8.2	4.8	10.9	13.6	11.3	8.4	6.3	5.7	7.2	8.7	6.7	4.6	4.1	3.4	2.3
Sweden	7.0	9.8	10.3	11.4	10.0	7.2	13.7	12.1	8.6	9.0	8.0	7.4	4.2	4.2	5.8	6.4	10.5	9.3	2.2
Switzerland	3.6	6.7	1.7	1.3	1.0	3.6	4.1	6.5	5.6	3.0	3.0	3.4	0.7	1.5	1.8	3.2	5.4	5.8	4.0
Total 9 countries	5.0	10.7	8.6	7.7	5.6	5.5	8.7	9.0	7.5	5.5	5.4	4.3	2.2	2.6	2.9	3.7	4.7	4.6	3.1
Total western Europe	4.9	13.3	10.6	10.1	7.2	9.2	12.7	11.2	9.5	7.2	6.0	5.2	2.6	2.8	3.2	4.6	5.1	4.6	3.7
Greece	2.9	13.4	13.3	12.1	12.6	19.0	24.9	24.5	21.0	20.2	18.5	19.3	23.0	16.4	13.5	13.7	20.5	18.9	14.4
Portugal [c,d]	6.4	15.3	21.0	27.4	22.0	24.2	16.6	20.0	22.4	25.5	29.3	19.2	11.8	10.2	9.6	12.7	13.5	11.3	9.3
Spain	5.6	16.9	14.9	24.5	19.8	15.7	15.5	14.5	14.4	12.2	11.3	8.8	8.7	5.3	4.8	6.8	6.7	6.0	5.9
Turkey [e]	8.1	20.1	15.3	28.4	49.5	56.5	116.6	35.9	27.1	31.4	48.4	44.9	34.6	38.9	73.7	63.3	60.3	66.0	70.1
Total southern Europe	5.9	17.1	15.3	24.1	25.4	25.5	38.3	20.7	18.5	18.3	21.5	18.6	16.2	14.2	20.9	20.1	20.3	20.7	20.9
Total Europe	5.0	13.7	11.1	11.5	9.1	10.8	15.3	12.2	10.5	8.3	7.6	6.6	4.0	4.0	5.0	6.2	6.6	6.3	5.4
United States [f]	5.9	9.1	5.8	6.4	7.6	11.3	13.5	10.3	6.2	3.2	4.3	3.6	1.9	3.7	4.1	4.8	5.4	4.2	1.5
Canada	3.3	10.8	7.5	8.0	9.0	9.1	10.2	12.3	10.9	5.7	4.4	3.9	4.2	4.4	4.0	5.0	4.8	5.6	3.0
North America	5.5	8.9	5.7	6.3	7.4	10.8	12.8	10.1	6.3	3.3	4.2	3.5	2.0	3.6	4.0	4.7	5.2	4.2	1.6
Total above	5.3	10.8	7.8	8.3	8.1	10.8	13.8	10.9	7.9	5.2	5.5	4.7	2.7	3.7	4.4	5.3	5.7	5.0	3.1
Cyprus	2.4	4.6	3.8	7.2	7.4	9.5	13.5	10.8	6.4	5.1	6.0	5.1	1.2	2.8	3.4	3.8	4.5	5.0	6.5
Israel	6.1	39.3	31.4	34.6	50.6	78.3	131.0	116.8	120.4	145.6	373.8	304.6	48.1	19.8	16.3	20.2	17.2	19.0	11.9
Malta	3.7	6.1	0.6	10.0	4.7	7.1	15.8	11.5	5.8	-0.9	-0.4	-0.3	2.0	0.5	0.9	0.9	3.0	2.5	..
Yugoslavia, SFR	10.6	24.3	11.7	14.9	14.3	20.6	30.2	40.9	32.3	42.9	50.0	60.0	87.5	126.7	194.1	1252.0	580.0	..	..

Sources: National statistics. Regional aggregates were obtained from time series in annual percentage change form, with weights taken from OECD *National Accounts*. (Private final consumption expenditure in US dollars for 1985 at current prices and exchange rates.)

[a] Cost-of-living index for Germany and Yugoslavia, retail price index for Cyprus, Malta, United Kingdom.
[b] Data refer to west Germany only.
[c] 1970–1976, Lisbon.
[d] Break in series after 1975.
[e] 1970–1982, Ankara; 1983 and thereafter, total urban areas.
[f] 1970–1978, urban wage earners and clerical workers; 1979 and thereafter, all urban consumers.

APPENDIX TABLE A.10
Average hourly earnings in manufacturing
(Annual percentage change)

	1970	1975	1976	1977	1978	1979	1980	1981	1982	1983	1984	1985	1986	1987	1988	1989	1990	1991
France [a]	10.8	17.3	14.0	12.7	12.9	13.0	15.1	14.5	15.3	11.1	7.7	5.8	3.9	3.2	3.1	3.8	4.5	4.3
Germany [b]	13.8	8.0	6.4	7.6	5.0	5.5	6.3	5.3	4.9	3.4	2.3	4.5	3.5	4.2	4.3	4.2	4.9	6.0
Italy [a]	22.4	26.6	20.8	27.9	16.2	19.0	18.5	23.1	21.4	14.8	11.5	11.2	4.8	6.5	6.1	6.1	7.2	9.8
United Kingdom [c]	13.1	26.3	16.6	10.3	14.4	15.7	17.4	13.4	11.2	8.9	8.8	9.1	7.7	8.0	8.5	8.7	9.4	8.2
Total 4 countries	13.8	14.4	11.0	11.2	9.7	10.7	12.0	11.5	11.1	8.3	6.7	7.2	4.9	5.4	5.5	5.7	6.5	7.1
Austria [d]	9.4	13.3	9.1	8.5	5.7	5.8	7.9	6.2	6.1	4.5	5.0	6.1	4.5	3.1	3.8	4.4	7.1	5.3
Belgium [e]	15.0	19.5	12.2	9.1	6.7	7.8	8.7	10.7	6.0	4.5	5.4	3.1	2.5	1.9	0.8	5.7	4.3	5.1
Denmark [f]	12.4	22.4	11.5	9.1	10.8	11.1	11.1	9.0	10.1	6.7	5.0	4.9	4.8	9.3	6.6	4.7	4.8	4.5
Finland	11.1	21.3	14.8	8.8	7.5	11.4	12.8	12.8	10.5	9.6	10.3	7.6	6.2	6.8	8.6	8.8	10.6	6.7
Ireland	11.1	25.0	16.0	20.7	14.3	15.0	21.7	16.1	15.1	11.5	10.4	8.6	6.9	5.7	5.3	4.8	5.4	5.6
Netherlands [a]	8.3	12.1	9.5	7.4	5.7	4.3	4.2	3.0	6.8	2.7	0.9	5.3	1.7	1.5	1.3	1.3	2.9	3.7
Norway [g]	10.0	20.0	14.3	12.5	7.4	3.4	8.3	10.8	9.7	8.9	8.1	7.5	10.0	16.4	5.5	5.2	5.6	5.3
Sweden [f]	15.6	14.9	19.3	6.8	7.7	8.2	8.7	11.1	7.1	8.4	9.3	7.5	7.4	6.4	8.0	10.0	9.4	4.6
Switzerland [h]	6.2	7.4	1.6	1.7	3.4	2.1	5.2	5.1	6.2	6.8	1.8	3.6	3.5	2.4	3.0	3.7	5.1	6.9
Total 9 countries	10.2	14.4	10.4	7.4	6.5	6.4	7.9	8.0	7.5	6.2	5.3	5.6	4.7	4.9	4.5	5.5	6.3	5.2
Total western Europe	12.9	14.4	10.9	10.2	8.9	9.7	11.1	10.7	10.3	7.8	6.4	6.8	4.8	5.3	5.3	5.6	6.5	6.7
Greece	6.1	24.4	27.3	21.4	23.5	23.8	26.9	24.2	34.1	20.0	25.8	20.5	13.0	9.7	17.7	20.5	19.3	17.1
Portugal [i]	29.4	51.0	22.9	19.2	15.2	21.1	26.2	20.0	20.6	18.6	18.8	21.1	16.8	14.0	11.3	14.8	13.7	14.5
Spain [j]	13.2	31.1	30.0	29.3	27.5	23.9	15.3	24.7	15.9	15.0	11.7	10.0	11.0	7.5	6.5	7.3	8.7	8.2
Turkey	:	:	:	:	:	:	:	:	:	:	:	:	:	:	:	:	:	:
Total southern Europe [k]	14.8	34.7	27.9	25.9	24.2	23.3	18.9	23.6	19.2	16.5	15.3	14.2	12.7	9.5	9.6	11.7	12.1	11.8
Total Europe [k]	12.9	15.2	11.7	11.1	9.8	10.7	11.7	11.8	11.1	8.7	7.3	7.7	5.8	5.8	5.9	6.5	7.3	7.5
United States	5.1	9.3	8.1	8.8	8.7	8.5	8.5	10.0	6.2	4.0	4.0	3.8	2.0	1.9	2.8	2.9	3.3	3.3
Canada	9.2	17.2	12.8	11.4	7.8	8.3	9.8	12.1	11.5	4.0	4.7	4.4	3.1	2.4	5.1	5.3	5.2	4.8
North America	5.3	9.8	8.5	9.0	8.7	8.5	8.6	10.1	6.6	4.0	4.1	3.9	2.1	1.9	3.0	3.1	3.5	3.4
Total above [k]	8.8	12.6	10.2	10.1	9.3	9.7	10.3	11.1	9.1	6.7	6.0	6.1	4.3	4.3	4.7	5.2	5.9	6.0
Cyprus [l]	:	:	:	20.4	21.4	26.9	19.8	14.6	17.3	7.3	4.0	7.2	3.1	7.6	6.5	10.0	9.2	9.0
Israel	:	:	:	:	:	:	:	:	:	:	:	:	:	:	:	:	:	:
Malta	:	:	:	17.9	19.1	20.7	24.2	37.0	27.7	27.5	45.3	76.9	105.4	105.0	171.21	580.6	:	:
Yugoslavia, SFR	16.3	22.4	14.2														406.3	

Sources: National statistics; OECD, *Economic Outlook – Historical Statistics, 1960-1990*; OECD, *Main Economic Indicators*, No.3, Paris, 1993. National data in annual percentage change form are aggregated by means of weights derived from manufacturing employment in 1985.

a Wage rates.
b Data refer to west Germany only.
c Weekly earnings of all employees in Great Britain.
d Monthly earnings in mining and manufacturing.
e Includes transport.
f Includes mining.
g Males only.
h Data refer to workers who had accidents during the relevant period.
i Daily earnings; for 1970-1973, wage bill for all activities.
j Data refer to all activities.
k Excluding Turkey.

Appendix A. Western Europe and North America

APPENDIX TABLE A.11
Total employment
(Annual percentage change)

	1970	1975	1976	1977	1978	1979	1980	1981	1982	1983	1984	1985	1986	1987	1988	1989	1990	1991	1992
France	0.9	-0.9	0.8	0.8	0.4	0.1	0.1	-0.6	0.2	-0.4	-0.9	-0.3	0.1	0.3	0.8	1.1	1.0	0.4	-0.5
Germany [a]	1.3	-2.7	-0.5	0.1	0.8	1.7	1.6	-0.1	-1.2	-1.4	0.2	0.7	1.4	0.7	0.8	1.5	3.0	2.6	0.8
Italy [b]	0.2	0.1	1.5	1.0	0.5	1.5	1.9	-	0.6	0.6	0.4	0.9	0.8	0.4	0.9	0.1	1.1	0.8	0.3
United Kingdom [c]	-0.4	-0.4	-0.8	0.1	0.6	1.5	-0.3	-3.9	-1.8	-1.2	2.6	1.3	0.1	2.1	3.3	3.4	0.5	-3.3	-2.5
Total 4 countries	0.5	-1.0	0.1	0.5	0.6	1.2	0.8	-1.2	-0.6	-0.6	0.6	0.7	0.6	0.9	1.5	1.6	1.4	0.1	-0.4
Austria	0.4	-0.4	0.5	0.9	0.2	0.7	0.4	0.1	-1.2	-0.8	0.1	0.2	0.4	-	0.6	1.4	1.9	2.0	1.6
Belgium [c]	1.8	-1.4	-0.7	-0.2	0.1	1.2	-0.1	-1.9	-1.3	-1.0	-0.2	0.6	0.6	0.5	1.5	1.6	1.4	0.3	-0.4
Denmark	0.7	-1.3	1.8	0.8	1.0	1.2	-0.4	-1.3	0.5	0.3	1.7	2.5	2.6	0.8	-0.6	-0.7	-0.5	-0.9	-0.2
Finland	2.1	-0.4	-1.3	-2.5	-1.1	2.2	2.9	1.0	0.7	0.4	0.3	-0.2	-0.5	0.2	0.2	0.5	-0.9	-5.3	-7.1
Ireland [d]	-1.2	-0.8	-0.8	1.8	2.5	3.2	1.0	-0.9	-	-1.9	-1.9	-2.2	2.0	-0.1	1.1	-0.2	3.3	-	-
Netherlands [e]	1.1	-0.7	-	0.2	0.7	1.3	0.7	-1.5	-2.5	-1.9	-0.1	1.5	3.0	1.4	1.4	1.7	2.1	1.3	0.4
Norway	1.5	1.9	3.3	2.6	1.8	1.5	2.3	1.0	0.1	-0.3	0.6	2.7	3.0	2.1	-0.8	-2.6	-0.6	-0.7	-0.2
Sweden	2.0	2.0	0.4	0.2	0.4	1.5	1.1	0.2	-0.2	0.2	0.8	1.0	0.6	0.8	1.4	1.5	0.9	-1.5	-4.0
Switzerland	1.5	-4.8	-3.0	0.4	1.0	1.1	2.3	1.3	-0.7	-1.3	4.4	1.9	1.4	1.2	1.2	1.2	1.3	-0.1	-2.0
Total 9 countries	1.3	-0.7	-0.1	0.3	0.6	1.4	1.0	-0.3	-0.8	-0.7	0.8	1.1	1.2	0.8	0.8	0.8	1.0	-0.3	-1.3
Total western Europe	0.7	-1.0	0.1	0.4	0.6	1.3	0.9	-1.0	-0.7	-0.7	0.6	0.8	0.7	0.9	1.3	1.4	1.3	-	-0.6
Greece	-0.1	0.1	1.2	0.8	0.4	1.1	1.4	5.2	-0.8	1.1	0.4	1.0	0.4	-0.1	1.7	0.4	1.3	-1.7	-0.7
Portugal	-0.7	-1.4	0.2	-0.1	-0.3	-0.5	2.2	0.5	-0.1	4.0	-0.1	-0.4	0.2	2.6	3.1	2.2	2.3	3.0	0.1
Spain	1.0	-1.8	-1.1	-0.8	-2.7	-0.4	-3.0	-3.0	-1.3	-1.1	-1.8	-0.9	2.2	3.1	2.9	4.0	2.6	0.2	-1.3
Turkey [f]	-0.3	1.7	2.3	1.9	1.0	0.1	-0.1	0.9	2.0	1.8	2.5	2.3	3.1	3.0	1.4	1.3	2.2	0.2	1.4
Total southern Europe	0.2	-0.2	0.7	0.5	-0.6	-0.1	-0.7	-0.1	0.3	1.0	0.5	0.8	2.2	2.6	2.1	2.2	2.3	0.4	0.1
Total Europe	0.6	-0.8	0.2	0.5	0.3	1.0	0.5	-0.8	-0.4	-0.3	0.6	0.8	1.1	1.3	1.5	1.6	1.6	0.1	-0.4
United States [b]	-0.8	-2.1	2.8	3.5	5.0	3.2	0.2	0.9	-1.7	1.0	4.9	2.4	1.7	2.9	2.8	2.3	1.2	-1.6	0.5
Canada	1.1	1.7	2.1	1.8	3.5	4.1	3.0	2.7	-3.5	0.5	2.4	2.6	2.8	2.9	3.2	2.0	0.7	-1.8	-0.8
North America	-0.6	-1.7	2.7	3.3	4.9	3.3	0.5	1.1	-1.8	1.0	4.6	2.4	1.8	2.9	2.9	2.3	1.2	-1.6	0.4
Total above	0.1	-1.2	1.2	1.5	2.1	1.9	0.5	-	-1.0	0.2	2.2	1.5	1.4	2.0	2.1	1.9	1.4	-0.7	-0.1
Cyprus	..	..	..	..	4.7	4.8	4.6	3.2	1.7	2.5	3.8	3.7	1.1	3.1	4.7	4.0	2.8	0.7	5.0
Israel	..	0.9	1.3	2.9	4.4	2.3	1.1	2.0	1.4	3.2	1.5	0.7	1.4	2.6	3.5	0.5	2.1	6.1	3.9
Malta	..	..	..	..	..	..	-	-3.1	-4.1	0.1	0.6	1.3	4.1	5.9	2.5	0.9	0.8	1.0	..
Yugoslavia, SFR [g]	3.9	5.5	3.6	4.5	4.5	4.3	3.2	2.9	2.3	2.0	2.1	2.5	2.9	2.1	0.2	-0.3	-3.8	..	..

Sources: National statistics: OECD, *National accounts*, detailed tables, vol.II, 1978-1990; OECD, *Labour force statistics 1970-1990*; *Quarterly labour force statistics*, No.3, 1992; ECE secretariat estimates. National data are aggregated by adding the annual data on persons engaged taken from the national accounts statistics, where available. Otherwise data refer to annual labour force surveys.

[a] Data refer to west Germany only.
[b] Data refers to full-time equivalent data.
[c] June of each year.
[d] April of each year.
[e] Man-years.
[f] Civilian employment.
[g] Socialist sector.

APPENDIX TABLE A.12
Annual unemployment rates [a]
(Percentage of total labour force)

	1970	1975	1976	1977	1978	1979	1980	1981	1982	1983	1984	1985	1986	1987	1988	1989	1990	1991	1992
France	2.5	4.0	4.4	4.9	5.2	5.9	6.3	7.4	8.1	8.3	9.7	10.2	10.4	10.5	10.0	9.4	9.0	9.5	10.3
Germany [b]	0.8	3.6	3.7	3.6	3.5	3.2	2.9	4.2	5.9	7.7	7.1	7.1	6.4	6.2	6.2	5.5	4.9	4.4	4.8
Italy	5.3	5.8	6.6	7.0	7.1	7.6	7.5	7.8	8.4	8.8	9.4	9.6	10.5	10.9	11.0	10.9	10.3	9.9	9.9
United Kingdom	3.0	4.3	5.6	6.0	5.9	5.0	6.4	9.8	11.3	12.4	11.7	11.2	11.2	10.3	8.5	7.1	6.8	8.7	10.0
Total 4 countries	2.8	4.4	5.0	5.3	5.4	5.3	5.7	7.3	8.4	9.4	9.5	9.5	9.6	9.4	8.8	8.1	7.6	8.0	8.6
Austria	1.1	1.5	1.5	1.4	1.7	1.7	1.5	2.1	3.1	3.7	3.8	3.6	3.1	3.8	3.6	3.1	3.2	3.5	3.6
Belgium	2.1	5.0	6.4	7.4	7.9	8.2	8.8	10.8	12.6	12.1	12.1	11.3	11.2	11.0	9.7	8.0	7.2	7.1	7.8
Denmark	1.3	5.3	5.3	6.4	7.3	6.2	7.0	9.2	9.8	10.4	10.1	9.0	7.8	7.8	8.5	9.2	9.5	10.4	11.2
Finland	1.9	2.2	3.8	5.8	7.2	5.9	4.6	4.8	5.3	5.4	5.2	5.0	5.3	5.0	4.5	3.4	3.4	7.5	13.0
Ireland	5.8	8.3	9.2	9.0	8.2	7.3	8.1	10.0	12.1	13.7	15.4	16.8	17.4	17.7	16.2	14.7	13.4	14.9	16.1
Netherlands	1.0	5.2	5.5	5.3	5.3	5.4	6.0	8.5	11.4	12.0	11.8	10.6	9.9	9.6	9.2	8.3	7.5	7.0	6.8
Norway	1.6	2.3	1.7	1.4	1.8	2.0	1.6	2.0	2.6	3.4	3.1	2.6	2.0	2.1	3.2	4.9	5.2	5.5	5.9
Sweden	1.2	1.3	1.3	1.4	1.8	1.7	1.6	2.1	2.6	2.9	2.6	2.4	2.2	1.9	1.6	1.4	1.5	2.7	4.8
Switzerland	-	0.3	0.7	0.4	0.3	0.4	0.2	0.2	0.4	0.9	1.1	1.0	0.8	0.8	0.7	0.6	0.6	1.3	3.0
Total 9 countries	1.4	3.3	3.7	4.1	4.4	4.2	4.3	5.6	6.8	7.2	7.2	6.7	6.3	6.3	6.0	5.5	5.2	5.9	7.1
Total western Europe	2.5	4.1	4.7	5.0	5.1	5.1	5.4	6.9	8.1	8.9	9.0	8.9	8.8	8.7	8.2	7.5	7.1	7.5	8.3
Greece	4.2	2.3	1.9	1.7	1.8	1.9	2.8	4.0	5.8	7.8	8.1	7.8	7.4	7.4	7.7	7.4	7.0	8.2	9.1
Portugal	2.2	3.5	5.8	7.1	7.9	8.2	8.0	7.7	7.5	7.9	8.4	8.5	8.5	7.0	5.7	5.0	4.6	4.1	4.1
Spain	0.9	3.6	4.5	5.1	6.8	8.4	11.1	13.8	15.6	17.0	19.7	21.1	20.8	20.1	19.1	16.9	15.9	16.0	18.1
Turkey	7.8	8.7	7.9	7.5	7.8	9.7	11.6	11.6	12.3	12.1	11.8	11.3	10.5	9.5	9.8	10.2	10.0	11.5	11.8
Total southern Europe	4.4	5.7	5.9	6.1	6.9	8.3	10.2	11.2	12.3	12.9	13.8	14.0	13.5	12.7	12.4	11.7	11.1	11.9	12.9
Total Europe	3.0	4.5	5.0	5.3	5.6	5.8	6.5	7.9	9.0	9.8	10.1	10.1	9.9	9.6	9.1	8.5	8.0	8.6	9.4
United States	4.8	8.3	7.6	6.9	6.0	5.8	7.0	7.5	9.5	9.5	7.4	7.1	6.9	6.1	5.4	5.2	5.4	6.6	7.2
Canada	5.6	6.9	7.1	8.0	8.3	7.4	7.4	7.5	10.9	11.8	11.2	10.4	9.5	8.8	7.7	7.5	8.1	10.2	11.2
North America	4.9	8.2	7.6	7.0	6.2	6.0	7.0	7.5	9.6	9.7	7.8	7.4	7.2	6.4	5.6	5.4	5.7	6.9	7.6
Total above	3.8	6.1	6.1	6.0	5.8	5.9	6.7	7.7	9.3	9.8	9.1	8.9	8.7	8.2	7.6	7.2	7.0	7.9	8.6
Cyprus	1.0	10.9	8.6	3.1	2.0	1.7	2.0	2.6	2.8	3.3	3.3	3.3	3.7	3.4	2.8	2.3	1.8	3.0	..
Israel	3.8	3.1	3.6	3.9	3.6	2.9	4.8	5.1	5.0	4.5	5.9	6.7	7.1	6.1	6.4	8.9	9.6	10.6	..
Malta	..	..	4.2	4.3	3.5	2.7	3.3	4.8	8.7	8.7	8.7	8.2	6.9	4.4	4.0	3.7	3.8	3.6	..
Yugoslavia, SFR	4.0	6.5	7.5	8.1	8.4	8.6	8.7	8.9	9.4	9.7	10.3	10.7	11.1	10.9	11.2	11.8	12.8	..	..

Sources: OECD, *Labour force statistics 1970-1990*, OECD, *Quarterly labour force statistics*, No.3, 1992; *Main economic indicators*, No.3, 1993; Yugoslavia: ILO *Yearbook of Labour Statistics 1991*; ECE secretariat estimates. National data are aggregated from annual figures on the number of unemployed and total labour force, and the rates have been calculated as percentages of the total labour force.

Note: Comparisons with previous years are limited due to changes in methodology in Germany, F.R. of (1984), United Kingdom (1984), Italy (1983), Belgium (1983), Ireland (1983), Netherlands (1983, 1988), Portugal (1983), Finland (1982), Norway (1980), Sweden (1987).

[a] Adjusted for comparability between countries, except for Denmark, Greece, Switzerland, Turkey and Yugoslavia.
[b] Data refer to west Germany only.

Appendix B. Eastern Europe and countries of the former Soviet Union

Data for this section were compiled from national and international statistical sources.[761]

---

[761] CMEA, UN, IMF.

APPENDIX TABLE B.1.A
Economies in transition: Selected indicators, 1990 or 1991

Country/units	Territory (thousand km2)	Population mid-1991 [a] (millions)	Physical indicator estimates of 1985 GDP [b] Total (billion dollars)	Physical indicator estimates of 1985 GDP [b] Per capita (dollars)	Total electricity consumption (billion kWh)	Employment Total (millions)	Employment Agriculture (per cent of total)	Employment Industry (per cent of total)	Unemployment (end-1991, per cent)	CPI change [c] 1991 average (per cent)
Albania	28.7	3.321	7	2 509	2.83	1.433	47.0	23.7	..	36.0
Bulgaria	111.0	8.982	64	7 122	46.78	3.466	19.0	35.1	9.6	338.5
Czechoslovakia	127.9	15.583	119	7 700	85.19	7.339	9.0	36.7	6.6	57.9
Czech Republic	78.9	10.306	84	8 105	61.29	5.059	8.2	38.4	4.1	..
Slovak Republic	49.0	5.277	36	6 888	23.90	2.280	10.7	33.0	11.8	..
Hungary	93.0	10.345	77	7 259	30.22	5.300	15.8	29.5	7.5	35.0
Poland	313.0	38.245	209	5 630	132.12	16.511	27.9	11.8	70.3	
Romania	238.0	23.185	108	4 735	73.79	10.919	27.6	34.9	3.1	165.5
Yugoslavia (SFR)	255.8	23.928	117	5 064	76.98	6.470	..	41.0	19.6	118.0
Bosnia-Hercegovina	51.1	4.551	15 [d]	3 437 [d]	12.56	1.026	..	45.2	17.0	..
Croatia	56.5	4.689	30 [d]	6 382 [d]	13.82	1.510	..	37.1	..	124.2
Macedonia	25.7	2.152	7 [d]	3 254 [d]	5.37	0.507	..	40.6	..	..
Slovenia	20.3	1.957	21 [d]	10 724 [d]	10.51	0.839	..	46.1	10.1	117.7
Yugoslavia (FR)	102.2	10.579	45 [d]	4 441 [d]	35.06	2.696	..	39.8 [d]	15.7	121.8
Former Soviet Union	22 403.0	290.077	1 839	6 627	1 692.70	138.500	18.4	28.2	..	..
CIS	22 031.2	276.976			1 625.10	130.000			..	89.1
Armenia	29.8	3.376	19 [d]	5 822 [d]	10.20	1.690	18.9	41.0	..	95.6
Azerbaijan	86.6	7.137	34 [d]	5 170 [d]	23.70	2.839	32.9	16.3	..	88.6
Belarus	207.6	10.271	74 [d]	7 388 [d]	49.90	5.008	19.1	41.6	..	81.5
Kazakhstan	2 717.3	16.793	76 [d]	4 776 [d]	101.60	7.288	16.5	20.1	..	86.0
Kyrgyzstan	198.5	4.422	14 [d]	3 512 [d]	8.10	1.748	32.8	27.8	..	89.6
Moldova	33.7	4.367	22 [d]	5 314 [d]	10.60	2.071	32.5	22.0	..	100.7
Russia	17 075.4	148.650	1 120 [d]	7 804 [d]	1 076.60	74.382	13.4	42.8	..	91.8
Tajikistan	143.1	5.358	14 [d]	3 096 [d]	17.40	1.938	42.9	13.5	..	91.9
Turkmenistan	488.1	3.714	13 [d]	4 070 [d]	9.40	1.585	42.4	10.6	..	86.7
Ukraine	603.7	52.000	298 [d]	5 853 [d]	263.80	19.886	19.5	30.8	..	84.6
Uzbekistan	447.4	20.708	64 [d]	3 523 [d]	53.79	8.100	..	..	..	85.0
Georgia	70.0	5.464	33 [d]	6 391 [d]	17.20	..	..	..	..	..
Estonia	45.2	1.566	11 [d]	7 163 [d]	9.86	0.799	12.0	26.2	0.1	155.8
Latvia	64.5	2.662	20 [d]	7 720 [d]	9.87	1.405	..	30.4	..	180.0
Lithuania	65.2	3.756	24 [d]	6 663 [d]	16.40	1.853	17.0	29.8	..	..
Ex-GDR Länder	108.0	15.915	160	9 604	124.90	8.886	10.8	37.6	12.7	17.3 [e]

Source: ECE secretariat estimates based on national and international statistics.

[a] End-1990 for CIS countries and Georgia.
[b] Physical indicator (PI) estimates of GDP for eastern countries were first computed by the ECE secretariat for the purpose of obtaining approximate weights for the share of centrally planned economies in the world economy (see United Nations Economic Commission for Europe, *Economic Bulletin for Europe*, Vol.31, No.2, New York, 1980). The present data stem from a follow-up study for the bench-mark year 1985 (I. Borenstein, *Comparative GDP Levels – Physical Indicators, Phase III*, United Nations Economic Commission for Europe, Economic Studies No.4, New York (forthcoming), Supplementary table 6).
[c] Retail price index of goods only in case of most CIS countries.
[d] Entries for the Czech and Slovak, Yugoslav and Soviet republics were obtained by allocating the PI estimates of GDP for Czechoslovakia, SFR Yugoslavia and the Soviet Union proportionally to the republics' shares in total NMP or GNP.
[e] Second half of 1991 compared to same period in 1990.

Appendix B. Eastern Europe and countries of the former Soviet Union

APPENDIX TABLE B.1.B

Economies in transition: Selected indicators, 1990 or 1991

Country/unit	Exports (billion dollars)	Imports (billion dollars)	Exports per capita (dollars)	Imports per capita (dollars)	Gross foreign debt (end-1991, billion dollars)	Foreign currency reserves (end-1991, billion dollars)	Indicators of living standards per 1,000 inhabitants			Infant mortality (per 1,000 live births)
							passenger cars	telephones	doctors	
Albania	0.10	0.26	31	77	..	..	..	..	..	39.0
Bulgaria	3.44	2.71	383	302	11.4	0.3	..	304.0	..	16.7
Czechoslovakia	11.32	10.96	726	703	9.3	3.0	200	273.0	3.6	11.5
Czech Republic	8.00	7.27	776	705	..	..	..	..	3.7	10.4
Slovak Republic	3.32	3.69	628	700	..	..	..	..	3.5	13.2
Hungary	10.23	11.45	988	1 107	22.7	3.9	187	194.0	3.8	15.1
Poland	14.91	15.55	390	407	48.4	3.6	127 [a]	128.0 [a]	2.1 [a]	16.0
Romania	4.25	5.67	183	245	1.9	0.4	..	..	..	23.0
Yugoslavia (SFR)	14.85	16.27	621	680	14.5	2.7	147	204.0	2.4	23.0
Bosnia-Hercegovina	2.03*	1.61*	447*	353*	..	..	97	159.0	1.8	15.3
Croatia	3.29	3.84	702	818	..	..	184	234.0	2.6	10.7
Macedonia	0.97*	1.15*	448*	535*	..	..	108	165.0	2.5	31.6
Slovenia	3.86	4.14	1 971	2 113	..	..	294	325.0	2.7	8.4
Yugoslavia (FR)	4.70*	5.55*	445*	524*	..	..	128	173.0	2.5	22.8
Former Soviet Union	46.66	45.41	161	157	65.3	8.8	..	87.4	4.4	23.0
CIS	45.68	40.80	165	147	..	..	..	..	..	..
Armenia	0.06	0.80	17	238	..	..	..	..	4.3	23.0
Azerbaijan	0.34	0.80	48	112	..	..	..	..	3.9	28.0
Belarus	1.66	1.95	162	190	..	..	..	108.2	4.1	13.0
Kazakhstan	0.80	1.66	48	99	..	..	..	..	4.1	28.0
Kyrgyzstan	0.06	0.57	13	130	..	..	..	..	3.7	36.0
Moldova	0.17	0.63	39	144	..	..	..	..	4.0	24.0
Russia	36.79	25.37	248	171	..	..	63	..	4.7	19.0
Tajikistan	0.29	0.46	53	86	..	..	..	..	2.7	46.0
Turkmenistan	0.11	0.40	31	108	..	..	..	..	3.6	54.0
Ukraine	4.81	6.65	93	128	..	..	..	..	4.4	13.6
Uzbekistan	0.63	1.32	30	64	..	..	..	..	3.6	43.0
Georgia	0.23	1.49	42	272	..	..	..	..	..	22.0
Estonia	0.17	0.40	109	254	..	..	123	225.0	4.6	15.0
Latvia	0.11	0.46	43	171	..	..	103	279.0	4.6	15.6
Lithuania	..	..	..	..	..	..	128	222.0	..	12.0
Ex-GDR Länder	18.55	21.07	1 166	1 324	..	..	..	..	2.4	7.6

Sources: ECE secretariat estimates based on national and international statistics.

[a] 1988.

Appendix C. International trade and payments

Data for this section were compiled from international [762] and national statistical sources, as indicated in the notes to individual tables. Regional aggregates for tables C.2-C.3 were obtained by means of weights representing 1985 shares in the US dollar value of trade.

---

[762] United Nations COMTRADE data base, IMF, IBRD, OECD, BIS.

Appendix C. International trade and payments

APPENDIX TABLE C.1
World trade: Value, by region
(Billion US dollars)

	1970	1975	1976	1977	1978	1979	1980	1981	1982	1983	1984	1985	1986	1987	1988	1989	1990	1991
Exports																		
Developed market economies	215.1	560.1	623.2	707.6	849.1	1042.2	1228.7	1210.1	1145.9	1132.9	1208.9	1251.1	1463.1	1710.9	1954.4	2090.5	2415.4	2462.5
North America	59.4	142.9	157.4	166.7	194.3	244.7	293.3	311.4	287.7	282.4	314.2	309.8	317.5	352.3	439.5	485.6	521.2	548.9
Western Europe [a]	131.9	348.0	383.4	443.1	535.4	667.4	771.5	713.7	685.1	669.1	684.3	721.8	887.3	1067.1	1181.6	1255.5	1513.8	1501.3
Southern Europe [b]	4.6	13.3	15.1	16.7	21.2	27.8	33.4	33.4	34.7	34.5	40.7	42.4	47.6	60.2	68.4	75.4	92.9	97.9
Japan	19.3	55.8	67.3	81.1	98.2	102.3	130.4	151.5	138.4	147.0	169.7	177.2	210.8	231.3	264.9	273.9	287.6	314.4
Developed market economies [c]	50.6	213.0	259.1	292.8	310.0	448.5	573.0	428.0	461.2	438.3	393.4	499.1	574.0	640.5	705.2	726.5	792.4	798.8
Oil-exporting countries [c]	17.4	120.8	146.5	159.3	155.6	236.4	300.8	178.8	173.1	154.8	100.9	121.2	120.6	148.5	177.6	158.6	178.1	164.2
Non-oil developing countries	33.2	92.2	112.7	133.5	154.4	212.1	272.3	249.2	288.1	283.5	292.5	377.9	453.4	492.0	527.6	567.9	614.3	634.6
Eastern Europe and the Soviet Union	32.6	87.0	71.2	82.5	92.3	111.9	128.0	132.3	135.9	134.1	134.4	131.5	137.5	143.8	144.0	142.5	138.5	104.8
Eastern Europe [d]	19.8	51.8	42.0	46.9	52.4	62.0	70.0	71.9	71.7	70.5	72.0	74.2	77.4	80.4	82.0	80.5	79.4	58.1
Soviet Union	12.8	35.2	29.2	35.5	39.8	49.9	57.9	60.4	64.2	63.6	62.4	57.3	60.0	63.4	62.0	62.3	59.1	46.7
Total above	298.3	860.1	953.5	1082.9	1251.4	1602.6	1929.7	1770.4	1743.0	1705.3	1736.6	1881.7	2174.6	2495.2	2803.6	2959.8	3346.3	3366.1
Memorandum item:																		
ECE region	228.4	591.3	627.1	709.0	843.2	1051.8	1226.2	1190.9	1143.4	1120.1	1173.5	1205.5	1389.8	1623.4	1833.5	1959.3	2266.3	2252.9
Imports																		
Developed market economies	224.6	589.0	679.7	770.1	888.6	1140.4	1368.9	1296.0	1215.0	1199.3	1310.0	1344.7	1523.3	1805.5	2037.0	2199.0	2541.8	2560.8
North America	56.6	142.0	172.7	202.5	232.3	278.9	319.5	343.4	313.0	334.7	424.2	433.1	467.8	517.0	572.3	612.7	640.2	633.7
Western Europe [a]	139.8	359.0	409.2	461.0	540.0	705.0	846.3	750.1	711.0	682.1	693.2	722.0	860.9	1047.0	1173.2	1254.6	1511.3	1527.6
Southern Europe [b]	9.2	30.2	32.9	35.2	36.4	46.7	61.8	59.6	59.9	56.1	56.9	59.0	67.1	90.4	104.2	122.0	154.9	162.6
Japan	18.9	57.9	64.9	71.3	79.9	109.8	141.3	142.9	131.5	126.4	136.2	130.5	127.6	151.0	187.4	209.7	235.4	236.8
Developed market economies	53.4	180.8	193.5	233.1	281.5	344.1	456.1	517.1	478.2	451.4	451.7	425.6	417.8	461.1	559.1	597.7	657.4	753.5
Oil-exporting countries [c]	8.9	50.7	63.1	83.9	94.7	97.8	131.3	158.1	160.5	142.3	126.0	105.1	91.8	91.1	104.1	102.7	118.2	128.2
Non-oil developing countries	44.5	130.1	130.4	149.2	186.8	246.3	324.8	359.0	317.7	309.1	325.7	320.5	326.0	370.0	455.0	495.0	539.2	625.3
Eastern Europe and the Soviet Union	33.1	100.2	81.3	89.3	101.5	117.8	133.4	134.8	127.1	122.0	120.7	125.2	132.7	132.8	135.9	143.1	149.2	106.5
Eastern Europe [d]	21.3	61.4	50.6	57.1	63.0	73.4	81.2	77.9	70.0	66.7	66.7	70.5	77.7	79.0	77.8	78.1	84.2	61.1
Soviet Union	11.7	38.8	30.8	32.2	38.5	44.4	52.2	56.9	57.1	55.4	53.9	54.8	55.0	53.8	58.0	65.0	65.0	45.4
Total above	311.0	870.1	954.6	1092.5	1271.6	1602.2	1958.5	1947.8	1820.7	1772.7	1882.8	1895.5	2073.8	2399.5	2732.0	2939.7	3348.5	3420.7
Memorandum item:																		
ECE region	238.8	631.4	696.1	788.1	910.2	1148.3	1361.0	1287.9	1211.0	1194.9	1294.9	1339.4	1528.4	1787.3	1985.5	2132.3	2455.6	2430.5

Sources: IMF, *International Financial Statistics*, March 1993 and ECE secretariat calculations, based on national publications for the countries of eastern Europe and the Soviet Union, using consistent rouble/dollar crossrates for recalculation of trade flows denominated in transferable roubles, as explained in the note to Appendix table C.4.

[a] Austria, Belgium-Luxembourg, Denmark, Finland, France, the Federal Republic of Germany, Ireland, Italy, the Netherlands, Norway, Sweden, Switzerland and the United Kingdom.
[b] Greece, Portugal, Spain and Turkey.
[c] IMF definition covering Algeria, Indonesia, the Islamic Republic of Iran, Iraq, Kuwait, Libyan Arab Jamahiriya, Nigeria, Qatar, Saudi Arabia, the United Arab Emirates and Venezuela.
[d] Including Yugoslavia, the former German Democratic Republic (until 1989), but excluding Albania.

APPENDIX TABLE C.2

World trade: Volume change, by region
(Annual percentage change)

	1970	1975	1976	1977	1978	1979	1980	1981	1982	1983	1984	1985	1986	1987	1988	1989	1990	1991
Exports																		
Developed market economies	10.3	-4.4	11.3	4.5	6.7	6.2	4.6	2.4	-2.1	2.3	9.6	3.9	1.2	5.4	7.9	7.1	5.2	3.3
North America	8.9	-3.2	5.3	1.8	11.3	7.3	7.5	-2.0	-9.3	-2.5	9.8	0.4	1.0	10.2	15.5	8.1	7.3	5.2
Western Europe	10.1	-5.8	12.6	5.0	5.5	6.8	0.8	2.8	1.1	2.9	7.6	5.1	2.0	4.1	5.7	7.0	3.9	2.1
Southern Europe[a]	17.5	-1.9	14.2	6.7	12.0	8.4	7.8	7.5	12.9	7.1	17.6	6.6	-3.3	12.4	1.7	12.2	9.5	9.3
Japan	15.8	-0.2	21.1	8.6	0.7	0.2	17.1	10.7	-2.3	8.7	15.8	4.9	-0.6	0.4	4.4	4.1	5.6	2.5
Developed market economies[b]	9.6	-8.7	13.9	2.6	0.6	5.1	-9.0	-5.5	-6.9	1.7	5.5	0.4	6.2	12.7	12.0	7.1	4.9	7.6
Oil-exporting countries[c]	18.1	-14.0	13.4	0.4	-4.2	2.8	-15.8	-13.9	-14.1	-4.7	-0.7	-4.1	1.5	5.7	11.0	10.6	1.7	2.7
Non-oil developing countries	5.0	-	14.6	5.1	6.3	9.3	2.1	4.7	0.2	6.7	10.4	3.6	8.1	15.6	12.6	5.9	6.2	9.5
Eastern Europe and the Soviet Union	7.8	4.4	8.3	9.5	4.8	3.8	2.3	2.3	4.9	5.5	4.6	-1.0	4.4	2.4	4.3	-0.9	-11.6	..
Eastern Europe[d]	10.7	7.0	7.9	8.0	6.7	7.9	3.1	2.7	5.3	8.0	7.0	2.5	-1.2	1.4	3.6	-1.9	-7.9	..
Soviet Union	6.1	2.5	8.6	10.6	3.4	0.6	1.6	1.9	4.5	3.3	2.5	-4.3	10.0	3.3	4.8	-	-13.1	..
Total above	10.0	-5.0	11.8	4.3	4.9	5.7	0.7	0.2	-3.0	2.3	8.1	2.7	2.8	7.2	8.8	6.6	4.1	..
Memorandum item																		
ECE region	9.8	-4.0	10.3	4.7	7.2	6.6	2.9	1.7	-0.8	1.9	8.2	3.3	1.8	5.8	7.9	6.7	3.4	..
Imports																		
Developed market economies	9.9	-7.7	14.8	4.0	6.2	7.3	-1.7	-2.0	-0.8	4.6	11.6	6.0	7.2	7.1	7.8	6.8	4.9	2.8
North America	2.0	-10.5	18.5	8.5	8.8	2.2	-6.8	2.6	-7.5	10.5	23.2	9.0	10.1	3.3	5.7	4.2	1.8	0.9
Western Europe	12.2	-5.8	14.4	1.7	5.1	9.4	1.1	-4.7	2.0	2.6	6.1	5.4	5.6	6.7	7.2	7.1	5.6	3.6
Southern Europe[a]	17.4	0.7	11.1	8.9	-0.1	5.7	6.9	6.1	7.9	3.4	10.6	6.4	-	38.5	9.5	16.1	13.6	7.2
Japan	18.7	-12.2	8.9	2.5	6.6	11.4	-5.1	-2.2	-0.6	1.2	10.5	-	10.5	9.0	16.6	7.9	6.0	1.9
Developed market economies[b]	8.8	6.5	5.7	10.9	6.3	2.0	9.3	8.0	-4.3	-3.3	1.2	1.0	-2.3	7.3	12.0	8.0	5.4	9.3
Oil-exporting countries[c]	6.6	45.2	23.6	22.4	0.3	-10.2	17.2	22.7	-1.8	-12.6	-6.6	-10.5	-17.6	-4.8	5.6	11.5	1.8	6.3
Non-oil developing countries	9.2	-1.6	0.7	6.9	8.8	6.4	6.8	2.2	-5.0	1.5	6.6	3.3	5.8	11.4	13.9	7.0	6.4	10.0
Eastern Europe and the Soviet Union	11.0	8.5	6.8	3.4	8.4	2.0	3.9	0.4	1.7	4.0	4.1	5.3	-0.6	1.0	3.6	5.0	-5.1	..
Eastern Europe[d]	14.0	4.3	6.7	5.2	5.0	2.7	1.3	-4.3	-5.3	4.0	3.9	5.8	4.8	3.4	3.3	1.2	-8.6	..
Soviet Union	7.8	14.8	7.0	0.9	13.3	1.1	7.5	6.4	9.7	4.0	4.4	4.7	-6.0	-1.6	4.0	9.3	-1.4	..
Total above	9.7	-3.5	12.2	5.5	6.3	5.8	1.2	0.4	-1.4	2.8	8.8	4.8	4.6	6.8	8.5	7.0	4.4	..
Memorandum item																		
ECE region	9.0	-5.9	14.9	4.4	6.4	6.2	-1.0	-1.4	-0.8	5.4	11.7	6.6	6.3	6.5	6.5	6.4	3.8	..

Sources: IMF, *International Financial Statistics*, March 1993 for the developed market economies; and IMF, *World Economic Outlook*, October 1992 for developing countries; ECE secretariat calculations, based on national sources for southern Europe and the countries of eastern Europe and the Soviet Union. For market economies and eastern Europe, weights for aggregations are US dollar trade shares in 1985, as reflected in Appendix table C.1.

[a] Includes ECE, OECD or World Bank estimates for certain sub-periods for Portugal and Turkey.
[b] IMF definitions for developing countries.
[c] See the definition applied in Appendix table C.1.
[d] Excludes Yugoslavia and Albania, but includes the former German Democratic Republic until 1990.

Appendix C. International trade and payments 281

APPENDIX TABLE C.3
Western Europe and North America: Trade volume change
(Annual percentage change) [a]

	1970	1975	1976	1977	1978	1979	1980	1981	1982	1983	1984	1985	1986	1987	1988	1989	1990	1991
Exports																		
France	15.9	-4.2	9.1	6.5	6.1	10.0	2.1	2.9	-2.9	3.4	5.4	2.7	0.5	3.7	8.7	7.8	5.0	4.7
Germany [b]	14.7	-11.2	18.6	3.9	3.2	4.9	1.7	6.6	3.3	-0.3	9.1	5.9	1.3	2.9	6.7	8.1	1.5	-0.1
Italy	8.0	3.7	11.7	7.6	10.8	7.8	-7.9	4.2	0.5	3.6	6.5	7.5	1.8	2.0	6.0	8.6	3.3	0.1
United Kingdom	2.3	-2.5	9.0	8.8	2.7	3.4	1.2	-1.3	3.3	2.2	8.4	5.8	4.0	5.6	2.5	4.3	6.6	1.4
Total 4 countries	10.7	-5.4	13.0	6.2	5.0	6.2	-	3.6	1.4	1.7	7.7	5.4	1.8	3.5	6.1	7.3	3.7	1.4
Austria	7.7	-7.0	15.7	3.2	10.1	12.9	2.9	5.1	1.1	4.4	9.3	10.1	0.8	2.5	12.1	13.6	6.1	5.7
Belgium	8.5	-8.3	13.6	2.7	3.9	7.5	2.3	-	2.3	3.3	4.3	3.1	5.0	6.7	6.3	7.6	3.1	3.0
Denmark	6.7	-3.3	3.4	8.3	6.3	10.3	8.0	2.5	2.4	4.7	6.7	5.3	-	4.0	4.8	7.3	1.7	7.6
Finland	8.0	-17.2	17.0	9.7	7.4	9.6	8.7	3.4	-2.2	3.4	9.9	-	-1.0	2.0	1.9	-	3.8	-9.2
Ireland	8.4	7.3	3.8	17.7	12.0	8.1	5.4	2.3	6.2	12.3	17.9	5.3	4.0	14.2	7.0	11.3	8.6	5.5
Netherlands	11.9	-5.6	13.2	-1.3	3.9	8.9	-	1.2	-1.1	4.7	7.8	3.1	3.0	5.8	8.3	4.2	4.9	6.2
Norway	4.5	4.0	15.3	-3.4	23.3	5.5	5.8	-2.0	1.3	7.4	9.1	3.2	-4.1	8.6	-2.0	15.0	7.0	6.7
Sweden	13.3	-9.9	4.7	1.5	7.4	8.2	-2.5	1.3	2.6	12.5	7.8	3.1	3.0	3.9	2.8	1.8	0.9	-1.8
Switzerland	6.3	-7.9	12.1	11.5	4.9	1.9	2.0	4.4	-3.6	-	6.9	13.0	2.1	0.5	-2.5	5.2	4.5	-1.4
Total 9 countries	9.2	-6.6	11.7	2.7	6.6	7.8	2.2	1.5	0.5	5.1	7.5	4.6	2.4	5.2	5.0	6.5	4.2	3.3
Total western Europe	10.1	-5.8	12.6	5.0	5.5	6.8	0.8	2.8	1.1	2.9	7.6	5.1	2.0	4.1	5.7	7.0	3.9	2.1
United States	8.5	-2.1	3.5	-0.2	11.8	8.9	10.0	-3.2	-11.7	-5.5	6.8	-2.0	-0.5	13.1	18.0	10.6	8.1	6.6
Canada	10.0	-7.3	11.9	9.0	9.9	1.8	-1.4	2.8	-0.5	7.5	18.4	6.5	4.5	3.6	9.3	1.4	4.7	1.0
Total above	9.7	-4.9	10.0	3.9	7.4	6.9	3.0	1.1	-2.5	1.2	8.3	3.7	1.7	5.9	8.7	7.4	5.0	3.1
Imports																		
France	6.6	-7.1	20.8	0.8	5.2	11.7	6.2	-3.3	3.6	-1.9	2.3	4.1	3.3	7.1	6.5	8.7	5.1	2.3
Germany [b]	18.4	-0.3	17.8	2.3	6.8	7.5	-	-5.0	1.4	3.9	5.3	4.2	6.1	5.4	6.4	7.3	11.9	12.7
Italy	15.6	-10.7	15.6	-0.4	7.5	13.1	2.8	-6.6	-	-	9.0	8.8	4.5	9.5	6.9	8.3	4.5	4.5
United Kingdom	4.9	-8.5	7.1	1.8	6.5	8.2	-4.7	-3.9	5.9	8.6	10.7	3.6	7.2	7.1	13.7	7.9	1.3	-2.9
Total 4 countries	11.9	-5.9	15.5	1.3	6.5	9.7	1.0	-4.7	2.6	2.7	6.5	4.9	5.4	6.9	8.2	8.0	6.3	5.1
Austria	18.1	-6.8	23.0	9.9	-1.6	10.5	5.0	-4.1	-2.1	7.1	8.7	4.8	5.2	5.2	8.2	14.0	8.1	3.0
Belgium	10.2	-6.3	13.5	2.4	3.5	7.9	1.0	-3.1	-	-	4.3	2.0	8.0	7.4	6.0	-6.5	4.6	3.6
Denmark	9.7	-6.5	19.4	-2.3	4.8	5.7	-8.6	-5.9	3.7	4.8	6.9	7.5	-1.0	6.1	-1.0	1.9	0.9	4.7
Finland	21.1	-	-3.5	-8.4	-5.3	19.4	11.6	-6.3	1.1	3.3	-	6.4	6.0	7.5	9.6	10.4	-3.6	-16.5
Ireland	3.6	-10.1	15.3	12.6	17.0	13.8	-4.9	1.2	-4.0	3.4	9.5	3.3	3.0	6.2	4.7	12.9	7.0	0.7
Netherlands	13.2	-5.2	12.3	1.2	6.0	6.8	-2.1	-7.6	1.2	2.3	6.8	6.4	4.0	6.7	7.2	2.5	5.7	3.1
Norway	12.7	2.7	11.4	7.9	-11.3	5.3	10.8	-4.0	4.2	-4.0	13.5	12.0	14.8	-1.8	-9.9	-4.9	9.9	2.6
Sweden	11.5	-18.2	5.3	-3.8	-5.2	15.1	2.4	-7.0	6.3	3.5	4.5	8.7	4.0	7.7	5.4	6.8	0.8	-6.3
Switzerland	13.7	-	16.7	9.5	9.1	7.0	7.2	-1.6	-2.1	5.0	0.7	11.6	8.3	6.5	7.7	5.3	2.7	-1.3
Total 9 countries	12.5	-5.7	12.4	2.3	2.7	8.8	1.4	-4.8	0.9	2.5	5.5	6.4	5.8	6.3	5.3	5.6	4.2	0.6
Total western Europe	12.2	-5.8	14.4	1.7	5.1	9.4	1.1	-4.7	2.0	2.6	6.1	5.4	5.6	6.7	7.2	7.1	5.6	3.6
United States	3.2	-12.0	21.8	10.7	10.3	0.1	-7.1	2.6	-5.1	10.4	24.0	8.7	10.5	2.6	3.8	3.9	2.2	0.6
Canada	-3.1	-5.4	7.8	0.5	3.1	11.1	-5.6	2.8	-16.4	11.0	19.7	10.4	8.1	6.2	14.2	5.4	0.1	2.4
Total above	8.7	-7.4	15.7	4.0	6.4	6.8	-1.6	-2.4	-1.2	5.1	11.8	6.7	7.2	5.4	6.7	6.0	4.2	2.6

Sources: IMF, *International Financial Statistics*, March 1993. IMF, *World Economic Outlook*, December 1992.

Note: National data are aggregated by means of weights derived from 1985 US dollar trade shares.

[a] Calculated from rounded index numbers (1985 = 100) for Belgium, Denmark, Finland, Netherlands, Norway, Sweden and Switzerland. Comparisons with the previous year are limited due to changes in methodology in France (1973, 1975, 1986), Italy (1970, 1981), United Kingdom (1970, 1973), Austria (1979,1988), Belgium (1974), Denmark (1971, 1974, 1985), Finland (1970, 1977), Greece (1977), Ireland (1975, 1985), Norway (1970, 1980), Sweden (1975, 1983), Netherlands (1985), United States (1989), Switzerland (1979 imports only, 1988) and Canada (1981).
[b] Data refer to west Germany only.

APPENDIX TABLE C.4

Eastern Europe and the Soviet Union: Exports by main directions

(Value, billion US dollars)

	1970	1975	1976	1977	1978	1979	1980	1981	1982	1983	1984	1985	1986	1987	1988	1989	1990	1991
Bulgaria																		
World	2.00	5.17	3.76	4.43	5.11	6.22	7.16	7.41	7.51	7.09	7.23	7.39	7.60	7.84	7.55	6.65	5.23	3.44
ECE-East	1.51	3.95	2.56	3.00	3.32	3.70	3.87	3.85	4.02	4.05	3.91	4.15	5.00	4.90	4.60	4.14	2.79	1.89
ECE-West	0.32	0.56	0.67	0.72	0.83	1.49	1.84	1.63	1.46	1.43	1.43	1.32	1.03	1.11	1.15	1.36	1.24	0.90
Other	0.17	0.67	0.53	0.72	0.96	1.03	1.44	1.92	2.02	1.61	1.90	1.92	1.57	1.82	1.80	1.16	1.20	0.65
Czechoslovakia																		
World	3.79	8.70	6.40	7.29	8.24	9.36	10.48	10.40	10.34	10.09	10.00	10.55	12.16	12.36	12.38	11.99	10.73	11.33
ECE-East	2.43	5.78	3.76	4.20	4.72	5.19	5.41	5.52	5.63	5.36	5.32	5.60	6.67	6.69	6.26	5.59	3.97	3.72
ECE-West	0.92	1.99	1.85	2.16	2.41	2.96	3.58	3.27	3.22	3.16	3.18	3.34	3.74	4.05	4.46	4.91	5.53	6.45
Other	0.43	0.93	0.79	0.93	1.11	1.21	1.48	1.61	1.49	1.57	1.49	1.61	1.75	1.62	1.66	1.49	1.23	1.16
ex-GDR																		
World	4.58	11.05	8.35	8.76	9.85	11.42	12.52	14.03	15.17	15.59	15.52	17.64	16.86	16.47	15.81	16.00	16.54	...
ECE-East	3.13	7.74	4.68	5.28	6.02	6.91	6.65	6.79	6.79	6.60	6.37	6.73	7.48	7.28	6.53	6.03	5.84	...
ECE-West	1.10	2.53	2.93	2.68	2.83	3.36	4.42	5.69	6.52	7.31	7.63	9.06	7.68	7.52	7.90	8.63	9.49	...
Other	0.36	0.78	0.73	0.80	1.00	1.15	1.45	1.55	1.87	1.68	1.52	1.84	1.70	1.67	1.38	1.34	1.21	...
Hungary																		
World	2.32	6.06	4.93	5.82	6.35	7.93	8.61	8.73	8.86	8.77	8.62	8.47	9.17	9.58	10.00	9.67	9.55	10.23
ECE-East	1.44	4.10	2.73	3.25	3.45	4.14	4.33	4.65	4.62	4.32	4.17	4.43	4.95	4.79	4.46	3.96	2.98	1.96
ECE-West	0.69	1.45	1.68	1.97	2.22	2.91	3.27	2.90	2.93	3.19	3.28	2.90	3.19	3.74	4.31	4.64	5.57	7.20
Other	0.19	0.52	0.52	0.60	0.68	0.88	1.01	1.18	1.31	1.25	1.17	1.13	1.04	1.05	1.23	1.07	1.00	1.07
Poland																		
World	3.55	11.11	8.59	9.60	10.75	12.53	13.07	10.08	10.53	10.44	10.90	11.10	13.13	14.09	14.57	14.67	18.29	14.91
ECE-East	2.13	6.60	3.93	4.46	5.07	5.80	5.21	4.37	4.88	4.80	4.90	5.18	6.47	6.63	6.27	5.99	6.10	2.51
ECE-West	1.09	3.52	3.69	4.03	4.55	5.31	6.09	4.13	3.96	4.12	4.49	4.48	4.61	5.63	6.52	6.94	9.55	11.20
Other	0.33	0.99	0.96	1.10	1.14	1.42	1.77	1.58	1.69	1.53	1.51	1.44	2.05	1.84	1.78	1.74	2.64	1.20
Romania																		
World	1.85	5.66	5.09	5.79	6.47	8.01	9.22	10.29	8.97	8.60	9.47	8.38	8.16	8.58	8.97	8.08	4.57	4.25
ECE-East	0.92	2.31	1.47	1.86	1.98	2.10	2.43	2.62	2.30	2.09	2.07	2.27	2.68	2.44	2.28	1.99	1.09	1.20
ECE-West	0.66	2.05	2.34	2.31	2.86	3.86	4.32	4.13	3.45	3.72	4.16	3.79	3.38	4.09	4.30	4.24	2.60	2.08
Other	0.27	1.30	1.29	1.63	1.62	2.05	2.47	3.54	3.22	2.79	3.24	2.32	2.10	2.04	2.39	1.85	0.89	0.97
Eastern Europe																		
World	18.10	47.77	37.12	41.69	46.77	55.46	61.06	60.94	61.38	60.59	61.74	63.53	67.07	68.93	69.29	67.06	64.84	46.66
ECE-East	11.57	30.48	19.13	22.04	24.56	27.85	27.90	27.81	28.24	27.22	26.74	28.35	33.24	32.73	30.41	27.70	23.79	9.23
ECE-West	4.78	12.10	13.15	13.87	15.70	19.88	23.54	21.75	21.54	22.95	24.17	24.90	23.63	26.15	28.64	30.72	33.85	27.82
Other	1.75	5.19	4.83	5.78	6.51	7.74	9.63	11.38	11.60	10.43	10.83	10.27	10.20	10.05	10.24	8.64	8.22	9.61
Soviet Union																		
World	12.80	35.20	29.20	35.53	39.81	49.92	57.94	60.42	64.20	63.60	62.36	57.32	60.04	63.41	62.02	62.29	59.06	...
ECE-East	6.76	17.99	11.04	13.03	14.87	16.68	17.81	19.01	18.66	17.76	17.53	18.09	21.93	19.90	17.07	15.17	11.10	...
ECE-West	2.78	9.74	11.23	13.00	13.81	20.85	26.70	26.14	27.90	28.23	27.96	23.81	19.83	23.81	25.34	27.44	31.05	...
Other	3.26	7.47	6.93	9.50	11.14	12.40	13.43	15.28	17.64	17.60	16.87	15.42	18.29	19.71	19.61	19.67	16.91	...
Eastern Europe and the Soviet Union																		
World	30.90	82.97	66.32	77.22	86.58	105.39	119.00	121.36	125.58	124.19	124.10	120.84	127.12	132.34	131.31	129.34	123.90	...
ECE-East	18.33	48.47	30.17	35.07	39.42	44.53	45.71	46.81	46.91	44.98	44.27	46.44	55.17	52.64	47.48	42.88	34.89	...
ECE-West	7.56	21.83	24.39	26.87	29.51	40.72	50.24	47.89	49.43	51.18	52.13	48.71	43.46	49.94	53.97	58.16	64.90	...
Other	5.01	12.66	11.76	15.28	17.65	20.13	23.05	26.66	29.24	28.03	27.69	25.68	28.49	29.76	29.85	28.30	25.13	...

Source: Secretariat of the United Nations Economic Commission for Europe, based on national foreign trade statistics.

Notes: Dollar values of intra-eastern and total trade were adjusted to avoid the distortions stemming from mutually inconsistent national rouble/dollar crossrates, as explained in the following paragraph. Partner country groupings: ECE-East – Soviet Union and east European former member countries of the CMEA; ECE-West – ECE market economies and Japan; Other – all remaining countries.

As an approximation to a consistent dollar valuation of rouble-denominated intra-group trade flows, the national-currency data on trade with the market economies were revalued, in national currency terms, at a common rouble-dollar crossrate and reaggregated with the data on trade with the "socialist" countries to obtain new trade totals (see United Nations Economic Commission for Europe, *Economic Bulletin for Europe*, Vol.43(1991), New York, 1991, pp.52-53 and 58-62). To generate the longest revalued series obtainable, Hungarian crossrates (available from 1976) were used. There is thus a substantial discontinuity in the time series for 1975/76. While data for prior years may overvalue rouble-denominated trade flows, they are at least internally consistent as all countries used similar crossrates during the early period. The procedure must be considered an approximation to the desired standard of consistent valuation because intra-group trade flows contained convertible-currency components (which in 1990 were significant).

Appendix C. International trade and payments 283

APPENDIX TABLE C.5
Eastern Europe and the Soviet Union: Imports by main directions
(Value, billion US dollars)

	1970	1975	1976	1977	1978	1979	1980	1981	1982	1983	1984	1985	1986	1987	1988	1989	1990	1991
Bulgaria																		
World	1.83	5.91	4.00	4.46	5.15	5.74	6.32	7.17	7.26	6.96	6.83	7.57	8.68	8.22	8.13	7.33	5.58	2.71
ECE-East	1.33	4.20	2.59	3.05	3.59	3.95	4.09	4.31	4.44	4.29	4.09	4.24	5.00	4.82	4.07	3.37	2.58	1.33
ECE-West	0.38	1.36	1.10	1.06	1.18	1.38	1.74	2.24	1.99	1.79	1.83	2.11	2.42	2.61	2.61	2.65	1.97	0.92
Other	0.13	0.36	0.32	0.35	0.39	0.42	0.49	0.63	0.82	0.87	0.91	1.22	1.26	0.89	1.44	1.31	1.03	0.46
Czechoslovakia																		
World	3.70	9.45	7.04	8.07	8.90	10.18	10.62	10.13	10.01	9.62	9.53	10.22	12.28	12.50	12.18	11.77	11.81	10.96
ECE-East	2.33	6.19	3.83	4.42	5.01	5.62	5.62	5.63	5.85	5.71	5.76	5.97	7.12	6.82	6.04	5.63	4.52	4.11
ECE-West	1.01	2.54	2.57	2.76	3.09	3.67	4.00	3.54	3.27	3.03	2.82	3.18	3.87	4.49	4.81	4.73	5.96	5.85
Other	0.35	0.72	0.65	0.89	0.80	0.90	1.01	0.96	0.90	0.88	0.95	1.07	1.28	1.20	1.33	1.41	1.33	1.00
ex-GDR																		
World	4.85	12.27	10.08	10.76	11.05	12.77	14.25	14.18	13.64	13.83	14.11	15.16	16.67	16.79	16.62	16.52	19.66	..
ECE-East	3.19	8.02	5.00	5.90	6.23	6.58	6.79	7.01	6.82	6.18	6.01	7.05	8.08	7.26	6.34	5.66	3.92	..
ECE-West	1.35	3.52	4.34	3.95	3.91	5.23	6.10	6.19	5.75	6.42	6.82	6.62	7.11	8.27	9.16	9.69	14.69	..
Other	0.30	0.73	0.73	0.90	0.91	0.96	1.37	0.98	1.07	1.23	1.28	1.49	1.48	1.25	1.12	1.16	1.05	..
Hungary																		
World	2.51	7.15	5.53	6.52	7.94	8.68	9.19	9.16	8.87	8.55	8.13	8.18	9.59	9.86	9.37	8.86	8.62	11.45
ECE-East	1.56	4.48	2.81	3.23	3.86	4.33	4.31	4.30	4.33	4.12	3.91	4.04	4.87	4.66	4.10	3.47	2.91	2.55
ECE-West	0.72	2.03	2.09	2.52	3.20	3.45	3.88	3.93	3.48	3.23	3.12	3.41	3.87	4.26	4.34	4.69	4.74	7.91
Other	0.22	0.64	0.64	0.77	0.88	0.83	1.01	0.93	1.06	1.21	1.10	0.73	0.85	0.94	0.93	0.69	0.97	0.99
Poland																		
World	3.61	13.31	11.47	11.88	12.65	14.00	14.71	11.39	9.50	9.38	9.72	10.43	12.31	12.69	12.99	12.94	12.62	15.53
ECE-East	2.36	6.18	3.90	4.63	5.17	5.64	5.90	5.66	5.29	5.13	5.27	5.51	7.08	6.50	5.59	4.92	4.26	2.95
ECE-West	0.98	6.34	6.85	6.36	6.44	6.76	6.82	4.60	3.32	3.24	3.36	3.80	4.15	4.75	5.96	6.57	6.19	10.90
Other	0.27	0.80	0.72	0.90	1.05	1.60	1.99	1.13	0.89	1.01	1.09	1.12	1.08	1.44	1.44	1.44	2.16	1.68
Romania																		
World	1.96	5.65	5.07	5.81	7.34	9.15	11.06	10.09	7.27	6.16	6.43	6.70	6.41	6.36	5.36	5.83	6.89	5.67
ECE-East	0.94	2.22	1.52	1.86	1.97	2.23	2.32	2.69	2.19	2.00	1.94	2.18	2.94	2.55	2.23	2.18	1.92	1.36
ECE-West	0.81	2.40	2.28	2.62	3.50	4.01	4.26	3.61	2.01	1.43	1.58	1.55	1.55	1.12	1.02	1.03	2.24	2.45
Other	0.21	1.02	1.27	1.33	1.87	2.92	4.48	3.79	3.07	2.72	2.91	2.97	1.92	2.68	2.11	2.62	2.73	1.86
Eastern Europe																		
World	18.45	53.75	43.19	47.49	53.03	60.53	66.14	62.13	56.55	54.50	54.75	58.26	65.95	66.41	64.65	63.25	65.08	45.41
ECE-East	11.72	31.30	19.64	23.08	25.82	28.35	29.01	29.59	28.92	27.44	26.98	28.98	35.10	32.61	28.38	25.24	21.28	8.77
ECE-West	5.25	18.18	19.22	19.26	21.32	24.51	26.79	24.10	19.82	19.14	19.54	20.67	22.96	25.41	27.90	29.37	35.81	27.62
Other	1.48	4.27	4.32	5.15	5.89	7.68	10.34	8.43	7.81	7.92	8.24	8.61	7.89	8.39	8.37	8.64	9.18	9.02
Soviet Union																		
World	11.73	38.77	30.79	32.21	38.48	44.37	52.22	56.91	57.12	55.39	53.95	54.76	55.02	53.79	58.04	64.98	64.96	45.41
ECE-East	6.63	17.15	10.29	11.82	14.72	15.73	16.20	16.82	17.26	16.77	16.51	17.59	20.34	19.87	18.15	17.12	15.04	8.77
ECE-West	3.02	14.20	14.55	13.78	16.70	20.55	24.47	26.55	27.11	25.94	25.00	24.28	23.43	22.63	27.53	33.34	35.38	27.62
Other	2.08	7.42	5.95	6.60	7.06	8.09	11.49	13.81	12.74	12.67	12.44	12.90	11.24	11.30	12.36	14.52	14.54	9.02
Eastern Europe and the Soviet Union																		
World	30.18	92.52	73.98	79.70	91.51	104.90	118.36	119.03	113.67	109.89	108.70	113.02	120.97	120.21	122.69	128.24	130.04	..
ECE-East	18.35	48.45	29.94	34.90	40.53	44.07	45.27	46.13	46.19	44.22	43.49	46.57	55.43	52.48	46.53	42.37	36.33	..
ECE-West	8.27	32.38	33.77	34.90	38.02	45.06	51.26	50.65	46.93	45.08	44.54	44.94	46.40	48.04	55.43	62.71	71.19	..
Other	3.56	11.69	10.27	11.75	12.95	15.77	21.83	22.25	20.55	20.59	20.67	21.51	19.13	19.69	20.73	23.16	23.72	..

Source: Secretariat of the United Nations Economic Commission for Europe, based on national foreign trade statistics.

Notes: Dollar values of intra-eastern and total trade were adjusted to avoid the distortions stemming from mutually inconsistent national rouble dollar crossrates, as explained in the following paragraph. Partner country groupings: ECE-East – Soviet Union and east European former member countries of the CMEA; ECE-West – ECE: market economies and Japan; Other – all remaining countries.

As an approximation to a consistent dollar valuation of rouble-denominated intra-group trade flows, the national-currency data on trade with the market economies were revalued, in national currency terms, at a common rouble-dollar crossrate and reaggregated with the data on trade with the "socialist" countries to obtain new trade totals (see United Nations Economic Commission for Europe, *Economic Bulletin for Europe*, Vol.43(1991), New York, 1991, pp.52-53 and 58-62). To generate the longest revalued series obtainable, Hungarian crossrates (available from 1976) were used. There is thus a substantial discontinuity in the time series for 1975/76. While data for prior years may overvalue rouble-denominated trade flows, they are at least internally consistent as all countries used similar crossrates during the early period. The procedure must be considered an approximation to the desired standard of consistent valuation because intra-group trade flows contained convertible-currency components (which in 1990 were significant).

APPENDIX TABLE C.6

East-west trade: Value of western exports, by country of origin
(Billion US dollars)

	1970	1975	1976	1977	1978	1979	1980	1981	1982	1983	1984	1985	1986	1987	1988	1989	1990	1991
Austria	0.48	1.51	1.50	1.68	1.98	2.45	2.46	2.08	2.00	1.92	1.95	2.08	2.43	2.52	3.01	3.20	4.36	4.51
Belgium and Luxembourg	0.21	0.89	0.83	0.82	0.94	1.12	1.35	1.17	0.98	1.11	0.98	1.13	1.08	1.10	1.06	1.33	1.31	1.71
Denmark	0.11	0.31	0.27	0.29	0.34	0.36	0.34	0.29	0.22	0.22	0.27	0.30	0.41	0.33	0.43	0.58	0.84	1.14
Finland	0.35	1.28	1.45	1.65	1.71	1.74	2.76	3.63	3.71	3.41	2.72	3.11	3.51	3.36	3.43	3.62	3.67	1.57
France	0.71	2.78	2.86	3.14	3.31	4.42	5.04	4.01	2.99	3.49	3.18	3.12	2.93	3.28	3.62	3.62	3.94	4.27
Germany [a]	1.93	8.29	7.83	8.78	10.25	11.88	12.31	9.86	9.57	9.75	9.06	9.47	12.03	13.16	14.69	16.93	19.61	26.90
Greece	0.14	0.31	0.31	0.39	0.42	0.40	0.59	0.40	0.45	0.37	0.33	0.37	0.32	0.31	0.25	0.45	0.52	0.59
Iceland	0.01	0.04	0.04	0.06	0.06	0.07	0.09	0.08	0.06	0.06	0.07	0.07	0.06	0.06	0.08	0.07	0.05	0.01
Ireland	0.01	0.04	0.02	0.03	0.04	0.08	0.11	0.08	0.07	0.09	0.05	0.08	0.11	0.11	0.09	0.15	0.25	0.20
Italy	1.08	2.86	2.43	2.94	3.18	3.75	3.83	3.41	3.34	3.65	3.33	3.74	4.07	4.83	4.95	5.97	7.78	7.00
Netherlands	0.23	0.81	0.79	0.93	1.02	1.22	1.49	1.44	1.10	1.15	0.95	1.04	1.20	1.35	1.29	1.64	1.87	2.32
Norway	0.06	0.24	0.25	0.25	0.33	0.25	0.30	0.37	0.24	0.25	0.17	0.17	0.18	0.22	0.31	0.32	0.43	0.44
Portugal	0.01	0.04	0.08	0.08	0.07	0.09	0.09	0.08	0.09	0.08	0.09	0.09	0.08	0.09	0.10	0.15	0.10	0.04
Spain	0.09	0.27	0.31	0.29	0.37	0.57	0.62	0.65	0.41	0.52	0.57	0.68	0.45	0.55	0.49	0.65	0.68	0.90
Sweden	0.30	1.07	0.98	0.96	1.02	1.27	1.33	1.12	0.89	0.77	0.77	0.85	0.91	0.96	1.10	1.24	1.33	1.32
Switzerland	0.25	0.91	0.91	1.05	1.27	1.28	1.26	1.03	0.92	0.91	0.89	0.98	1.29	1.54	1.67	1.77	2.04	1.62
Turkey	0.08	0.12	0.17	0.18	0.33	0.30	0.49	0.35	0.31	0.24	0.28	0.33	0.30	0.32	0.58	1.01	0.97	1.14
United Kingdom	0.66	1.43	1.33	1.67	2.09	2.30	2.86	2.26	1.68	1.57	1.85	1.68	1.87	2.04	2.29	2.41	2.65	2.21
Western Europe	6.71	23.21	22.36	25.19	28.72	33.55	37.31	32.31	29.00	29.53	27.52	29.29	33.23	36.16	39.43	45.11	52.39	57.89
Canada	0.16	0.64	0.76	0.57	0.76	1.01	1.82	1.96	2.08	1.56	1.78	1.32	1.08	0.72	1.08	0.72	1.11	1.44
United States	0.49	3.09	3.73	2.85	3.97	6.05	4.12	4.61	3.87	3.30	4.47	3.74	2.42	2.59	4.06	5.68	4.72	5.03
North America	0.65	3.73	4.49	3.42	4.73	7.06	5.95	6.57	5.95	4.85	6.25	5.06	3.49	3.31	5.14	6.39	5.83	6.47
Japan	0.47	2.26	2.85	2.77	3.25	3.12	3.57	3.95	4.34	3.29	2.90	3.25	3.78	3.16	3.87	3.77	3.45	3.04
Developed market economies	7.83	29.19	29.70	31.38	36.70	43.72	46.82	42.83	39.29	37.68	36.67	37.59	40.51	42.63	48.45	55.28	61.67	67.40

Sources: United Nations commodity trade data (COMTRADE); OECD, *Monthly Statistics of Foreign Trade*, Series A, Paris, February 1993.

Note: Data cover reported balances (f.o.b.-f.o.b.) with seven east European countries (Albania, Bulgaria, Czechoslovakia, Hungary, Poland, Romania, Yugoslavia) and the Soviet Union.

[a] Including trade between the Federal Republic of Germany and the former German Democratic Republic through 1989.

Appendix C. International trade and payments

APPENDIX TABLE C.7
East-west trade: Value of western imports, by country of destination
(Billion US dollars)

	1970	1975	1976	1977	1978	1979	1980	1981	1982	1983	1984	1985	1986	1987	1988	1989	1990	1991
Austria	0.34	0.93	1.08	1.24	1.37	1.75	2.33	2.40	2.10	1.98	2.27	2.22	2.24	2.27	2.43	2.52	3.27	3.41
Belgium and Luxembourg	0.15	0.55	0.52	0.66	0.75	0.95	1.49	1.32	1.72	1.42	2.10	1.55	1.41	1.70	1.74	1.71	2.16	1.13
Denmark	0.12	0.45	0.51	0.42	0.49	0.67	0.67	0.45	0.51	0.44	0.54	0.52	0.50	0.51	0.51	0.54	0.70	0.78
Finland	0.39	1.43	1.48	1.64	1.64	2.42	3.57	3.53	3.48	3.44	3.06	2.93	2.59	3.06	2.80	3.18	3.03	2.25
France	0.43	1.58	1.87	2.15	2.40	3.17	5.09	4.85	3.99	3.90	3.73	3.66	4.25	4.41	4.73	4.69	5.90	5.52
Germany [a]	1.28	3.71	4.64	5.13	6.34	8.91	9.52	8.43	8.66	8.69	9.46	9.39	10.30	11.20	11.74	13.07	17.43	23.60
Greece	0.11	0.24	0.39	0.42	0.62	0.61	0.62	0.51	0.51	0.46	0.74	0.73	0.57	0.73	0.65	0.78	0.94	1.07
Iceland	0.01	0.05	0.06	0.07	0.06	0.09	0.10	0.08	0.09	0.08	0.08	0.07	0.06	0.08	0.08	0.08	0.10	0.07
Ireland	0.03	0.08	0.08	0.11	0.12	0.17	0.13	0.10	0.12	0.13	0.14	0.14	0.17	0.17	0.16	0.19	0.22	0.22
Italy	0.94	2.03	2.65	2.83	3.16	4.07	5.58	4.94	5.49	5.73	6.62	5.53	5.54	6.30	7.05	8.03	8.89	8.56
Netherlands	0.18	0.74	0.90	1.01	1.17	1.86	2.17	2.41	3.04	3.06	2.76	2.88	1.76	2.19	1.98	2.29	2.72	2.30
Norway	0.08	0.23	0.31	0.37	0.25	0.36	0.32	0.33	0.40	0.27	0.33	0.32	0.26	0.38	0.42	0.46	0.67	0.64
Portugal	0.01	0.08	0.15	0.14	0.12	0.18	0.21	0.24	0.11	0.10	0.08	0.07	0.08	0.08	0.09	0.10	0.09	0.12
Spain	0.06	0.41	0.39	0.32	0.35	0.51	0.70	0.73	0.76	0.77	0.81	0.64	0.57	1.23	1.52	1.71	1.81	1.30
Sweden	0.28	0.90	0.96	0.99	0.93	1.54	1.37	1.03	1.16	1.36	1.21	1.16	1.02	1.44	1.47	1.53	1.61	1.38
Switzerland	0.14	0.35	0.51	0.61	0.86	1.16	1.43	1.23	1.15	1.03	0.94	0.88	0.83	0.74	0.68	0.73	0.81	1.04
Turkey	0.10	0.22	0.28	0.32	0.35	0.57	0.77	0.79	0.39	0.70	0.88	0.60	0.81	0.87	1.03	1.42	2.09	1.91
United Kingdom	0.56	1.36	1.71	1.97	2.05	2.66	2.68	1.47	1.72	1.74	2.14	1.84	2.08	2.68	2.80	2.82	3.09	2.89
Western Europe	5.24	15.35	18.50	20.40	23.03	31.66	38.77	34.84	35.40	35.30	37.87	35.13	35.02	40.05	41.90	45.84	55.53	58.18
Canada	0.07	0.17	0.20	0.19	0.20	0.27	0.25	0.24	0.17	0.18	0.21	0.20	0.22	0.26	0.46	0.45	0.49	0.44
United States	0.29	0.90	1.14	1.14	1.48	1.57	1.66	1.82	1.30	1.60	2.37	2.28	2.33	2.51	2.79	2.63	2.82	2.59
North America	0.35	1.07	1.33	1.32	1.68	1.84	1.90	2.06	1.46	1.77	2.57	2.48	2.55	2.78	3.25	3.07	3.31	3.03
Japan	0.46	1.21	1.22	1.47	1.48	1.99	1.90	1.58	1.39	1.53	1.63	1.47	1.74	2.31	2.93	3.21	3.47	3.47
Developed market economies	6.05	17.62	21.05	23.20	26.20	35.48	42.58	38.48	38.25	38.60	42.07	39.08	39.32	45.14	48.07	52.13	62.31	64.68

Sources: United Nations commodity trade data (COMTRADE); OECD, *Monthly Statistics of Foreign Trade*, Series A, Paris, February 1993.

Note: Data cover reported balances (f.o.b.-f.o.b.) with seven east European countries (Albania, Bulgaria, Czechoslovakia, Hungary, Poland, Romania, Yugoslavia) and the Soviet Union.

[a] Including trade between the Federal Republic of Germany and the former German Democratic Republic through 1989.

APPENDIX TABLE C.8

East-west trade: Western trade balances by western country

(Billion US dollars)

	1970	1975	1976	1977	1978	1979	1980	1981	1982	1983	1984	1985	1986	1987	1988	1989	1990	1991
Austria	0.13	0.58	0.42	0.44	0.61	0.69	0.12	-0.32	-0.10	-0.06	-0.32	-0.14	0.19	0.25	0.58	0.68	1.09	1.11
Belgium and Luxembourg	0.06	0.34	0.31	0.16	0.19	0.17	-0.15	-0.14	-0.74	-0.31	-1.13	-0.41	-0.33	-0.60	-0.68	-0.37	-0.86	0.58
Denmark	-0.02	-0.14	-0.24	-0.13	-0.16	-0.31	-0.34	-0.17	-0.29	-0.23	-0.27	-0.22	-0.09	-0.18	-0.08	0.04	0.13	0.35
Finland	-0.04	-0.15	-0.03	0.01	0.07	-0.68	-0.82	0.11	0.23	-0.03	-0.34	0.18	0.92	0.30	0.63	0.44	0.64	-0.68
France	0.28	1.20	0.99	0.99	0.91	1.25	-0.05	-0.83	-1.00	-0.41	-0.55	-0.54	-1.32	-1.14	-1.11	-1.08	-1.96	-1.25
Germany [a]	0.65	4.58	3.19	3.65	3.91	2.97	2.79	1.42	0.90	1.06	-0.40	0.08	1.73	1.96	2.95	3.87	2.18	3.30
Greece	0.02	0.08	-0.08	-0.03	-0.20	-0.21	-0.02	-0.11	-0.07	-0.09	-0.41	-0.36	-0.25	-0.42	-0.40	-0.33	-0.42	-0.48
Iceland	-	-0.01	-0.02	-	-	-0.02	-0.01	-0.01	-0.03	-0.02	-0.01	-0.01	-	-0.02	-	-0.01	-0.05	-0.05
Ireland	-0.02	-0.05	-0.06	-0.07	-0.08	-0.09	-0.02	-0.02	-0.05	-0.04	-0.08	-0.06	-0.05	-0.06	-0.07	-0.04	0.04	-0.02
Italy	0.14	0.83	-0.22	0.11	0.02	-0.32	-1.76	-1.53	-2.15	-2.08	-3.29	-1.78	-1.47	-1.46	-2.11	-2.06	-1.11	-1.56
Netherlands	0.05	0.07	-0.11	-0.08	-0.15	-0.64	-0.68	-0.97	-1.95	-1.91	-1.81	-1.84	-0.55	-0.83	-0.69	-0.65	-0.86	0.02
Norway	-0.02	0.01	-0.06	-0.13	0.08	-0.11	-0.03	0.04	-0.16	-0.02	-0.16	-0.15	-0.08	-0.16	-0.12	-0.14	-0.24	-0.20
Portugal	-	-0.03	-0.07	-0.06	-0.05	-0.09	-0.12	-0.16	-0.02	-0.02	0.02	0.02	-	0.01	0.01	0.05	0.01	-0.08
Spain	0.03	-0.14	-0.08	-0.03	0.02	0.07	-0.08	-0.08	-0.35	-0.25	-0.24	0.03	-0.11	-0.67	-1.03	-1.07	-1.13	-0.40
Sweden	0.02	0.16	0.01	-0.03	0.08	-0.27	-0.04	0.09	-0.28	-0.59	-0.45	-0.32	-0.11	-0.48	-0.38	-0.28	-0.28	-0.05
Switzerland	0.11	0.57	0.40	0.43	0.41	0.12	-0.17	-0.20	-0.23	-0.12	-0.05	0.10	0.46	0.80	0.99	1.04	1.23	0.58
Turkey	-0.01	-0.10	-0.12	-0.15	-0.02	-0.27	-0.28	-0.45	-0.08	-0.46	-0.60	-0.27	-0.51	-0.54	-0.45	-0.41	-1.11	-0.78
United Kingdom	0.10	0.07	-0.38	-0.30	0.04	-0.36	0.18	0.79	-0.04	-0.17	-0.29	-0.16	-0.21	-0.64	-0.51	-0.41	-0.45	-0.67
Western Europe	1.48	7.86	3.86	4.79	5.68	1.89	-1.46	-2.53	-6.40	-5.77	-10.35	-5.84	-1.79	-3.89	-2.46	-0.73	-3.14	-0.29
Canada	0.09	0.47	0.56	0.38	0.56	0.74	1.58	1.72	1.91	1.38	1.57	1.11	0.86	0.46	0.62	0.27	0.62	0.99
United States	0.20	2.19	2.59	1.72	2.49	4.48	2.47	2.79	2.57	1.70	2.10	1.46	0.08	0.08	1.28	3.05	1.89	2.45
North America	0.29	2.66	3.16	2.10	3.05	5.22	4.04	4.51	4.48	3.08	3.67	2.58	0.94	0.54	1.90	3.32	2.52	3.44
Japan	0.01	1.05	1.63	1.29	1.77	1.13	1.67	2.37	2.95	1.76	1.27	1.78	2.04	0.85	0.94	0.56	-0.01	-0.44
Developed market economies	1.78	11.57	8.65	8.18	10.50	8.24	4.25	4.35	1.04	-0.93	-5.41	-1.49	1.19	-2.51	0.38	3.16	-0.64	2.71

Sources: United Nations commodity trade data (COMTRADE); OECD, *Monthly Statistics of Foreign Trade*, Series A, Paris, February 1993.

Note: Data cover reported balances (f.o.b.-f.o.b.) with seven east European countries (Albania, Bulgaria, Czechoslovakia, Hungary, Poland, Romania, Yugoslavia) and the Soviet Union.

[a] Including trade between the Federal Republic of Germany and the former German Democratic Republic through 1989.

APPENDIX TABLE C.9

East-west trade: Western exports, imports and balances by eastern country
(Billion US dollars)

	1970	1975	1976	1977	1978	1979	1980	1981	1982	1983	1984	1985	1986	1987	1988	1989	1990	1991
Western exports to:																		
Albania	0.02	0.06	0.04	0.06	0.07	0.09	0.12	0.13	0.15	0.11	0.11	0.10	0.08	0.08	0.11	0.16	0.19	0.26
Bulgaria	0.33	1.10	0.94	0.90	1.10	1.23	1.61	1.85	1.55	1.56	1.44	1.83	2.19	2.34	2.37	2.39	1.56	1.66
Czechoslovakia	0.78	1.86	2.06	2.06	2.32	2.74	2.94	2.30	2.12	1.91	1.87	2.12	2.70	3.28	3.49	3.54	4.75	6.18
Hungary	0.62	1.83	1.81	2.31	2.99	2.96	3.25	3.19	2.85	2.57	2.50	2.80	3.45	3.87	3.95	4.60	5.37	6.59
Poland	0.87	5.41	5.43	4.94	5.51	5.92	6.41	4.19	3.15	2.81	2.84	3.02	3.26	3.84	4.73	6.00	7.55	12.46
Romania	0.70	1.99	1.99	2.29	2.97	3.75	3.87	3.00	1.65	1.26	1.34	1.38	1.59	1.22	1.17	1.13	2.29	2.23
Yugoslavia, SFR	1.95	4.79	4.13	5.54	6.45	8.39	8.16	6.88	5.72	5.51	5.28	6.01	7.43	7.91	8.68	9.81	13.57	10.33
Eastern Europe	5.26	17.03	16.42	18.11	21.42	25.08	26.36	21.53	17.20	15.73	15.38	17.26	20.68	22.54	24.49	27.63	35.28	39.72
Soviet Union	2.57	12.16	13.28	13.27	15.28	18.64	20.46	21.30	22.09	21.95	21.29	20.34	19.83	20.08	23.96	27.65	26.39	27.68
Eastern Europe and Soviet Union	7.83	29.19	29.70	31.38	36.70	43.72	46.82	42.83	39.29	37.68	36.67	37.59	40.51	42.63	48.45	55.28	61.67	67.40
German Democratic Republic	1.09	2.72	2.99	3.08	3.77	4.96	5.39	4.90	4.32	4.66	4.02	4.19	5.36	6.67	7.00	7.39	15.21	..
Western imports (f.o.b.) from:																		
Albania	0.01	0.05	0.04	0.06	0.05	0.11	0.11	0.12	0.13	0.11	0.11	0.08	0.10	0.09	0.12	0.14	0.13	0.10
Bulgaria	0.22	0.37	0.45	0.48	0.55	0.85	0.94	0.81	0.76	0.69	0.70	0.67	0.71	0.72	0.71	0.75	0.94	1.21
Czechoslovakia	0.67	1.53	1.60	1.77	2.05	2.59	3.05	2.58	2.53	2.48	2.58	2.51	2.96	3.34	3.63	3.92	4.63	6.31
Hungary	0.50	1.18	1.36	1.59	1.80	2.41	2.69	2.38	2.16	2.22	2.43	2.49	2.87	3.55	3.92	4.30	5.44	6.43
Poland	0.98	2.98	3.40	3.63	4.04	4.75	5.18	3.41	3.13	3.10	3.70	3.69	3.98	4.67	5.42	5.81	8.45	9.45
Romania	0.51	1.56	1.85	1.79	2.19	3.05	3.30	3.35	2.44	2.59	3.51	3.28	3.43	3.88	3.82	3.68	2.59	2.23
Yugoslavia, SFR	0.86	1.72	2.24	2.46	2.82	3.55	3.87	3.20	3.33	3.85	4.27	4.62	5.89	7.44	8.66	9.45	11.44	10.74
Eastern Europe	3.74	9.38	10.95	11.78	13.50	17.31	19.13	15.85	14.47	15.03	17.30	17.33	19.95	23.70	26.28	28.06	33.62	36.47
Soviet Union	2.32	8.24	10.10	11.42	12.69	18.17	23.45	22.63	23.79	23.57	24.78	21.75	19.37	21.44	21.80	24.07	28.69	28.21
Eastern Europe and Soviet Union	6.05	17.62	21.05	23.20	26.20	35.48	42.58	38.48	38.25	38.60	42.07	39.08	39.32	45.14	48.07	52.13	62.31	64.68
German Democratic Republic	0.92	2.33	2.55	2.78	3.27	4.04	5.06	4.76	5.00	5.00	4.94	4.83	5.51	6.20	6.47	6.58	7.20	..
Western trade balances with:																		
Albania	0.01	0.02	0.01	0.01	0.03	-0.02	0.01	0.00	0.02	-0.00	-0.00	0.02	-0.03	-0.01	-0.01	0.02	0.06	0.17
Bulgaria	0.11	0.73	0.48	0.42	0.56	0.38	0.67	1.04	0.79	0.87	0.74	1.16	1.48	1.61	1.67	1.64	0.62	0.46
Czechoslovakia	0.11	0.33	0.47	0.29	0.27	0.14	-0.11	-0.27	-0.41	-0.56	-0.71	-0.39	-0.26	-0.07	-0.14	-0.38	0.12	-0.13
Hungary	0.13	0.65	0.45	0.72	1.19	0.55	0.57	0.81	0.70	0.35	0.06	0.30	0.57	0.33	0.03	0.30	-0.07	0.16
Poland	-0.11	2.43	2.03	1.31	1.47	1.18	1.23	0.78	0.02	-0.29	-0.86	-0.67	-0.73	-0.82	-0.69	0.19	-0.91	3.01
Romania	0.19	0.43	0.14	0.50	0.78	0.70	0.57	-0.35	-0.78	-1.33	-2.17	-1.89	-1.85	-2.66	-2.65	-2.55	-0.30	-0.00
Yugoslavia, SFR	1.09	3.06	1.89	3.08	3.63	4.84	4.30	3.68	2.39	1.66	1.01	1.39	1.54	0.47	0.02	0.36	2.13	-0.41
Eastern Europe	1.53	7.65	5.47	6.33	7.91	7.77	7.24	5.68	2.73	0.70	-1.92	-0.08	0.73	-1.16	-1.78	-0.43	1.66	3.24
Soviet Union	0.25	3.92	3.18	1.85	2.58	0.47	-2.99	-1.33	-1.69	-1.62	-3.49	-1.41	0.46	-1.35	2.16	3.58	-2.30	-0.53
Eastern Europe and Soviet Union	1.78	11.57	8.65	8.18	10.50	8.24	4.25	4.35	1.04	-0.93	-5.41	-1.49	1.19	-2.51	0.38	3.16	-0.64	2.71
German Democratic Republic	0.17	0.39	0.44	0.30	0.49	0.92	0.33	0.14	-0.68	-0.34	-0.92	-0.64	-0.16	0.46	0.53	0.81	8.01	..

Source: United Nations commodity trade data (COMTRADE); OECD, *Monthly Statistics of Foreign Trade*, Series A, Paris, February 1993.

APPENDIX TABLE C.10
Eastern Europe and the Soviet Union: Balance of payments in convertible currencies
(Billion US dollars)

	1970	1975	1976	1977	1978	1979	1980	1981	1982	1983	1984	1985	1986	1987	1988	1989	1990	1991
Bulgaria																		
Merchandise export	0.5	1.0	1.1	1.2	1.5	2.4	3.3	3.4	3.1	2.7	3.3	3.3	2.7	3.3	3.5	3.1	2.6	3.7
Merchandise import	0.5	1.8	1.5	1.5	1.5	1.8	2.5	3.1	2.6	2.7	3.0	3.7	3.5	4.2	4.5	4.3	3.4	3.8
Balance	-	-0.8	-0.4	-0.3	0.1	0.5	0.8	0.3	0.5	0.1	0.3	-0.4	-0.8	-1.0	-1.0	-1.2	-0.8	-
Invisibles	-	-0.1	-	0.1	-	0.1	0.1	0.3	0.3	0.2	0.4	0.3	0.1	0.2	0.1	-0.1	-0.4	-
Current account	-	-0.9	-0.4	-0.3	0.1	0.6	0.9	0.6	0.8	0.3	0.7	-0.1	-0.7	-0.8	-0.8	-1.3	-1.2	-0.1
Czechoslovakia																		
Merchandise export	0.8	2.1	2.1	2.5	2.9	3.5	4.4	4.2	4.1	4.0	4.0	3.9	4.3	4.5	5.0	5.4	6.0	8.3
Merchandise import	0.9	2.5	2.8	3.1	3.3	4.1	4.4	3.9	3.4	3.2	3.1	3.2	4.1	4.7	5.1	5.0	6.8	8.8
Balance	-0.1	-0.4	-0.7	-0.6	-0.4	-0.5	-	0.3	0.7	0.8	0.9	0.7	0.2	-0.1	-0.1	0.4	-0.8	-0.4
Invisibles	0.1	0.1	0.1	-	-	-	-0.3	-0.3	-0.2	-	0.2	0.1	0.2	0.2	0.2	-0.4	-0.3	0.8
Current account	-	-0.3	-0.6	-0.6	-0.4	-0.6	-0.3	-	0.4	0.9	1.1	0.7	0.4	0.1	0.1	0.3	-1.1	0.4
Hungary																		
Merchandise export	0.6	2.2	2.3	2.7	3.2	4.1	4.9	4.9	4.8	4.8	4.9	4.2	4.2	5.0	5.5	6.4	6.3	9.3
Merchandise import	0.7	2.5	2.5	3.0	4.0	4.2	4.6	4.4	4.2	4.1	4.0	4.1	4.7	5.0	5.0	5.9	6.0	9.1
Balance	-0.1	-0.3	-0.2	-0.4	-0.8	-0.2	0.3	0.4	0.7	0.8	0.9	0.1	-0.5	-	0.5	0.5	0.3	0.2
Invisibles	-	-0.2	-0.2	-0.4	-0.5	-0.7	-0.6	-1.2	-1.0	-0.7	-0.8	-1.0	-1.0	-0.9	-1.3	-2.0	-0.2	0.1
Current account	-0.1	-0.5	-0.4	-0.8	-1.2	-0.8	-0.4	-0.7	-0.3	0.1	0.1	-0.8	-1.5	-0.9	-0.8	-1.5	0.1	0.3
Poland																		
Merchandise export	1.1	4.1	4.3	4.7	5.3	6.1	7.4	5.0	4.5	4.8	5.3	5.1	5.3	6.2	7.2	7.6	10.9	12.8
Merchandise import	1.0	6.9	7.0	6.6	7.4	8.2	8.2	5.8	4.3	3.9	3.9	4.0	4.3	5.1	6.3	7.3	8.6	12.7
Balance	0.1	-2.8	-2.7	-1.9	-2.1	-2.1	-0.8	-0.8	0.3	0.9	1.4	1.1	1.0	1.0	0.9	0.2	2.2	0.1
Invisibles	0.1	-0.2	-0.1	-0.3	-0.4	-1.2	-1.8	-2.4	-2.5	-2.3	-2.1	-1.7	-1.7	-1.4	-1.5	-2.1	2.0	-1.4
Current account	0.1	-3.0	-2.8	-2.1	-2.4	-3.4	-2.5	-3.2	-2.3	-1.4	-0.8	-0.6	-0.7	-0.4	-0.6	-1.8	0.7	-1.4
Romania																		
Merchandise export	0.7	2.8	3.4	3.7	4.0	5.4	6.5	7.2	6.2	6.6	6.9	6.3	6.0	5.1	6.0	5.5	3.4	3.5
Merchandise import	0.8	2.9	3.3	3.8	4.6	6.5	8.0	7.0	4.7	4.6	4.7	4.8	4.0	2.8	2.5	3.0	5.1	4.9
Balance	-0.1	-0.1	0.1	-0.1	-0.6	-1.2	-1.5	0.2	1.5	1.7	2.2	1.4	1.9	2.3	3.5	2.6	-1.7	-1.4
Invisibles	-0.1	-0.1	-	-0.2	-0.2	-0.5	-0.9	-1.0	-0.9	-0.8	-0.6	-0.5	-0.5	-0.1	-	0.3	0.1	-
Current account	-0.2	-0.3	-0.1	-0.3	-0.8	-1.7	-2.4	-0.8	0.7	0.9	1.5	0.9	1.4	2.2	3.5	2.9	-1.6	-1.4
Yugoslavia, SFR																		
Merchandise export	-	-	4.9	5.2	5.8	6.8	9.1	10.4	10.5	6.3	6.6	6.5	7.2	8.6	9.6	10.5	11.8	12.3
Merchandise import	-	-	6.8	9.0	9.6	12.9	14.0	13.5	12.5	8.1	7.8	8.3	9.7	9.6	10.2	12.0	16.5	12.8
Balance	-	-	-1.9	-3.8	-3.8	-6.1	-4.9	-3.2	-2.0	-1.8	-1.2	-1.8	-2.5	-1.0	-0.6	-1.5	-4.7	-0.5
Invisibles	-	-	1.7	2.5	2.5	2.4	2.6	2.0	1.5	2.1	2.0	2.1	2.7	2.1	2.8	3.5	2.0	-0.7
Current account	-	-	0.2	-1.3	-1.3	-3.7	-2.3	-1.0	-0.5	0.3	0.9	0.3	0.2	1.1	2.2	2.0	-2.7	-1.2
Eastern Europe																		
Merchandise export	3.7	12.3	18.1	20.0	22.7	28.2	35.5	35.0	33.3	28.9	31.0	29.2	29.7	32.8	37.0	38.6	41.0	49.9
Merchandise import	3.9	16.7	10.3	27.0	30.4	37.8	41.7	37.7	31.7	26.4	26.6	28.1	30.3	31.5	33.7	37.5	46.4	52.1
Balance	-0.2	-4.4	-5.8	-7.0	-7.6	-9.6	-6.2	-2.7	1.6	2.5	4.5	1.2	-0.6	1.3	3.3	1.1	-5.4	-2.1
Invisibles	-0.1	-0.5	1.4	1.7	1.5	-	-0.9	-2.6	-2.8	-1.4	-1.0	-0.7	-0.2	-	0.4	-0.5	3.1	-1.3
Current account a	-0.2	-4.9	-4.4	-5.3	-6.1	-9.5	-7.0	-5.3	-1.2	1.1	3.5	0.5	-0.8	1.3	3.6	0.6	-2.3	-3.4
Soviet Union																		
Merchandise export	4.8	14.2	16.6	20.7	21.8	30.9	38.2	39.1	43.4	44.2	43.3	36.9	26.8	31.3	33.4	35.2	33.6	37.7
Merchandise import	4.3	18.7	19.3	18.8	20.9	26.7	34.8	39.9	39.1	38.0	36.6	36.2	23.2	23.1	28.7	35.4	35.3	35.3
Balance	0.4	-4.5	-2.7	1.9	0.9	4.2	3.4	-0.7	4.3	6.2	6.7	0.7	3.6	8.2	4.8	-0.1	-1.7	2.4
Invisibles	0.5	0.3	0.1	-	0.1	0.1	-0.4	-0.9	-0.7	-0.4	-	-0.7	-1.8	-1.7	-3.3	-3.8	-5.6	-6.6
Current account a	0.9	-4.2	-2.6	1.9	1.0	4.3	3.0	-1.6	3.6	5.8	6.7	0.1	2.7	7.3	2.9	0.8	-4.8	-0.8
Eastern Europe and the Soviet Union																		
Merchandise export	8.5	26.5	34.7	40.7	45.5	59.1	73.8	74.2	76.7	73.1	74.4	66.2	56.5	64.1	70.4	73.9	74.6	87.6
Merchandise import	8.2	35.3	29.6	45.9	52.2	64.5	76.5	77.5	70.8	64.4	63.2	64.3	53.5	54.6	62.4	72.9	81.7	87.3
Balance	0.2	-8.9	-8.5	-5.1	-6.7	-5.3	-2.8	-3.3	5.9	8.7	11.2	1.9	3.0	9.5	8.0	0.9	-7.1	0.3
Invisibles	0.4	-0.2	1.5	1.7	1.5	0.1	-1.3	-3.5	-3.5	-1.8	-1.0	-1.4	-2.0	-1.7	-3.0	-4.4	-2.5	-7.9
Current account a	0.7	-9.1	-7.0	-3.5	-5.1	-5.3	-4.0	-7.0	2.4	6.9	10.2	0.5	1.9	8.5	6.5	1.3	-10.6	-4.2

Sources: ECE secretariat Common Data Base. National statistics for Bulgaria, Czechoslovakia, Hungary (revised data for 1982-1991); IMF, *Balance of Payment Statistics*, 1971-1981 for Poland, Romania and Yugoslavia; IMF, *The economy of the former USSR in 1991*, April 1992. ECE estimates for the German Democratic Republic and the Soviet Union. For these two countries trade balances reflect trade with all developed and developing market economies (non-socialist countries *plus* Yugoslavia) based on national foreign trade statistics.

Note: Substantial discontinuities in the series are indicated by ‖.

a Includes gold sales 1986-1991.

APPENDIX TABLE C.11

Eastern Europe and the Soviet Union: Gross debt, foreign currency reserves and net debt in convertible currencies
(Billion US dollars)

	1970	1975	1976	1977	1978	1979	1980	1981	1982	1983	1984	1985	1986	1987	1988	1989	1990	1991
Gross debt																		
Bulgaria	0.7	2.7	3.3	3.8	4.4	4.6	4.9	4.1	3.5	3.1	2.9	4.1	5.5	7.4	9.1	10.7	10.0	11.4
Czechoslovakia	0.3	1.0	1.7	2.4	3.0	3.8	‖ 6.8	6.3	5.8	5.2	4.7	4.6	5.6	6.7	7.3	7.9	8.1	9.3
Hungary	0.8	4.2	5.2	6.3	9.5	10.5	11.5	10.7	‖ 10.2	10.7	11.0	14.0	16.9	19.6	19.6	20.4	21.3	22.7
Poland	1.2	8.4	12.1	14.9	18.6	23.7	24.1	25.9	26.3	26.4	26.9	29.3	33.5	39.2	39.2	40.8	48.5	48.4
Romania	1.0	2.9	2.8	3.6	5.1	7.2	9.6	10.2	9.8	8.9	7.2	6.6	6.4	5.7	1.9	0.7	0.4	1.9
Yugoslavia, SFR	2.1	6.0	7.5	10.1	12.5	16.0	18.5	20.6	19.9	19.0	18.8	18.4	19.2	20.5	18.9	17.3	16.5	14.5
Eastern Europe	6.0	25.2	32.7	41.1	53.0	65.7	75.3	77.8	75.5	73.2	71.6	77.0	87.1	99.0	96.0	97.8	104.8	108.2
Soviet Union	1.6	15.4	20.9	22.7	24.4	26.1	‖ 25.2	29.0	28.4	26.9	25.6	31.4	37.4	40.2	‖ 49.4	58.5	61.1	65.3
Eastern Europe and the Soviet Union	7.6	40.6	53.6	63.8	77.4	91.9	100.5	106.8	103.9	100.1	97.2	108.4	124.5	139.2	145.4	156.3	165.9	173.5
Addendum: GDR	1.1	5.2	6.0	7.5	9.3	11.1	‖ 13.6	14.4	12.6	12.1	11.6	13.6	16.1	19.1	20.2	20.6	‖ 33.0 [a]	..
Foreign currency reserves [b]																		
Bulgaria	-	0.4	0.4	0.5	0.6	0.7	0.8	0.8	1.0	1.2	1.4	2.1	1.4	1.1	1.8	..	..	0.3
Czechoslovakia	0.3	0.3	0.4	0.5	0.7	1.0	‖ 1.8	1.0	0.8	0.8	1.0	0.9	1.1	1.4	1.6	1.1	1.1	3.0
Hungary	0.2	0.9	1.2	1.1	0.9	1.2	1.4	0.9	0.7	1.1	1.6	2.2	2.3	1.6	1.5	1.2	1.1	3.9
Poland	0.3	0.6	0.8	0.4	0.8	0.6	0.1	0.3	0.6	0.8	1.1	0.9	0.7	1.5	2.1	2.3	4.5	3.6
Romania	-	0.5	0.5	0.2	0.4	0.5	0.3	0.4	0.4	0.5	0.7	0.2	0.6	1.4	0.8	1.8	0.4	0.4
Yugoslavia, SFR	..	0.8	2.0	2.0	2.3	1.2	1.4	1.5	1.0	0.9	1.2	1.1	1.5	0.7	2.3	4.1	5.5	2.7
Eastern Europe	0.8	3.5	5.3	4.8	5.7	5.2	5.8	4.9	4.3	5.3	6.9	7.3	7.5	7.7	10.0	10.6	12.5	14.0
Soviet Union	1.0	3.1	4.7	4.4	6.1	8.8	8.6	8.5	10.0	10.9	11.3	13.1	14.8	14.1	15.3	14.7	8.6	8.8
Eastern Europe and the Soviet Union	1.8	6.6	10.0	9.2	11.8	14.0	14.4	13.3	14.3	16.2	18.3	20.3	22.4	21.8	25.3	25.3	21.1	22.8
Addendum: GDR	0.2	1.6	0.8	0.9	1.3	2.0	2.0	2.2	1.9	3.4	4.5	6.2	7.5	9.0	9.5	9.5	9.6	..
Total assets [c]																		
Hungary	0.6	2.0	2.3	2.4	3.1	3.3	3.7	3.2	‖ 2.9	3.8	4.4	5.9	6.2	5.9	5.6	5.5	5.3	8.1
Romania	..	..	..	..	..	..	1.7	1.9	2.3	2.9	3.6	3.0	3.1	3.2	3.3	3.1	3.2	-
Net debt (reflecting foreign currency reserves)																		
Bulgaria	0.7	2.3	2.9	3.3	3.8	3.9	4.1	3.3	2.5	1.9	1.5	2.0	4.1	6.3	7.3	10.7	10.0	11.1
Czechoslovakia	-	0.7	1.3	1.9	2.3	2.8	‖ 5.0	5.4	5.0	4.4	3.8	3.8	4.5	5.3	5.7	6.8	7.0	6.3
Hungary	0.6	3.3	4.0	5.2	8.6	9.3	10.1	9.8	‖ 9.5	9.6	9.4	11.8	14.6	18.0	18.2	19.1	20.2	18.8
Poland	0.9	7.8	11.3	14.5	17.8	23.2	24.0	25.6	25.7	25.6	25.8	28.4	32.8	37.7	37.1	38.5	44.0	44.8
Romania	1.0	2.3	2.3	3.3	4.7	6.6	9.2	9.8	9.3	8.4	6.5	6.4	5.8	4.3	1.1	-1.1	-	-1.5
Yugoslavia, SFR	2.1	5.2	5.6	8.1	10.2	14.8	17.1	19.1	19.1	18.1	17.7	17.3	17.7	19.8	16.6	13.2	11.1	11.8
Eastern Europe	5.2	21.7	27.4	36.4	47.4	60.6	69.5	73.0	71.1	67.9	64.6	69.8	79.5	91.3	86.1	87.2	92.3	94.2
Soviet Union	0.6	12.3	16.2	18.3	18.3	17.3	16.6	20.5	18.4	16.0	14.2	18.3	22.5	26.1	34.1	43.8	52.5	56.5
Eastern Europe and the Soviet Union	5.8	34.0	43.6	54.7	65.7	77.9	86.1	93.5	89.5	83.9	78.9	88.1	102.1	117.4	120.1	131.0	144.8	150.7
Addendum: GDR	0.9	3.6	5.2	6.6	8.1	9.1	‖ 11.6	12.2	10.7	8.7	7.1	7.4	8.6	10.1	10.7	11.1	‖ 23.4	..
Net debt (reflecting total assets)																		
Hungary	0.2	2.2	3.0	3.9	6.3	7.2	7.8	7.5	7.3	7.0	6.5	8.0	10.7	13.7	14.0	14.9	16.0	14.6
Romania	..	..	..	..	..	5.2	7.8	8.2	7.5	6.0	3.6	3.6	3.3	2.5	-1.4	-2.4	-2.8	1.9

Sources: ECE secretariat Common Data Base. National statistics for Bulgaria (1980-1990), GDR (1989), Hungary (revised data), Poland, Romania (1970-1990), Yugoslavia (1980-1990) and Soviet Union (1988, 1990). Bulgaria, GDR and the Soviet Union (including CMEA banks) prior to discontinuity: BIS/OECD *Statistics on External Indebtedness: Bank and Trade Related Non-Bank External Claims on Individual Borrowing Countries and Territories*, Paris and Basle (various years). BIS/OECD figures are adjusted here to include gross claims of the Federal Republic of Germany *vis-à-vis* the GDR arising from clearing exchanges). For these three countries data reflect convertible currency debt *vis-à-vis* reporting institutions only and thus exclude any claims of developing countries. IMF, *International Financial Statistics*, February 1993; BIS, *International Banking and Financial Market Developments* (various issues); Yugoslavia: World Bank, *World tables 1989-1990*, prior to discontinuity.

Note: Substantial discontinuities in the series are indicated by '‖'.

[a] End-June 1990.
[b] For Bulgaria, GDR and Soviet Union, assets with BIS reporting banks; also for other countries prior to discontinuity in series. Hungarian gold reserves at national valuation: $275/oz. (1982-1985); $320/oz. (1986-1991). Romania's gold reserves valued at SDR 35/oz.
[c] International reserves *plus* other assets (mainly trade credits).

ECE SECRETARIAT

EMPLOYMENT OPPORTUNITIES

Although there is a freeze on recruitment at present, vacancies for professional posts are likely to occur in the future in a number of areas of the Commission's work in economics and statistics. Candidates should have an advanced university degree in economics, econometrics or statistics and at least three years relevant professional experience. Candidates for the more senior posts in economic analysis will be expected to have a record of professional publications, preferably in empirical economics, and in statistics experience of managing large projects and of international comparison work is required.

The emphasis in the Commission's work is increasingly on the problems of the transition to a market economy in eastern Europe and the countries of the former Soviet Union. Some experience in this area together with a knowledge of Slavic and other east European languages is highly desirable.

Vacancies are open to men and women but, because of their under-representation in the secretariat, applications from qualified women will be especially welcome.

Those interested in being placed on ECE's roster of candidates for professional posts should apply, stating area of interest and enclosing a detailed curriculum vitae, to:

Office of the Executive Secretary
United Nations Economic Commission for Europe
Palais des Nations
CH-1211 Geneva 10
Switzerland

Letters of application should refer to this notice, quoting the reference DEAP.

In addition to its regular review of recent economic developments in Europe and North America, the Division for Economic Analysis and Projections of the United Nations Economic Commission for Europe, in Geneva, also publishes research studies with a longer time perspective.

Recent publications include:

"East-west agricultural and food trade: development and prospects", *Economic Bulletin for Europe,* vol.40, No. 3, Published by Pergamon Press for the United Nations, 1988, chapter 3

"Long-term prospects in ECE countries: National and international aspects", *Economic Bulletin for Europe,* vol.40, No. 4, Published by Pergamon Press for the United Nations, 1988

"The effects of west European integration on imports of manufactures from eastern and southern Europe" *Economic Survey of Europe in 1988-1989,* New York, 1989, Sales No. E.89.II.E.1
"Retail trade in eastern Europe and the Soviet Union", ibid.

"A note on recent developments in east-west trade in services", *Economic Bulletin for Europe,* vol.41, (1989), Sales No. E.89.II.E.26
"East-west trade in investment goods, 1970-1987", ibid.

"Economic reform in the east: a framework for western support", *Economic Survey of Europe in 1989-1990,* New York, 1989, Sales No. E.90.II.E.1
"The broader policy framework for 1990 and beyond", ibid.
"Developments in the service sector", ibid.
"International initiative in support of eastern reforms", ibid.
"Some implications of a German monetary union", ibid.
"Economic integration and the export performance of west European countries outside the EC", ibid.
"Europe's trade in engineering goods: Specialization and technology", ibid.
"The broader policy framework for 1990 and beyond", ibid.

"The unification of Germany", *Economic Bulletin for Europe,* vol.42, 1990, Sales No. E.90.II.E.37
"The Free Trade Agreement between Canada and the United States", ibid.

"Explaining unemployment in the market economies: theories and evidence", *Economic Survey of Europe in 1990-1991,* New York, 1991, Sales No. E.91.II.E.1
"The hard road to the market economy: problems and policies", ibid.
"Developments in the service sector", ibid.

"Reforms in foreign economic relations of eastern Europe and the Soviet Union", *Economic Studies No. 2,* Sales No. E.91.II.E.5

"External economic relations of the Baltic States", *Economic Bulletin for Europe,* vol.43(1991), Sales No. E.91.II.E.39

"International support for eastern transformation", *Economic Survey of Europe in 1991-1992,* New York, 1992, Sales No. E.92.II.E.1
"On property rights and privatization in the transition economies", ibid.
"Migration from east to west", ibid.

"International support for transition countries", *Economic Bulletin for Europe,* vol.44(1992), Sales No. E.93.II.E.3
"Finland's trade with the Soviet Union: its impact on the Finnish economy", ibid.

"Economic growth in the market economies, 1950-2000", *Discussion Papers, Volume 1 (1991), No. 1;* Sales No. GV.E.91-0-15
"Five years of *Perestroika:* Results, problems, prospects", *Discussion Papers, Volume 1 (1991), No. 2,* Sales No. GV.E.91-0-22
"Managing reforms in the east European countries", *Discussion Papers, Volume 1 (1991), No. 3,* Sales No. GV.E-92-0-1
"Economic reforms and their significance for all-European cooperation", *Discussion Papers, Volume 1 (1991), No. 4,* Sales No. GV.E.92-0-2
"Personal and collective services: An international perspective", *Discussion Papers, Volume 2 (1992), No. 1,* Sales No. GV.E.92-0-22
"The conditions for economic recovery in central and eastern Europe", *Discussion Papers, Volume 2 (1992), No. 2,* Sales No. GV.E.92-0-24
"The scope for macroeconomic policy to alleviate unemployment in western Europe", *Discussion Papers, Volume 2 (1992), No. 3,* Sales No. GV.E.92-0-27
"Economics and environment in the former Soviet Union and Czechoslovakia", *Discussion Papers, Volume 2 (1992), No. 4,* Sales No. GV.E.92-0-11

Back issues of the *Economic Survey of Europe* (from 1948) and the *Economic Bulletin for Europe* (from 1949) are available in microfiche form (which can be supplied as printed pages).

Your source of information: *United Nations publications*

DATE DUE

FEB 27 1995			
FEB 15 1995			

Demco, Inc. 38-293